Electrospun Nanofibers for Biomedical Applications

Electrospun Nanofibers for Biomedical Applications

Special Issue Editors

Albino Martins
Rui L. Reis
Nuno M. Neves

MDPI • Basel • Beijing • Wuhan • Barcelona • Belgrade • Manchester • Tokyo • Cluj • Tianjin

Special Issue Editors

Albino Martins
3B's Research Group;
I3Bs - Research Institute on
Biomaterials, Biodegradables
and Biomimetics of the
University of Minho
Portugal

Rui L. Reis
3B's Research Group;
I3Bs - Research Institute on
Biomaterials, Biodegradables
and Biomimetics of the
University of Minho
Portugal

Nuno M. Neves
3B's Research Group;
I3Bs - Research Institute on
Biomaterials, Biodegradables
and Biomimetics of the
University of Minho
Portugal

Editorial Office
MDPI
St. Alban-Anlage 66
4052 Basel, Switzerland

This is a reprint of articles from the Special Issue published online in the open access journal *Nanomaterials* (ISSN 2079-4991) (available at: https://www.mdpi.com/journal/nanomaterials/special_issues/electrospun_nano_fiber).

For citation purposes, cite each article independently as indicated on the article page online and as indicated below:

LastName, A.A.; LastName, B.B.; LastName, C.C. Article Title. *Journal Name* **Year**, *Article Number*, Page Range.

ISBN 978-3-03928-774-1 (Pbk)
ISBN 978-3-03928-775-8 (PDF)

Cover image courtesy of Catarina Silva.

© 2020 by the authors. Articles in this book are Open Access and distributed under the Creative Commons Attribution (CC BY) license, which allows users to download, copy and build upon published articles, as long as the author and publisher are properly credited, which ensures maximum dissemination and a wider impact of our publications.

The book as a whole is distributed by MDPI under the terms and conditions of the Creative Commons license CC BY-NC-ND.

Contents

About the Special Issue Editors vii

Preface to "Electrospun Nanofibers for Biomedical Applications" ix

Mark A. Calhoun, Sadiyah Sabah Chowdhury, Mark Tyler Nelson, John J. Lannutti, Rebecca B. Dupaix and Jessica O. Winter
Effect of Electrospun Fiber Mat Thickness and Support Method on Cell Morphology
Reprinted from: *Nanomaterials* **2019**, *9*, 644, doi:10.3390/nano9040644 1

Naresh Kasoju, Julian George, Hua Ye and Zhanfeng Cui
Sacrificial Core-Based Electrospinning: A Facile and Versatile Approach to Fabricate Devices for Potential Cell and Tissue Encapsulation Applications
Reprinted from: *Nanomaterials* **2018**, *8*, 863, doi:10.3390/nano8100863 15

Ana Agustina Aldana, Laura Malatto, Muhammad Atiq Ur Rehman, Aldo Roberto Boccaccini and Gustavo Abel Abraham
Fabrication of Gelatin Methacrylate (GelMA) Scaffolds with Nano- and Micro-Topographical and Morphological Features
Reprinted from: *Nanomaterials* **2019**, *9*, 120, doi:10.3390/nano9010120 27

Jeong Hwa Kim, Ju Young Park, Songwan Jin, Sik Yoon, Jong-Young Kwak and Young Hun Jeong
A Microfluidic Chip Embracing a Nanofiber Scaffold for 3D Cell Culture and Real-Time Monitoring
Reprinted from: *Nanomaterials* **2019**, *9*, 588, doi:10.3390/nano9040588 39

Aochen Wang, Ming Hu, Liwei Zhou and Xiaoyong Qiang
Self-Powered Well-Aligned P(VDF-TrFE) Piezoelectric Nanofiber Nanogenerator for Modulating an Exact Electrical Stimulation and Enhancing the Proliferation of Preosteoblasts
Reprinted from: *Nanomaterials* **2019**, *9*, 349, doi:10.3390/nano9030349 51

Hyo-Geun Jeong, Yoon-Soo Han, Kyung-Hye Jung and Young-Jin Kim
Poly(vinylidene fluoride) Composite Nanofibers Containing Polyhedral Oligomeric Silsesquioxane–Epigallocatechin Gallate Conjugate for Bone Tissue Regeneration
Reprinted from: *Nanomaterials* **2019**, *9*, 184, doi:10.3390/nano9020184 67

Francesca Serio, Marta Miola, Enrica Vernè, Dario Pisignano, Aldo R. Boccaccini and Liliana Liverani
Electrospun Filaments Embedding Bioactive Glass Particles with Ion Release and Enhanced Mineralization
Reprinted from: *Nanomaterials* **2019**, *9*, 182, doi:10.3390/nano9020182 83

Chien-Chung Chen, Sheng-Yang Lee, Nai-Chia Teng, Hsin-Tai Hu, Pei-Chi Huang and Jen-Chang Yang
In Vitro and In Vivo Studies of Hydrophilic Electrospun PLA95/β-TCP Membranes for Guided Tissue Regeneration (GTR) Applications
Reprinted from: *Nanomaterials* **2019**, *9*, 599, doi:10.3390/nano9040599 99

Elisa Bacelo, Marta Alves da Silva, Cristina Cunha, Susana Faria, Agostinho Carvalho, Rui L. Reis, Albino Martins and Nuno M. Neves
Biofunctional Nanofibrous Substrate for Local TNF-Capturing as a Strategy to Control Inflammation in Arthritic Joints
Reprinted from: *Nanomaterials* **2019**, *9*, 567, doi:10.3390/nano9040567 111

A. Sandeep Kranthi Kiran, T.S. Sampath Kumar, Rutvi Sanghavi, Mukesh Doble and Seeram Ramakrishna
Antibacterial and Bioactive Surface Modifications of Titanium Implants by PCL/TiO$_2$ Nanocomposite Coatings
Reprinted from: *Nanomaterials* **2018**, *8*, 860, doi:10.3390/nano8100860 127

Yu Chen, Weipeng Lu, Yanchuan Guo, Yi Zhu and Yeping Song
Electrospun Gelatin Fibers Surface Loaded ZnO Particles as a Potential Biodegradable Antibacterial Wound Dressing
Reprinted from: *Nanomaterials* **2019**, *9*, 525, doi:10.3390/nano9040525 143

Shiao-Wen Tsai, Sheng-Siang Huang, Wen-Xin Yu, Yu-Wei Hsu and Fu-Yin Hsu
Fabrication and Characteristics of Porous Hydroxyapatite-CaO Composite Nanofibers for Biomedical Applications
Reprinted from: *Nanomaterials* **2018**, *8*, 570, doi:10.3390/nano8080570 157

Chuan Yin, Sélène Rozet, Rino Okamoto, Mikihisa Kondo, Yasushi Tamada, Toshihisa Tanaka, Hatsuhiko Hattori, Masaki Tanaka, Hiromasa Sato and Shota Iino
Physical Properties and In Vitro Biocompatible Evaluation of Silicone-Modified Polyurethane Nanofibers and Films
Reprinted from: *Nanomaterials* **2019**, *9*, 367, doi:10.3390/nano9030367 167

S. M. Shatil Shahriar, Jagannath Mondal, Mohammad Nazmul Hasan, Vishnu Revuri, Dong Yun Lee and Yong-Kyu Lee
Electrospinning Nanofibers for Therapeutics Delivery
Reprinted from: *Nanomaterials* **2019**, *9*, 532, doi:10.3390/nano9040532 187

Rafael Contreras-Cáceres, Laura Cabeza, Gloria Perazzoli, Amelia Díaz, Juan Manuel López-Romero, Consolación Melguizo and Jose Prados
Electrospun Nanofibers: Recent Applications in Drug Delivery and Cancer Therapy
Reprinted from: *Nanomaterials* **2019**, *9*, 656, doi:10.3390/nano9040656 219

Hana Kadavil, Moustafa Zagho, Ahmed Elzatahry and Talal Altahtamouni
Sputtering of Electrospun Polymer-Based Nanofibers for Biomedical Applications: A Perspective
Reprinted from: *Nanomaterials* **2019**, *9*, 77, doi:10.3390/nano9010077 243

Jolanta Wróblewska-Krepsztul, Tomasz Rydzkowski, Iwona Michalska-Pożoga and Vijay Kumar Thakur
Biopolymers for Biomedical and Pharmaceutical Applications: Recent Advances and Overview of Alginate Electrospinning
Reprinted from: *Nanomaterials* **2019**, *9*, 404, doi:10.3390/nano9030404 273

About the Special Issue Editors

Albino Martins (BSc, PhD) is an Assistant Researcher with an Investigator FCT starting grant since 2015, a member of the Services and Innovation Subunit and of the Scientific Council of the I3Bs Research Institute on Biomaterials, Biodegradables and Biomimetics of University of Minho. He is the author of 57 publications, 12 book chapters, 5 patents, and 125 scientific communications.

Rui L. Reis (CEng, MSc, PhD, DSc, MD h.c.) is the founder and Director of the I3Bs Research Group and a Full Professor of Tissue Engineering, Regenerative Medicine, Biomaterials, and Stem Cells at the University of Minho. He is a recognized world expert in the TERM and biomaterials fields and has edited several books and encyclopedias, being the author/inventor of more than 1250 publications and around 75 patents. He has been awarded several major international prizes.

Nuno M. Neves (CEng, MSc, PhD, DSc) graduated in Polymer Engineering, has a Master degree on Polymer Engineering, a PhD on Polymer Science and Engineering (1998) and a Doctor of Science (D.Sc.) in Tissue Engineering, Regenerative Medicine and Stem Cells at the University of Minho. Currently, he is an Associate Professor with Habilitation at the University of Minho and Vice-Director of the 3B's Research Group.

Preface to "Electrospun Nanofibers for Biomedical Applications"

Nanomaterial-based substrates and devices offer opportunities to develop promising therapies to face the increasing incidence of cardiovascular, respiratory, musculoskeletal, cancer, and neurodegenerative diseases. Nanofiber meshes have already been proposed as devices to be used in the treatment of many diseases, taking advantage of their specific properties. The versatility in their composition, the high specific surface area, the high porosity, and the interconnectivity of the porosity of electrospun nanofiber meshes allow exploring different biological functionalities. Additionally, drug delivery systems and tissue-engineered scaffolds have been developed to address specific needs of particular tissues and organs.

Electrospun nanofibrous meshes can physically mimic the native extracellular matrix of most connective tissues. However, they present three important drawbacks (i.e., limited thickness, pore size in the micrometer range, and mechanical properties more compatible to skin than to bone) which can hinder their use in biomedical applications, namely, in tissue engineering and regenerative medicine. Therefore, many efforts have been made to enhance the performance of electrospun nanofibrous meshes, ranging from morphological to compositional and biochemical cues.

Despite the many efforts to translate electrospun nanofibrous meshes into medical devices, only few have reached clinical practice. It is expected that in the near future, long-lasting and highly-effective solutions will be developed that are able to address relevant clinical needs, increasing the quality of life of millions of patients and reducing healthcare costs.

This Special Issue assembles a set of original and highly-innovative contributions showcasing advanced devices and therapies based on or involving electrospun meshes. It comprises 13 original research papers covering topics that span from biomaterial scaffolds structure and functionalization, nanocomposites, antibacterial nanofibrous systems, wound dressings, monitoring devices, electrical stimulation, bone tissue engineering to first in human clinical trials. This publication also includes 4 review papers focused on drug delivery and tissue engineering applications.

Albino Martins, Rui L. Reis, Nuno M. Neves
Special Issue Editors

Article

Effect of Electrospun Fiber Mat Thickness and Support Method on Cell Morphology

Mark A. Calhoun [1], Sadiyah Sabah Chowdhury [2], Mark Tyler Nelson [1], John J. Lannutti [3], Rebecca B. Dupaix [2] and Jessica O. Winter [1,4,*]

[1] Department of Biomedical Engineering, The Ohio State University, Columbus, OH 43210, USA; Calhoun.89@osu.edu (M.A.C.); Nelsonmt05@gmail.com (M.T.N.)
[2] Department of Mechanical and Aerospace Engineering, The Ohio State University, Columbus, OH 43210, USA; Chowdhury.68@osu.edu (S.S.C.); Dupaix.1@osu.edu (R.B.D.)
[3] Department of Materials Science and Engineering, The Ohio State University, Columbus, OH 43210, USA; Lannutti.1@osu.edu
[4] William G. Lowrie Department of Chemical and Biomolecular Engineering, The Ohio State University, 453 CBEC, 151 W. Woodruff Ave., Columbus, OH 43210, USA
* Correspondence: winter.63@osu.edu; Tel.: +1-614-247-7668

Received: 13 March 2019; Accepted: 10 April 2019; Published: 20 April 2019

Abstract: Electrospun fiber mats (EFMs) are highly versatile biomaterials used in a myriad of biomedical applications. Whereas some facets of EFMs are well studied and can be highly tuned (e.g., pore size, fiber diameter, etc.), other features are under characterized. For example, although substrate mechanics have been explored by several groups, most studies rely on Young's modulus alone as a characterization variable. The influence of fiber mat thickness and the effect of supports are variables that are often not considered when evaluating cell-mechanical response. To assay the role of these features in EFM scaffold design and to improve understanding of scaffold mechanical properties, we designed EFM scaffolds with varying thickness (50–200 µm) and supporting methodologies. EFM scaffolds were comprised of polycaprolactone and were either electrospun directly onto a support, suspended across an annulus (3 or 10 mm inner diameter), or "tension-released" and then suspended across an annulus. Then, single cell spreading (i.e., Feret diameter) was measured in the presence of these different features. Cells were sensitive to EFM thickness and suspended gap diameter. Overall, cell spreading was greatest for 50 µm thick EFMs suspended over a 3 mm gap, which was the smallest thickness and gap investigated. These results are counterintuitive to conventional understanding in mechanobiology, which suggests that stiffer materials, such as thicker, supported EFMs, should elicit greater cell polarization. Additional experiments with 50 µm thick EFMs on polystyrene and polydimethylsiloxane (PDMS) supports demonstrated that cells can "feel" the support underlying the EFM if it is rigid, similar to previous results in hydrogels. These results also suggest that EFM curvature may play a role in cell response, separate from Young's modulus, possibly because of internal tension generated. These parameters are not often considered in EFM design and could improve scaffold performance and ultimately patient outcomes.

Keywords: electrospun fiber mats; mechanobiology; glioblastoma; biomaterials; finite element modeling

1. Introduction

Electrospun fibers are used widely across a range of applications, including filtration [1], drug delivery [2] and tissue engineering [3]. This is a result of the high degree of tunability of their pore size [4], fiber diameter [5] and degradation rate [6]. Generally the most important aspect of electrospun fiber mats (EFMs) as biomedical scaffolds is their fibrous topography [7], which influences cell morphology [8], migration [9] and gene regulation [10]. This has enabled their use in vascular grafts [11], organ replacement [12] and cancer treatment [13]. Thus, EFM features, such as fiber diameter and pore size features [14–16] have been well studied. Further, many studies address the mechanics of EFMs, which are powerful regulators of cell phenotype [17,18]. However, few of these go beyond Young's modulus, which assumes linear elasticity that may be inconsistent with the polymers used in EFM scaffolds.

Further, EFMs can be presented in a variety of conformations that may alter their mechanical properties. For example, EFMs are often synthesized on the surface of a much stiffer solid support, such as tissue culture polystyrene [19]. These mechanical nuances could potentially have a large effect on observed cell response, as we and others have shown the presence of edge effects in similar hydrogel culture models supported on glass or polystyrene that influence cell morphology [20,21]. Additionally, EFMs may be synthesized across an annular gap with support on the outer edges only (i.e., suspended). Such a configuration increases internal tension in the fibers as the EFM curvature increases. The effects of this configuration have not been widely explored. Addressing mechanical nuances in EFM scaffolds represents a valuable opportunity to advance understanding and to enable the design of next generation of EFM biomaterials.

Here, we employed EFMs in different scaffold configurations to correlate features of the mechanical environment to changes in cell morphology, extending our studies beyond Young's modulus. In particular, similar to our previous study in hydrogels [20], we examined cell morphology as a function of interfacial mechanics by altering the EFM support material to determine if cells cultured on EFM supports can "feel" the underlying substrate. Cell morphology often relates to or precedes other cell behavior in a myriad of conditions [17,22–24], and thus is a critical characteristic reporter of cell behaviors. We also evaluated the effect of EFM presentation: such as suspension across a gap, which induces curvature and may increase deformability versus support on a solid material; and we investigated the effect of releasing EFM internal residual tension that occurs during the spinning process, which would also alter deformability, alignment and presentation of focal adhesion sites. In the latter cases, the EFM material stiffness, as measured by Young's modulus, remains relatively constant; however, deformability of the fibers is altered, permitting subtle mechanical effects to be observed. As a model system, glioblastoma cells were employed because of their highly invasive nature and dysregulation in cell signaling related to migration and morphology. These studies highlight the importance of considering factors beyond Young's modulus in materials design to more fully understand the interaction between substrate mechanics and cell response.

2. Materials and Methods

2.1. Preparation of Aligned Polycaprolactone Electrospun Fiber Mat Constructs

Aligned polycaprolactone (PCL) fiber mats were prepared by electrospinning onto a rotating mandrel, as described previously [25]. Briefly, 5 wt% PCL (Mn 70,000–90,000, Sigma-Aldrich, St. Louis, MO, USA) solution in 1,1,1,3,3,3-hexafluoro-2-propanol (HFP) (>99% purity; Oakwood Products, Inc., Columbia, SC, USA) was electrospun at 4 mL/h at a 20 cm needle-to-collector distance [26]. The rotating mandrel was set to maintain a linear velocity at the collecting surface of 15 m/s. Fiber mat thickness was varied from 50, 100 and 200 microns by spinning for ~45 min, ~1 h 30 min and ~4 h, respectively. Fiber mats were spun either directly onto a supporting material or across an annular gap (Figure 1). Fiber mats were rendered hydrophilic for cell culture by air plasma treatment (Harrick Plasma, Ithaca, NY, USA) under vacuum at ~700 mTorr and with a plasma radio frequency of 8–12 MHz for 3 min. Fiber mats were used immediately in experiments.

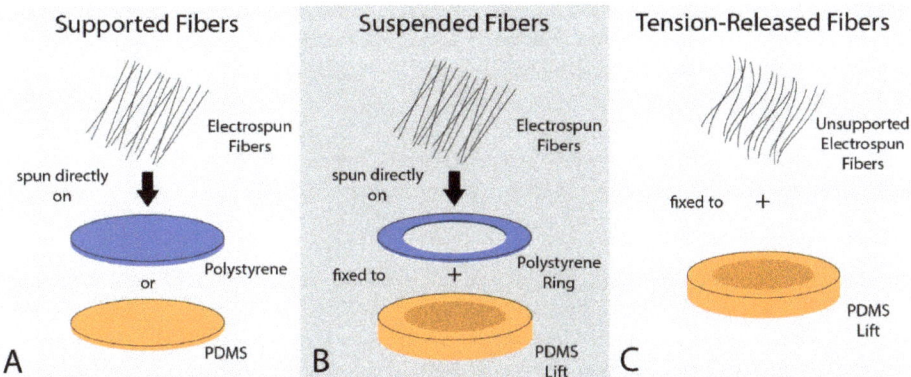

Figure 1. Schematic of Electrospun Fiber Mats. Electrospun fiber mats (EFMs) were either spun directly onto a stiff support (**A**), spun across a gap (**B**), or spun unto foil, removed, and then fixed across a gap (**C**).

2.1.1. Polydimethylsiloxane- and Polystyrene-Supported Fiber Mats

Supported fiber mats (Figure 1A) were synthesized by fixing polydimethylsiloxane (PDMS) and polystyrene (PS) wafers to the rotating mandrel with double sided tape. PS wafers (Multi-Plastics Inc., Lewis Center, OH, USA) were 0.2 mm thick and cut to a 10 mm diameter with an arch punch (Grainger, Columbus, OH, USA). PDMS was made from a combination of SylGard 184 and SylGard 527 (Dow Corning, Midland, MI, USA) to vary stiffness without changing surface charge or chemistry [27]. Stiffer "PDMS 100/0" was made of 100% SylGard 184, and less stiff "PDMS 50/50" was made of 50% SylGard 184 and 50% SylGard 527. Each was cured at 65 °C for 18 h. PDMS wafers were ~1 mm thick and 12 mm in diameter. Electrospinning on these substrates produced a layer of electrospun fibers that was irreversibly bound to the support, such that fiber-support adhesion was not considered to be a confounding variable.

2.1.2. Suspended Fiber Mats

Suspended fiber mats (Figure 1B) were produced by fixing PS rings of varying inner diameter to the rotating mandrel with double sided tape. The rings were cut so that the outer diameter of the ring was 16 mm and the inner gap diameter was either 3 or 10 mm. After electrospinning, a free, suspended EFM area was produced inside the annular ring.

2.1.3. Tension-Released Fiber Mats

Tension-released fiber mats (Figure 1C) were produced by electrospinning on non-stick aluminum foil attached to the rotating mandrel. EFM sections were carefully cut with a scalpel, removed from the foil, and fixed to PDMS annular rings (as above) with gap diameters of 10 mm using Sylastic (Dow Corning, Midland, MI, USA). This produced a layer of electrospun fibers irreversibly bound to the support.

2.2. PDMS Control Substrates

As a control for supported EFM studies, cell adhesion was also compared to bare PDMS wafers free of EFMs. PDMS was synthesized as above, then air plasma treated for 10 min without cracking the PDMS surface to render surfaces hydrophilic for cell adhesion [28].

2.3. Scanning Electron Microscopy

For examination of fiber alignment, each type of EFM construct was attached to an aluminum stub using carbon tape (Ted Pella, Inc., Redding, CA, USA), sputter coated with gold for 30 s (Model 3

Sputter Coater 91000, Pelco, Reading, CA, USA) and imaged using a scanning electron microscope (SEM) (Quanta 200 SEM, FEI Company, Hillsboro, OR, USA). Fast Fourier Transform (FFT) was used to quantitatively validate fiber alignment [29,30].

2.4. Mechanical Characterization

The mechanics of the EFM constructs were characterized by measuring the Young's modulus and calculating the indentation moduli using finite element modeling (FEM).

2.4.1. Elastic Stress Characterization

Tensile testing (RSA2, New Castle, DE, USA) was carried out on PDMS 100/0, PDMS 50/50, and PS supports as well as 50 µm and 200 µm thick EFMs. EFMs were oriented in the aligned fiber direction for testing. Each was cut to a standard dogbone shape and tested up to a strain of 10% at a strain rate of 0.6% per second. The linear portion of the stress–strain curve was used to determine the Young's modulus of the samples.

2.4.2. Finite Element Model Indentation Characterization

A Finite Element Model (FEM) was created using ABAQUS/CAE 6.13-1 software (Dassault Systèmes Simulia Corporation, Providence, RI, USA, 2013). This model featured a 10 µm diameter spherical indentation (E = 200 GPa, Poisson's ratio = 0.3) impressed into each EFM by 5 µm. Based on EFM morphology, we assume rigid, non-slip contact with the support. In the model, EFMs were either fixed along the base and outside edge (supported fiber mats) or solely on the outside edge (suspended fiber mats) (shown in Supplementary Figure S1 and further described in the Supplementary Material). We tested suspended EFM gap diameters of 3 and 10 mm, with EFM thicknesses of 50, 100 and 200 µm. An axisymmetric model captured the geometry of the EFMs and spherical indentation. The EFMs were assumed to be elastic and isotropic with a Young's modulus of 7.9 MPa and a Poisson's ratio of 0.35. Isotropy was assumed as shear modulus for PCL electrospun fibers is not statistically significantly different in perpendicular directions [31]. The indentation modulus of each EFM was determined by the Hertzian-contact method, which is valid for ideal elastic materials under infinitesimal deformation [32]. The elastic modulus of the substrate, given a rigid spherical indenter, is based on the load-displacement curve, as in Equation (1).

$$E = \sqrt{\frac{S^3(1-\nu^2)^2}{6RP}}, \qquad (1)$$

where E is the elastic modulus of the substrate, ν is Poisson's ratio of the substrate, R is the nominal radius of curvature of the indenter tip, P is the applied load, and S is the material stiffness (S = dP/dh) evaluated at P.

2.5. Cell Culture

Human glioblastoma U87 MG cells (ATCC) were used for all experiments because of their highly invasive nature. Cells were cultured at 37 °C in a 5% CO_2 atmosphere in DMEM/F12 (Sigma-Aldrich, St. Louis, MO, USA) supplemented with 10% fetal bovine serum (Fisher Scientific, Hampton, NH, USA), 1% penicillin/streptomycin (Invitrogen, Carlsbad, CA, USA) and 1% MycoZap (Life Technologies, Carlsbad, CA, USA). Cells were fed 2–3 times a week and passaged at confluence prior to use.

2.6. Analysis of Cell Morphology on PDMS Substrates and PCL Electrospun Fiber Mats

Supported fiber mats were fixed to the bottom of a 12 well plate using Sylastic to prevent floating. Suspended or tension-released EFMs were first fixed to a PDMS ring to prevent the unsupported area from touching the bottom of the well. Then, the PDMS ring was fixed to the bottom of the plate using Sylastic. PDMS wafers and EFMs were sterilized by soaking with 70% ethanol under UV for 20 min, then rinsed twice with PBS and once with media. Materials were then inoculated in cell culture media for 30 min. Cells were stained with CellTracker Green (Invitrogen, Carlsbad, CA, USA), then seeded at 40 k per well. After 18 h, cells were imaged using reflectance microscopy (Olympus IX-2, Shinjuku, Tokyo, Japan) near the center of the fiber mat to avoid potential edge effects. Feret diameter is the largest distance between any two points of an object and was measured using Image J Image Analysis Software (v1.52n, Bethesda, MD, USA) freely available from the National Institutes of Health.

2.7. Statistical Analysis

All data was analyzed using JMP statistical analysis software (JMP Pro 14, Cary, NC, USA). Statistical differences in cell response were detected with ANOVA and Tukey–Kramer HSD tests. In all cases, $p < 0.05$ was considered statistically significant.

3. Results

3.1. Fiber Mat Characterization

Correlating to standard EFM characterization methods, we first characterized EFM morphology and stiffness. Aligned EFMs were spun directly onto PS or PDMS to create supported (solid) or suspended (annulus) scaffolds (Figure 1A,B). Alternatively, EFMs were spun onto non-stick foil, cut and removed from foil to release internal tension arising from the electrospinning process, and fixed to a PDMS annulus to create tension-released scaffolds (Figure 1C). As-spun EFMs demonstrated excellent alignment (Figure 2A), some of which was preserved in the tension-released EFMs, though a clear non-linearity is introduced to the fibers (Figure 2B). In these representative EFM images, individual fiber diameter was consistent at 0.97 ± 0.04 µm and 1.00 ± 0.04 µm for 100 µm thick supported and tension-released EFMs, respectively ($n = 127$ total individual fibers). Fiber diameter distribution was normal (Gaussian) for PS-supported EFMs, but not for the tension-released EFMs (Supplementary Figure S2). Fiber densities in these EFMs were statistically different at 595.8 ± 17.6 mm^{-1} and 539.3 ± 11.7 mm^{-1} for 100 µm thick supported and tension-released EFMs, respectively ($p = 0.0318$, two-sided t-test). At the cellular level (~10 µm across), this is a difference between cells interacting with 5.96 fibers or 5.39 fibers. This may be negligible when considering that out-of-plane fibers were quantified, while cells largely interact with the top layer of fibers and do not penetrate the EFM. These fiber diameters and densities are very similar to our previously published results from the same electrospinning process [19].

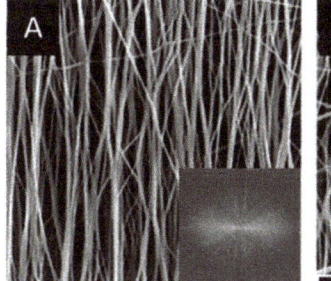 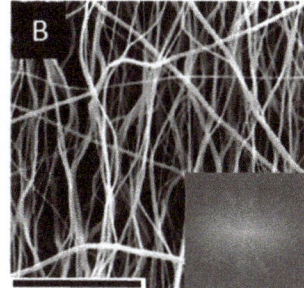

Figure 2. Representative scanning electron microscope (SEM) images of EFMs. Aligned (**A**) and tension-released (**B**) 100 µm thick EFMs showing differences in fiber morphology following release of internal tension. Insets: Fourier transform of the figure, indicating degree of alignment. Scale bar = 50 µm.

Mechanical properties of EFMs were first characterized by calculating an FEM-generated indentation modulus (Table 1). The indentation modulus is based on an axisymmetric model simulating a rigid steel indenter impressing an EFM at its center. The indentation modulus for PS-supported

EFMs was up to ~500×, or 8.233 MPa, higher than the indentation modulus for suspended EFMs of the same thickness. However, moduli for supported EFMs of different thicknesses were similar (e.g., 8.250 vs. 7.443 MPa). In contrast for suspended EFMs, the indentation modulus increased by ~50×, or 0.79 MPa, as the thickness of the suspended EFM increased from 50 to 200 μm at constant gap diameter (i.e., 10 mm). Finally, indentation modulus decreased by ~10×, or 0.145 MPa, as the gap diameter increased from 3 to 10 mm. Thus, the weight of each factor on indentation modulus from greatest to least was supported vs. suspended, EFM thickness (suspended), and gap diameter. Based on this model, supported presentations, increasing EFM thickness (suspended), and decreasing gap diameter all lead to a higher indentation modulus. Supported EFMs are generally stiffer than suspended EFMs. This suggests that decreasing gap diameter leads to a mechanical response more similar to a stiffer, supported EFM.

Table 1. Indentation modulus (MPa) for electrospun fiber mat constructs.

	50 μm	100 μm	200 μm
Supported, PS	8.250	7.814	7.443
Suspended, 3 mm gap diameter	0.162	1.069	3.932
Suspended, 10 mm gap diameter	0.017	0.118	0.807

Indentation modulus generated by finite element modeling.

A uniaxial tensile test was then performed on the EFM support materials (i.e., PS and PDMS) and on the unsupported EFMs. The PDMS 100/0 was made of 100% SylGard 184 and PDMS 50/50 was mixed with 50% SylGard 527 to make it less stiff while keeping surface charge and chemistry constant [27]. As expected, PDMS 50/50 was less stiff, by about 1.5×, than PDMS 100/0 (Table 2). For EFMs with different thicknesses, the mean Young's modulus increased with thickness, although results were not statistically significant. The 200 μm EFM was 1.4× stiffer than the 50 μm EFM (Table 2). These results are similar to indentation modulus calculations for supported EFMs (Table 1), which show that modulus is not a strong function of thickness. Thus, we conclude that EFM thickness does not play a strong role in the mechanical response as measured by Young's modulus.

Table 2. Young's modulus of electrospun fiber mats and support materials.

Material	Young's Modulus (MPa)
50 μm EFM	15.00 ± 1.01
200 μm EFM	20.76 ± 3.37
PS	2160.63 ± 51.35
PDMS 50/50	1.29 ± 0.10
PDMS 100/0	3.13 ± 0.35

Data are displayed as mean ± standard error.

3.2. The Role of the Support in EFM Scaffolds

In our previous work [20], we showed that the mechanics of an underlying support material can influence the response of cells cultured at close proximity by inducing edge effects in an interfacial region. To investigate how such edge effects can potentially be utilized in EFM scaffolds, cells were grown on EFMs of varying thickness and with differing support materials (Figure 1A). Cell morphology was assessed by the Feret diameter (Figure 3A, red lines), which is the longest distance between any two points of an object. We have previously shown that cells exhibit polarized morphologies on aligned EFMs and that Feret diameter correlates with migration speed for glioblastoma cells [19], the model system employed here. Increasing EFM thickness from 50 to 200 μm led to a statistically significant increase of ~19% in cell Feret diameter for PS-supported EFMs (Figure 3B). Thus, although a statistically significant difference in Young's modulus could not be detected, statistically significant cell responses were still observed.

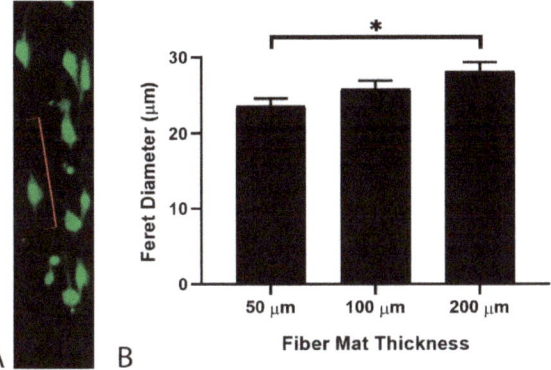

Figure 3. Glioblastoma cell Feret diameter (**A**) as a function of polystyrene (PS)-supported EFM thickness (**B**). Feret diameter is indicated for a representative cell (red lines). A total of 979 cells were analyzed across at least two independent experiments. Levels connected by a star (*) are statistically significant from each other ($p < 0.05$).

Cells were also cultured on EFMs at a thickness of 50 μm using support materials of decreasing Young's modulus (i.e., PS, PDMS 100/0 and PDMS 50/50) (Figure 4A). PS was selected because it exhibits a Young's modulus several orders of magnitude higher than that of the EFMs employed (i.e., ~2200 vs. ~15 MPa for EFMs), similar to our previous work examining Matrigel (e.g., 1 kPa) supported on glass (e.g., GPa) [20]. For comparison, two types of softer PDMS were employed; both with Young's moduli lower than that of the EFMs (3.13 and 1.29 MPa, respectively vs. 15.00 MPa for EFMs). We observed a ~26% decline in Feret diameter between the PS and the PDMS 50/50-supported 50 μm EFMs. In contrast, no difference in Feret diameter was detected between the two PDMS-supported models. However, when the cells were cultured directly on PDMS 100/0 and PDMS 50/50 support materials without an EFM, a significant difference in Feret diameter was seen (Figure 4B). This suggests that the mechanical response of the EFM scaffold is a combination of the support material and the EFM itself. Thus, tuning the mechanobiological response of cells to EFM scaffolds may be achieved either by altering the EFM thickness or by altering the stiffness of the underlying support material and that cells are able to sense the stiffer material in the interfacial region of composites.

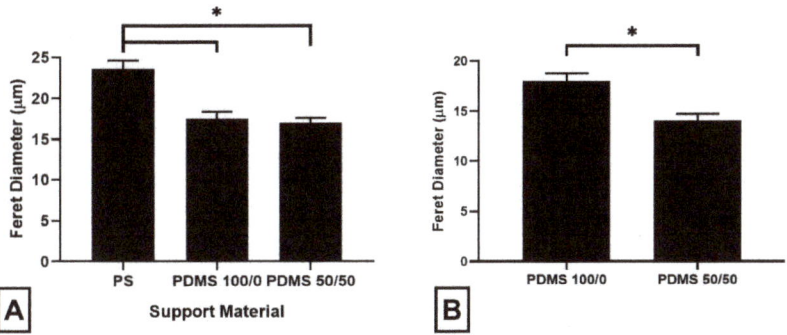

Figure 4. Glioblastoma cell Feret diameter in response to culture on 50 μm EFMs supported on substrates of declining stiffness (**A**) and on polydimethylsiloxane (PDMS) supports of declining stiffness with no EFM present (**B**). A total of 1623 cells were analyzed across at least two independent experiments. Levels connected by a star (*) are statistically significant.

3.3. The Role of Suspension in EFM Scaffolds

To further elucidate how mechanical nuances can influence cell behavior, we investigated EFMs lacking support in a central region, i.e., EFMs suspended across an annulus (Figure 1B). This architecture resembles tissue engineered constructs in which a support material may be missing or only partially present [12,13], such as artificial trachea constructs. PS disks were used to suspend EFMs of different thicknesses (i.e., 50, 100 and 200 μm) over two inner gap diameters (i.e., 3 and 10 mm). Compared to 50 μm PS-supported EFMS, cells cultured on 50 μm EFMs supported across a 10 mm annular gap displayed Feret diameters similar in magnitude, but cells on the 3 mm gap scaffold had a significantly higher mean Feret diameter than either support (Figure 5). According to FEM, the 3 mm gap scaffold would have an indentation modulus between that of the 10 mm gap scaffold and the PS support.

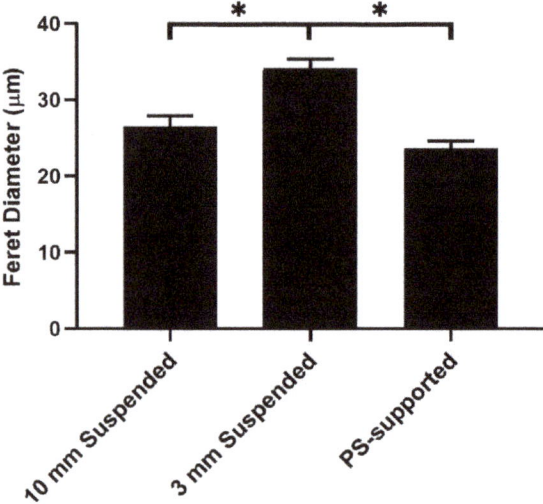

Figure 5. Glioblastoma cell Feret diameter on supported (PS) and suspended (10 mm, 3 mm) 50 μm EFMs. Scaffolds are arranged left to right by increasing indentation modulus. A total of 935 cells were analyzed across at least two independent experiments. Levels connected by a star (*) are statistically significant from each other ($p < 0.05$).

To gain additional information on these responses, we investigated the effect of EFM thickness coupled to annular diameter (Figure 6). As EFM thickness increased from 50 to 200 μm, Feret diameter decreased by ~37% for a fixed 3 mm gap diameter. At the larger 10 mm gap, this decrease was much less pronounced (~18%). In addition, there was a statistically significant difference in Feret diameter as a function of gap diameter for the 50 μm thick and 100 μm thick EFMs. This difference was abrogated at the 200 μm EFM thickness. Thus, increasing gap diameter correlates with a decreased indentation modulus and an observed decrease in Feret diameter. Interestingly, increased EFM thickness correlates with increased indentation moduli, but in this case caused a decreased Feret diameter. This highlights the mechanical complexity of EFM scaffolds, especially in the more nuanced scenario of a suspended EFM. These results are likely influenced by the role of curvature, which could be pronounced at the length scale of a cell. Alternatively, these responses may reflect internal residual tension that is introduced by deformation of the EFM in the central region.

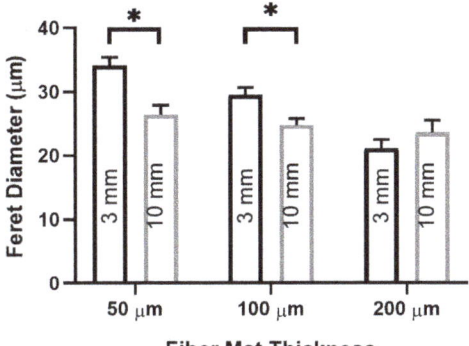

Figure 6. Glioblastoma cell Feret diameter measured on supports with 3 and 10 mm diameter annular gaps supporting EFMs of different thicknesses. A total of 1097 cells were analyzed across at least two independent experiments. Levels connected by a star (*) are statistically significant from each other ($p < 0.05$).

3.4. The Role of Tension in EFM Scaffolds

To further explore the role of internal tension, suspended EFMs were modified to reduce internal residual tension by electrospinning onto non-stick aluminum foil, cutting the fibers and removing them from the foil, and fixing free-standing EFMs to PDMS annular rings with an inner gap diameter of 10 mm (Figure 1C). These "tension-released" EFMs produced a statistically significant increase of 9.7% in Feret diameter compared to suspended EFM without released tension at the same gap diameter (Figure 7A). In addition, increasing EFM thickness led to lower Feret diameter (Figure 7B). This is the same trend observed for suspended EFMs. In this case, we have empirically reduced the stiffness of the EFMs by releasing internal tension, yet EFM thickness remains inversely correlated with Feret diameter in contrast to results on supported EFMs (Figure 3B). Taken together, this data suggests that the mechanical environment of EFM scaffolds is complex and that cells may react to this environment in different ways on depending on the stiffness and presentation of the support.

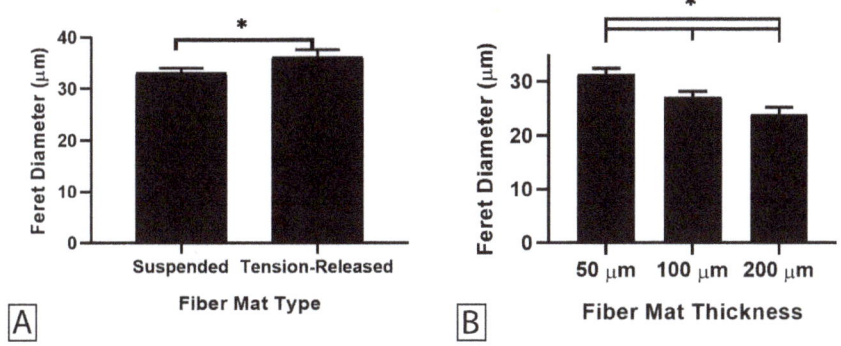

Figure 7. Average cell Feret diameter observed on tension-released and suspended EFM constructs, pooled data across all EFM thicknesses (**A**) and between tension-released mats at different fiber mat thicknesses (**B**). A total of 1651 cells were analyzed across at least two independent experiments. Levels connected by a star (*) are statistically significant from each other ($p < 0.05$).

4. Discussion

This work explores the impact of EFM mechanical parameters, beyond Young's modulus, on cell morphology as measured by the Feret diameter. In particular, the effect of the mechanics of the support, the presence or absence of the support, inherent tension in EFMs and EFM thickness were explored. For supported EFMs, changing the stiffness of the support material underlying the EFM caused a large change in cell morphology, but only if the support was much stiffer than the EFM (Figure 4A). This is not because cells could not sense differences in stiffness between the underlying substrates; indeed, Feret diameter was different when grown directly on softer flat PDMS supports without an EFM (Figure 4B). However, when the much stiffer EFM was added, these differences were abrogated. Although this could result from topography differences, EFMs are fibrous whereas PDMS supports are relatively smooth and flat, others have shown that cells can "feel" the underlying substrate when cultured on thin hydrogels. Experiments in which cells spread more on thin gels than on thicker gels demonstrate this interplay of support substrates [33].

We are the first to demonstrate such edge effects in EFMs. Buxboim et al. described a "threshold matrix thickness," the length scale at which cells respond not only to the stiffness of the matrix, but also to the rigidity of the underlying support for hydrogel models [21]. However, these length scales were on the order of 10–20 μm compared to our observed responses on 50–200 μm thick EFMs. Cells may be able to "feel" further when grown on EFMs than when grown on hydrogels because of the orders of magnitude difference in length scales of individual fibers. Electrospun fiber length can range upwards of 35–50 cm [34], whereas collagen fibers in hydrogels are only 0.5–3 μm long [35]. Other factors aside (e.g., fiber strength, fiber interconnectivity, etc.), this difference in length scale may lead to differences in how well the matrix transmits tension. This likely permits stress to be transferred over a larger distance.

Suspended EFMs produced perhaps the most intriguing results. From a design perspective, the 50 μm thick, 3 mm gap EFM scaffolds yielded a key finding; these substrates elicited the highest mean Feret diameter of all (Figure 5). Cells on a thin EFM suspended over a small gap spread more than cells on a thin EFM fixed to a rigid support with a higher indentation modulus (Table 1). In general, cells on stiffer substrates demonstrate higher contractility and spreading [36], i.e., Feret diameter. Engler et al. demonstrated cells grown on hydrogels show a bell curve-shaped response to ligand density, though ligand density and substrate stiffness are highly difficult to decouple in gels [37]. We have previously shown that this bell curve-shaped response also exists in EFMs for cell mechanics when ligand density and stiffness were decoupled through the use of core-shell electrospinning [19]. Here, we have extended upon that work by decoupling material properties from more complex composite and structural mechanics by suspending fiber mats across gaps of different diameter without changing surface ligand density or Young's modulus. We also interrogated focal adhesion kinase (FAK) expression through Western blotting and found no significant differences in expression (Supplementary Figure S3). Although FAK phosphorylation could play a role, this suggests that this mechanosensing mechanism may not be mediated by classical focal adhesion kinase machinery, which we have previously implicated in EFM mechanotransduction [19]. These data help to illustrate the complex interplay of substrate mechanics, cell adhesion molecules, and mechanosensing.

For suspended EFMs, changing gap diameter caused significant changes in cell morphology (Figure 6). Releasing the internal tension in EFMs also altered mean Feret diameter (Figure 7A). These results highlight the interplay of substrate mechanics and mechanosensing beyond Young's modulus. It is generally believed that cells spread more on stiffer substrates [36]. Our investigation into the effect of support material stiffness (Figure 4) generally corroborates this belief; Feret diameter correlated positively with stiffness, as measured by Young's modulus. However, these results were inverted with EFM thickness for both supported and suspended EFM presentations (Figures 3B and 6). Feret diameter inversely correlated with stiffness, as measured by indentation modulus. In supported and suspended EFMs, increasing EFM thickness had opposite effects on indentation modulus. Increasing supported EFM thickness results in only a slight decrease in indentation modulus; however, increasing suspended EFM thickness resulted in ~50× increase in indentation modulus (Table 1). For supported scaffolds, a given cell at the center of the EFM is only hundreds of microns from the rigid support (i.e., one EFM thickness away), whereas for suspended scaffolds the rigid support is millimeters away. This likely affects how the cell feels mechanics between these models. Suspending EFMs may also cause an impactful change in the curvature of the EFM. These complexities that arise from inverted cell response to stiffness and suspending EFMs are reflected in part by the bell curve-shaped response to indentation modulus when comparing suspended and supported presentations (Figure 5). Further experiments are warranted to decouple these variables to explain these differences.

In addition, improved characterization employing more advanced FEM models that consider individual fibers over different length scales or empirical data that recapitulates cell contractility would yield helpful insights. We employ an FEM-generated indentation modulus and an experimentally derived Young's modulus. Neither completely describes the results seen herein. Young's modulus is especially poorly-suited to provide insights because the most challenging data arise specifically from changes in structural mechanics, not changes in the materials themselves. Thus, systems have identical Young's moduli.

These data highlight a gap in understanding between cell sensing and response to the mechanical environment. Further, the large differences generated by small, nuanced changes in the mechanical environment, for example a ~37% change in morphology with a change in EFM thickness (Figure 6), suggest caution in interpreting data from mechanical studies that may lack scientific rigor. In addition, these data suggest new variables (i.e., support material, suspension diameter) that can be tuned to alter material mechanobiology effects. The field of biomaterials has historically progressed from early generation scaffolds that are bioinert to later generations that are bioactive or bioresorbable. For example, earlier generation bioinert alumina vs. more recent bioresorbable tri-calcium phosphate dental implants [38]. Whereas it is acceptable to "avoid" biological complexity early, this complexity also provides powerful opportunities to improve patient outcomes. Impending generations of biomedical scaffolds will need to harness the powerful, yet relatively untapped realm of nuanced mechanobiology.

Supplementary Materials: The following are available online at http://www.mdpi.com/2079-4991/9/4/644/s1, Figure S1: Boundary conditions and interactions for indentation FE models, Figure S2: Individual fiber diameter distributions. Frequency distributions for PS-supported (A) and tension-released (B) EFMs, Figure S3: Focal adhesion kinase expression.

Author Contributions: Conceptualization, M.A.C., R.B.D., J.O.W.; methodology, M.A.C., S.S.C., M.T.N., R.B.D.; formal analysis, M.A.C.; investigation, M.A.C.; resources, J.J.L., R.B.D., J.O.W.; data curation, M.A.C., R.B.D.; writing—original draft preparation, M.A.C.; writing—review and editing, M.A.C., M.T.N., J.J.L., R.B.D., J.O.W.; funding acquisition, M.A.C., M.T.N., J.J.L., R.B.D., J.O.W.

Funding: This research was funded by the National Science Foundation (DGE0221678, CMMI0747252), and the National Institutes of Health R01HL132355.

Acknowledgments: The authors acknowledge support of the Pelotonia Cancer Research Fellowship (MAC, MTN). These are the words and opinions of the authors only and do not reflect those of the Pelotonia program.

Conflicts of Interest: The authors declare no conflict of interest. The funders had no role in the design of the study; in the collection, analyses, or interpretation of data; in the writing of the manuscript, or in the decision to publish the results.

References

1. Yun, K.M.; Hogan, C.J., Jr.; Matsubayashi, Y.; Kawabe, M.; Iskandar, F.; Okuyama, K. Nanoparticle filtration by electrospun polymer fibers. *Chem. Eng. Sci.* **2007**, *62*, 4751–4759. [CrossRef]
2. Kumbar, S.G.; Nair, L.S.; Bhattacharyya, S.; Laurencin, C.T. Polymeric nanofibers as novel carriers for the delivery of therapeutic molecules. *J. Nanosci. Nanotechnol.* **2006**, *6*, 2591–2607. [CrossRef] [PubMed]
3. Sill, T.J.; von Biomaterials, R.-H.A. Electrospinning: Applications in drug delivery and tissue engineering. *Biomaterials* **2008**, *29*, 1989–2006. [CrossRef] [PubMed]
4. Eichhorn, S.J.; Sampson, W.W. Statistical geometry of pores and statistics of porous nanofibrous assemblies. *J. R. Soc. Interface* **2005**, *2*, 309–318. [CrossRef]

5. Deitzel, J.M.; Kleinmeyer, J.; Harris, D.E.A.; Polymer, N.C.B. The effect of processing variables on the morphology of electrospun nanofibers and textiles. *Polymer* **2001**, *42*, 261–272. [CrossRef]
6. Kim, K.; Yu, M.; Zong, X.; Chiu, J.; Fang, D.; Seo, Y.S.; Hsiao, B.S.; Chu, B.; Hadjiargyrou, M. Control of degradation rate and hydrophilicity in electrospun non-woven poly(D,L-lactide) nanofiber scaffolds for biomedical applications. *Biomaterials* **2003**, *24*, 4977–4985. [CrossRef]
7. Sundararaghavan, H.G.; Saunders, R.L.; Hammer, D.A.; Burdick, J.A. Fiber alignment directs cell motility over chemotactic gradients. *Biotechnol. Bioeng.* **2013**, *110*, 1249–1254. [CrossRef] [PubMed]
8. Chew, S.Y.; Mi, R.; Hoke, A.; Biomaterials, L.-K.W. The effect of the alignment of electrospun fibrous scaffolds on Schwann cell maturation. *Biomaterials* **2008**, *29*, 653–661. [CrossRef] [PubMed]
9. Nelson, M.T.; Short, A.; Cole, S.L.; Gross, A.C.; Winter, J. Preferential, enhanced breast cancer cell migration on biomimetic electrospun nanofiber 'cell highways'. *BMC Cancer* **2014**, *14*, 825. [CrossRef]
10. Wang, Z.; Cui, Y.; Wang, J.; Yang, X.; Wu, Y.; Wang, K.; Biomaterials, G.-X. The effect of thick fibers and large pores of electrospun poly (ε-caprolactone) vascular grafts on macrophage polarization and arterial regeneration. *Biomaterials* **2014**, *35*, 5700–5710. [CrossRef] [PubMed]
11. Hasan, A.; Memic, A.; Annabi, N.; Hossain, M.; Paul, A.; Dokmeci, M.R.; Dehghani, F.; Khademhosseini, A. Electrospun scaffolds for tissue engineering of vascular grafts. *Acta Biomater.* **2014**, *10*, 11–25. [CrossRef] [PubMed]
12. Ajalloueian, F.; Lim, M.L.; Lemon, G.; Haag, J.C.; Gustafsson, Y.; Sjoqvist, S.; Beltran-Rodriguez, A.; Del Gaudio, C.; Baiguera, S.; Bianco, A.; et al. Biomechanical and biocompatibility characteristics of electrospun polymeric tracheal scaffolds. *Biomaterials* **2014**, *35*, 5307–5315. [CrossRef]
13. Jain, A.; Betancur, M.; Patel, G.D.; Valmikinathan, C.M.; Mukhatyar, V.J.; Vakharia, A.; Pai, B.S.; Brahma, B.; MacDonald, T.J.; Bellamkonda, R.V. Guiding intracortical brain tumour cells to an extracortical cytotoxic hydrogel using aligned polymeric nanofibres. *Nat. Mater.* **2014**, *13*, 308–316. [CrossRef]
14. Pham, Q.P.; Sharma, U.; Mikos, A.G. Electrospinning of polymeric nanofibers for tissue engineering applications: A review. *Tissue Eng.* **2006**. [CrossRef]
15. Nam, J.; Huang, Y.; Agarwal, S.; Lannutti, J. Improved cellular infiltration in electrospun fiber via engineered porosity. *Tissue Eng.* **2007**, *13*, 2249–2257. [CrossRef] [PubMed]
16. Nisbet, D.R.; Forsythe, J.S.; Shen, W.; Finkelstein, D.I.; Horne, M.K. A review of the cellular response on electrospun nanofibers for tissue engineering. *J. Biomater. Appl.* **2009**, *24*, 7–29. [CrossRef]
17. Engler, A.J.; Sen, S.; Sweeney, L.H.; Discher, D.E. Matrix Elasticity Directs Stem Cell Lineage Specification. *Cell* **2006**, *126*, 677–689. [CrossRef] [PubMed]
18. Shin, J.-W.; Discher, D.E. Cell culture: Soft gels select tumorigenic cells. *Nat. Mater.* **2012**, *11*, 662–663. [CrossRef] [PubMed]
19. Rao, S.S.; Nelson, M.T.; Xue, R.; DeJesus, J.K.; Viapiano, M.S.; Lannutti, J.J.; Sarkar, A.; Winter, J.O. Mimicking white matter tract topography using core-shell electrospun nanofibers to examine migration of malignant brain tumors. *Biomaterials* **2013**, *34*, 5181–5190. [CrossRef]
20. Rao, S.S.; Bentil, S.; DeJesus, J.; Larison, J.; Hissong, A.; Dupaix, R.; Sarkar, A.; Winter, J.O. Inherent interfacial mechanical gradients in 3D hydrogels influence tumor cell behaviors. *PLoS ONE* **2012**, *7*. [CrossRef] [PubMed]
21. Buxboim, A.; Rajagopal, K.; Brown, A.E.X.; Discher, D.E. How deeply cells feel: Methods for thin gels. *J. Phys. Condens. Matter* **2010**, *22*, 194116. [CrossRef] [PubMed]
22. Zhao, M.; Bai, H.; Wang, E.; Forrester, J.V.; McCaig, C.D. Electrical stimulation directly induces pre-angiogenic responses in vascular endothelial cells by signaling through VEGF receptors. *J. Cell Sci.* **2004**, *117*. [CrossRef] [PubMed]
23. Czeisler, C.; Short, A.; Nelson, T.; Gygli, P.; Ortiz, C.; Catacutan, F.; Stocker, B.; Cronin, J.; Lannutti, J.; Winter, J.; et al. Surface topography during neural stem cell differentiation regulates cell migration and cell morphology. *J. Comp. Neurol.* **2016**, *524*, 3485–3502. [CrossRef] [PubMed]
24. Tse, J.M.; Cheng, G.; Tyrrell, J.A.; Wilcox-Adelman, S.A.; Boucher, Y.; Jain, R.K.; Munn, L.L. Mechanical compression drives cancer cells toward invasive phenotype. *Proc. Natl. Acad. Sci. USA* **2012**, *109*, 911–916. [CrossRef] [PubMed]
25. Johnson, J.; Nowicki, M.O.; Lee, C.H.; Chiocca, E.A.; Viapiano, M.S.; Lawler, S.E.; Lannutti, J.J. Quantitative analysis of complex glioma cell migration on electrospun polycaprolactone using time-lapse microscopy. *Tissue Eng. Part C Methods* **2009**. [CrossRef] [PubMed]

26. Gaumer, J.; Prasad, A.; Lee, D.; Lannutti, J. Structure-function relationships and source-to-ground distance in electrospun polycaprolactone. *Acta Biomater.* **2009**, *5*, 1552–1561. [CrossRef] [PubMed]
27. Palchesko, R.N.; Zhang, L.; Sun, Y.; Feinberg, A.W. Development of Polydimethylsiloxane Substrates with Tunable Elastic Modulus to Study Cell Mechanobiology in Muscle and Nerve. *PLoS ONE* **2012**, *7*. [CrossRef]
28. Bodas, D.; Khan-Malek, C. Formation of more stable hydrophilic surfaces of PDMS by plasma and chemical treatments. *Microelectron. Eng.* **2006**, *83*, 1277–1279. [CrossRef]
29. Wang, H.B.; Mullins, M.E.; Cregg, J.M.; Hurtado, A.; Oudega, M.; Trombley, M.T.; Gilbert, R.J. Creation of highly aligned electrospun poly-L-lactic acid fibers for nerve regeneration applications. *J. Neural Eng.* **2009**, *6*, 016001. [CrossRef]
30. Ayres, C.E.; Jha, B.S.; Meredith, H.; Bowman, J.R.; Bowlin, G.L.; Henderson, S.C.; Simpson, D.G. Measuring fiber alignment in electrospun scaffolds: A user's guide to the 2D fast Fourier transform approach. *J. Biomater. Sci. Polym. Ed.* **2008**, *19*, 603–621. [CrossRef]
31. Driscoll, T.P.; Nerurkar, N.L.; Jacobs, N.T.; Elliott, D.M.; Mauck, R.L. Fiber angle and aspect ratio influence the shear mechanics of oriented electrospun nanofibrous scaffolds. *J. Mech. Behav. Biomed. Mater.* **2011**, *4*, 1627–1636. [CrossRef] [PubMed]
32. Hertz, H. Uber die Beruhrung Fester Elastischer Korper (On the Contact of Elastic Solids). *J. Reine Angew. Math.* **1882**, *92*, 156–171.
33. Maloney, J.M.; Walton, E.B.; Bruce, C.M.; Van Vliet, K.J. Influence of finite thickness and stiffness on cellular adhesion-induced deformation of compliant substrata. *Phys. Rev. E* **2008**, *78*, 04192. [CrossRef]
34. Beachley, V.; Wen, X. Effect of electrospinning parameters on the nanofiber diameter and length. *Mater. Sci. Eng. C Mater. Biol. Appl.* **2009**, *29*, 663–668. [CrossRef]
35. Chevallay, B.; Herbage, D. Collagen-based biomaterials as 3D scaffold for cell cultures: Applications for tissue engineering and gene therapy. *Med. Biol. Eng. Comput.* **2000**, *38*, 211–218. [CrossRef] [PubMed]
36. Solon, J.; Levental, I.; Sengupta, K.; Georges, P.C.; Janmey, P.A. Fibroblast Adaptation and Stiffness Matching to Soft Elastic Substrates. *Biophys. J.* **2007**, *93*, 4453–4461. [CrossRef]
37. Engler, A.; Bacakova, L.; Newman, C.; Hategan, A.; Griffin, M.; Discher, D. Substrate Compliance versus Ligand Density in Cell on Gel Responses. *Biophys. J.* **2004**, *86*, 617–628. [CrossRef]
38. Heness, G.L.; Ben-Nissan, B. Innovative bioceramics. *Mater. Forum* **2004**, *27*, 104–114.

© 2019 by the authors. Licensee MDPI, Basel, Switzerland. This article is an open access article distributed under the terms and conditions of the Creative Commons Attribution (CC BY) license (http://creativecommons.org/licenses/by/4.0/).

Article

Sacrificial Core-Based Electrospinning: A Facile and Versatile Approach to Fabricate Devices for Potential Cell and Tissue Encapsulation Applications

Naresh Kasoju [1,2], Julian George [1], Hua Ye [1,*] and Zhanfeng Cui [1,*]

1. Institute of Biomedical Engineering, Department of Engineering Science, University of Oxford, Oxford OX3 7DQ, UK; naresh.kasoju@sctimst.ac.in (N.K.); julian.george@eng.ox.ac.uk (J.G.)
2. Current affiliation: Division of Tissue Culture, Department of Applied Biology, Biomedical Technology Wing, Sree Chitra Tirunal Institute for Medical Sciences and Technology, Thiruvananthapuram 695 012, India
* Correspondence: hua.ye@eng.ox.ac.uk (H.Y.); zhanfeng.cui@eng.ox.ac.uk (Z.C.); Tel.: +44-1865-617689 (H.Y.); +44-1865-617693 (Z.C.)

Received: 3 September 2018; Accepted: 19 October 2018; Published: 21 October 2018

Abstract: Electrospinning uses an electric field to produce fine fibers of nano and micron scale diameters from polymer solutions. Despite innovation in jet initiation, jet path control and fiber collection, it is common to only fabricate planar and tubular-shaped electrospun products. For applications that encapsulate cells and tissues inside a porous container, it is useful to develop biocompatible hollow core-containing devices. To this end, by introducing a 3D-printed framework containing a sodium chloride pellet (sacrificial core) as the collector and through post-electrospinning dissolution of the sacrificial core, we demonstrate that hollow core containing polyamide 66 (nylon 66) devices can be easily fabricated for use as cell encapsulation systems. ATR-FTIR and TG/DTA studies were used to verify that the bulk properties of the electrospun device were not altered by contact with the salt pellet during fiber collection. Protein diffusion investigations demonstrated that the capsule allowed free diffusion of model biomolecules (insulin, albumin and Ig G). Cell encapsulation studies with model cell types (fibroblasts and lymphocytes) revealed that the capsule supports the viability of encapsulated cells inside the capsule whilst compartmentalizing immune cells outside of the capsule. Taken together, the use of a salt pellet as a sacrificial core within a 3D printed framework to support fiber collection, as well as the ability to easily remove this core using aqueous dissolution, results in a biocompatible device that can be tailored for use in cell and tissue encapsulation applications.

Keywords: electrospinning; 3D printing; nanofibers; encapsulation; protein diffusion; *in vivo* tissue engineering; immuno-isolation; transplantation

---

1. Introduction

Electrospinning is a process in which a charged polymer jet is deposited on a grounded collector. Introduced in the early 1900s, the technology is now a well-established and an intensively investigated way of producing continuous fibers from a wide variety of polymer solutions, with diameters ranging from several microns down to several nanometers [1]. With a high surface area to volume ratio, tunable porosity and flexibility in surface modification, electrospun fibers have been successfully used in various applications including the fabrication of protective clothing, electronics, catalysis and pharmaceutics, as well as in the fields of filtration, environmental engineering and regenerative medicine. Typically, a syringe is used to load the polymer solution and a syringe pump controls the solution flow rate from a tip charged by a high-voltage power unit. The charged polymer jet takes a torturous path, elongating and drying as it is pulled towards a grounded collector. Many modifications to the electrospinning setup have been investigated over the last couple of decades [2,3]. For instance, a co-axial electrospinning system was developed to fabricate core-shell fibrous structures, wherein the inclusion of a sacrificial core can be used to yield hollow fibers or the inclusion of multiple cores can be used to yield multi-core fibers [4,5]. An axillary cylindrical electrode has also been used to facilitate multiple spinneret electrospinning that dramatically increases the rate of fiber production [6]. Similarly, conductive patterned collector-based electrospinning has been used to generate fibrous materials with controlled architectures [7].

Despite innovations in jet facilitation and control over path and fiber collection, the most commonly fabricated electrospun forms are flat membranes produced using flat collector plates, and tubular structures produced using rod-like collectors. In the context of biomedical applications, flat membranes are widely used as substrates for culturing a variety of cells and engineering tissues, such as skin [8], whereas the tubular structures are used to engineer tissues such as blood vessels and neural conduits [9]. Beyond these two forms, there is demand and scope for novel products and devices such as encapsulation systems that expand the reach of electrospinning applications in regenerative medicine. For example, a common approach used to treat type 1 diabetes mellitus is to infuse pancreatic islets into the hepatic portal vein; however, this implantation site has proven to be sub-optimal. The development of devices that encapsulate cells within a protective environment would facilitate the choice of alternative sites. In the field of tissue engineering, current *in vitro* approaches are unable to replicate native cell niche environments and hence there is interest in the development of *in vivo* bio-incubators that better replicate the tissue environment. Similarly, in the field of allo- or xeno-cell and tissue transplantation, the side effects of using immune suppressants are undesirable and there is increasing demand to develop suitable immuno-isolating devices [10,11].

In this report, we describe the fabrication of a macro-scale encapsulation device, produced through manipulation of the collection of electrospun fibers. It was hypothesized that the introduction of a water-soluble pellet (sacrificial core), such as a custom-made sodium chloride pellet, as the collector would allow for post-electrospinning dissolution of this sacrificial core, making it possible to fabricate a hollow polymeric macro-scale encapsulation device. Fabrication of capsules with varying size, shape and thickness would be possible by modifying the size, shape and thickness of the sacrificial core. It was further hypothesized that a framework would be required to prevent the collapse of the electrospun device upon sacrificial core dissolution, and therefore, it was proposed that a 3D printed framework would be used to support the sacrificial core during preparation and electrospinning, and that this framework would remain in place to prevent collapse following core dissolution. Polyamide 6,6 (PA66) was selected as the polymer of choice. This is a non-degradable and biocompatible polymer that has been widely explored in biomedical applications. The electrospun capsules produced were evaluated through the use of scanning electron microscopy (SEM), attenuated total reflectance Fourier transform infrared spectroscopy (ATR-FTIR) and simultaneous thermo gravimetric/differential thermal analysis (TG/DTA). Finally, protein diffusion studies using model biomolecules (insulin, albumin and Ig G) were performed to study the free diffusion of essential proteins through the capsule walls, and cell encapsulation studies with model cell types (fibroblasts and lymphocytes) were performed to investigate whether the device could support the viability of encapsulated cells whilst restricting immune cell access from the surrounding space.

2. Materials and Methods

2.1. Materials

Polyamide 6,6 (PA66, also known as nylon 6,6 or Poly(N,N′-hexamethyleneadipinediamide)), 2,2,2-Trifluoroethanol (TFE), formic acid (FA), 1,1,1,3,3,3-Hexafluoro-2-propanol (HFIP), human insulin, human Ig G and bovine serum albumin (BSA) were purchased from Sigma-Aldrich, Dorset, UK, and the protein determination kit was purchased from Cayman Chemical, Ann Arbor, MI, USA. Dulbecco's Modified Eagle Medium (DMEM), Roswell Park Memorial Institute 1640 (RPMI-1640) medium, fetal bovine serum (FBS), trypsin-Ethylenediaminetetraacetic acid (EDTA), phosphate-buffered saline (PBS, 1×), Penicillin-Streptomycin (Pen-Strep, 100×) and Alamar blue were purchased from Thermo Fisher Scientific, Waltham, MA, USA.

2.2. Preparation of Macroscale Capsules by Sacrificial Core Electrospinning

2.2.1. Preparation of a 3D Printed Supporting Framework

An online computer-assisted design tool (Tinkercad, AUTODESK, San Rafael, CA, USA) was used to prepare a 3D stereolithographic design in STL (stereolithography) file format. As shown in Figure 1a, a typical framework contains an external thick skeletal structure with a thin internal lattice of crisscross elements used to create a luminal space. An opening injection port to one side allows access to the internal space. The design was uploaded to a computer-assisted 3D printer manufacturing system (Form 2, FormLabs, Somerville, MA, USA). The 3D printer uses a laser to cure solid isotropic parts from a liquid photopolymer resin. A biocompatible resin was used (EN-ISO 10993-1:2009/AC:2010, USP Class VI, Dental SG, FormLabs, Somerville, MA, USA). Removal of non-cured resin was achieved by post-fabrication washing in absolute ethanol for 15 min, as per the supplier's recommendation.

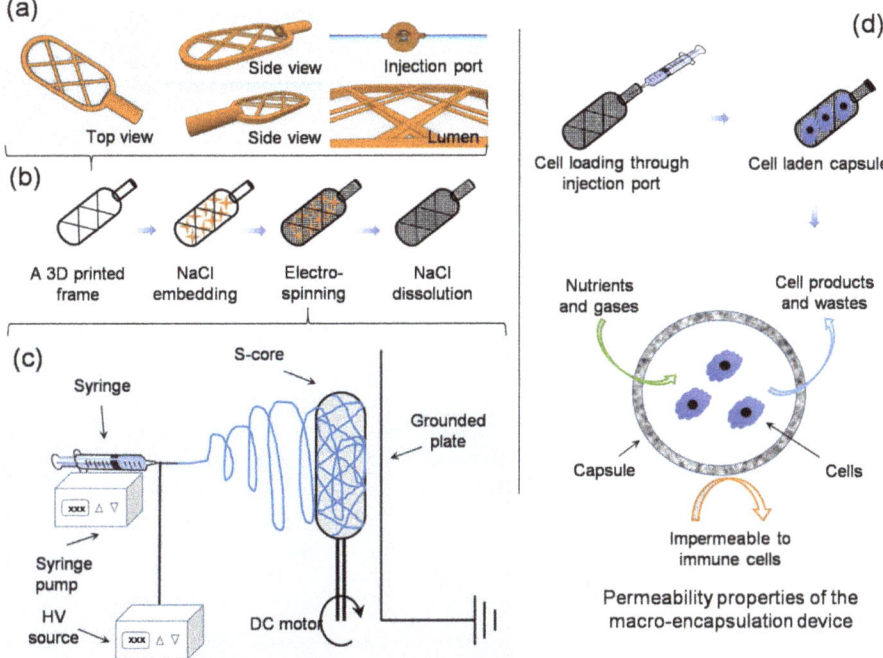

Figure 1. Schematic showing sacrificial core electrospinning capsule fabrication and use: (**a**) Images show the CAD 3D model of the supporting framework (frame). (**b**) Work flow showing the steps involved in device assembly, detailing supporting framework production, salt embedding, electrospinning and salt dissolution. (**c**) Electrospinning setup, highlighting how the salt aggregate contacting supporting frame was placed directly in front of the grounded collecting plate and rotated during fiber collection. (**d**) Loading of the electrospun capsule with cells for use in cell and tissue encapsulation applications, highlighting how the capsule allows free diffusion on nutrients, gases, cellular products and waste, whilst blocking access to immune cell penetration.

2.2.2. Embedding Salt into the Printed Framework

The printed framework was kept on a flat surface and sodium chloride crystals were packed into the lumen of the framework. Water was used to wet the packed crystals, partially solubilizing the salt at the crystal boundaries, and fusing the crystals into a single aggregate. After allowing to dry at room temperature on an open bench for 24 h, the salt pellet containing the framework was ready to be used as a collector during electrospinning (Figure 1b).

2.2.3. Sacrificial Core Electrospinning

PA66 pellets were suspended in different organic solvents at various concentrations (see Table 1) and stirred continuously until completely dissolved. The solution was electrospun at ambient temperature (20–22 °C) in a low humidity environment (25 ± 5% RH – relative humidity, controlled by purging the chamber with nitrogen gas). The sacrificial core electrospinning setup included a high-voltage power supply, a digital syringe pump and a specialized collecting unit, composed of the sacrificial core and printed framework connected to a DC (direct current) motor and placed directly in front of a grounded metal plate (Figure 1c). The entire setup was enclosed inside a custom-made plastic chamber with fume extraction and safety interlocks. To reduce variability in fiber diameter and packing density, the DC motor was run at a constant speed during the fabrication of all capsules. Systematic optimization was performed by varying the processing parameters, as detailed in Table 1.

Table 1. Details of electrospinning parameters varied during optimization.

Parameter	Test Conditions	Constant Variables
Solvent type	FA, TFE and HFIP	Polymer conc. = 7.5% (w/v); Voltage = 15 kV; Distance = 15 cm; Flow rate = 0.05 mL/min
Polymer concentration	5, 7.5 and 10% (w/v)	Solvent = HFIP; Voltage = 15 kV; Distance = 15 cm; Flow rate = 0.05 mL/min
Voltage	10, 15, 20 and 30 kV	Polymer conc. = 7.5% (w/v); Solvent = HFIP; Distance = 15 cm; Flow rate = 0.05 mL/min
Tip to collector distance	10, 15, 20 and 30 cm	Polymer conc. = 7.5% (w/v); Solvent = HFIP; Voltage = 15 kV; Flow rate = 0.05 mL/min
Voltage: Tip to collector distance	1:2 (15 kV:30 cm), 1:1 (15 kV:15 cm) and 2:1 (30 kV:15 cm)	Polymer conc. = 7.5% (w/v); Solvent = HFIP; Flow rate = 0.05 mL/min
Flow rate	0.025, 0.050, 0.075 and 0.100 mL/min	Polymer conc. = 7.5% (w/v); Solvent = HFIP; Voltage = 15 kV; Distance = 15 cm

2.3. Characterization Studies

The morphological features of the electrospun PA66 materials were imaged and analyzed using SEM (Evo LS15, Carl Zeiss, Oberkochen, Germany). For this purpose, the samples were air dried in a chemical fume hood. They were then mounted on the specimen holder with double adhesive electro-conductive carbon tape, sputter-coated with platinum for 90 seconds under argon gas in a coating unit (Polaron range SC7620, Quorum Technologies Ltd, East Sussex, UK), and imaged. Fiber diameter was calculated manually by analyzing the images using Image J software (NIH, Bethesda, MD, USA).

Changes to the chemistry of PA66 after electrospinning were tracked using Attenuated Total Reflection-Fourier-transform infrared spectroscopy (ATR-FTIR, Tensor 37, Bruker, Billerica, MA, USA). For each measurement, 16 scans were recorded with a resolution of 1 cm^{-1} and wave numbers ranging from 800 to 4000 cm^{-1}. During the measurements, the instrument was continuously purged with nitrogen using blow-off from a liquid nitrogen tank to eliminate the spectral contributions of atmospheric water vapor.

The thermal properties of PA66 before and after electrospinning were analyzed by simultaneous thermogravimetry/differential thermal analysis (TG/DTA, Diamond, PerkinElmer, Waltham, MA, USA). In each measurement, the sample was heated from 40 to 600 °C at a rate of 10 °C/min. During the measurements, the instrument was continuously purged with nitrogen gas at a rate of 20 mL/min to eliminate the spectral alterations due to the presence of ambient air.

2.4. Protein Diffusion Assay

PA66 capsules (1 cm width × 2 cm length × 1 mm thickness) made from 7.5% (w/v) polymer solution with 15 kV voltage, 15 cm distance and 0.5 mL/min flow rate, were sterilized by autoclaving and saturated with 1× PBS for 3 h in an incubator (5% CO_2, 37 °C). Capsules were placed in a 6-well plate and simultaneously loaded with insulin, Ig G or BSA solution (1 mg/mL). The capsule loading port was tied tightly with a sterile nylon thread and kept in a CO_2 incubator overnight. The contents were manually mixed at regular intervals. The spent protein solution released from the capsule was collected and assayed by the Bradford method following the kit instructions (Cayman Chemical, Ann Arbor, MI, USA). Briefly, 100 µL of diluted protein solution was mixed with equal amounts of assay reagent. The contents were incubated at room temperature for 5 min and the absorbance was read at 595 nm in a multi-well plate reader (Spectra Max i3x, Molecular Devices, San Jose, CA, USA). Results were compared with a standard curve made using known concentrations of insulin, Ig G and BSA.

2.5. Cell Encapsulation Studies

Human dermal fibroblasts (HDF, Thermo Fisher Scientific, Waltham, MA, USA) and Jurkat cells (ATCC, Manassas, VA, USA) were routinely cultured in DMEM and RMPI-1640 medium respectively, in a CO_2 incubator. The medium was supplemented with 10% FBS and 1% Pen-Strep. PA66 capsules were prepared as described in Section 2.4. The capsules were saturated in complete medium overnight, and 5.0×10^5 cells/mL of exponentially growing cells were loaded into each capsule through the injection port. The injection port was closed tightly with a sterile nylon thread, and they were cultured in a CO_2 incubator. After 3 d of incubation, the cell-laden capsules were transferred to a fresh plate with fresh medium. The presence of cells, both inside and outside the capsules, was measured using the metabolic Alamar blue assay, comparing the medium in the spent and receiving plates. In the case of HDF cells, if any cells migrate out of the capsule, they would attach to the bottom of the plate; the spent medium was exchanged directly with the fresh medium. In the case of Jurkat cells, the spent medium was centrifuged to collect cells at the bottom of the well prior to replacing the spent medium with fresh medium. Briefly, 1 mL of fresh medium together with 100µL of Alamar blue reagent was added to both the wells containing the cell-laden capsules and the wells of the spent plate. The contents were mixed well and incubated in a CO_2 incubator. After 4 h of incubation, the absorbance was measured at a wavelength of 570 nm, using a 600 nm wavelength as a reference. An equal number of cells cultured in a standard tissue culture treated polystyrene (TCPS) plate, under similar conditions, were treated as the positive control. Cell-free Alamar blue reagent, under similar conditions, was treated as the negative control. The cell viability was measured using the following Equation:

$$\text{Cell viability\%} = \frac{\text{Absorbance of Test Sample}}{\text{Absorbance of Control Sample}} \times 100 \qquad (1)$$

2.6. Statistical Analysis

The quantitative values were averaged and expressed as mean ± standard deviation. One-way ANOVA was performed to compare significance across groups, and the differences were denoted with *, # and ‡ representing $p < 0.05$, $p < 0.01$ and $p < 0.001$ respectively.

3. Results and Discussion

Inspired by techniques such as freeze-drying [12] and salt-leaching [13] where ice crystals or salt crystals are used as porogens, it was hypothesized that it should be possible to use a similar approach to modify the macroscale form of fibers collected using conventional electrospinning. This setup, termed "sacrificial core electrospinning", could be used to produce devices for potential cell and tissue encapsulation applications. As presented in Figure 1, the setup was simple and extends conventional systems though inclusion of a sacrificial core (a sodium chloride pellet) rotated directly in front of a grounded plate. Fibers deposited onto the pellet conform to the shape and size of the pellet. After electrospinning, the sacrificial core pellet can be dissolved by suspending in ultrapure water. Whilst this results in the creation of a hollow capsule with a port for cell or tissue injection, the electrospun fibers do not provide sufficient internal support and the device is susceptible to collapse (Figure S1). To overcome this issue, a custom-made 3D printed supporting frame was developed (Figure S2) and used both as a former for the sacrificial core and to provide internal support once the core has been dissolved (it is worthwhile to mention here that use of 3D printed frame without salt embedding was not efficient - check Figure S3). Using this approach, it was possible to successfully fabricate devices with varying length, width (Figure 2a), shape (Figure 2b) and thickness (Figure 2c).

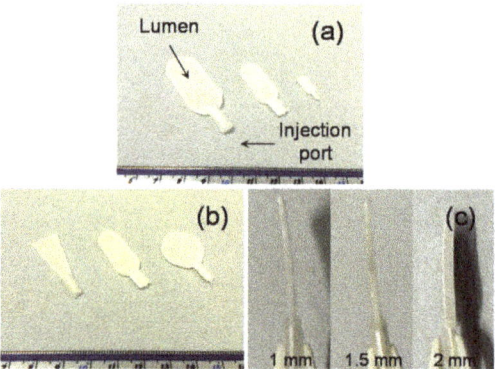

Figure 2. Sacrificial core electrospun products: With sacrificial core-based electrospinning, it was possible to prepare PA66 macro-capsules with variable size (**a**), shape (**b**), and thickness (**c**).

Whilst conventional electrospinning techniques enable control to be gained over the microscale properties of fiber diameter and density, the use of a printed framework and salt core facilities the ability to control the macro-scale features of the fabricated construct. Of the 21 parameters commonly observed to influence fiber formation and the properties of the resultant product, we studied the effects of solvent type, polymer concentration, flow rate, applied voltage, tip-to-collector distance and voltage–to–distance ratio using polyamide 66 (PA66 or nylon 66) (Table 1 and Figure 3) [14]. In accordance with the literature, electrospunPA66 dissolved in formic acid (FA) resulted in thinner fibers (121 ± 12 nm), in comparison to PA66 electrospun in 2,2,2-Trifluoroethanol (TFE, 388 ± 127 nm) or 1,1,1,3,3,3-Hexafluoro-2-propanol (HFIP, 967 ± 127 nm) (Figure 3a). Differences in the evaporation rate of the solvents, as well as the conductivity and viscosity of the resultant solutions may have also contributed to the differences observed in fiber diameter [14]. Similarly, an increase in the PA66 concentration from 5 to 7.5 to 10% (w/v) resulted in an increase in fiber diameter from 486 ± 159 nm to 967 ± 127 nm to 1908 ± 258 nm respectively (Figure 3b). The viscosity and elastic modulus of the solution increases with polymer concentration and this may be attributable for the changes in the fiber diameter observed [14]. The change in fiber diameter in relation to voltage was found to follow a slight inverse relationship, with fiber diameters of 1113 ± 258 nm, 967 ± 127 nm and 904 ± 247 nm recorded at 10, 15 and 20 kV (Figure 3c). These changes may be attributed to the differences in volumetric charge density associated with the applied voltage, which would significantly affect the initial jet diameter at the tip of the Taylor cone [14]. Whilst the distance (Figure 3d) and the voltage–to–distance ratio (Figure 3e) do not seem to have any significant influence on fiber diameter, we noticed a change in fiber morphology in relation to the voltage–to–distance ratio. Finally, the fiber diameters were found to be 937 ± 189 nm, 967 ± 127 nm, 1165 ± 412 nm and 1292 ± 194 nm, when the flow rates were 0.025, 0.050, 0.075 and 0.100 mL/min respectively (Figure 3f). It has previously been noted that, with other parameters constant, the lowest flow rate typically results in a small jet radius that yields the thinnest fibers, whereas higher flow rates typically result in a larger jet radius and yield thicker fibers [14]. Our observations were in agreement with earlier reports [14–16].

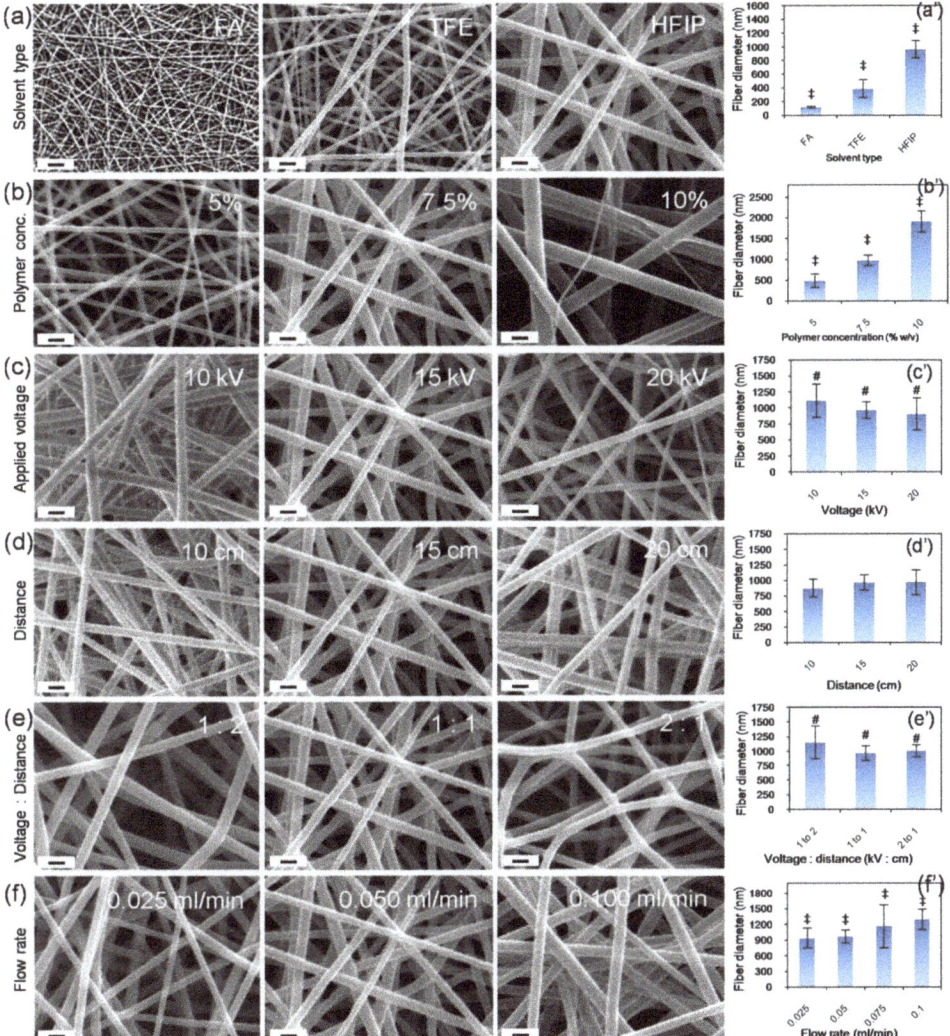

Figure 3. Effect of electrospinning parameters on fiber morphology: scanning electron microscopy (SEM) images of electrospun PA66 samples showing the effects of solvent type (**a**), polymer concentration (**b**), voltage (**c**), distance (**d**), voltage–to–distance ratio (**e**), and flow rate (**f**). Corresponding plots of fiber diameter are presented in (**a'**) to (**f'**) respectively. Fiber width was analyzed by manually measuring the width of individual fibers against a given scale using Image J software (n = 25). # and ‡ denotes statistical differences at $p < 0.01$ and $p < 0.001$ respectively. Scale bar = 2 μm.

The bulk properties of the polymer were compared before and after the electrospinning process. Electricity, a physical property, does not typically affect the material properties of the polymer unless reactive additives are present in the polymer solution. In the current approach, the sodium chloride core was in direct contact with the deposited fibers within the electric field. Attenuated total reflectance—Fourier transform infrared spectroscopy (ATR-FTIR) revealed similar spectral patterns with characteristic peaks related to N–H stretching at ~3330 cm^{-1}, CH$_2$ stretching at ~2840 cm^{-1}, amide I at ~1650 cm^{-1}, amide II at ~1545 cm^{-1}, and amide III at ~1370 cm^{-1} [15,17] (Figure 4a). A simultaneous TG/DTA investigation also suggested comparable heat flow patterns with a characteristic peak melting temperature (T_m) at ~264.5 °C for pristine and ~262.5 °C for electrospun PA66, and decomposition patterns with the onset decomposition temperature of ~380 °C for pristine and ~373 °C for electrospun PA66 samples (Figure 4b,c). The spectral peak values were in good agreement with previously reported values for PA66 [16,18]. Collectively, ATR-FTIR and TG/DTA studies confirmed that the sacrificial core electrospinning was similar to the typical electrospinning process and did not influence the bulk properties of the polymer used in this study.

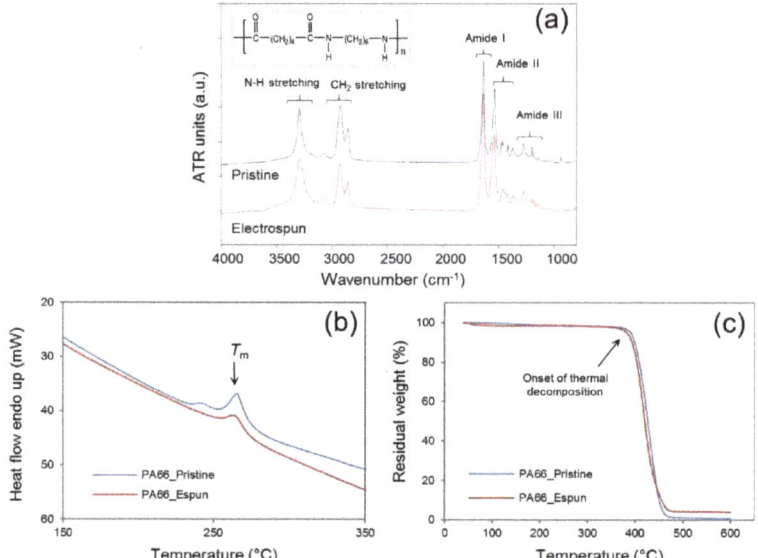

Figure 4. Material characteristics before and after sacrificial core electrospinning: Attenuated Total Reflection-Fourier-transform infrared spectroscopy (ATR-FTIR) (**a**), and thermogravimetry/differential thermal analysis (TG/DTA) (**b**—DSC, **c**—TGA) confirmed that the use of sodium chloride within the collector has no significant influence on the bulk properties of the polymer (PA66).

Since the devices fabricated using the sacrificial core electrospinning technique are intended to be used in cell and tissue encapsulation applications, it is of prime importance to verify the free movement of biomolecules, such as cytokines and growth factors, across the capsule walls. To investigate this, fabricated capsules were independently loaded with three model proteins: insulin, albumin and immunoglobulin G (Ig G) (1 mg/mL in PBS), with molecular weights of 5.8, 66 and 150 kDa, respectively, and incubated in protein-free PBS. Following overnight incubation at 37 °C, the capsule was removed, and the solution taken from around the capsule was assayed using the Bradford method. The presence of protein was detected in the solutions from all tested capsules (see Figure 5). However, minor differences were observed between the solutions from the 5, 7.5 and 10% (w/v) PA66 capsules. This may be due to the differences in the total surface area of the fibrous constructs. However, these differences were negligible. The results suggested that, in contrast to conventional hydrogel-based encapsulation systems, where the rate of biomolecule diffusion is restricted, the electrospun membranes produced in this study allow free diffusion of biomolecules. This may be beneficial for cell and tissues encapsulation applications that require rapid biomolecule transport [19].

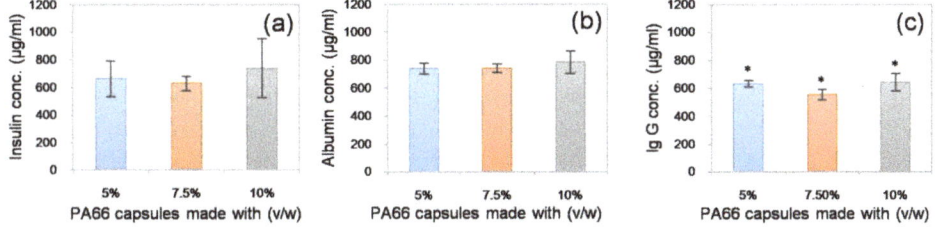

Figure 5. Protein diffusion study: Biomolecules such as insulin (**a**), albumin (**b**), and Ig G (**c**) loaded (1 mg/mL) into the sacrificial core electrospun capsules, were found to diffuse through the membrane into the incubation buffer. * denotes statistical differences at $p < 0.05$.

Finally, the ability of the electrospun capsule to support cells was assessed *in vitro* using two different model cell types: adherent fibroblasts (from human dermis) and non-adherent lymphocytes (Jurkat cells from human peripheral blood). In each case, 5.0×10^5 cells were loaded into the capsule and the cell-laden capsule was incubated in a CO_2 incubator. After 3 d of incubation, cell viability was indirectly measured using the metabolic Alamar blue assay. Cells cultured in a well plate under similar conditions were treated as the positive control. The purpose of the study was to investigate whether the porous structure of the capsule posed a barrier to diffusion that would affect the cell viability. Cell viability remained unchanged across all capsules tested (Figure 6). These results suggest that the electrospun membranes posed no limitation to the exchange of gases, nutrients and metabolic wastes essential for cell survival. We also collected the spent medium and performed further Alamar blue assays to determine if the encapsulated cells were released through the porous wall of the capsule. Both the fibroblasts and lymphocytes were not detected in the spent media, indicating that the porosity of the device (Figure S4) was efficient enough to restrict access to cells (Figure 6). The results indicate that the resultant electrospun device could be potentially used in various cell and tissue encapsulation applications. Considering the lymphocyte-based results, the capsule could also be used as an immuno-isolating system; however, detailed experiments are required to prove the immuno-protective properties of the device.

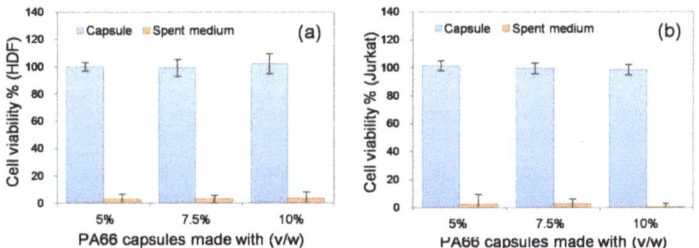

Figure 6. Cell encapsulation study: (**a**) Adherent cells (human dermis fibroblasts, HDF) and (**b**) non-adherent cells (lymphocytes from human peripheral blood, Jurkat cells) were encapsulated in the electrospun capsules. The number of cells present at day 3 remained unaffected in comparison to cultures on standard tissue culture-treated polystyrene dishes. Spent media showed no detectable cell activity, suggesting that the capsules effectively compartmentalized the cells. No statistical differences were found within the capsule and spent medium groups.

Previously, Lathuilière et al., generated flat sheet devices consisting of an outer polymeric frame, porous membranes and a reinforcement mesh assembled using ultrasonic welding [20]. These devices were used for the implantation of genetically engineered allogeneic cells and passive immunization against amyloid-β. Similarly, Nyitray et al., reported a two-step heat-sealing process to fabricate a polycaprolactone-based thin-film device and used this device to transplant pancreatic islets [21], whilst Park et al., described the assembly of a bacterial cellulose, collagen and alginate-based composite device for encapsulation of neuronal cells [22]. Recently, David et al., showed how a Polytetrafluoroethylene-based bilaminar device (Theracyte) can be used in the encapsulation of ovarian allograft to restore ovarian endocrine function [23]. While these reports suggest that there are a wide range of potential applications for bio-encapsulation technology, they also reveal the unmet need for an efficient macro-encapsulation device and illustrate the complexity of fabrication processes currently being investigated. In contrast, the sacrificial core electrospinning approach presented in this report is relatively simple as it follows well-known electrospinning principles and there is no complicated assembly or other processing steps involved. It is also versatile as it allows the fabrication of a device with a wide range of macro- and micro-scale features. The resultant device can be used as a cell-based drug delivery device or as a cell reservoir to treat conditions such as diabetes, neurological disorders and sensory diseases; however, the true efficiency of the device can only be determined by further *in vivo* studies.

4. Conclusions

In conclusion, this report describes a technological advancement in the field of electrospinning. While retaining all the benefits of the electrospinning process, a sodium chloride-based pellet (sacrificial core) has been introduced to act as a rotating collector, placed directly in front of a grounded plate. Subsequently, the dissolution of the salt pellet results in a hollow core containing capsule. With this technique, we have developed nylon (PA66)-based devices with varying macroscale features such as size (2×4, 1×2 and 0.5×1 cm), shape (rectangular, triangular and circular) and thickness (1, 1.5 and 2 mm). Additionally, it was possible to control the morphological properties of the capsule, such as fiber diameter by altering the conventional electrospinning parameters, including solvent type, polymer concentration, voltage, distance, voltage–to–distance ratio and flow rate. The results from the ATR-FTIR and TG/DTA studies performed suggest that the use of the sodium chloride pellet as the sacrificial core had no influence on the bulk properties of the polymer. Protein diffusion studies revealed that the capsules allowed free diffusion of biomolecules such as insulin, albumin and Ig G. Cell encapsulation studies with HDF and Jurkat cells revealed that the optimized capsules could be used to support the viability of encapsulated cells, whilst the walls of the capsules would restrict the ingress of immune cells. Further studies to unravel the potential of this system in various cell and tissue encapsulation applications are in progress.

Supplementary Materials: The following are available online at http://www.mdpi.com/2079-4991/8/10/863/s1, Figure S1: Physical state of the sacrificial core electrospun device without a supporting framework; Figure S2: A representative image of a 3D-printed supporting framework; Figure S3: Electrospinning onto 3D printed framework without inclusion of a NaCl pellet; Figure S4: Pore area calculations.

Author Contributions: Conceptualization, N.K., H.Y. and Z.C.; Methodology, N.K. and J.G.; Data Curation, N.K., J.G. and H.Y.; Writing-Original Draft Preparation, N.K.; Writing-Review & Editing, H.Y. and Z.C.

Funding: This research was funded by China Regenerative Medicine International Limited, Hong Kong.

Acknowledgments: We extend our thanks to our colleague *Dr. Nguyen T.B. Linh* for assistance with ATR-FTIR studies, *Dr. Colin Johnston* (Department of Materials) for assistance with TG/DTA analysis and *Dr. Michelle Kümin* for critical reading of the manuscript.

Conflicts of Interest: The authors declare no conflict of interest. The funders had no role in the design of the study, in the collection, analyses, or interpretation of data, in the writing of the manuscript, and in the decision to publish the results.

References

1. Teo, W.E.; Ramakrishna, S.A. Review on electrospinning design and nanofibre assemblies. *Nanotechnology* **2006**, *17*, R89–R106. [CrossRef] [PubMed]
2. Al-Enizi, A.M.; Zagho, M.M.; Elzatahry, A.A. Polymer-based electrospun nanofibers for biomedical applications. *Nanomaterials* **2018**, *8*, 259. [CrossRef] [PubMed]

3. Fang, Y.; Xu, L.; Wang, M. High-throughput preparation of silk fibroin nanofibers by modified bubble-electrospinning. *Nanomaterials* **2018**, *8*, 471. [CrossRef] [PubMed]
4. Sun, Z.; Zussman, E.; Yarin, A.L.; Wendorff, J.H.; Greiner, A. Compound core–shell polymer nanofibers by co-electrospinning. *Adv. Mater.* **2003**, *15*, 1929–1932. [CrossRef]
5. Zhao, Y.; Cao, X.; Jiang, L. Bio-mimic multichannel microtubes by a facile method. *J. Am. Chem. Soc.* **2007**, *129*, 764–765. [CrossRef] [PubMed]
6. Kim, G.; Cho, Y.S.; Kim, W.D. Stability analysis for multi-Jets electrospinning process modified with a cylindrical electrode. *European Polym. J.* **2006**, *42*, 2031–2038. [CrossRef]
7. Kai, Z.; Xuefen, W.; Dazheng, J.; Yin, Y.; Meifang, Z. Bionic Electrospun Ultrafine fibrous poly(L-Lactic Acid) scaffolds with a multi-Scale structure. *Biomed. Mater.* **2009**, *4*, 035004.
8. Mouthuy, P.A.; El-Sherbini, Y.; Cui, Z.; Ye, H. Layering PLGA-based Electrospun Membranes and Cell Sheets for Engineering Cartilage-Bone Transition. *J. Tissue Eng. Regen. Med.* **2016**, *10*, E263–E274. [CrossRef] [PubMed]
9. Braghirolli, D.I.; Helfer, V.E.; Chagastelles, P.C.; Dalberto, T.P.; Gamba, D.; Pranke, P. Electrospun Scaffolds Functionalized With Heparin And Vascular Endothelial Growth Factor Increase The Proliferation of Endothelial Progenitor Cells. *Biomed. Mater.* **2017**, *12*, 025003. [CrossRef] [PubMed]
10. Kasoju, N.; Kubies, D.; Fabryova, E.; Kriz, J.; Kumorek, M.M.; Sticova, E.; Rypacek, F. In Vivo Vascularization of Anisotropic Channeled Porous Polylactide-Based Capsules for Islet Transplantation: The Effects of Scaffold Architecture And Implantation Site. *Physiol. Res.* **2015**, *64*, S75–S84. [PubMed]
11. Headen, D.M.; Aubry, G.; Lu, H.; Garcia, A.J. Microfluidic-Based Generation of Size-Controlled, Biofunctionalized Synthetic Polymer Microgels for Cell Encapsulation. *Adv. Mater.* **2014**, *26*, 3003–3008. [CrossRef] [PubMed]
12. Kasoju, N.; Kubies, D.; Sedlacik, T.; Janouskova, O.; Kumorek, M.M.; Rypacek, F. Polymer Scaffolds with No Skin-Effect for Tissue Engineering Applications Fabricated By Thermally Induced Phase Separation. *Biomed. Mater.* **2016**, *11*, 015002. [CrossRef] [PubMed]
13. Chiu, Y.C.; Larson, J.C.; Isom, A.; Brey, E.M. Generation of Porous Poly(EthyleneGlycol) Hydrogels by Salt Leaching. *Tissue Eng. Part C Methods* **2010**, *16*, 905–912. [CrossRef] [PubMed]
14. Thompson, C.J.; Chase, G.G.; Yarin, A.L.; Reneker, D.H. Effects of Parameters on Nanofiber Diameter Determined from Electrospinning Model. *Polymer* **2007**, *48*, 6913–6922. [CrossRef]
15. Abbasi, A.; Nasef, M.M.; Takeshi, M.; Faridi-Majidi, R. Electrospinning Of Nylon-6,6 Solutions into Nanofibers: Rheology and Morphology Relationships. *Chin. J. Polym.Sci.* **2014**, *32*, 793–804. [CrossRef]
16. Carrizales, C.; Pelfrey, S.; Rincon, R.; Eubanks, T.M.; Kuang, A.; McClure, M.J.; Bowlin, G.L.; Macossay, J. Thermal And Mechanical Properties of Electrospun PMMA, PVC, Nylon 6, And Nylon 6,6. *Polym. Adv. Technol.* **2008**, *19*, 124–130. [CrossRef]
17. Pramanik, N.K.; Alam, M.S.; Khandal, R.K. Electron Beam Irradiation of Nylon 66: Characterization by IR Spectroscopy and Viscosity Studies. *Int. J. Innovative Res. Sci. Eng. Technol.* **2015**, *4*, 18547–18555. [CrossRef]
18. Zhang, G.; Watanabe, T.; Yoshida, H.; Kawai, T. Phase Transition Behavior of Nylon-66, Nylon-48, and Blends. *Polym. J.* **2003**, *35*, 173–177. [CrossRef]
19. Shoichet, M.S.; Li, R.H.; White, M.L.; Winn, S.R. Stability of Hydrogels Used In Cell Encapsulation: An In Vitro Comparison of Alginate and Agarose. *Biotech. Bioeng.* **1996**, *50*, 374–381. [CrossRef]
20. Lathuiliere, A.; Cosson, S.; Lutolf, M.P.; Schneider, B.L.; Aebischer, P. A High-Capacity Cell Macroencapsulation System Supporting the Long-Term Survival of Genetically Engineered Allogeneic Cells. *Biomaterials* **2014**, *35*, 779–791. [CrossRef] [PubMed]
21. Nyitray, C.E.; Chang, R.; Faleo, G.; Lance, K.D.; Bernards, D.A.; Tang, Q.; Desai, T.A. Polycaprolactone Thin-Film Micro- and Nanoporous Cell-Encapsulation Devices. *ACS Nano* **2015**, *9*, 5675–5682. [CrossRef] [PubMed]
22. Park, M.; Shin, S.; Cheng, J.; Hyun, J. Nanocellulose Based Asymmetric Composite Membrane for the Multiple Functions in Cell Encapsulation. *Carbohydr. Polym.* **2017**, *158*, 133–140. [CrossRef] [PubMed]
23. David, A.; Day, J.R.; Cichon, A.L.; Lefferts, A.; Cascalho, M.; Shikanov, A. Restoring Ovarian Endocrine Function with Encapsulated Ovarian Allograft in Immune Competent Mice. *Ann. Biomed. Eng.* **2017**, *45*, 1685–1696. [CrossRef] [PubMed]

© 2018 by the authors. Licensee MDPI, Basel, Switzerland. This article is an open access article distributed under the terms and conditions of the Creative Commons Attribution (CC BY) license (http://creativecommons.org/licenses/by/4.0/).

Article

Fabrication of Gelatin Methacrylate (GelMA) Scaffolds with Nano- and Micro-Topographical and Morphological Features

Ana Agustina Aldana [1], Laura Malatto [2], Muhammad Atiq Ur Rehman [3,4], Aldo Roberto Boccaccini [3,*] and Gustavo Abel Abraham [1]

[1] Instituto de Investigaciones en Ciencia y Tecnología de Materiales, INTEMA (UNMdP-CONICET), Av. Juan B. Justo 4302, Mar del Plata B7608FDQ, Buenos Aires, Argentina; aaldana@fi.mdp.edu.ar (A.A.A.); gabraham@fi.mdp.edu.ar (G.A.A.)
[2] Instituto Nacional de Tecnología Industrial, Centro de Micro y Nanoelectrónica del Bicentenario (INTI-CMNB), Av. Gral. Paz 5445, San Martin B1650KNA, Buenos Aires, Argentina; laura@inti.gob.ar
[3] Institute of Biomaterials, Department of Materials Science and Engineering, University of Erlangen-Nuremberg, 91058 Erlangen, Germany; muhammad.rehman.ur@fau.de
[4] Department of Materials Science and Engineering, Institute of Space Technology Islamabad, 1, Islamabad Highway, Islamabad 44000, Pakistan
* Correspondence: aldo.boccaccini@ww.uni-erlangen.de; Tel.: +49-9131-85-28601

Received: 12 December 2018; Accepted: 12 January 2019; Published: 18 January 2019

Abstract: The design of biomimetic biomaterials for cell culture has become a great tool to study and understand cell behavior, tissue degradation, and lesion. Topographical and morphological features play an important role in modulating cell behavior. In this study, a dual methodology was evaluated to generate novel gelatin methacrylate (GelMA)-based scaffolds with nano and micro topographical and morphological features. First, electrospinning parameters and crosslinking processes were optimized to obtain electrospun nanofibrous scaffolds. GelMA mats were characterized by SEM, FTIR, DSC, TGA, contact angle, and water uptake. Various nanofibrous GelMA mats with defect-free fibers and stability in aqueous media were obtained. Then, micropatterned molds produced by photolithography were used as collectors in the electrospinning process. Thus, biocompatible GelMA nanofibrous scaffolds with micro-patterns that mimic extracellular matrix were obtained successfully by combining two micro/nanofabrication techniques, electrospinning, and micromolding. Taking into account the cell viability results, the methodology used in this study could be considered a valuable tool to develop patterned GelMA based nanofibrous scaffolds for cell culture and tissue engineering.

Keywords: biomimetic scaffolds; gelatin; electrospinning; micromolding; biomaterials

1. Introduction

The design of biomimetic biomaterials as scaffolds for cell culture is a powerful tool for studying and understanding fundamental cell behavior, specific tissue environment, degradation, and reasons for tissue damage [1]. Scaffolding structures should mimic not only biological properties of extracellular matrix (ECM), but also morphological and topographical features [2–4]. The ECM directs and modulates cell behavior, is composed of fibrous proteins (mainly collagens and elastin), glycosaminoglycans (GAGs), proteoglycans, and glycoproteins [5,6].

In order to emulate the natural structure of the ECM, different technologies have been developed. Current techniques for generating topographical features on polymeric scaffolds for cell culture, especially those with nano-scale resolution, are typically complex and expensive. Thus, a simple and tunable fabrication method for the production of patterned biomimetic scaffold is still pending. By using electrospinning technology, it is possible to obtain micro- and/or nano-fibrous mats that mimic ECM [7,8]. The optimization of different processing parameters and the use of post-processing treatments allow handling dimensions, porosity, morphology, and the spatial arrangement of nanofibers. A huge variety of natural and/or synthetic polymeric solutions has been electrospun. Composition and processing techniques determine the scaffold architecture, mechanical performance, degradation rate, and cell-material interactions. Aligned and randomly oriented electrospun mats have been also developed to study how morphology affects cell behavior. Gao et al. studied the influence of aligned and randomly oriented fibrous gelatin/PLLA scaffolds to guide the growth of corneal stroma cells [9]. The aligned scaffold not only increased cell viability more significantly than that in a randomly oriented scaffold, but it also provided an external stimulus for the orderly arrangement of cells. Similar results were observed by Shalumon et al., who prepared aligned and randomly PLLA/gelatin nanofibrous scaffolds [10]. In these structures, an increase in viability and proliferation of human umbilical vein endothelial cells (HUVECs) and smooth muscle cells (SMCs) was observed.

On the other hand, substrates with various micro- and nano-features such as lines, wells, and holes among others have been explored to introduce significant effects on cell behavior [11]. Most of these reports relate the topographical features with cell orientation, migration, morphology, proliferation, cell gene expression, and differentiation [12–14].

The main goal of this work is to design gelatin-based scaffolds with micro and nano-topographical and morphological features, achieving a high resolution, and performance with low cost. In addition, gelatin, a biocompatible and biofunctional polymer, and benign solvents (Class 3 according to ICH guidelines) [15] are used for scaffold fabrication. Gelatin is an inexpensive biomacromolecule obtained from denatured collagen, and presents integrin cell-binding motifs, such as RGD and matrix metalloproteinases (MMP) degradable sites [16,17]. Compared to native collagen, gelatin has lower antigenicity and less batch-to-batch variation due to the denaturation process, in which tertiary protein structures are removed. Functionalization of amine-containing groups of gelatin with methacrylate groups was used to provide a photopolymerizable biomaterial named GelMA that has been widely investigated for cell-based studies and tissue engineering applications [18–22]. Crosslinking of the methacrylic side groups results in hydrogels with stiffness and density that can be controlled by varying the polymer dry mass, degree of functionalization, photo-initiator concentration, ultraviolet (UV) intensity, and exposure time. In this work, electrospinning and photolithography techniques were used to design 3D scaffolds with novel topographical features in micro- and nanoscale, while UV exposition time was varied. Moreover, the use of micropatterned molds with different sizes as collectors in the electrospinning process is proposed to produce electrospun fibrous mats.

2. Materials and Methods

2.1. Materials

Gelatin type A from porcine skin Gel Strength 300, methacrylic anhydride (MAA), and glacial acetic acid (AA) were purchased from Aldrich (Darmstadt, Germany). The photoinitiator 1-[4-(2-hydroxyethoxy)phenyl]-2-hydroxy-2-methyl-1-propan-1-one (Irgacure®2959) was kindly provided by BASF (Nienburg, Germany). Phosphate-buffered saline (PBS) was freshly prepared in the laboratory.

2.2. Synthesis of GelMA

The preparation of GelMA has been recently described by the authors [23]. Briefly, gelatin was dissolved in PBS (pH 7.4) at a concentration of 1% (wt/v) and 50 °C. Then, a predetermined amount of methacrylic anhydride was added, under vigorously magnetic stirring conditions. The mixture was left to react for one hour at 50 °C and was afterwards dialyzed by using a 12–14 kDa cutoff membrane against distilled water for several days. Finally, functionalized gelatin was frozen and freeze-dried.

The degree of methacrylation, defined as the percent of amine groups converted to methacrylamide groups, was determined by Habeeb's test [24]. Finally, GelMA was characterized by ^{1}H-NMR (Bruker 400 MHz NMR spectrometer, Bruker Biospin, Rheinstetten, Germany) and FTIR spectroscopy (Mattson Instruments Inc., model Genesis II, Madison, WI, USA).

2.3. Fabrication of Electrospun GelMA Nanofibers

A predetermined amount of GelMA was completely dissolved in acetic acid at a 250 mg/mL concentration. Then, Irgacure 2959 was added to GelMA solution. The solution was electrospun through a blunt 18-gauge stainless steel needle onto an aluminum collector plate 10 cm away. A solution flow rate of 0.2 mL/h and an applied high-voltage of 12 kV were used. All experiments were carried out at room temperature and a relative humidity of 50%.

2.4. Photocrosslinking of GelMA Nanofibers

The obtained electrospun GelMA scaffolds were crosslinked by UV irradiation. Fibrous meshes were initially wetted in anhydrous ethanol and then exposed to UV light (UVL-28 lamp, 365 nm), during different time periods (0, 6, 9 and 12 min) at 2.5 cm (named NG-0UV, NG-6UV, NG-9UV and NG-12UV, respectively). After crosslinking treatment, samples were removed from aluminum foil and dried at room temperature in a vacuum oven.

2.5. Fabrication of Micropatterned Molds

Master mold layouts were designed using L-EditTM from Tanner EDA and printed on polyester-based film using a 3600 DPI printer. Molds were then created on silicon wafers by spin-coating SU8 2000 negative photoresist (10 s at 500 rpm and 30 s at 3000 rpm, Microchem Inc., Westborough, MA, USA). Wafers were previously cleaned with the standard piranha solution. The coated wafers were then soft-baked on a hot plate for 20 min at 65 °C, followed by 25 min at 95 °C, before exposing the photoresist through the photomask. Near UV (365 nm) was applied with an exposure dose of 650 mJ/cm^2 (EVG 620, EV Group). Post exposure bake conditions were 5 min at 65 °C, followed by 10 min at 95 °C. Molds were finished by 10 min of immersion on the developer (MicroChem's SU-8 Developer) with strong agitation, rinsed with isopropyl alcohol, and dried with nitrogen stream.

2.6. Fabrication of Micropatterned Nanofibrous GelMA Scaffolds

The electrospinning of micropatterned GelMA scaffolds was carried out by using the same procedure as described above except that the collector plate consisted in a micropatterned mold obtained by photolithography. After electrospinning, the meshes were crosslinked by UV irradiation for 9 min just as it was previously described.

2.7. Morphology Characterization

Surface morphology of the nanofibrous scaffolds before and after crosslinking was examined by scanning electron microscopy (SEM). Samples were placed on double-sided graphite tape, attached onto a metal surface, and sputter-coated with gold for 10 s. SEM micrographs were acquired with different magnifications using a SEM (Jeol USA Inc., model JSM-6460LV, Peabody, MA, USA). The average fiber diameter and fiber diameter distribution were estimated from SEM images, using ImagePro-Plus 6.0® software.

2.8. Infrared Analysis

Chemical composition of the gelatin scaffolds was assessed by Fourier transform infrared spectroscopy with attenuated total reflectance (FTIR-ATR) using a Nicolet 6700 Thermo Scientific (Waltham, MA, USA) spectrometer equipped with a diamond crystal at a nominal incidence angle of

45° and ZnSe lens. Spectra were recorded in the range of 600–4000 cm^{-1} at 32 scans with a resolution of 4 cm^{-1}.

2.9. Contact Angle Measurements

Water contact angles of the fibrous scaffolds were measured with a Ramé-hart goniometer using the sessile drop method. The samples were attached to a glass slide and placed in the sample stage. A droplet of deionized water (10 µL) was automatically dispersed onto the sample surface and its evolution with time was recorded using a CCD video camera attached to the equipment. From the film frames, the water contact angles along time were automatically calculated by the equipment software.

2.10. Water Uptake Measurements

Electrospun GelMA scaffolds discs (1 cm in diameter) were cut and weighted (Wo). Then, samples were immersed in PBS at room temperature. After predetermined immersion times, mats were retrieved and weighted (Wt). The water retention (WR) at time t was calculated according to the following equation:

$$WR(\%) = (Wt - Wo)/Wo \times 100, \qquad (1)$$

2.11. Tensile Testing

Tensile properties of the electrospun scaffolds were tested by using an Instron 4467 universal testing machine (Instron, Norwood, MA, USA). Prior to uniaxial tensile testing, the electrospun fibrous sheets were cut into rectangular shapes (50 mm × 10 mm). Samples were secured between opposing clamps which were approximately 30 mm apart from each other. For tensile testing, samples were stretched until failure at 10 mm/min.

2.12. Thermal Analysis

Thermal properties were determined by thermogravimetric analysis (TGA) and differential scanning calorimetry (DSC). TGA were performed on a Shimadzu TGA-50 analyzer from ambient temperature to 300 °C at 10 °C/min under nitrogen atmosphere. DSC thermograms were obtained in a Perkin-Elmer Pyris 1 calorimeter (PerkinElmer Inc., Waltham, MA, USA). Scans were carried out from 25 to 300 °C at a heating rate of 10 °C/min under a nitrogen atmosphere.

2.13. X-ray Diffraction

The XRD patterns were obtained using a X-ray diffractometer (PANalytical Model X'pert PRO, Royston, UK). Film samples with dimensions of 4.0 cm × 1.5 cm were cut and fixed in a circular clamp of the instrument. The analysis was carried out directly and the conditions were as follows: (i) voltage and current: 40 kV and 40 mA, respectively; (ii) scan range from 3° to 30°; (iii) step: 0.1° and (iv) speed 1°/min, equipped with a secondary monochromator of graphite beam. The samples were stored at 25 °C and 50% Relative humidity (RH) and analyzed in triplicate.

2.14. Roughness Measurements

Roughness was measured by using a laser profilometer (UBM™, ISC-2). A measurement length of 5–7 mm was used with a scanning velocity of 400 points per second. The roughness was calculated using the LMT Surface View UBM™ software.

2.15. Cell Viability

For cell culture studies MG-63 osteoblast-like cells (Sigma-Aldrich, Darmstadt, Germany) as an adequate model for bone cells were used [25]. Culture medium Dulbecco's modified Eagle's medium (DMEM, Gibco, Darmstadt, Germany) supplemented with 10% (v/v) fetal bovine serum (FBS, Merck, Darmstadt, Germany) and 1% (v/v) penicillin/streptomycin (PS, Merck, Darmstadt, Germany) was chosen. The electrospun GelMA samples were cut into pieces at a diameter of 10 mm and then placed onto the bottom of the culture plates, followed by sterilization under UV light. MG-63 cells were seeded onto the samples in 24-well plate at a density of 1×10^4 cells per well, and conserved into an incubator at 37 °C in a humidified atmosphere of 95% air and 5% CO_2 for 48 h. After cell culture, the cell viability was determinate by the enzymatic conversion of tetrazolium salt (WST-8 assay, Sigma-Aldrich) to formazan. A volume of 1 mL of a solution of 1% WST-8 assay n cell culture medium was added to each sample, which were incubated for 4 h. The absorbance at 450 nm was measured with a plate reader (typo Phomo, Anthos Mikrosysteme GmbH, Krefeld, Germany). As a blank value, the cell media containing 1% WST-8 without contact to a sample, was used and measured after 4 h of incubation.

3. Results

3.1. GelMA Nanofibrous Mats

3.1.1. Synthesis of GelMA

GelMA was synthesized according to previously reported methods, in which methacrylate functional groups were grafted onto the gelatin backbone through reactions between methacrylic anhydride and lysine residues [21]. GelMA infrared spectrum showed peaks at 1645, 1526, and 1240 cm^{-1} related to the C=O stretching (amide I), N–H bending (amide II), and C–N stretching plus N–H bending (amide III), respectively. Moreover, a N–H stretching (amide A) could be observed at 3284 cm^{-1}. The modification of lysine residues with methacrylate groups was confirmed by a decrease in the lysine signal at 2.9 ppm, and the appearance of the methacrylate group signal at 5.4 ppm and 5.7 ppm and the methyl group signal at 1.8 ppm. A degree of methacrylation of 71% was calculated using Habeeb's test [24].

3.1.2. Fabrication of Electrospun GelMA Matrices

Electrospun GelMA mats were prepared by electrospinning technique. Defect-free nanofibrous matrices were obtained as shown in Figure 1 (NG-0UV). Then, electrospun mats were successfully crosslinked by UV irradiation. As it is well-known, photo-chemical crosslinking takes place only in the presence of a photoinitiator, and thus hydrogel formation is triggered by the external source of UV light. Irgacure 2959 is usually chosen as photoinitiator due to its low toxicity at the used concentration as demonstrated in previous studies [23]. To evaluate the UV irradiation process time, GelMA mats were immersed in ethanol and exposed at different irradiation times. Figure 1 shows SEM micrographs of GelMA mats with 0, 6, 9 and 12 min of UV irradiation. In a first step of UV irradiation, fibers increased their diameter as result of their swelling in polar solvents. After UV irradiation, mats still exhibited their fibrous structures and their diameters were reduced with respect to the diameter of NG-0UV. Probably, this decrease in fiber size is related to the crosslinking of GelMA. The increase of diameter in NG-12UV could be ascribed to the increase in temperature produces by UV device.

The mean fiber diameter is summarized in Table 1.

Table 1. Fiber diameter, contact angle, water uptake (WR), and DSC results, melting temperature (Tm), and the onset of glass transition temperature (Tg), of GelMA mats.

Matrix.	Mean Diameter (μm)	Contact Angle (°)	WR (%)	Tg_{onset} (°C)	Tm_{onset} (°C)
NG-0UV	1.05 ± 0.17	0	n/d	57.6	74.1
NG-6UV	0.70 ± 0.10	43.2	390 ± 21	52.3	66.5
NG-9UV	0.60 ± 0.12	66.1	405 ± 10	52.9	71.2
NG-12UV	0.80 ± 0.12	56.0	290 ± 15	53.9	68.6

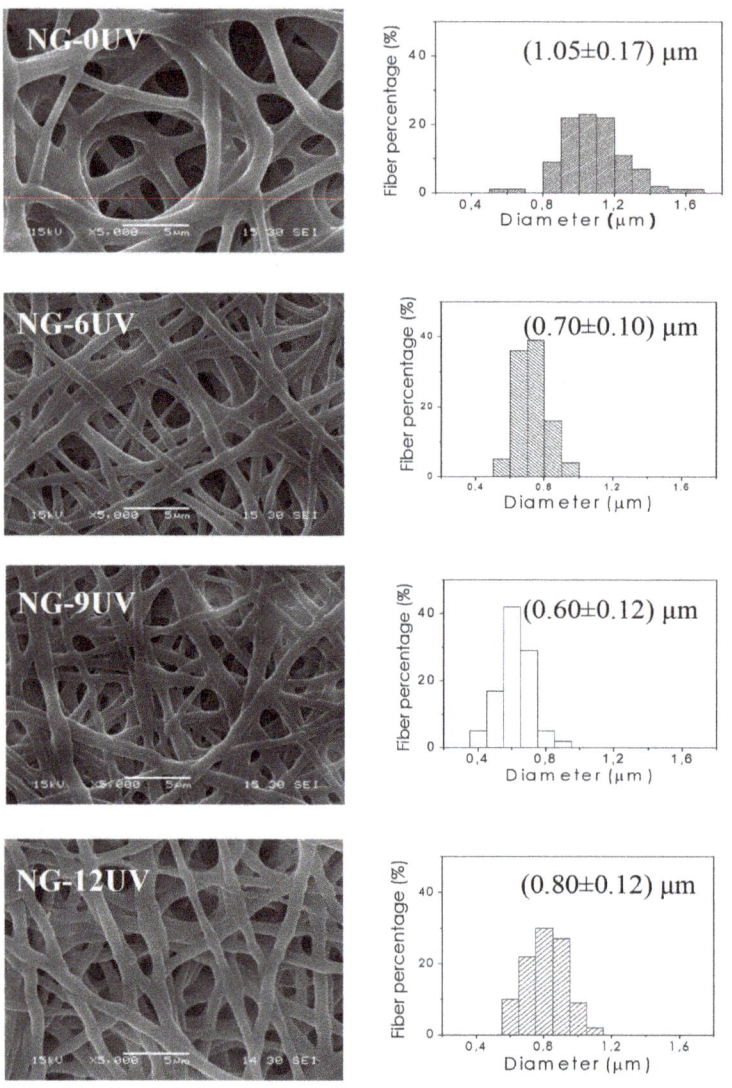

Figure 1. SEM micrographs of samples at different irradiation times: 0 (NG-0UV), 6 (NG-6UV), 9 (NG-9UV) and 12 min (NG-12UV).

The FTIR spectra of GelMA mats showed typical peaks of gelatin (Figure 2a) as described for GelMA. On the other hand, TGA thermograms in Figure 2b show the thermal degradation behavior of GelMA mats. All samples showed a small loss of mass in the range of 50 °C to 120 °C, probably caused by the loss of water molecules. At temperatures above 150 °C, the degradation of biopolymer started for all samples. In addition, their onset points of degradation increased slightly with the time exposed to UV irradiation. Probably, it is due to an increase in the crosslinking degree.

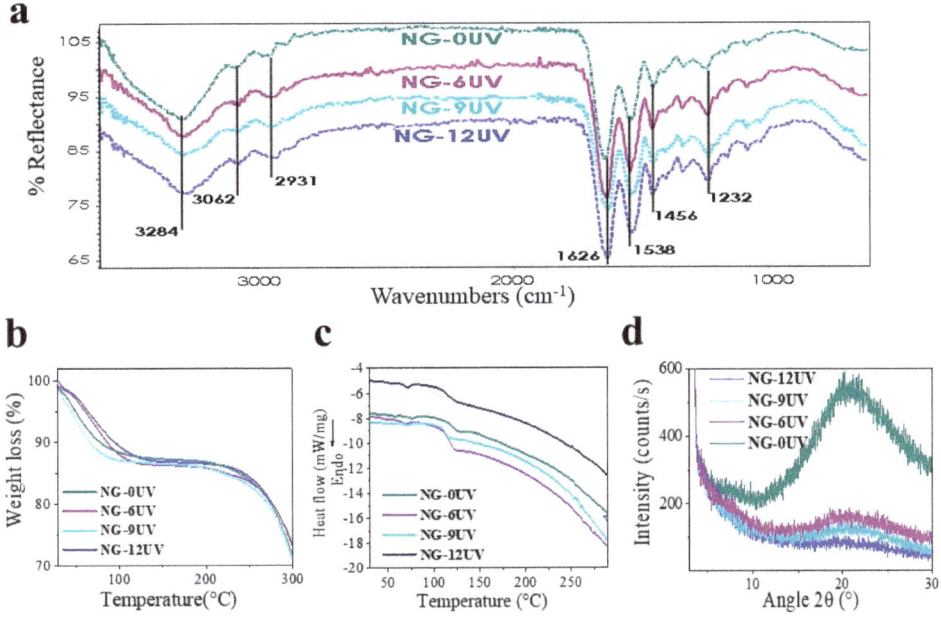

Figure 2. Characterization of samples by (**a**) ATR-FTIR, (**b**) TGA, (**c**) DSC, (**d**) XRD.

Differential scanning calorimetry (DSC) curves of electrospun GelMA mats are shown in Figure 2c. All samples displayed two endothermic peaks at around 50 and 70 °C. The first peak could be attributed to glass transition of amino acid blocks in the peptide chain relating to the amorphous regions of gelatin. Uncrosslinked gelatin showed a higher Tg than the observed in crosslinked gelatin due to peptide chains, could interact physically by intermolecular bonds. In addition, Tg of GelMA increased with the crosslinking percentage because of the reduced polymeric chain mobility. As suggested in the literature, the endothermic peak at around 70 °C is attributed to denaturation protein [26]. The high Tm of uncrosslinked gelatin could be associated to physical crosslinking of peptide chains. All samples showed endotherm falling at around 97 °C; this is attributed to water loss and degradation of protein.

In order to correlate thermal behavior with structural organization, XRD patterns of samples were obtained (Figure 2d). In spite, gelatin has a crystalline structure originated from α-helix and triple helical; the amorphous structures were observed for all the samples. Probably, the electrospinning processing hinders the re-crystallization of gelatin and the XRD pattern showed an amorphous halo. This peak could be associated with the short range order of protein chains. In addition, the intensity of amorphous peak decreased with the increase in crosslinking degree. Probably, the crosslinking hindered the peptide chains re-ordering.

The hydrophilicity of the electrospun meshes was evaluated by measuring their water contact angles (Figure 3a, Table 1). The contact angle of uncrosslinked gelatin mat could not be measured because it was absorbed instantaneously when the PBS solution was in contact. However, crosslinked gelatin nanofibers are stable at these conditions. As expected, the CA values of crosslinked gelatin confirm the hydrophilic character of matrices. UV irradiation time could affect not only crosslinking extension, but could also modify the structure, as can be seen in SEM images. Thus, CA values are determined by both chemical and morphological properties.

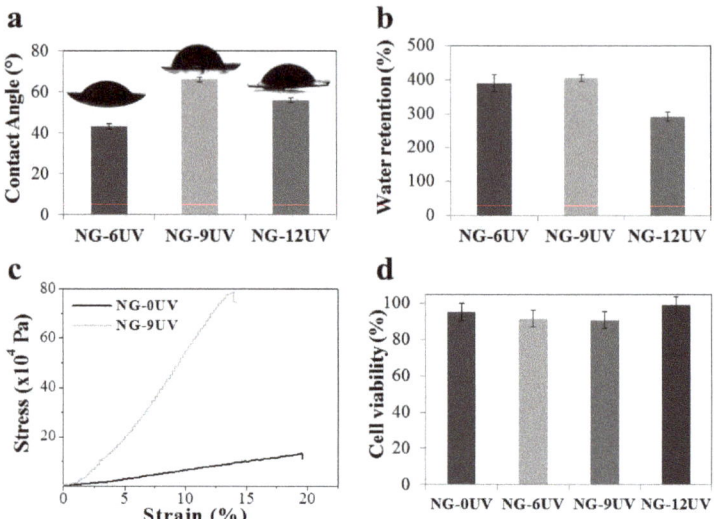

Figure 3. Results of (**a**) CA, (**b**) WR, (**c**) Tensile properties, (**d**) cell viability of GelMA matrices.

Although water uptake measurements were carried out for all samples, only crosslinked gelatin electrospun mats were stable in the PBS solution (Figure 3b, Table 1). Swelling percentages of the matrices were around 300%, demonstrating the high hydrophilic character and wettability of matrices. There were no significant differences between water uptake values of NG-6UV and NG-9UV samples. However, NG-12UV showed the lowest swelling degree. This fact can be probably ascribed to the major crosslinking extension of this sample, which clearly affects the water uptake capability.

In order to confirm the crosslinking, the mechanical properties of NG-9UV and NG-0UV nanofiber mats were tested (Figure 3c). A uniaxial tensile testing was performed and the Young's modulus (YM) was determined for both matrices. YM value of NG-9UV sample was seven times higher than the measured for NG-0UV (0.717 ± 0.001 MPa and 4.89 ± 0.03 MPa, respectively). Thus, the crosslinking methodology used in this work for GelMA curing was successful. Moreover, a relationship between the morphological/chemical properties and UV irradiation time was observed. Thus, a UV irradiation time of 9 min was chosen for further studies, according to the obtained results.

3.2. Micropatterned Nanofibrous GelMA Scaffolds

3.2.1. Fabrication of Micropatterned Molds

3D structures were created by optical lithography of epoxy-based photoresist on silicon wafers, obtaining 115 μm thick molds. Designs included patterns of 50 μm, 100 μm, 200 μm, and 400 μm. Figure 4 shows a microscope image (BX 51, Olympus) of a chip with 200 μm structures.

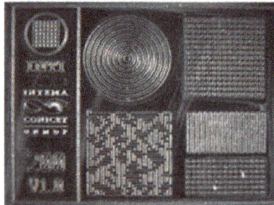

Figure 4. Microscope images of micropatterned mold.

3.2.2. Fabrication of Micropatterned Nanofibrous GelMA Scaffolds

Once molds were fabricated, micropatterned nanofibrous scaffolds were obtained. The electrospun mats were collected over micropatterned molds and then they were peeled out carefully. GelMA nanofibrous mats were obtained using five different types of surface topographies: concentric-circles, dot, parallel-lines, random-lines (maze), and squares. After crosslinking, samples were examined by SEM. The electrospun mats successfully reproduced the micropatterned design of the molds, keeping the nanofibrous structure when the size of structures are 200 μm or more. Two of the micropatterned designs are shown in Figure 5.

Surface roughness is among the key parameters for biomaterials, which affect the cell attachment and proliferation. Roughness measurements were carried out on the patterned GelMA electrospun mats (Table 2). Fibers produced with 400 μm of thickness (NG-P400) were rougher than the sample produced with 200 μm of thickness (NG-P200). The depth of the patterns on the mats can be estimated from the roughness values recorded at Z-axis. The depth of the step on the mat NG-P400 was around 4 μm, whereas the depth of the step on the NG-P200 was estimated to be 2.5 μm. For all GelMA fiber mats with different topographies, the measured values of roughness showed an increase of surface patterning.

Table 2. Roughness values [1] of patterned GelMA electrospun mats.

Sample	Ra (μm)	Rz (μm)	Rmax (μm)
NG-P400	1.5 ± 0.5	4.0 ± 0.2	14 ± 1
NG-P200	0.7 ± 0.05	2.5 ± 0.4	6 ± 1

[1] Ra: arithmetical mean deviation of the assessed profile, Rz: arithmetical mean deviation of the assessed profile, and Rmax: maximum peak-to-valley height.

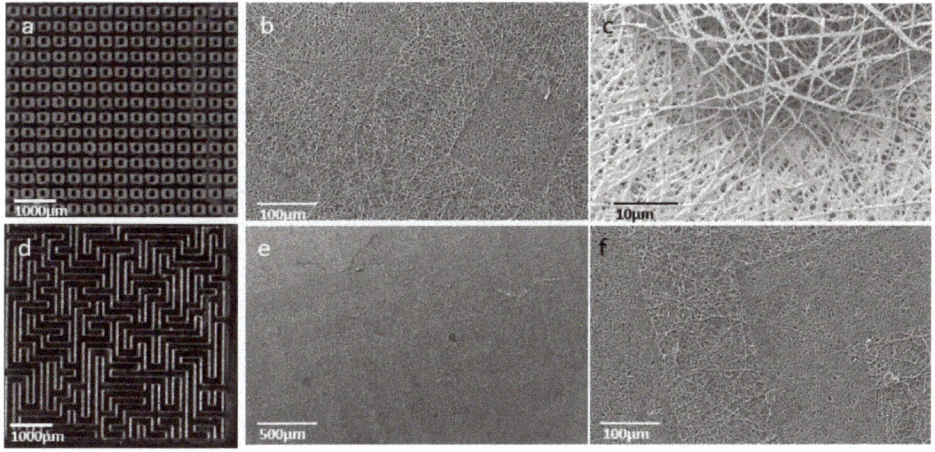

Figure 5. Optic images of 200 μm row and maze patterned molds (**a** and **d**, respectively) and SEM images of 200 μm patterned GelMA electrospun mats (**b** and **c**: row patterns; **e** and **f**: maze patterns).

3.3. Cell Viability

GelMA electrospun mats were tested in vitro to determine the biocompatible character of the process (Figure 3d). Cell viability percentages of all samples were above 90%. The high number of living cells indicates that the crosslinking process and the modified gelatin were compatible with the used cell-line MG-63. Thus, gelatin based electrospun mat with micro- and nanotopographical features have a great potential for tissue engineering.

4. Discussion

The use of gelatin as a biopolymer scaffolding material for tissue engineering applications is directly related to its high biocompatibility, hydrophilicity, and bioactivity associated with specific peptide sequences. Electrospinning is a very attractive, complex, and versatile technique to prepare advanced functional nanofibrous scaffolds for a variety of tissue engineering and cell culture applications. Current efforts are focused on preparing electrospun scaffolds with controlled multilevel hierarchical structures. GelMA electrospun scaffolds have been reported using synthetic polymers and/or hazard, expensive solvents [27–29]. However, a pure gelatin scaffold that mimics the important features of natural ECM obtained by electrospinning is still pending. In this work, GelMA nanofibers without defects were obtained using acetic acid, a low toxic potential solvent. Furthermore, in order to overcome the limitations of the low stability of gelatin in aqueous media as well as the possible toxicity of chemical crosslinkers, the as-spun scaffolds were crosslinked by UV irradiation at different exposition times. The same strategy, but while varying the photoinitiator concentration, was reported by Lai et al. for cultivation of limbal epithelial cells [30].

The GelMA matrix was soluble in PBS before UV irradiation (NG-0UV) while crosslinked GelMA nanofibers were stable in PBS (NG-6UV, NG-9UV and NG-12UV). The mechanical properties of UV irradiated mats (NG-9UV) increased 7 times related to mats unexposed to UV (NG-0UV), indicating the efficiency of crosslinking.

In the last few decades, surface engineering technologies have been used as important tools to clarify the effects of the microenvironment on cellular behavior [3,11–14]. The fabrication of the surface topographies with geometrical micro and nanopatterns, like channels, pillars and pits with controlled dimensions have been possible through the use of various methods as photolithography, electron beam lithography and microfluidics. These geometric and topographical factors can have an influence in cell adhesion, migration, differentiation, and the shape of cells. To modulate cell behavior through surface engineering methods is not only useful to stimuli stem cell differentiation, but also to generate a favorable response of the implant. In order to control substrate topography, the micropatterns molding with different geometries and dimensions, were developed to fabricate scaffolds. Molds were obtained by photolithography and they were used as collector in electrospinning process. Therefore, GelMA electrospun mats were obtained with nano structure and microroughness. The SEM images and roughness analysis showed that micropatterns were successfully copied over GelMA electrospun mats. To the best of our knowledge, this is the first report in literature in which GelMA nanofibrous scaffolds with micropattern topographical features were designed and prepared.

By combining molding and electrospinning processing, it was possible to obtain scaffolds which mimic ECM. Thus, this simple and easy technique can be useful for developing sophisticated and complex materials for tissue engineering and cell culture. In future works, we will study deeply how different patterns influence the behavior of specific cell lines.

5. Conclusions

In summary, an inexpensive and rapid method for the fabrication of well-defined micro- and nanotopographic features on uniform bead-free electrospun GelMA fibers were developed in order to design 3D scaffolds that mimic ECM. The synthesis of GelMA and its processing by electrospinning was optimized, so therefore, defect-free GelMA nanofibrous mats were obtained. The results showed that crosslinking took place successfully and it could modulate the final properties. Furthermore, the use of micropatterned molds obtained by photolithography as a collector in the electrospinning process allowed controlling the roughness of mats. Taking into account the cell viability results, the methodology used in this study is a valuable tool to develop patterned GelMA based nanofibrous scaffolds for cell culture and tissue engineering.

Author Contributions: Ana Agustina Aldana performed the experimental work and wrote the first drafts of the manuscript and final version, Laura Malatto designed and prepared micropatterned molds, Muhammad

Atiq Ur Rehman measured the roughness, Aldo Roberto Boccaccini advised during the project progress and collaborated in writing the manuscript and Gustavo Abel Abraham contributed to the conception of the project and collaborated in writing the manuscript.

Funding: This research was funded by CONICET (grant UE73) and MINCYT-DAAD (DA/16/02 Project). The APC was funded by University of Erlangen-Nuremberg.

Acknowledgments: The authors would like to thank Alina Grünewald and Vanesa Fuchs for technical support in cell viability and X-ray diffraction studies, respectively.

Conflicts of Interest: The authors declare no conflict of interest.

References

1. Chen, F.-M.; Liu, X. Advancing biomaterials of human origin for tissue engineering. *Prog. Polym. Sci.* **2016**, *53*, 86–168. [CrossRef] [PubMed]
2. Kim, H.N.; Kang, D.-H.; Kim, M.S.; Jiao, A.; Kim, D.-H.; Suh, K.-Y. Patterning methods for polymers in cell and tissue engineering. *Ann. Biomed. Eng.* **2012**, *40*, 1339–1355. [CrossRef] [PubMed]
3. Ishikawa, S.; Iijima, K.; Otsuka, H. Nanofabrication technologies to control cell and tissue function for biomedical applications. In *Nanobiomaterials*; Narayan, R., Ed.; Woodhead Publishing: Sawston, UK, 2018; pp. 385–409. ISBN 9780081007167.
4. Petrie, R.J.; Doyle, A.D.; Yamada, K.M. Random versus directionally persistent cell migration. *Nat. Rev. Mol. Cell Biol.* **2009**, *10*, 538–549. [CrossRef] [PubMed]
5. Wang, X.; Ding, B.; Li, B. Biomimetic electrospun nanofibrous structures for tissue engineering. *Mater. Today* **2013**, *16*, 229–241. [CrossRef] [PubMed]
6. Pelipenko, J.; Kocbek, P.; Kristl, J. Critical attributes of nanofibers: Preparation, drug loading, and tissue regeneration. *Int. J. Pharm.* **2015**, *484*, 57–74. [CrossRef] [PubMed]
7. Ngadiman, N.; Noordin, M.; Idris, A.; Kurniawan, D.; Fallahiarezoudar, E.; Sudin, I. Developments in tissue engineering scaffolding using an electrospinning process. In *Electrospinning and Electroplating: Fundamentals, Methods and Applications*; Jacobs, T., Ed.; Nova Science Publishers: Hauppauge, NY, USA, 2017; pp. 87–125. ISBN 9781536123890.
8. Senthamizhan, A.; Balusamy, B.; Uyar, T. Electrospinning: A versatile processing technology for producing nanofibrous materials for biomedical and tissue-engineering applications. In *Electrospun Materials for Tissue Engineering and Biomedical Applications*; Uyar, T., Kny, E., Eds.; Woodhead Publishing: Sawston, UK, 2017; pp. 3–41. ISBN 9780081010228.
9. Gao, Y.; Yan, J.; Cui, X.-J.; Wang, H.-Y.; Wang, Q. Aligned fibrous scaffold induced aligned growth of corneal stroma cells in vitro culture. *Chem. Res. Chin. Univ.* **2012**, *28*, 1022–1025.
10. Shalumon, K.T.; Deepthi, S.; Anupama, M.S.; Nair, S.V.; Jayakumar, R.; Chennazhi, K.P. Fabrication of poly (I-lactic acid)/gelatin composite tubular scaffolds for vascular tissue engineering. *Int. J. Biol. Macromol.* **2015**, *72*, 1048–1055. [CrossRef]
11. Martínez, E.; Engel, E.; Planell, J.A.; Samitier, J. Effects of artificial micro- and nano-structured surfaces on cell behavior. *Ann. Anat.* **2009**, *191*, 126–135. [CrossRef]
12. Ermis, M.; Antmen, E.; Hasirci, V. Micro and Nanofabrication methods to control cell-substrate interactions and cell behavior: A review from the tissue engineering perspective. *Bioact. Mater.* **2018**, *3*, 355–369. [CrossRef]
13. Zheng, L.; Jiang, J.; Gui, J.; Zhang, L.; Liu, X.; Sun, Y.; Fan, Y. Influence of Micropatterning on Human Periodontal Ligament Cells' Behavior. *Biophys. J.* **2018**, *114*, 1988–2000. [CrossRef] [PubMed]
14. Rosenfeld, D.; Levenberg, S. Effect of Matrix Mechanical Forces and Geometry on Stem Cell Behavior. In *Biology and Engineering of Stem Cell Niches*; Vishwakarma, A., Karp, J.M., Eds.; Academic Press: London, UK, 2017; pp. 233–243. ISBN 9780128027349.
15. ICH. Q3C Guideline for Residual Solvents (R5). In *International Conference Harmon Tech Requir Regist Pharm Hum Use 29*; ICH: Geneva, Switzerland, 2011.
16. Aldana, A.A.; Abraham, G.A. Current advances in electrospun gelatin-based scaffolds for tissue engineering applications. *Int. J. Pharm.* **2017**, *523*, 441–453. [CrossRef]

17. Chou, S.-F.; Luo, L.-J.; Lai, J.-Y.; Ma, D.H.-K. Role of solvent-mediated carbodiimide cross-linking in fabrication of electrospun gelatin nanofibrous membranes as ophthalmic biomaterials. *Mater. Sci. Eng. C* **2017**, *71*, 1145–1155. [CrossRef]
18. Van den Bulcke, A.I.; Bogdanov, B.; Cornelissen, M.; Schacht, E.H.; de Rooze, N.; Berghmans, H. Structural and rheological properties of methacrylamide modified gelatin hydrogels. *Biomacromolecules* **2000**, *1*, 31–38. [CrossRef] [PubMed]
19. Hutson, C.B.; Nichol, J.W.; Aubin, H.; Bae, H.; Yamanlar, S.; Al-Haque, S.; Koshy, S.T.; Khademhosseini, A. Synthesis and characterization of tunable poly(ethylene glycol): Gelatin methacrylate composite hydrogels. *Tissue Eng. Part A* **2011**, *17*, 1713–1723. [CrossRef]
20. Nichol, J.W.; Koshy, S.T.; Bae, H.; Hwang, C.M.; Yamanlar, S.; Khademhosseini, A. Cell-laden microengineered gelatin methacrylate hydrogels. *Biomaterials* **2010**, *31*, 5536–5544. [CrossRef]
21. Aldana, A.A.; Rial-Hermida, M.I.; Abraham, G.A.; Concheiro, A.; Alvarez-Lorenzo, C. Temperature-sensitive biocompatible IPN hydrogels based on poly(NIPA-PEGdma) and photocrosslinkable gelatin methacrylate. *Soft Mater.* **2017**, *15*, 341–349. [CrossRef]
22. Kim, J.W.; Kim, M.J.; Ki, C.S.; Kim, H.J.; Park, Y.H. Fabrication of bi-layer scaffold of keratin nanofiber and gelatin-methacrylate hydrogel: Implications for skin graft. *Int. J. Biol. Macromol.* **2017**, *105*, 541–548. [CrossRef] [PubMed]
23. Williams, C.G.; Malik, A.N.; Kim, T.K.; Manson, P.N.; Elisseeff, J.H. Variable cytocompatibility of six cell lines with photoinitiators used for polymerizing hydrogels and cell encapsulation. *Biomaterials* **2005**, *26*, 1211–1218. [CrossRef]
24. Habeeb, A.F.S.A. Determination of free amino groups in proteins by trinitrobenzenesulfonic acid. *Anal. Biochem.* **1966**, *14*. [CrossRef]
25. Schumacher, M.; Uhl, F.; Detsch, R.; Deisinger, U.; Ziegler, G. Indirect rapid prototyping of biphasic calcium phosphate scaffolds as bone substitutes: Influence of phase composition, macroporosity and pore geometry on mechanical properties. *J. Mater. Sci. Mater. Med.* **2010**, *21*, 3039–3048. [CrossRef] [PubMed]
26. Mukherjee, I.; Rosolen, M.A. Thermal transitions of gelatin evaluated using DSC sample pans of various seal integrities. *J. Therm. Anal. Calorim.* **2013**, *114*, 1161–1166. [CrossRef]
27. Ferreira, P.; Santos, P.; Alves, P.; Carvalho, M.P.; de Sá, K.D.; Miguel, S.P.; Correia, I.J.; Coimbra, P. Photocrosslinkable electrospun fiber meshes for tissue engineering applications. *Eur. Polym. J.* **2017**, *97*, 210–219. [CrossRef]
28. Zhao, X.; Sun, X.; Yildirimer, L.; Lang, Q.; Lin, Z.Y.; Zheng, R.; Zhang, Y.; Cui, W.; Annabi, N.; Khademhosseini, A. Cell infiltrative hydrogel fibrous scaffolds for accelerated wound healing. *Acta Biomater.* **2017**, *49*, 66–77. [CrossRef] [PubMed]
29. Coimbra, P.; Santos, P.; Alves, P.; Miguel, S.P.; Carvalho, M.P.; de Sá, K.D.; Correia, I.J.; Ferreira, P. Coaxial electrospun PCL/Gelatin-MA fibers as scaffolds for vascular tissue engineering. *Colloids Surf. B Biointerfaces* **2017**, *159*, 7–15. [CrossRef] [PubMed]
30. Lai, J.-Y.; Luo, L.-J. Effect of riboflavin concentration on the development of photo-cross-linked amniotic membranes for cultivation of limbal epithelial cells. *RSC Adv.* **2015**, *5*, 3425–3434. [CrossRef]

© 2019 by the authors. Licensee MDPI, Basel, Switzerland. This article is an open access article distributed under the terms and conditions of the Creative Commons Attribution (CC BY) license (http://creativecommons.org/licenses/by/4.0/).

Article

A Microfluidic Chip Embracing a Nanofiber Scaffold for 3D Cell Culture and Real-Time Monitoring

Jeong Hwa Kim [1], Ju Young Park [2], Songwan Jin [3], Sik Yoon [4], Jong-Young Kwak [5] and Young Hun Jeong [6],*

- [1] Department of Mechanical Engineering, Graduate School, Kyungpook National University, Daegu 41566, Korea; qhfekrn89@gmail.com
- [2] Department of Mechanical Engineering, Pohang University of Science and Technology (POSTECH), Pohang 37673, Korea; juyoung1489@postech.ac.kr
- [3] Department of Mechanical Engineering, Korea Polytechnic University, Siheung 15073, Korea; songwan@kpu.ac.kr
- [4] Department of Anatomy, Pusan National University School of Medicine, Yangsan 50612, Korea; sikyoon@pusan.ac.kr
- [5] Department of Pharmacology, Ajou University School of Medicine, Suwon 16499, Korea; jykwak@ajou.ac.kr
- [6] School of Mechanical Engineering, Kyungpook National University, Daegu 41566, Korea
- * Correspondence: yhjeong@knu.ac.kr; Tel.: +82-53-950-5577

Received: 20 February 2019; Accepted: 1 April 2019; Published: 10 April 2019

Abstract: Recently, three-dimensional (3D) cell culture and tissue-on-a-chip application have attracted attention because of increasing demand from the industries and their potential to replace conventional two-dimensional culture and animal tests. As a result, numerous studies on 3D in-vitro cell culture and microfluidic chip have been conducted. In this study, a microfluidic chip embracing a nanofiber scaffold is presented. A electrospun nanofiber scaffold can provide 3D cell culture conditions to a microfluidic chip environment, and its perfusion method in the chip can allow real-time monitoring of cell status based on the conditioned culture medium. To justify the applicability of the developed chip to 3D cell culture and real-time monitoring, HepG2 cells were cultured in the chip for 14 days. Results demonstrated that the cells were successfully cultured with 3D culture-specific-morphology in the chip, and their albumin and alpha-fetoprotein production was monitored in real-time for 14 days.

Keywords: nanofibers; microfluidic chip; electrospinning; live assay; hepatocellular carcinoma cells

1. Introduction

Over the last century, two-dimensional (2D) in-vitro cell culture has been used in studying the responses to stimulation from biological and biochemical materials, such as drugs, toxic materials, and detoxification. Traditional in-vitro cell tests are based on 2D culture on a flat surface, and the 2D culture environment is completely different from the human body. Cells inside the human body are surrounded by extracellular matrix (ECM) and tissue fluid in a complex three-dimensional (3D) space. Thus, cells are difficult to activate in 2D environments to maintain differentiation and expression of tissue-specific physiological functions and physiological activity [1,2].

Tissue-on-a-chip is a recapitulation of the biological and mechanochemical environment of human body tissues using a small device chip [1]. This technique includes cells, chemical and physical environments, and the microenvironment. It is used in the development of in-vitro disease models, drug screening, toxicity testing, and disease research, by providing a cellular environment that better mimics human physiological conditions [3–5]. Typical tissue-on-a-chip consists of an integrated microscale engineering system, several types of cells, and culture medium. The microelectromechanical system (MEMS) allows the use of microfluidic devices to reconstitute distinct features of the tissue–tissue interface, physiological movements, and biochemical environment similar to the human body. Marino et al. [6] presented a microfluidic system to mimic the blood–brain barrier (BBB). The system consisted of porous microtubes fabricated using two-photon lithography and enabled co-culturing brain microcapillary cells and functioning.

Recently, various 3D environmental features, such as microporous membrane [7,8], hydrogel [9–11], and nanofibers [12,13], have been introduced to tissue-on-a-chip applications to provide the similar structures and functions of the human body to cell culture. The hydrogel is highly permeable and an excellent biocompatible material. Annabi et al. coated microfluidic channels with synthesized photo-crosslinkable gelatin and tropoelastin-based hydrogel to improve cardiomyocyte culture in a polydimethylsiloxane (PDMS) surface [9]. Au et al. fabricated hepatic tissue models by encapsulating HepG2 and NIH-3T3 cells in a hydrogel. Their platform showed better results than 2D cell culture systems in drug screening [14]. Gumuscu et al. [15] proposed a microfluidic cell culture platform composed of 3D collagen hydrogel compartments, and they applied their system for co-culture of human intestinal cells and drug screening in preliminary level. Porous membranes are considered mimic basal membranes of barrier tissue, such as tissue–tissue, tissue–liquid, and tissue–air interfaces in organ-on-a-chip [16,17].

Nanofibers have diameters ranging from tens to hundreds of nanometers and similar morphology to the extracellular matrix of the human body. In particular, nanofibers are well suited for cell nutrient exchange, communication, and efficient cellular responses because of large surface areas, high porosity, and spatial interconnectivity [18,19]. As a result, nanofibers have been extensively applied to a variety of applications such as porous membrane and scaffold in biotechnology [20–22]. In addition, cells are easily attached and better proliferated in electrospun composite nanofiber than in a conventional 2D culture environment, such as a petri dish [23].

Here, we present a microfluidic chip with a nanofiber scaffold, which can provide a 3D human body ECM-like environment to cell culture and monitor cell status and activity using a conditioned culture medium. The nanofiber scaffold was electrospun so that it composed various diameter fibers, thereby providing highly porous morphology to cells under the well-defined microfluidic chip conditions. In particular, a perfusion method, which enables real-time monitoring cell status, was demonstrated. The developed chip was applied to 3D culture of HepG2 cells, which has various functions, such as cell growth and secretion of proteins. Our results demonstrated that HepG2 cells were cultured with 3D culture-specific morphology (i.e., large spheroids) [3,24], and their protein production was successfully monitored for 14 days.

2. Materials and Methods

2.1. Concept of Microfluidic Chip with Nanofiber Mat

Figure 1 shows the schematic concept of the proposed microfluidic chip embracing nanofibers, which consists of a nanofiber scaffold, a microfluidic chip structure, perfusion environment, and cells. The chip was designed to mimic the dynamic microenvironment of the human body, support perfusion-based long-term culture, and allow real-time monitoring of secretions and functionalities of the cultured cells in vitro. An electrospun nanofiber scaffold was introduced to a microfluidic chip to provide a 3D extracellular matrix (ECM)-like environment, because the nanofibers have similar morphology to the human body's ECM [25]. The nanofiber scaffold was located on the chamber bottom. The microfluidic chip has a simple structure composed of micro-channels, a chamber, and gate holes, and it is made of Polydimethylsiloxane (PDMS). The cell suspension was introduced into the top opening of the chamber before closing the opening with a cover slip. Fresh culture medium was supplied to the cell-seeded nanofiber scaffold via the inlet microchannel to provide supplied oxygen and nutrients to the cells. The cell culture-conditioned medium through the outlet microchannel was collected to monitor cell activity. Fluidic connections to the microfluidic chip were made with tubing inserted through the inlet and outlet holes. Culture medium perfusion/flow in the chip was established with the help of a perfusion environment consisting of a fresh media reservoir at the side of the inlet and a syringe pump at the side of the outlet. Therefore, medium flow was derived by negative pressure. The collected conditioned medium in a syringe was easily transferred for analysis, such as enzyme-linked immunosorbent assay (ELISA).

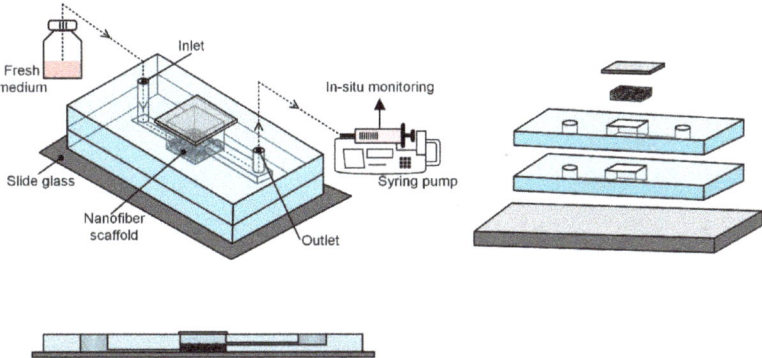

Figure 1. Concept of a microfluidic chip with a nanofiber scaffold.

As shown in Figure 1, the chip is composed of four layers: a supportive plate (slide glass), two microfluidics layers (PDMS), and a window to chamber. The microfluidic layer has a microchannel on the bottom plane and a rectangular hole as a chamber at the center of the layer. A thin coverslip was used as the window to chamber in this study. After assembling all the layers, the rectangular holes formed the center chamber with the supportive plate and window to chamber.

Flow rate of the culture medium through the chip was determined by considering the required amount of culture medium for cell growth, 2D culture condition, and allowable shear stress on the cells induced by culture medium flow in microfluidics. Information on the required amount of culture medium can be obtained from the culture product company, whereas the shear stress limit can be obtained from a previous study [26]. The amount of media supplement per hour (flow rate) can be determined from the following equation:

$$\frac{M}{N_R \times T} \times N_{chip} \leq q \leq \frac{\tau_{max} b h^2}{6\mu}$$

where M is the amount (volume) of culture medium supply for the number (M_R) of cells at a given culture area. T is the medium exchange period. M, N_R, and T are the recommended values under 2D culture condition. N_{chip} is the number of cells seeded in the chip, τ_{max} is the allowable shear stress loaded to cells to avoid cellular damage, μ is the viscosity of medium (g/cm·s), q is the flow rate through the chip (cm^3/s), and b and h are the width (cm) and height (cm) of the microchannel, respectively. The maximum limit of the flow rate, which was calculated from the allowable shear stress, was 1.54 mL/h. The minimum flow rate, which was calculated from the 2D culture protocol [27] considering the number of seeded cells in a scaffold and scaffold area, was 0.001 mL/h. We set the flow rate at 0.1 mL/h between the maximum and minimum limits.

2.2. Fabrication of Microfluidic Chip Structure

The microfluidic chip structure was made of PDMS using well-established soft lithography and micromolding [28,29] at a preliminary level. The master molds were prepared by cutting adhesive tape (3M) according to the channel shape, and removing the tape from all areas except the channel shape on a slide glass (ThermoFisher Scientific, Waltham, MA, USA)). PDMS (Dow Corning, Midland,

MI, USA) was cast on the master molds using the pre-polymer (base) to curing agent weight ratio of 10:1, and the molds were cured in a dry oven at 60 °C for 3 h.

In the microfluidic layer, the microchannels had a cross-section of about 300 × 100 μm and a length of about 10 mm, whereas the chamber hole had a rectangular shape with a size of about 12 × 12 mm and a height of about 3 mm. The microchannels were patterned using soft lithography (Figure 2). After curing, a rectangular chamber hole was made by cutting a rectangle in the cured PDMS slab. A hole to the inlet microchannel was made by using a biopsy puncher.

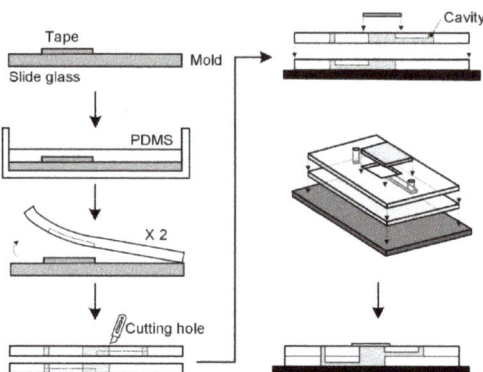

Figure 2. Fabrication of patterned microchannels using soft lithography.

All the PDMS structures and slide glass were autoclaved for sterilization and bonded together after corona treatment (Electro-Technic Products Inc., Chicago, IL, USA) to secure tight bonding (Figure 1). Before inserting the nanofiber scaffold into the chamber, the scaffold was sterilized in 70% ethanol and thoroughly washed three times with PBS. A nanofiber scaffold was inserted into the chamber inside the bonded PDMS. The assembled chip samples were stored on a clean bench for a while. After seeding the cell suspension onto the nanofiber scaffold, the chamber was covered with glass coverslips. The microfluidic device was immobilized with two polycarbonate (PC) plates and screws to ensure no leakage.

2.3. Fabrication of Nanofiber Scaffold

In this study, 8 wt% polycaprolactone (PCL) solution was prepared by dissolving PCL with an average molecular weight (Mw) of 70,000–90,000 (PCL, Sigma-Aldrich, St.Louis, MO, USA) in 99.5% chloroform (Samchun Pure Chemical Co., Seoul, Republic of Korea). The mixture was then homogenized with a magnetic stirrer for 12 h at room temperature.

The electrospinning setup consisted of a high-voltage power supply, syringe pump, nozzle spinneret with an inner diameter of 210 μm, and a grounded solid drum collector, which rotated at a speed of about 30 rpm. The nozzle spinneret tip-to-collector distance was maintained at 100.0 mm. The electrical potential between the nozzle spinneret and the drum collector was 10.0 kV. The flow rate of the PCL solution was controlled by a syringe pump at 0.1 mL/h. The thickness uniformity of the nanofibers was improved by providing the nozzle spinneret with a traveling motion speed of 1 mm/s along the drum axis direction. The nanofibers were electrospun until they reached a thickness of 100 μm (actually, 113.7 ± 2.7 μm), which was measured using an ultra-precision micrometer (Mitutoyo) with a constant compression of 0.5 N. During electrospinning, the temperature and relative humidity were maintained at 19–20 °C and 50–55%, respectively. The electrospun nanofibers were provided with 5 min of heat treatment at 60 °C using a dry oven to improve nanofiber interconnectivity. The nanofibers were then cut into the same size of the chamber (12 × 12 mm) with a scalpel and carefully placed into the microfluidic chamber using a tweezer. The morphology of the electrospun nanofibers was observed by scanning electron microscopy (SEM; Hitachi, Tokyo, Japan). The porosity of the nanofibers was measured using a mercury porosimeter (Micromeritics Instrument Co., Norcross, GA, USA). The filling pressure of mercury was 1.23 psi, and the equilibrium time was 10 s.

2.4. Cell Culture and Flow Experiments

Human liver hepatocellular carcinoma cells (HepG2) were obtained from Korean Cell Line Bank. HepG2 cells in suspension containing 1×10^4 cell/chamber were seeded on the nanofiber scaffold in the microfluidic chip under static conditions, and allowed to form a stable attachment for 24 h. After the cells were allowed to settle, the medium was continuously perfused to the chip at a flow rate of 0.1 mL/h using a commercial syringe pump (New Era Pump Systems, Farmingdale, NY, USA). The perfused culture medium was Dulbecco's modified Eagle's medium (Life Technologies, ThermoFisher Scientific, Waltham, MA, USA) supplemented with 10% fetal bovine serum (Life Technologies, ThermoFisher Scientific, Waltham, MA, USA) and 1% penicillin streptomycin. The cultures were stored for 14 days in a humidified atmosphere of 5% CO_2 in air at 37 °C. We could not find any deterioration in the device operation nor change in the nanofiber morphology during the culture. Moreover, the nanofibers stuck tightly in the chamber and embedded well on the chamber bottom.

2.5. Functional Assays

To verify hepatocyte viability of the developed microfluidic chip, the cells on the scaffold after settling down were stained with Calcein-AM to indicate live cells and ethidium homodimer-1 to indicate dead cells, according to the manufacturer's instructions (Life Technologies, ThermoFisher Scientific, Waltham, MA, USA). The live and dead cells were observed using a confocal microscope (Carl Zeiss, Oberkochen, Germany). We investigated cell proliferation behavior under static and perfusion conditions to justify the effect of medium perfusion with microfluidic chip. Unlike cell culture under perfusion condition, the medium in the chamber was exchanged with fresh medium daily by using a pipet to compare cell proliferation under both conditions. Cell proliferation was assessed by measuring the DNA content of the cultured cells in each chip. The DNA content was measured on days 1 and 14. DNA from cultured cells was extracted using a commercial kit (Qiagen, Hilden, Germany). The DNA contents for both static and perfusion conditions were quantified by spectrophotometers (Thermo Fisher Scientific, Waltham, MA, USA).

Here, we quantified the secretion of albumin and alpha-fetoprotein (AFP) using enzyme-linked immunosorbent assay (ELISA) to demonstrate the applicability of the developed microfluidic chip to the 3D cell culture and real-time monitoring with HepG2 cells. Therefore, the conditioned culture medium was continuously collected at a rate of 0.1 mL/h and harvested daily by changing the syringe in the syringe pump. The amount of albumin secreted from the HepG2 cells was quantified every 2 days until day 14 using a human albumin ELISA kit (Bethyl Laboratories, Montgomery, TX, USA). AFP secretion was measured using an AFP ELISA kit (CUSABIO, Wuhan, China) every 3 days, from day 9 to day 14, according to the manufacturer's instructions. The required volume per sample of the conditioned medium for the assays was 100 µL, and the volume of the daily collected medium (2.4 mL) was sufficient to analyze various secreted proteins, up to 12 types, after duplication.

The morphology of the cells cultured in the microfluidic chip was investigated by SEM and fluorescent imaging. Cells cultured in the chips for 1 and 14 days were harvested with phosphate buffered saline (PBS) and then fixed with 4% paraformaldehyde solution. The samples for SEM were dried in a freeze-dryer (Labconco, Kansas City, MO, USA). Sample morphology was observed using SEM (Hitachi Hitachi, Tokyo, Japan). The cell morphology in the chip was assessed by staining F-actin with the phalloidin-fluorescein isothiocyanate (Phalloidin-FITC, Sigma-Aldrich, St. Louis, MO, USA) and nuclei with DAPI (4',6-diamidino-2-phenylindole, Life Technologies, ThermoFisher Scientific, Waltham, MA, USA). The cells were then visualized using a confocal laser scanning microscope (Carl Zeiss, Oberkochen, Germany).

2.6. Statistical Analysis

All experiments were carried out with 5 samples (N = 5) except DNA content assay with four samples. All the samples were duplicated in assays. The experimental data are presented as means ± standard deviation (SD). ANOVA was conducted when we compared more than two groups for statistical comparison. The differences between the mean values of each group were evaluated by Student's *t*-test, in which a *p* value less than 0.001 was considered significant.

3. Results and Discussion

3.1. Microfluidic Chip Embracing a Nanofiber Scaffold

Figure 3a shows the SEM image of the electrospun nanofiber scaffold for the proposed chip. First, the electrospun fibers were randomly oriented. Moreover, the nanofiber scaffold in this study had various diameter fibers due to the combination effect of the chloroform solvent, with faster evaporation, and the process conditions [30,31]. Some fibers had diameters ranging from several to 10 µm, whereas others appeared fine with diameters in the nanometer scale. Results of detailed analysis of fiber diameter distribution are in Figure 3b. The peak frequency diameter of fine fibers (i.e., nanofibers) was between 600 and 1000 nm, and a significant number of microfibers had a diameter of 2 to 6 µm in the scaffold. This nanofiber scaffold possibly contained larger pores than the scaffold consisting of small diameter nanofibers, because the microfibers can introduce larger pores than nanofibers [32]. Figure 3c shows the differential intrusion curve with respect to pore size of the nanofiber scaffold. From the figure, the pore diameter (equivalently, size) ranged between 0.3 and 100 µm and the peak frequency pore diameter was about 3 µm. The porosity of the nanofiber scaffold was about 76%. Therefore, the fabricated nanofiber scaffold may allow the cells to infiltrate into the deep regions of the scaffold. Figure 4a shows an assembled microfluidic chip embracing a nanofiber scaffold. The developed chip looks simple and is appropriate for use in a real PDMS fabrication environment. The microchannel of the chip had a width of 307.6 ± 4.5 µm and depth of 97.2 ± 4.9 µm, as shown in Figure 4b. The microchannel had a flat and smooth bottom plane, while its side walls had a rough surface because of the cut edge of tapes. However, the rough wall surface of the microchannel did not affect the device operation quality because the microchannel engaged only in the delivery of culture medium with slow velocity.

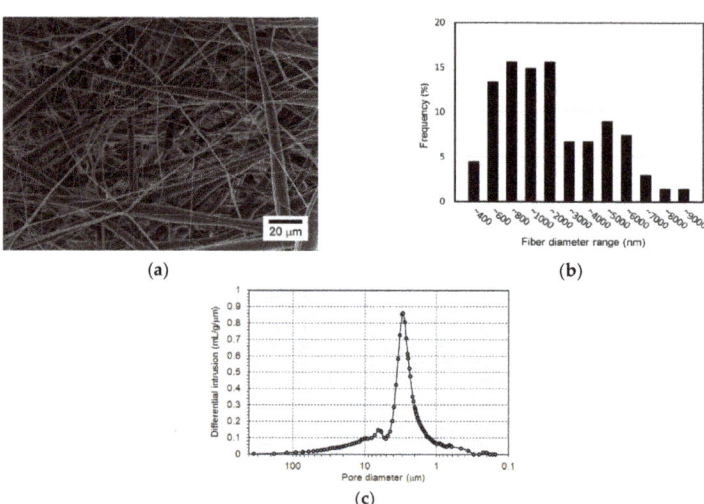

Figure 3. Electrospun nanofiber scaffold: (**a**) SEM images; (**b**) diameter distribution of the electrospun fibers; (**c**) porosity analysis result.

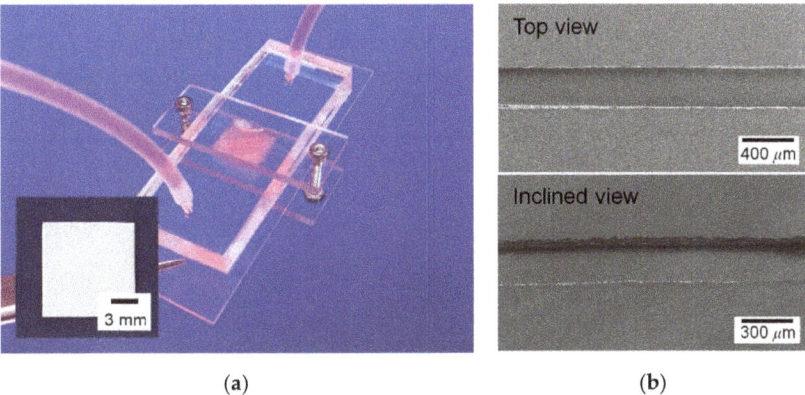

Figure 4. Microfluidic chip with a nanofiber scaffold (**a**) and SEM images of the microchannel of the chip (**b**).

3.2. Application to 3D Cell Culture and Real-Time Monitoring

The cell viability of a newly built culture system should be determined. Here, we investigated HepG2 cell viability of the microfluidic chip, embracing a nanofiber scaffold via live and dead cell assay. Figure 5 shows the assay results, in which the live cells were stained green, and the red ones corresponded to dead cells. The cell counting assay revealed that the developed microfluidic chip had sufficiently high viability for HepG2 cells (>95%).

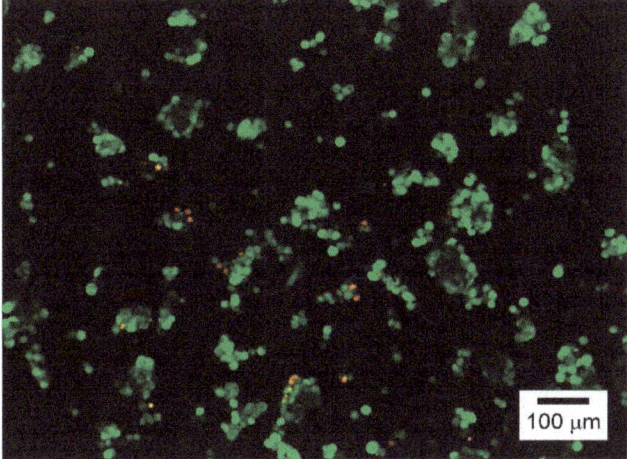

Figure 5. Live and dead cell assay results of the developed microfluidic chip for HepG2 cells (day 1).

Figure 6 shows the comparison results of the proliferation of HepG2 cells cultured under static and perfused conditions for 1 and 14 days. As shown in Figure 6, the cells cultured for 14 days in the chip successfully proliferated regardless of the medium supply method used, although the standard deviations looked relatively large. The large standard deviation possibly resulted from the variation in cell seeding quality because of the innate uneven fibrous morphology. Even though the cells cultured under static condition revealed excellent proliferation for 14 days (6.6-fold), the perfused condition promoted cell proliferation more. The cells cultured under perfused condition proliferated by more than 12-fold during the same period. Thus, perfusion culture condition in the developed chip gave rise to about two times higher cell proliferation than the static culture condition ($p < 0.001$).

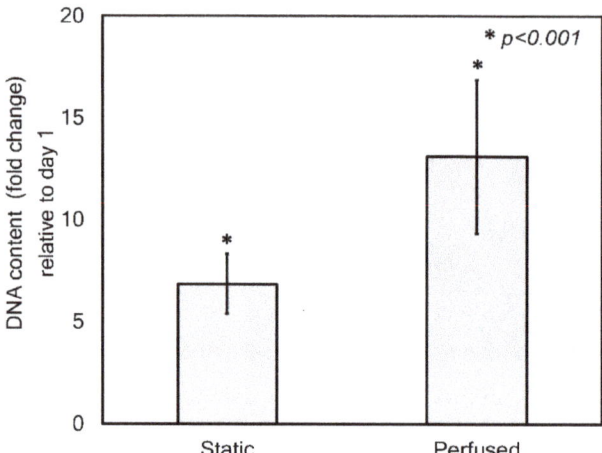

Figure 6. DNA content analysis of the cells cultured under static and perfusion conditions for 1 and 14 days in the chip.

The results demonstrated that the perfusion culture in the developed chip could provide cell-friendly conditions to the cells. To justify the applicability of the developed chip, 3D cell culture and real-time monitoring of the cells' status using the developed chip were carried out. Figure 7 shows the HepG2 cells cultured for 1 and 14 days in the chip with perfusion. As shown in Figure 7a,b, the cells with the cell length of 16.47 ± 4.84 µm successfully attached and hung on the nanofibers in the form of small spheroids, with diameters ranging from 20 to 50 µm on day 1. Moreover, the cytoskeleton organization, visualized by actin staining, revealed that actin filaments stretched along the nanofibers (Figure 7c). After 14 days, HepG2 cells re-established cell–cell contact and formed about 300 µm-diameter aggregates (or large spheroid) engulfed on nanofibers (Figure 7d–f), which has been reported to be the more desirable formation of HepG2 cells because of higher hepatocyte synthetic functions, higher response for drug toxicity, and better mimicking of in-vivo oxygen gradient in the hepatic lobule [3,24,33]. These results were far from that under 2D culture conditions based on a flat surface. In 2D conditions, HepG2 cells grow as monolayers with flat morphology, which results in short-lived canalicular-like structures [24]. Therefore, the developed microfluidic chip with nanofiber scaffold could successfully provide a 3D cell culture environment.

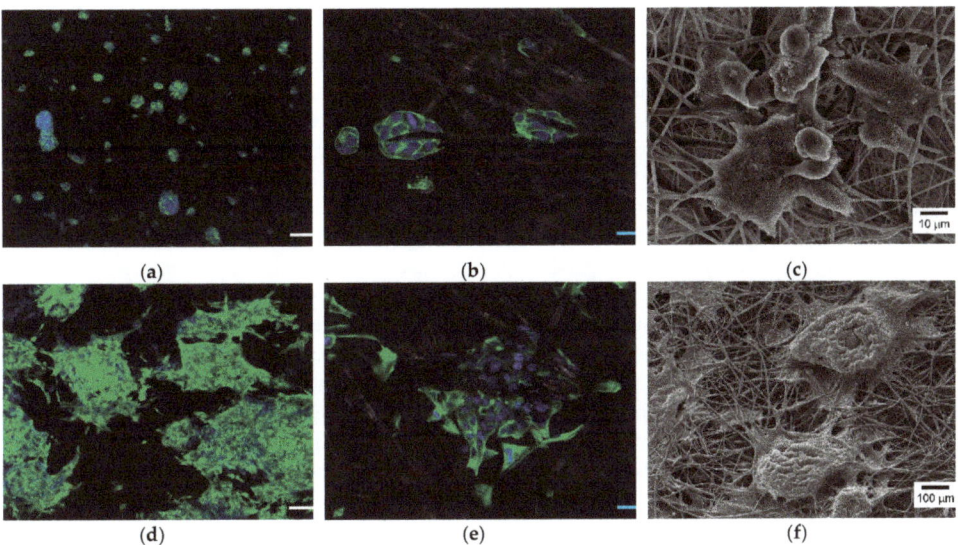

Figure 7. Confocal microscopy and SEM images of the cells cultured on the nanofiber chamber with microfluidic system for 1 (**a,b,e**) and 14 days (**c,d,f**). Especially, (**b,e**) are enlarged images of (**a,d**). The scale bars in the confocal microscopy images are 100 μm (**a,d**) and 20 μm (**b,e**).

The typical specific functionality of HepG2 cells was evaluated through real-time monitoring of the concentration of secreted albumin and AFP in the conditioned medium for 14 days. Albumin, which is secreted from HepG2 cells from human hepatocytes, is a marker of metabolic activity of hepatocytes in vitro, and it can indicate liver-specific functions [24]. Thus, the amount of albumin secreted from the cell cultured for 14 days in the chip was examined every two days (Figure 8a). The albumin production rate consistently increased from 3.07 ± 0.61 ng/mL at day 1 to 42.09 ± 1.91 ng/mL at day 14. Moreover, we monitored AFP secretion from the cells because AFP is known as a tumor-associated protein and important indicator of the differentiation of hepatoma cells and disease progression [34]. As shown in Figure 8b, the increase in secreted AFP level was similar to that in albumin secretion for the same duration. The AFP concentration in the conditioned medium collected at day 2 was 15.45 ± 5.13 ng/mL, and this value increased to 118.54 ± 3.19 ng/mL at day 14. As shown in these monitoring results, the developed microfluidic chip embracing nanofiber scaffold enabled the real-time monitoring of cell status based on the conditioned culture medium.

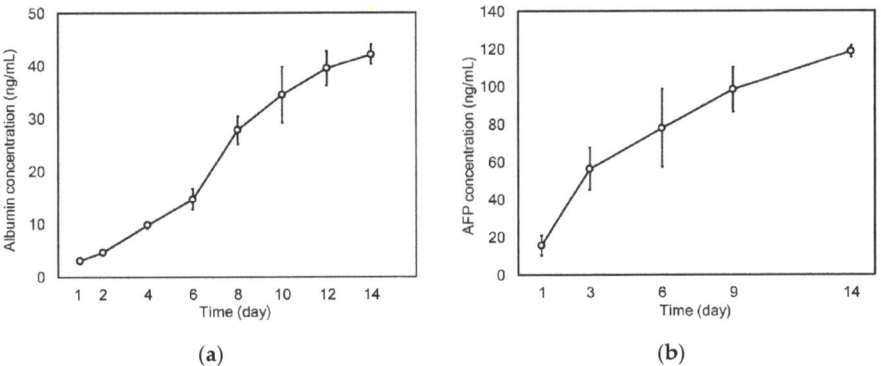

Figure 8. Real-time monitoring of secretion of (**a**) albumin and (**b**) AFP for 14 days.

4. Conclusions

In this study, we presented a microfluidic chip with a nanofiber scaffold. Microfluidics can introduce dynamic conditions, and a nanofiber scaffold provides a 3D environment to an in-vitro cell culture. Moreover, the perfusion method in the chip allows the real-time monitoring of cell status with the conditioned culture medium. The microfluidic chip structure was fabricated using well-established soft-lithography at a preliminary level, and a nanofiber scaffold was prepared using electrospinning. In particular, the nanofiber scaffold had various diameter fibers. It was highly porous and heterogeneous in morphology, making it appropriate to provide a human body's ECM-like 3D environment. To justify the developed chip, HepG2 cells with various protein syntheses and 3D culture-specific morphology were cultured in the chip for 14 days [35]. Results demonstrated that the developed chip had excellent viability (higher than 95%) for HepG2 cells through live and dead cell assay. Additionally, the perfusion-based dynamic culture had a positive effect on the proliferation of HepG2 cells, which was about two-times higher than that under static culture condition. Finally, we presented the application of the chip to 3D cell culture and real-time monitoring based on conditioned culture medium. The cells cultured in the chip successfully formed engulfed aggregates (or large spheroid), which are a 3D culture-specific morphology of the cells. Moreover, albumin levels and AFP secretion from the cells were successfully monitored for 14 days. We expect the concept of the developed chip to be useful for studying 3D cell cultures, live assay or real-time monitoring of cell activity, and in-vitro drug screening and toxicity testing.

Author Contributions: Methodology, S.J.; validation, S.Y. and J.-Y.K.; formal analysis, J.H.K. and J.Y.P.; investigation, J.H.K.; writing—original draft preparation, J.H.K. and Y.H.J.; writing—review and editing, J.H.K. and Y.H.J.; supervision, Y.H.J.

Funding: This work was supported by the National Research Foundation of Korea (NRF) grant funded by the Korea government (MSIT) (No. NRF-2012-0009665, 2015R1A2A2A01005515, 2018R1A2B2009540).

Conflicts of Interest: The authors declare no conflict of interest.

References

1. Huh, D.; Hamilton, G.A.; Ingber, D.E. From 3D cell culture to organs-on-chips. *Trends Cell Biol.* **2011**, *21*, 745–754. [CrossRef] [PubMed]
2. Lovitt, C.J.; Shelper, T.B.; Avery, V.M. Advanced cell culture techniques for cancer drug discovery. *Biology* **2014**, *3*, 345–367. [PubMed]
3. Bhise, N.S.; Ribas, J.; Manoharan, V.; Zhang, Y.S.; Polini, A.; Massa, S.; Dokmeci, M.R.; Khademhosseini, A. Organ-on-a-chip platforms for studying drug delivery systems. *J. Control. Release* **2014**, *190*, 82–93. [CrossRef] [PubMed]
4. Esch, E.W.; Bahinski, A.; Huh, D. Organs-on-chips at the frontiers of drug discovery. *Nat. Rev. Drug Discov.* **2015**, *14*, 248. [CrossRef]
5. Ho, C.M.B.; Ng, S.H.; Yoon, Y.-J. A review on 3D printed bioimplants. *Int. J. Precis. Eng. Manuf.* **2015**, *16*, 1035–1046. [CrossRef]
6. Marino, A.; Tricinci, O.; Battaglini, M.; Filippeschi, C.; Mattoli, V.; Sinibaldi, E.; Ciofani, G. A 3D Real-Scale, Biomimetic, and Biohybrid Model of the Blood-Brain Barrier Fabricated through Two-Photon Lithography. *Small* **2018**, *14*, 1702959. [CrossRef]
7. Jang, K.-J.; Suh, K.-Y. A multi-layer microfluidic device for efficient culture and analysis of renal tubular cells. *Lab Chip* **2010**, *10*, 36–42. [CrossRef] [PubMed]
8. Jang, K.-J.; Mehr, A.P.; Hamilton, G.A.; McPartlin, L.A.; Chung, S.; Suh, K.-Y.; Ingber, D.E. Human kidney proximal tubule-on-a-chip for drug transport and nephrotoxicity assessment. *Integr. Biol.* **2013**, *5*, 1119–1129. [CrossRef]
9. Annabi, N.; Selimović, Š.; Cox, J.P.A.; Ribas, J.; Bakooshli, M.A.; Heintze, D.; Weiss, A.S.; Cropek, D.; Khademhosseini, A. Hydrogel-coated microfluidic channels for cardiomyocyte culture. *Lab Chip* **2013**, *13*, 3569–3577. [CrossRef] [PubMed]

10. Kolesky, D.B.; Homan, K.A.; Skylar-Scott, M.A.; Lewis, J.A. Three-dimensional bioprinting of thick vascularized tissues. *Proc. Natl. Acad. Sci. USA* **2016**, *113*, 3179–3184. [CrossRef] [PubMed]
11. Verhulsel, M.; Vignes, M.; Descroix, S.; Malaquin, L.; Vignjevic, D.M.; Viovy, J.-L. A review of microfabrication and hydrogel engineering for micro-organs on chips. *Biomaterials* **2014**, *35*, 1816–1832. [CrossRef] [PubMed]
12. Yang, X.; Li, K.; Zhang, X.; Liu, C.; Guo, B.; Wen, W.; Gao, X. Nanofiber membrane supported lung-on-a-chip microdevice for anti-cancer drug testing. *Lab Chip* **2018**, *18*, 486–495. [CrossRef] [PubMed]
13. Lee, K.H.; Kwon, G.H.; Shin, S.J.; Baek, J.Y.; Han, D.K.; Park, Y.; Lee, S.H. Hydrophilic electrospun polyurethane nanofiber matrices for hMSC culture in a microfluidic cell chip. *J. Biomed. Mater. Res. Part A Off. J. Soc. Biomater. Jpn. Soc. Biomater. Aust. Soc. Biomater. Korean Soc. Biomater.* **2009**, *90*, 619–628. [CrossRef]
14. Au, S.H.; Chamberlain, M.D.; Mahesh, S.; Sefton, M.V.; Wheeler, A.R. Hepatic organoids for microfluidic drug screening. *Lab Chip* **2014**, *14*, 3290–3299. [CrossRef] [PubMed]
15. Gumuscu, B.; Albers, H.J.; Van Den Berg, A.; Eijkel, J.C.; Van Der Meer, A.D. Compartmentalized 3D Tissue Culture Arrays under Controlled Microfluidic Delivery. *Sci. Rep.* **2017**, *7*, 3381. [CrossRef] [PubMed]
16. Sakolish, C.M.; Esch, M.B.; Hickman, J.J.; Shuler, M.L.; Mahler, G.J. Modeling barrier tissues in vitro: Methods, achievements, and challenges. *EBioMedicine* **2016**, *5*, 30–39. [CrossRef] [PubMed]
17. Bhatia, S.N.; Ingber, D.E. Microfluidic organs-on-chips. *Nat. Biotechnol.* **2014**, *32*, 760. [CrossRef] [PubMed]
18. Agarwal, S.; Wendorff, J.H.; Greiner, A. Progress in the field of electrospinning for tissue engineering applications. *Adv. Mater.* **2009**, *21*, 3343–3351. [CrossRef] [PubMed]
19. Ma, Z.; Kotaki, M.; Inai, R.; Ramakrishna, S. Potential of nanofiber matrix as tissue-engineering scaffolds. *Tissue Eng.* **2005**, *11*, 101–109. [CrossRef] [PubMed]
20. Ramakrishna, S. *An Introduction to Electrospinning and Nanofibers*; World Scientific: Singapore, 2005.
21. Liu, W.; Thomopoulos, S.; Xia, Y. Electrospun nanofibers for regenerative medicine. *Adv. Healthc. Mater.* **2012**, *1*, 10–25. [CrossRef]
22. DeFrates, K.; Moore, R.; Borgesi, J.; Lin, G.; Mulderig, T.; Beachley, V.; Hu, X. Protein-based fiber materials in medicine: A review. *Nanomaterials* **2018**, *8*, 457. [CrossRef]
23. Das, S.; Sharma, M.; Saharia, D.; Sarma, K.K.; Muir, E.M.; Bora, U. Electrospun silk-polyaniline conduits for functional nerve regeneration in rat sciatic nerve injury model. *Biomed. Mater.* **2017**, *12*, 045025. [PubMed]
24. Bokhari, M.; Carnachan, R.J.; Cameron, N.R.; Przyborski, S.A. Culture of HepG2 liver cells on three dimensional polystyrene scaffolds enhances cell structure and function during toxicological challenge. *J. Anat.* **2007**, *211*, 567–576. [CrossRef] [PubMed]
25. Kolambkar, Y.M.; Dupont, K.M.; Boerckel, J.D.; Huebsch, N.; Mooney, D.J.; Hutmacher, D.W.; Guldberg, R.E. An alginate-based hybrid system for growth factor delivery in the functional repair of large bone defects. *Biomaterials* **2011**, *32*, 65–74. [CrossRef] [PubMed]
26. Tanaka, Y.; Yamato, M.; Okano, T.; Kitamori, T.; Sato, K. Evaluation of effects of shear stress on hepatocytes by a microchip-based system. *Meas. Sci. Technol.* **2006**, *17*, 3167.
27. *Surface Areas and Recommended Medium Volumes for Corning®Cell Culture Vessels*; Application Note; CLS-AN-209; Corning Incorporated: Midland, MI, USA, 2012.
28. Luo, L.W.; Teo, C.Y.; Ong, W.L.; Tang, K.C.; Cheow, L.F.; Yobas, L. Rapid prototyping of microfluidic systems using a laser-patterned tape. *J. Micromech. Microeng.* **2007**, *17*, N107. [CrossRef]
29. Shrirao, A.B.; Hussain, A.; Cho, C.H.; Perez-Castillejos, R. Adhesive-tape soft lithography for patterning mammalian cells: Application to wound-healing assays. *Biotechniques* **2012**, *52*, 315–318. [CrossRef]
30. Hsu, C.M.; Shivkumar, S. N,N-Dimethylformamide Additions to the Solution for the Electrospinning of Poly (ε-caprolactone) Nanofibers. *Macromol. Mater. Eng.* **2004**, *289*, 334–340.
31. Qin, X.; Wu, D. Effect of different solvents on poly (caprolactone)(PCL) electrospun nonwoven membranes. *J. Therm. Anal. Calorim.* **2011**, *107*, 1007–1013. [CrossRef]
32. Kim, T.-E.; Kim, C.G.; Kim, J.S.; Jin, S.; Yoon, S.; Bae, H.-R.; Kim, J.-H.; Jeong, Y.H.; Kwak, J.-Y. Three-dimensional culture and interaction of cancer cells and dendritic cells in an electrospun nano-submicron hybrid fibrous scaffold. *Int. J. Nanomed.* **2016**, *11*, 823.
33. Chua, K.-N.; Lim, W.-S.; Zhang, P.; Lu, H.; Wen, J.; Ramakrishna, S.; Leong, K.W.; Mao, H.-Q. Stable immobilization of rat hepatocyte spheroids on galactosylated nanofiber scaffold. *Biomaterials* **2005**, *26*, 2537–2547. [CrossRef] [PubMed]

34. Vucenik, I.; Tantivejkul, K.; Zhang, Z.S.; Cole, K.E.; Saied, I.; Shamsuddin, A.M. IP6 in treatment of liver cancer. I. IP6 inhibits growth and reverses transformed phenotype in HepG2 human liver cancer cell line. *Anticancer Res.* **1998**, *18*, 4083–4090. [PubMed]
35. Rabouille, C.; Spiro, R.G. Nonselective utilization of the endomannosidase pathway for processing glycoproteins by human hepatoma (HepG2) cells. *J. Biol. Chem.* **1992**, *267*, 11573–11578. [PubMed]

 © 2019 by the authors. Licensee MDPI, Basel, Switzerland. This article is an open access article distributed under the terms and conditions of the Creative Commons Attribution (CC BY) license (http://creativecommons.org/licenses/by/4.0/).

Article

Self-Powered Well-Aligned P(VDF-TrFE) Piezoelectric Nanofiber Nanogenerator for Modulating an Exact Electrical Stimulation and Enhancing the Proliferation of Preosteoblasts

Aochen Wang *, Ming Hu *, Liwei Zhou and Xiaoyong Qiang

School of Microelectronics, Tianjin University, Tianjin 300072, China; 13602176911@163.com (L.Z.); shawn_q@tju.edu.cn (X.Q.)
* Correspondence: aochen_wang@tju.edu.cn (A.W.); huming@tju.edu.cn (M.H.); Tel.: +86-18622273980 (A.W. & M.H.)

Received: 24 January 2019; Accepted: 19 February 2019; Published: 3 March 2019

Abstract: Electric potential plays an indispensable role in tissue engineering and wound healing. Piezoelectric nanogenerators based on direct piezoelectric effects can be self-powered energy sources for electrical stimulation and have attracted extensive attention. However, the accuracy of piezoelectric stimuli on piezoelectric polymers membranes in vitro during the dynamic condition is rarely studied. Here, a self-powered tunable electrical stimulation system for assisting the proliferation of preosteoblasts was achieved by well-aligned P(VDF-TrFE) piezoelectric nanofiber membrane (NFM) both as a nanogenerator (NG) and as a scaffold. The effects of electrospinning and different post-treatments (annealing and poling) on the surface wettability, piezoelectric β phase, ferroelectric properties, and sensing performance of NFMs were evaluated here. The polarized P(VDF-TrFE) NFM offered an enhanced piezoelectric value (d_{31} of 22.88 pC/N) versus pristine P(VDF-TrFE) NFM (d_{31} of 0.03 pC/N) and exhibited good sensing performance. The maximum voltage and current output of the P(VDF-TrFE) piezoelectric nanofiber NGs reached −1.7 V and 41.5 nA, respectively. An accurate electrical response was obtained in real time under dynamic mechanical stimulation by immobilizing the NGs on the flexible bottom of the culture plate, thereby restoring the real scene of providing electrical stimulation to the cells in vitro. In addition, we simulated the interaction between the piezoelectric nanofiber NG and cells through an equivalent circuit model. To verify the feasibility of P(VDF-TrFE) nanofiber NGs as an exact electrical stimulation, the effects of different outputs of P(VDF-TrFE) nanofiber NGs on cell proliferation in vitro were compared. The study realized a significant enhancement of preosteoblasts proliferation. This work demonstrated the customizability of P(VDF-TrFE) piezoelectric nanofiber NG for self-powered electrical stimulation system application and suggested its significant potential application for tissue repair and regeneration.

Keywords: well-aligned nanofibers; P(VDF-TrFE); piezoelectric nanogenerator; preosteoblasts electrospinning

1. Introduction

Electrical stimulation is widely used to compensate for the altered electrical communication in diseased tissue and thus improve tissue regeneration [1–3]. External electric fields can improve physiological strength and can guide cell-orientated growth and influence cell proliferation and differentiation including for nerves, cardiac cells, and osteoblasts [4–10]. However, traditional electrical stimulator requires invasive microelectrodes, an external power supply, and electrical wires. This is very uncomfortable, inconvenient, and unreliable. Thus noninvasive, wireless, portable, self-powered, and wearable electronic devices are urgently needed for electrical stimulation system. Recent work in nanogenerators (NGs) has exhibited significant progress in noninvasive and self-powered electrical stimulation [8,11]. In the framework, piezoelectric nanogenerators (PENGs) based on piezoelectric polymers that can generate electric surface charges under external mechanical vibration and thus achieve cordless electrical stimulation, have attracted a lot of attentions [12–15]. The electrical output produced by the PENG acts as an electrical stimulation signal whose value is related to the piezoelectric property of the material [16,17]. One such material is poly(vinylidene fluoride-trifluoroethylene) (P(VDF-TrFE)) with outstanding piezoelectric properties due to steric hindrance from the extra fluorine atoms in the TrFE inducing an all-trans stereochemical configuration [18,19]. It is well-known that the β phase is the most highly polar crystalline phase of PVDF and its copolymers [20]. Accordingly, P(VDF-TrFE) with a high content of piezoelectric β-crystalline offers excellent piezoelectric properties.

The electrospinning process can produce piezoelectric fibers by stretching as well as in situ poling during the fabrication process, paving the way for piezoelectric nanofiber NGs [21–24]. These electrospun aligned fibers have a higher content of β-crystalline phase versus random fibers [25]. In addition, various post-treatments, such as annealing and poling treatments, can produce piezoelectricity via dipole orientations; however, the influence of poling treatments on the β-crystalline phase content is rarely studied [26]. Therefore, to further improve the piezoelectricity of electrospun fibers and thus enhance the performance of PENG, it is critical to understand the specific significance of post-treatments on the piezoelectric performance [27,28].

In addition to its good piezoelectric properties, P(VDF-TrFE) also offers easy processing and nanomaterial formability. Therein, electrospinning has been widely employed to fabricate the micro-/nanofibers in ordered, random, and specific patterns. The electrospun fibrous films have a high surface-area-to-volume ratio similar to the structure and characteristics of extracellular matrix (ECM) [29]. In this framework, research has demonstrated that the alignment of nanofibers can structurally mimic the parallel orientation of the tissues and modulate cell adhesion, migration, proliferation, and differentiation [30–33].

Kai et al. prepared electrospun aligned and randomly oriented poly(e-caprolactone)/gelatin (PG) scaffolds. They found that the aligned PG scaffold could enhance the cells attachment and alignment [34]. Hitscherich et al. reported that mouse embryonic stem cell-derived cardiomyocytes (mES-CM), cultured on the aligned P(VDF-TrFE), were aligned along the fibers and expressed classic cardiac-specific markers [35]. Therefore, based on its piezoelectric property and processability, P(VDF-TrFE) has been widely used for biomedical scaffolds for tissue engineering and electrical stimulation. Genchi et al. fabricated P(VDF-TrFE)/$BaTiO_3$ composite films as a substrate for piezoelectric stimulation to enhance the differentiation of neuroblastoma cells [36]. Later Deng's group used the $BaTiO_3$/P(VDF-TrFE) nanocomposite membrane and leveraged the piezoelectric properties to promote bone regeneration [37]. However, the exact value of electrical stimulation induced by the piezoelectric substrate is not clear when external mechanical vibration is also applied to the cells in vitro. Thus, to better understand the effect of piezoelectric regulation on cell behavior, there is a need to measure this electrical stimulation.

In this work, we explored the effects of the exact electrical signals generated by the P(VDF-TrFE) piezoelectric nanofibers NGs on the proliferation fate of preosteoblasts. Here, the fabrication of self-powered piezoelectric nanofiber NG used as cell scaffold was based on electrospun well-aligned P(VDF-TrFE) nanofiber membranes (NFMs). The effects of annealing and poling post-treatments on the surface wettability, piezoelectric β phase, piezoelectricity, and sensing performance of P(VDF-TrFE) NFMs were investigated. In order to study the dependence of the electrical outputs of NGs on the degree of polarization, two kinds of NGs processed by different poling electrical fields were prepared. In particular, they were fixed to the flexible bottom of the culture plate, and the accurate electrical response was measured in real time under dynamic mechanical stimulation, thereby restoring the real scene of the electrical stimulation of the cells in vitro. In addition, we simulated the interaction between piezoelectric nanofiber NG and cells through an equivalent circuit model. In order to study the role of NFMs as a scaffold, the effects of well-aligned and random interfaces of NFMs on the morphology of preosteoblasts were investigated. The well-aligned nanofibrous platforms could guide and elongate the cells. Finally, we compared the effects of different outputs stimulation of P(VDF-TrFE) nanofiber NGs on cell proliferation in vitro by applying a dynamic piezoelectric stimulus. This work demonstrates a significant potential of P(VDF-TrFE) piezoelectric nanofiber NG as self-powered electrical stimulation system for assisting tissue repair and regeneration.

2. Materials and Methods

2.1. Electrospinning of Nanofiber Membranes (NFMs)

The P(VDF-TrFE) (75/25 mol%, Piezotech Inc., Pierre-Bénite, France) nanofibers were prepared as described previously [38]. Briefly, P(VDF-TrFE) powders were dissolved in the N,N-dimethylformamide (DMF) and acetone mixture solution (6:4 v/v) at 20% (w/v). The P(VDF-TrFE) spinning solution was injected into a 5-ml syringe fitted with a 22 G needle. A syringe pump (KDS101, KD Scientific, Holliston, MA, USA) was used to supply a constant flow rate of 1 mL/h. A high voltage of 15 kV was applied between the tip of the syringe needle and the grounded roller collector at a distance of 10 cm. The thickness of the electrospun NFM was regulated at ~30 μm by controlling the electrospinning time. The electrospun NFMs were dried at 65 °C for 10 h to volatilize the residual solvent. The annealed NFMs were kept in a vacuum oven at 135 °C for 4 h. The poled samples were pressed via a powder compression machine (BJ-15, BOJUNKEJI Inc., Tianjin, China) after annealing, in order to ensure no conduction between subsequent sputtering electrodes. Next, the samples were prepared by sputtering gold electrodes on both the top and bottom surfaces, then placed in a fixture and immersed completely in silicon oil. Afterwards, the silicon bath was heated to 115 °C by a hot plate (HCT basic, IKA Inc., Staufen, Germany) and then a polarization electric field of 80 to 100 MV/m was applied. After 30 min of thermal poling treatment, the polarization voltage was kept constant, cooled down to room temperature and finally the voltage was removed. Next, pristine P(VDF-TrFE) NFMs without any postprocessing (annealing and poling) were labeled as U-NFM, and the samples treated by annealing were coded as A-NFM. The poled samples were denoted as P-NFM. The NFMs poled with the electric field of 80 MV/m and 100 MV/m were labeled as P80-NG and P100-NG, respectively.

2.2. Characterization and Measurements of NFMs

The morphology of electrospun NFMs was observed with a scanning electron microscope (SEM, SU8020, Hitachi Ltd., Tokyo, Japan) at an accelerating voltage of 5 kV. ImageJ software (National Institutes of Health, Bethesda, USA) was used to analyze the mean fiber diameter. Tensile testing was done with a tensile test machine (ESM301, Mark-10, Copiague, NY, USA) at room temperature with a cross-head velocity of 10 mm/min. The sample was cut to dumbbell shape (10 mm long and 5 mm wide). The contact angles were recorded employing contact angle goniometer (XG-CAMB1, Xuanzhun co., Ltd, Shanghai, China) by sessile drop method at room temperature. A droplet of deionized water was dropped from the capillary mouth to stop on the membrane surface and the angle of the droplet on the upper surface of the membrane was collected and analyzed. X-ray diffraction (XRD) patterns were done on an X-ray diffractometer (X'pert^3 Powder, PANalytical Ltd., Almelo, The Netherlands) and recorded over an angular range from 10° to 50°. Infrared spectra were recorded on a Fourier transform infrared spectrometer (FTIR, VERTEX80v, Bruker Corp., Billerica, MA, USA) from 400 cm^{-1} to 1600 cm^{-1}. The polarization-electric field (P-E) hysteresis loops were obtained by precision multiferroic and ferroelectric test systems (Radiant Technologies Inc., Alpharetta, GA, USA) under a unipolar electric field at a measurement frequency of 10 Hz.

The dynamic piezoelectric coefficient d_{31} was determined with a homemade measurement system nearly identical to setup described previously [39]. The output voltage of the samples during the process of stretching–relaxing was recorded with a DSP lock-in amplifier (SR830, Stanford Research

Systems, Sunnyvale, CA, USA). The piezoelectricity coefficient d_{33} was measured using a quasi-static d_{33} measuring instrument (Institute of Acoustics, Chinese Academy of Science, ZJ-4AN, Beijing, China).

2.3. Measurements and Stimulation of Piezoelectric Nanofiber Nanogenerators (NGs)

The generated voltage, current, and charge of the NGs generated via dynamic mechanical stimulus from a speaker (8 Ω, 1 W) were assessed by a Tektronix Keithley electrometer 6514. The speakers were driven by sinusoidal signals including a 4 V amplitude at different frequencies (2 Hz, 3 Hz, 4 Hz, and 5 Hz) generated by function generator (DS345, Stanford Research Systems, Sunnyvale, CA, USA).

To quantitatively analyze the deformation and potential distribution of P(VDF-TrFE) nanofibers based on NG by a uniaxial stress (~0.5 Mpa), we performed finite element modeling (FEM) by COMSOL multiphysics software (5.3, COMSOL Inc., Stockholm, Sweden).

From the analysis of electrical characteristics, the cell membrane can be equivalently modeled as a resistor capacitor with electrical properties such as extracellular medium resistance (R_{cm}), membrane resistance (R_m), membrane capacitance (C_m) and ion equilibrium potential (V_m). According to these four characteristics, an equivalent circuit can be constructed [40,41]. The external electrical stimulation is generated by a P(VDF-TrFE) piezoelectric nanofiber NG under a dynamic mechanical vibration, which can be represented as a voltage source. By using the equivalent circuit model, the expected behavior of the effective voltage and current applied to the cell membrane by NG was evaluated. The circuit was simulated by Multism software (14.0, National Instruments Co., Austin, TX, USA).

2.4. Cell Culture

MC3T3-E1 cells (Subclone 14, mouse preosteoblasts, Innochem Ltd., Beijing, China) were cultured in Alpha Minimum Essential Medium (α-MEM, Gibco) with 2 Mm L-glutamine (Gibco) and 1 mM sodium pyruvate (Gibco) supplemented with 10% fetal bovine serum (Gibco), 100 U/mL penicillin, and streptomycin (Gibco) in a humidified atmosphere with 5% CO_2 in air at 37 °C.

2.5. Cell Alignment Quantification

The cell alignment on the NFM was quantified by a two-dimensional fast Fourier transform (2-D FFT) image analysis method [42]. Briefly, the fluorescent cell photographs were converted to 8-bit grayscale and then cropped a 1024 × 1024 pixel square. This was then masked with a transparent circular pattern, and the corners were filled with black. The processed photograph was analyzed by FFT function in ImageJ software [43]. Pixel intensities along the radian were summed using the oval profile plug-in and normalized by the lowest intensity value [44]. The normalized results represent the percentage of cells aligned along a certain direction.

2.6. Piezoelectric Stimulation and Cell Proliferation Assay

The dynamic piezoelectric stimulus used the custom-made speakers that provided uniform mechanical vibration to cells in monolayer cultures. Specialized flexible-bottomed culture plates (BF-3001U, Flexcell Int. Co., Austin, TX, USA) made of silicone elastomer membrane were used to culture the cells. The strain applied to the silicone elastomer membrane was directly transmitted to the NGs to generate piezoelectricity. The synthesized function generator and power amplifier were used to control the frequency and amplitude of deformation applied to the culture plate (experimental vibration frequency: 2 Hz; amplified voltage: 4 V).

To clarify the feasibility of P(VDF-TrFE) nanofiber NGs as exact electrical stimulation and demonstrate the effects on the proliferation of MC3T3-E1 cells, P100-NG and P80-NG were selected as the experimental groups, and the nonpiezoelectric A-NFM served as the control group. The cells were seeded at a density of 2×10^4 cells per well on the various NGs. The piezoelectric stimulation was applied to MC3T3-E1 cells for 30 min per day for 1 day, 3 days, or 5 days. The proliferation of the cultured MC3T3-E1 cells was measured using the cell count kit-8 (CCK-8, Dojindo Molecular Technology). The culture medium was first replaced with 1.5 mL α-MEM medium plus 10% CCK-8 solution. After 4 h incubation at 37 °C, the production of water-soluble formazan dye was determined using a microplate reader (MULTISKA NMK3, Thermo Fisher Scientific, Waltham, MA, USA) at a wavelength of 450 nm. The culture medium was changed every 2 days. Three parallel replicates were examined each time for each group.

To observe the cell morphology on the NFMs, the MC3T3-E1 cells were fixed with 4% paraformaldehyde solution in PBS (Sigma) for 10 min and then washed three times with warm 1× PBS and blocked with 1% bovine serum albumin (BSA, Sigma) solution for 60 min. The cytoskeleton was stained with Phalloidin (Invitrogen) conjugated to Alexa Fluor 488 (1:200 diluted) for 2 h at 37 °C, and the nucleus was stained with 4′,6-diamidino-2-phenylindole (DAPI, 300 nM, Life Technology) for 10 min.

2.7. Statistical Analysis

The data were expressed as the mean ±standard deviation (SD). Statistical analysis was determined using one-way ANOVA. Statistical differences were tested with a one-way ANOVA using the *t*-test (Tukey test) for independent samples. Statistical significance was accepted at * $p < 0.05$ and ** $p < 0.01$.

3. Results and Discussion

3.1. Morphology and Characterization of NFMs

Figure 1 shows a schematic for the electrospinning process and post-treatment process. The P(VDF-TrFE) nanofibers were fabricated with the optimized electrospun parameters using the electrospinning setup. They had a strongly aligned and uniform morphology (Figure 1b). The mean diameter of electrospun nanofibers was 590 nm ± 26 nm with further size analysis in Figure S1a. After annealing (Figure 1c and Figure S1b), several microvoids appeared clearly on the surface of the nanofibers and the surface of single fiber became rough. Here, the A-NFM exhibited a porous structure and a higher surface-area-to-volume ratio than the U-NFM, which is beneficial for the migration of implanted cells. Although the P-NFM exhibited the flat surface owing to the mechanical pressing to make the nanofibers flattened, it still had nanofibrous surface. As shown in the insets of Figure 1b–d, the average contact angle of U-NFM was 129.13°. Although the single nanofiber surface of A-NFM was rougher than that of U-NFM, the contact angle of A-NFM decreased to 113.38°. This phenomenon can be explained as follows. In addition to the influence of surface roughness, the contact angle is also related to the surface free energy [45]. The annealing treatment resulted in the transformation of some nonpolar α phases into polar β phase. The presence of polar β phase increased the dipolar interaction between the NFM and water molecules, which increased surface energy and reduced the contact angle [46,47]. The influence of high surface energy on the contact angle of A-NFM is greater than that of surface roughness. The P-NFM had the smallest contact angle (91.14°) due to the combined effects of nanostructure flattening and polar β phase, demonstrating the surface wettability of P(VDF-TrFE) NFM can be improved by poling treatment. In addition to the highly-aligned and porous properties, the fibers of the polarized P-NFM samples were arranged more closely, and the fibers were bonded to each other. This improves the material's mechanical properties (Figure 1d) [48]. Representative stress–strain plots are displayed in Figure S2. The mean elastic modulus of U-NFM, A-NFM, and P-NFM were 0.148 GPa, 0.426 GPa, and 0.876 GPa, respectively. This result proves that the poling treatment can significantly affect the mechanical robust. A well-aligned and uniform P-NFM with favorable mechanical properties shows its potential as a scaffold in tissue engineering.

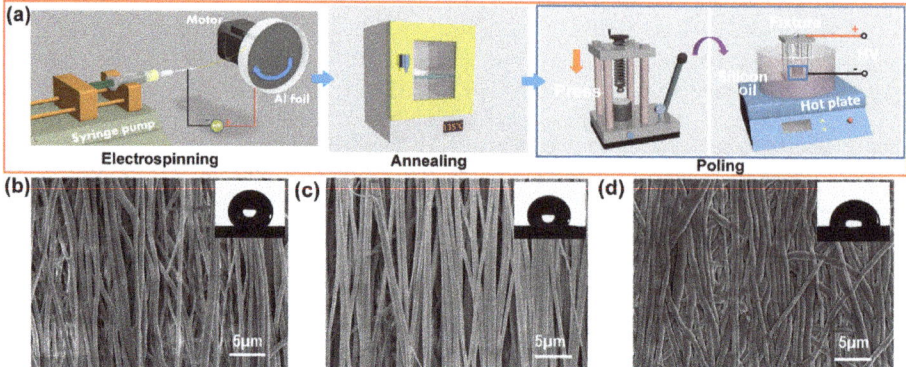

Figure 1. The morphology and contact angle of electrospun nanofiber membranes (NFMs). (**a**) Schematic illustration of the preparation and treatment of different samples. Scanning electron microscope (SEM) micrographs of (**b**) U-NFM, (**c**) A-NFM, and (**d**) P-NFM. The insets represent the contact angles corresponding to U-NFM, A-NFM, and P-NFM, respectively. (U-NFM represented pristine poly(vinylidene fluoride-trifluoroethylene)(P(VDF-TrFE)) NFM without any postprocessing; A-NFM represented annealed P(VDF-TrFE) NFM; P-NFM represented the poled samples.)

The crystallinity of P(VDF-TrFE) NFMs treated with different postprocessing steps was determined via the XRD patterns (Figure 2a). There was a distinct reflection peak from 19 to 21 degrees for all NFMs corresponding to the diffraction of plane (200)/(110) of the β phase crystal [49,50]. The broad shoulder at ~18 degrees for U-NFM is associated with the amorphous phase. After annealing treatment, the shoulder completely disappeared for both A-NFM and P-NFM, but the diffraction intensity of the (200)/(110) plane increased. In addition, the increase in diffraction peak intensity of P-NFM versus A-NFM demonstrates that the β phase is enhanced after poling. By fitting the XRD patterns [51], the diffraction curve could be resolved into three regions: amorphous as well as α- and β-crystalline phases. Thus, α and β phases could be measured. The percentage of α phase in the U-NFM was 23.3% while that of the A-NFM was 26.5%. The P-NFM had less α phase—this might be due to the phase transformation from α-crystal phase to β-crystal phase after poling treatment. The percentage of β phase in the U-NFM and A-NFM was 43.1% and 46.6%, respectively; it was 69.2% in P-NFM.

The crystal phase structures could also be characterized by FTIR spectra (Figure 2b). The characteristic absorption bands [52,53] at 506, 840, 1285, and 1430 cm^{-1} are recognized as β phase structures whereas the absorbance peaks of the α phase structure appears at 532, 614, 765, 870, and 976 cm^{-1}. Versus U-NFM, the intensity of characteristic bands corresponding to β phase increased for the A-NFM, which suggests that the annealing treatment improves the β phase content. Furthermore, the highest peak intensity of β phase was seen with P-NFM. This phenomenon is mainly due to thermal poling that increases with the degree of dipole orientation and phase transition of the β phase. There are fewer crystalline defects and enhanced β-crystallinity.

The XRD patterns and FTIR spectra indicate that an annealing treatment can slightly increase the β phase content while the thermal poling treatment can significantly improve the β phase crystallinity.

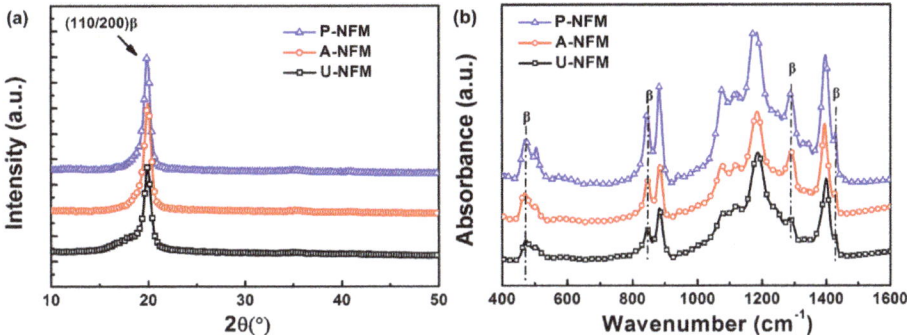

Figure 2. Crystalline characterization of P(VDF-TrFE) NFMs treated with different postprocessing steps. (**a**) X-ray diffraction (XRD) patterns and (**b**) Fourier transform infrared spectroscopy (FTIR) spectra.

3.2. Effect of Postprocessing on NFM Piezoelectric Properties

The polarization-electric field hysteresis loops (P-E loops) of U-NFM, A-NFM, and P-NFM at various electric fields are presented in Figure 3 and illustrate the ferroelectric behavior of nanofiber membranes treated with different postprocessing steps. The remnant polarization (Pr) and the saturated polarization (Ps) of U-NFM was 17.1 mC/m^2 and 37.9 mC/m^2, respectively (Figure 3a). After annealing, the Pr and Ps of A-NFM could reach 26.9 mC/m^2 and 43.1 mC/m^2, respectively (Table 1 and Figure 3b), indicating that the annealing process can increase the crystallinity. The Pr is mainly associated with a highly polar β-crystalline phase. Table 1 and Figure 3c show that a higher Pr of 32.6 mC/m^2 could be obtained by P-NFM under a polarizing electric field of 160 MV/m. In addition, Pr was closer to Ps in sample P-NFM, and the P-E loops tended to be saturated. This suggested that the ferroelectric domain trends toward a single-domain. In particular, the higher Pr mainly originated from a β phase crystal domain reflecting the better ferroelectric properties. This result suggests that thermal poling can increase the β-crystalline phase content. In addition, the coercive electric field (E_c) increased from 60.9 MV/m for U-NFM and 65.2 MV/m for A-NFM to 88.1 MV/m for P-NFM, suggesting that the ferroelectric domain of the β-crystal phase is not oriented easily. Consequently, the P-NFM had a strong ability to maintain polarization and possessed excellent piezoelectric property.

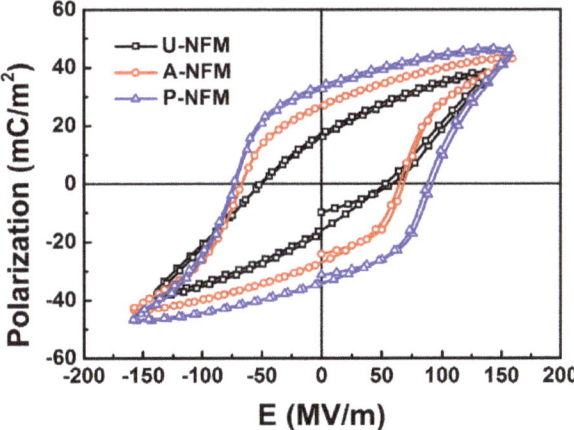

Figure 3. The P-E hysteresis loops of P(VDF-TrFE) NFMs treated with different postprocessing at a polarization electric field of 160 MV/m.

The relationship between charge density and applied stress is plotted in Figure 4 and can evaluate the piezoelectric coefficient d_{31}. The slope of the line represents the piezoelectric coefficient d_{31} (Figure 4). The calculated d_{31} and the corresponding linear regression correlation coefficients (R^2) are summarized in Table 1. The data illustrate that there is better agreement between the theoretical and experimental values of d_{31} for A-NFM and P-NFM than for U-NFM. The A-NFM and P-NFM showed good sensing performances. It means that P-NFM can provide accurate electrical stimulation under varying external stress and that P-NFM is more reliable in the application of NGs that provide electrical stimulation.

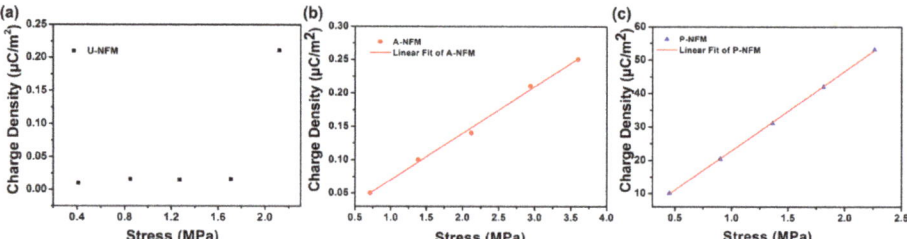

Figure 4. Experimental and theoretical studies of piezoelectric coefficient d31 of nanofiber membranes. (**a**) Experimental charge density–stress curve of U-NFM. Experimental (symbols) and linear fitting (lines) charge density–stress curves of (**b**) A-NFM and (**c**) P-NFM.

Table 1. Electric properties of P(VDF-TrFE) NFMs with different postprocessing steps at an electric field of 160 MV/m.

Sample	Ec (MV/m)	Ps (mC/m^2)	Pr (mC/m^2)	d_{33} (pC/N)	d_{31} (pC/N)	R^2
U-NFM	60.9	37.9	17.1	0	0.03	0.3496
A-NFM	65.2	43.1	26.9	0	0.07	0.9948
P-NFM	88.1	44.1	32.6	−31	22.88	0.9997

In addition, there was almost no piezoelectricity in A-NFM (d_{31} = 0.07 pC/N) and U-NFM (d_{31} = 0.03 pC/N); however, P-NFM showed the highest piezoelectric coefficient (d_{31} = 22.88 pC/N, d_{33} = −31 pC/N) among all samples (Table 1). This is due to the fact that U-NFM has many amorphous crystalline phases, some nonpolar α phases and a small amount of polar β phases. With an annealing treatment, the amorphous phase of the nanofibers decreases, the total crystallinity increases and some nonpolar α phases are transformed to polar β phase, which increases the content of β phase and thus the remnant polarization was increased compared to U-NFM, nevertheless, the dipole orientation is disorderly, so the total spontaneous polarization is zero and the piezoelectric property is weak. After poling treatment, the transformation of the α phase to the β phase is further promoted leading to high remnant polarization and most electric dipoles are oriented along the direction of the externally applied electric field, underscoring the high piezoelectric performance. With the favorable mechanical and ferroelectric properties of P-NFM confirmed, we next used this material as piezoelectric nanofiber NG and evaluated the electrical response as well as the arrangement and proliferation of cells cultured on this material.

3.3. Effect of Mechanical Stimulus on Electrical Performances

To study the dependence of the electrical outputs of NGs on the degree of polarization, two kinds of NGs processed by different poling electrical fields were measured. Figure 5a shows the illustrative diagram of the experiment setup for imitating the real scene of the electrical stimulation of the cells in vitro during the dynamic mechanical vibration. The NG was deformed periodically via a self-designed setup driven with an amplitude of 4 V (peak-to-peak value). The resulting curve exhibited a periodic alternation of negative and positive responses corresponding to the deformed and released states of a piezoelectric nanofiber NG, respectively (Figure 5b). As the frequency increases from 2 Hz to 5 Hz, the generated peak-to-peak piezoelectric current increased from 18.1 nA to 39.7 nA in the P80-NG sample (Figure 5c upper); the P100-NG sample was modulated from 23.1 nA to 41.5 nA (Figure 5c lower). These results confirm that the charge transfer is kept equal at different vibration frequencies, but the output is due to rapid electron flow.

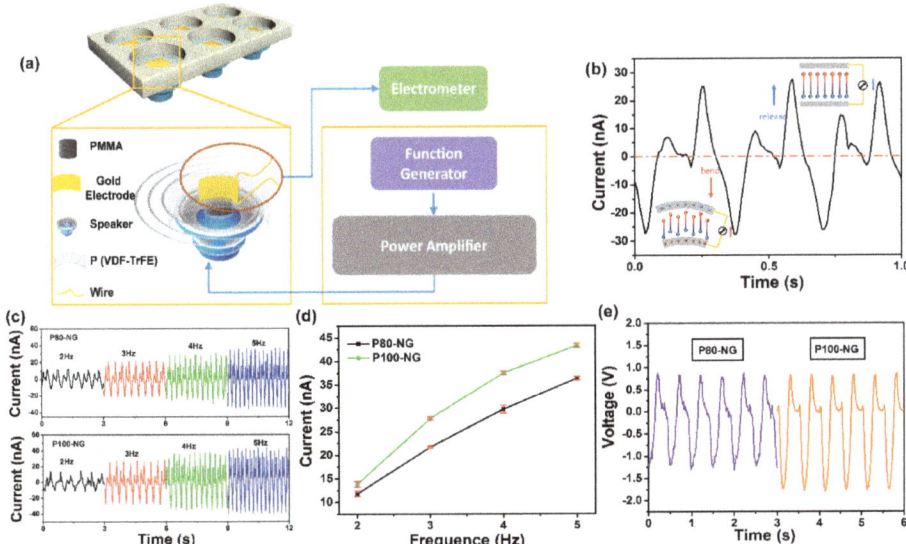

Figure 5. Effect of mechanical stimulus on electrical performances of different NGs. (**a**) Schematic diagram of the home-designed experimental shaker for providing periodic mechanical vibrations. (**b**) Piezoelectric output currents recorded during bending and releasing. The insets are the schematics of NG under mechanical bending deformation and releasing state, respectively, (P100-NG at a frequency of 3 Hz). (**c**) Current outputs of P80-NG (above) and P100-NG (below) in the frequency range of 2 to 5 Hz. (**d**) Comparison of measured current of P80-NG and P100-NG under different frequency from 2 to 5 Hz. (**e**) Voltage outputs of P80-NG and P100-NG at a frequency of 2 Hz.

Figure 5d shows that the measured current increased with increasing vibration frequency in both P80-NG and P100-NG. The slope of the P100-NG sample was 1.2-fold higher than that of P80-NG sample under the same vibration force. Figure 5e shows the results of induced voltage under a frequency of 2 Hz. The output voltage of P80-NG and P100-NG reached −1.3 V and −1.75 V, respectively. These results indicate that the electrical performance of the piezoelectric nanofiber NG are affected by poling treatment and the outputs can be modulated and optimized by adjusting the polarization treatment. In the following in vitro assay, piezoelectric nanofiber NG as an electrical stimulator can provide an exact stimulation to cells in real time during the dynamic status.

3.4. Theoretical Modeling of NG-Cell Interaction

To estimate the expected behavior of the effective voltage and current applied to the cell membrane by NG, an equivalent circuit was used to model the interaction of the piezoelectric nanofiber NG and cells. Figure 6a shows the diagram of NG-cell and corresponding equivalent circuit model [54,55]. Under external mechanical vibration, the NG deformation produced an external voltage excitation that could be modeled as a voltage source (V_{NG}). The voltage reached the cell membrane through the culture medium, the conductivity of the medium would affect the voltage that stimulated the cell. The cell membrane is composed of a phospholipid bilayer, which can be regarded as an insulator.

The extracellular fluid, cell membrane and intracellular fluid constitute a capacitor (C_m) [56]. The initial potential of ion channel is represented as V_m [57]. Figure 6b shows the simulation results of voltage and current transmitted to the cell membrane through the circuit due to sinusoidal voltage stimulation. When an excitation of 1.3 V was input, the voltage and current delivered to the cell was ~0.8 V and 0.4 mA, respectively. According to the basic principles of the circuit, when V_{NG} is active, the capacitor C_m is charged (corresponding to the current that is stimulating cell growth), and the voltage across the capacitor (the voltage applied to the cell) V_c (t) (t is time, hereinafter abbreviated as V_c), which is expressed as

$$V_c = -\frac{R_m V_{NG}}{R_{cm} + R_m} \cdot e^{-\frac{R_{cm}+R_m}{R_{cm}R_m C_m}t} + \frac{R_{cm} V_m + R_m V_{NG}}{R_{cm} + R_m}$$

The result calculated by the formula of the voltage applied to the cell was consistent with that of the circuit simulation.

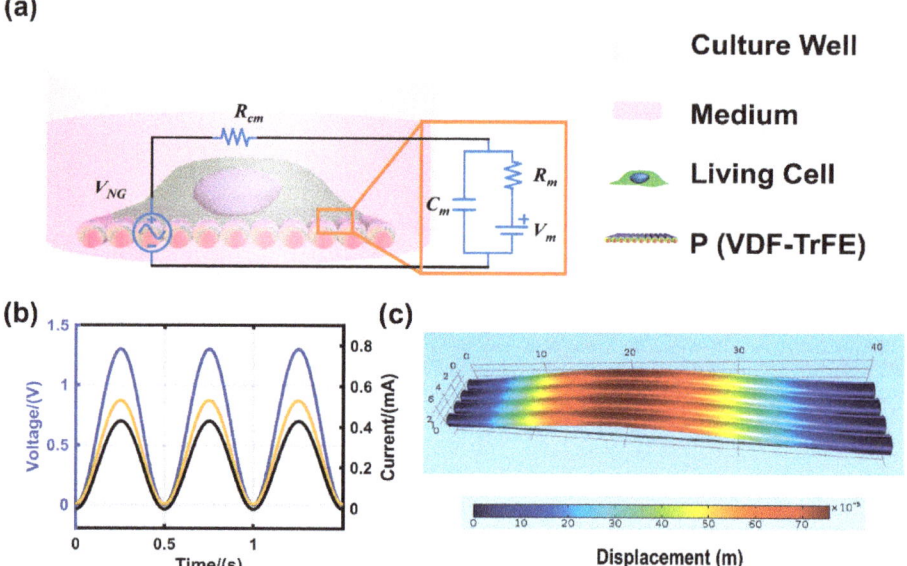

Figure 6. Electrical model of the NG-cell and FEM model of NG. (**a**) The diagrammatic sketch of NG-cell and corresponding equivalent circuit model. (**b**) The voltage (yellow) and current (black) applied to the cell membrane due to sinusoidal voltage stimulation (blue). (**c**) FEM stimulation of deformation distribution of nanofibers on NG at the stress amplitude of 0.5 Mpa.

Since a large deformation of the substrate will hinders cell adhesion, the deformation of aligned nanofibers based on NG under 0.5 MPa stress was analyzed by using finite element model (Figure 6c). The resulting maximum deformation was ~7 μm.

3.5. Cell Morphology on NG without Piezoelectric Stimulaton

For tissue engineering, the scaffold material should have excellent cytocompatibility to support cell growth and proliferation [58]. The cell viability data were validated by live/dead kit (Figure S3). The viability was similar between NG and control, suggesting that the P(VDF-TrFE) piezoelectric nanofiber NG has good cytocompatibility. Furthermore, the highly aligned micro-/nanostructure of the fiber-based scaffolds can provide morphological cues for cell attachment and behavioral modulation [59–61]. To study the effect of nanofiber morphology on the alignment of cells, the high aligned and random P(VDF-TrFE) nanofibers were used for cell culture. Figure 7a,c shows that the MC3T3-E1 cells attached nicely on both the aligned and random P(VDF-TrFE) nanofibers. The cell cytoskeleton and nucleus showed an elongated morphology on the direction of nanofibers alignment of P(VDF-TrFE) NG, while the MC3T3-E1 cells seeded on random nanofibers displayed a random orientation. This was further verified in that the aligned P(VDF-TrFE) NGs not only have an excellent cytocompatibility but can also guide cell elongation and orientation.

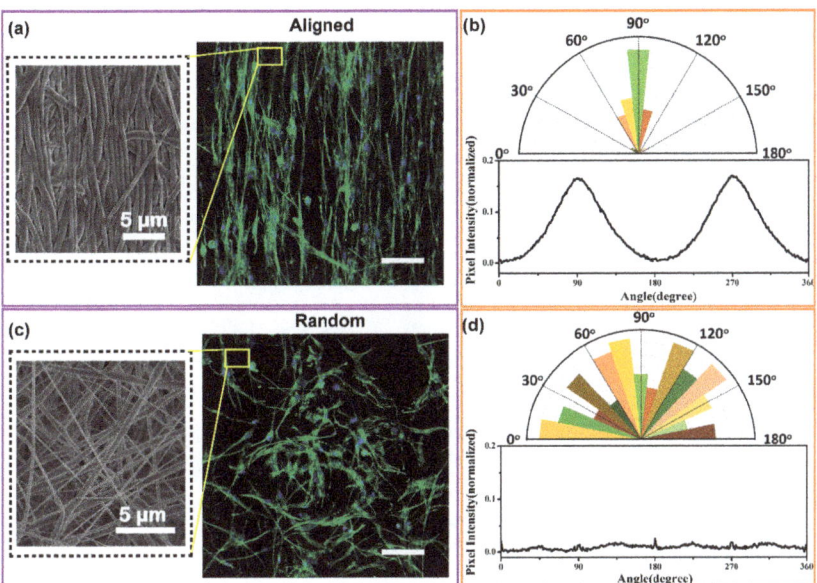

Figure 7. Attachment and alignment of MC3T3-E1 cells on well-aligned and random P(VDF-TrFE) nanofibers after 3 days of culture. (**a**) The SEM image of well-aligned nanofiber substrate (left column) and the confocal fluorescence micrographs of MC3T3-E1 cells (right column). (**b**) 2-D FFT image analysis of cell nuclei (above) and cytoskeleton alignment (below) on well-aligned P(VDF-TrFE) nanofibers. (**c**) SEM image of random nanofiber substrate (left column) and the confocal fluorescence micrographs of MC3T3-E1 cells (right column). (**d**) 2-D FFT image analysis of nuclei (above) and cytoskeleton alignment (below) on random P(VDF-TrFE) nanofibers. F-actin was stained by Alexa Fluor 488-labeled phalloidin (green); cell nuclei were stained by DAPI (blue). The scale bar for confocal fluorescence micrographs is 100 μm.

The results of representative 2-D FFT image are shown in Figure 7b,d. There were two significant symmetrical peaks at 90° and 270° in the plot of aligned P(VDF-TrFE) NGs illustrating that the direction of actin filaments is the same as the nanofibers (Figure 7b below) [62]. The frequency distributions of nuclei were preferentially concentrated along 90° (Figure 7b above) suggesting a specific nuclei orientation. In comparison, the 2-D FFT plot of random samples showed no obvious peaks; the actin filaments were randomly arranged (Figure 7d blow). The arrangement angles of nuclei on the random nanofibers were disordered as the angular histogram exhibits. This confirms that the cell cytoskeleton orientation is caused by the well-aligned surface topography of P(VDF-TrFE) NGs. The results demonstrate that P(VDF-TrFE) NGs can provide not only electrical stimulation signals but also morphologic cues.

3.6. Effect of Piezoelectric Stimulation Induced by NG on MC3T3-E1 Cells

We next verified the effect of the piezoelectric response of the P(VDF-TrFE) NG acted as electrical stimulus on the cell behavior under dynamic mechanical vibration. To further illustrate the effect of the accurate electrical stimulation modulated by NG on cell proliferation. Here, P100-NG and P80-NG were the experimental groups, and A-NFM was the control. The vibration frequency was set at 2 Hz for mimicking low-frequency biomechanics. The cell proliferation was quantitatively assessed by the CCK-8 assay to estimate the metabolic activity of the total number of MC3T3-E1 cells. Figure 8a shows that the MC3T3-E1 had an elongated morphology along the direction of the nanofibers for all the samples. There was no obvious difference in morphology between cells grown on the two membranes.

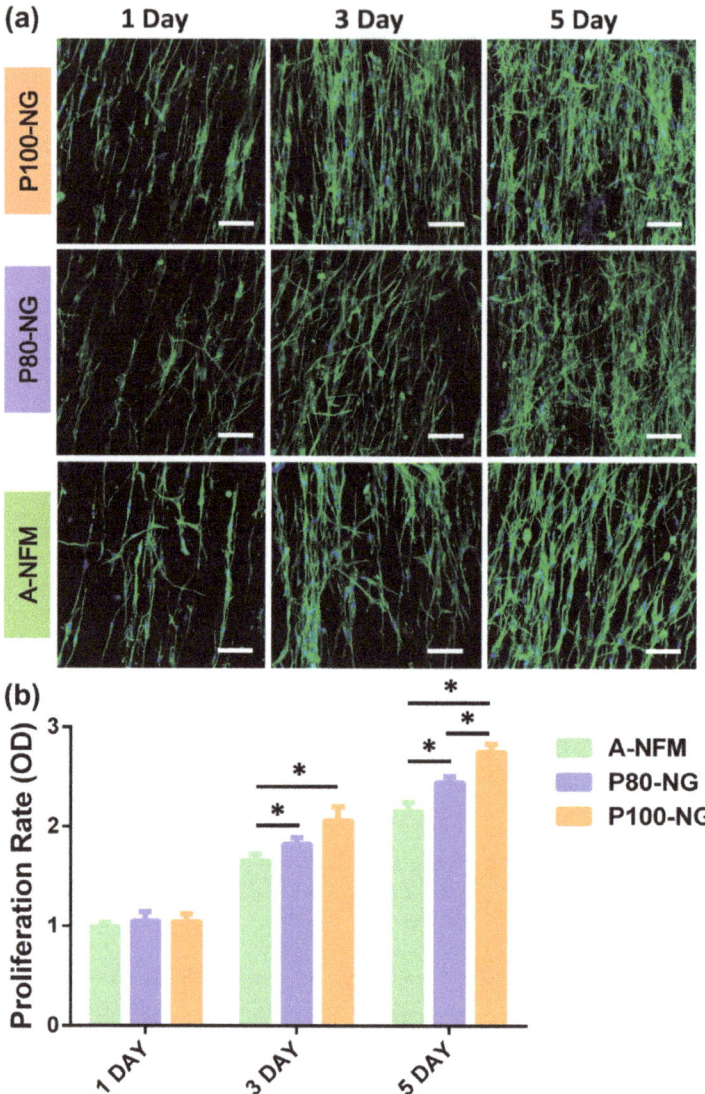

Figure 8. Proliferation of MC3T3 cells on P80-NG, P100-NG, and control A-NFM. (**a**) Fluorescence microscopy images of MC3T3 cells on A-NFM, P80-NG, and P100-NG. (**b**) MC3T3 cells proliferation analyzed by ImageJ software after 1, 3, and 5 days culture. All data represent the mean standard deviation ($n = 3$, * $p < 0.05$). The scale bar is 100 μm.

The cell counts of all groups increased on days 1, 3, and 5. The CCK-8 assay showed that the cells grown on both the P100-NG and P80-NG had the higher proliferation rate than those grown on A-NFM illustrating that the piezoelectricity increased cell proliferation (Figure 8b). On day 3, the MC3T3-E1 cells proliferation on the P100-NG and P80-NG was enhanced by 1.24- and 1.10-fold versus that of A-NFM, respectively. No significant difference was observed between the P100-NG and P80-NG. There were significant statistical differences in proliferation rate between cells on P100-NG, P80-NG and A-NFM on day 3 and 5. On day 5, the cells proliferation on the P100-NG and P80-NG was 1.27- and 1.13-fold versus that of A-NFM, respectively; there were significant statistical differences in proliferation rate between cells on P100-NG and P80-NG. A preliminary in vitro assay suggested that piezoelectric stimulations induced by the P(VDF-TrFE) NGs are suitable for the biopotential of MC3T3-E1 cells and can significantly promote cell growth. These results indicate that the combination of customizable exact piezoelectric stimulation and aligned nanostructured NG show the potential application to meet the demand of electrical stimulation according to specific tissue repair.

4. Conclusions

In conclusion, we introduced a promising strategy of a self-powered well-aligned P(VDF-TrFE) piezoelectric nanofiber nanogenerator as an exact piezoelectric stimulator for bone tissue engineering. We investigated the specific effects of post-treatments on the properties of NG. Poling post-treatment could effectively improve the mechanical, piezoelectric, and sensing performances of NG. We also measured the accurate piezoelectric response of NG and emulated the real scene of the electrical stimulation of the cells in vitro during the dynamic status. The well-aligned piezoelectric P(VDF-TrFE) NGs with different encouraged the MC3T3-E1 cells to proliferate in vitro under a sustainable piezoelectric stimulus. To clarify the effect of the exact electrical stimulation on the proliferation fate of preosteoblasts, two different output voltages of NG as stimulators were compared. Our work provides additional insights into the application of P(VDF-TrFE) piezoelectric nanofiber NG as self-powered electrical stimulation system for assisting tissue repair and regeneration.

Supplementary Materials: The following are available online at http://www.mdpi.com/2079-4991/9/3/349/s1, Figure S1: The diameter distribution and morphology, Figure S2: The strain-stress plots of the different NFMs, Figure S3: The cell viability of MC3T3-E1 on P80-NG, P100-NG and control.

Author Contributions: Conceptualization, A.W. and M.H.; methodology, A.W., L.Z.; software, A.W.; validation, A.W.; formal analysis, A.W.; investigation, A.W.; resources, M.H.; data curation, A.W.; writing—original draft preparation, A.W.; writing—review and editing, L.Z. and X.Q.; visualization, X.Q.; supervision, M.H.; project administration, M.H.; funding acquisition, M.H.

Funding: This work was supported by the National Natural Science Foundation of China (grant Nos. 61271070 and 60771019).

Acknowledgments: We would like to thank LetPub (for providing linguistic assistance during the preparation of this manuscript.

Conflicts of Interest: The authors declare no conflict of interest.

References

1. Zhao, M.; Song, B.; Pu, J.; Wada, T.; Reid, B.; Tai, G.; Wang, F.; Guo, A.; Walczysko, P.; Gu, Y.; et al. Electrical Signals Control Wound Healing through Phosphatidylinositol-3-OH Kinase-Gamma and PTEN. *Nature* **2006**, *442*, 457–460. [CrossRef] [PubMed]
2. Zhao, M. Electrical Fields in Wound Healing—An Overriding Signal That Directs Cell Migration. *Semin. Cell Dev. Biol.* **2009**, *20*, 674–682. [CrossRef] [PubMed]
3. Sundelacruz, S.; Li, C.; Choi, Y.J.; Levin, M.; Kaplan, D.L. Bioelectric Modulation of Wound Healing in A 3D in Vitro Model of Tissue-Engineered Bone. *Biomaterials* **2013**, *34*, 6695–6705. [CrossRef] [PubMed]
4. Hinkle, L.; McCaig, C.D.; Robinson, K.R. The Direction of Growth of Differentiating Neurones and Myoblasts From Frog Embryos in An Applied Electric Field. *J. Physiol.* **1981**, *314*, 14. [CrossRef]

5. Zheng, Q.; Zou, Y.; Zhang, Y.; Liu, Z.; Shi, B.; Wang, X.; Jin, Y.; Ouyang, H.; Li, Z.; Wang, Z.L. Biodegradable Triboelectric Nanogenerator as A Life-Time Designed Implantable Power Source. *Sci. Adv.* **2016**, *2*, e1501478. [CrossRef] [PubMed]
6. Kim, S.J.; Cho, K.W.; Cho, H.R.; Wang, L.; Park, S.Y.; Lee, S.E.; Hyeon, T.; Lu, N.; Choi, S.H.; Kim, D.H. Stretchable and Transparent Biointerface Using Cell-Sheet-Graphene Hybrid For Electrophysiology and Therapy of Skeletal Muscle. *Adv. Funct. Mater.* **2016**, *26*, 3207–3217. [CrossRef]
7. Lee, J.H.; Jeon, W.Y.; Kim, H.H.; Lee, E.J.; Kim, H.W. Electrical Stimulation by Enzymatic Biofuel Cell to Promote Proliferation, Migration and Differentiation of Muscle Precursor Cells. *Biomaterials* **2015**, *53*, 358–369. [CrossRef] [PubMed]
8. Guo, W.; Zhang, X.; Yu, X.; Wang, S.; Qiu, J.; Tang, W.; Li, L.; Liu, H.; Wang, Z.L. Self-Powered Electrical Stimulation for Enhancing Neural Differentiation of Mesenchymal Stem Cells on Graphene-Poly(3,4-ethylenedioxythiophene) Hybrid Microfibers. *ACS Nano* **2016**, *10*, 5086–5095. [CrossRef] [PubMed]
9. Tandon, N.; Cannizzaro, C.; Chao, P.H.G.; Maidhof, R.; Marsano, A.; Au, H.T.H.; Radisic, M.; Vunjak-Novakovic, G. Electrical Stimulation Systems for Cardiac Tissue Engineering. *Nat. Protoc.* **2009**, *4*, 155–173. [CrossRef] [PubMed]
10. Liu, Y.; Cui, H.; Zhuang, X.; Wei, Y.; Chen, X. Electrospinning of Aniline Pentamer-Graft-Gelatin/PLLA Nanofibers for Bone Tissue Engineering. *Acta Biomater.* **2014**, *10*, 5074–5080. [CrossRef] [PubMed]
11. Wagner, T.; Valero-Cabre, A.L. A Noninvasive Human Brain Stimulation. *Ann. Rev. Biomed. Eng.* **2007**, *9*, 527–565. [CrossRef] [PubMed]
12. Fukada, E. Piezoelectric Properties of Biological Polymers. *Q. Rev. Biophys.* **2009**, *16*, 59. [CrossRef]
13. Rajabi, A.H.; Jaffe, M.; Arinzeh, T.L. Piezoelectric Materials for Tissue Regeneration: A Review. *Acta Biomater.* **2015**, *24*, 12–23. [CrossRef] [PubMed]
14. Shimono, T.; Matsunaga, S.; Fukada, E.; Hattori, T.; Shikinami, Y. The Effects of Piezoelectric Poly-L-lactic Acid Films in Promoting Ossification in Vivo. *In Vivo* **1996**, *10*, 471–476. [PubMed]
15. Ribeiro, C.; Sencadas, V.; Correia, D.M.; Lanceros-Mendez, S. Piezoelectric Polymers as Biomaterials for Tissue Engineering Applications. *Colloids Surface B* **2015**, *136*, 46–55. [CrossRef] [PubMed]
16. Liu, C.; Yu, A.; Peng, M.; Ming, S.; Wei, L.; Yang, Z.; Zhai, J. Improvement in Piezoelectric Performance of ZnO Nanogenerator by A Combination of Chemical Doping and Interfacial Modification. *J. Phys.Chem. C* **2016**, *120*, 6971–6977. [CrossRef]
17. Zhang, Y.; Liu, C.; Liu, J.; Xiong, J.; Liu, J.; Zhang, K.; Liu, Y.; Peng, M.; Yu, A.; Zhang, A. Lattice Strain Induced Remarkable Enhancement in Piezoelectric Performance of ZnO Based Flexible Nanogenerators. *ACS Appl. Mater. Int.* **2015**, *8*, 1381. [CrossRef] [PubMed]
18. Garcia-Iglesias, M.; de Waal, B.F.; Gorbunov, A.V.; Palmans, A.R.; Kemerink, M.; Meijer, E.W. A Versatile Method for The Preparation of Ferroelectric Supramolecular Materials via Radical End-Functionalization of Vinylidene Fluoride Oligomers. *J. Am. Chem. Soc.* **2016**, *138*, 6217–6223. [CrossRef] [PubMed]
19. Dey, S.; Purahmad, M.; Ray, S.S.; Yarin, A.L.; Dutta, M. Investigation of PVDF-TrFE Nanofibers for Energy Harvesting. In Proceedings of the Nanotechnology Materials and Devices Conference, Waikiki Beach, HI, USA, 9–12 October 2012; pp. 21–24.
20. Prateek, T.V.; Gupta, R.K. Recent Progress on Ferroelectric Polymer-Based Nanocomposites for High Energy Density Capacitors: Synthesis, Dielectric Properties, and Future Aspects. *Chem. Rev.* **2016**, *116*, 4260–4317. [CrossRef] [PubMed]
21. Wang, X.; Sun, F.; Yin, G.; Wang, Y.; Liu, B.; Dong, M. Tactile-Sensing Based on Flexible PVDF Nanofibers via Electrospinning: A Review. *Sensors* **2018**, *18*, 330. [CrossRef] [PubMed]
22. Reneker, D.H.; Yarin, A.L.; Fong, H.; Koombhongse, S. Bending Instability of Electrically Charged Liquid Jets of Polymer Solutions in Electrospinning. *J. Appl. Phys.* **2000**, *87*, 4531–4547. [CrossRef]
23. Reneker, D.H.; Yarin, A.L. Electrospinning Jets and Polymer Nanofibers. *Polymer* **2008**, *49*, 2387–2425. [CrossRef]
24. Saeid, L.; Claire, G.; Ata, Y.; Kumar, T.V.; Yazdani, N.H. Electrospun Piezoelectric Polymer Nanofiber Layers for Enabling in Situ Measurement in High-Performance Composite Laminates. *ACS Omega* **2018**, *3*, 8891–8902.

25. Kang, S.B.; Won, S.H.; Im, M.J.; Kim, C.U.; Park, W.I.; Baik, J.M.; Choi, K.J. Enhanced Piezoresponse of Highly Aligned Electrospun Poly(vinylidene fluoride) Nanofibers. *Nanotechnology* **2017**, *28*, 395402. [CrossRef] [PubMed]
26. Lovinger, A.J. Ferroelectric Polymers. *Science* **1983**, *220*, 1115–1121. [CrossRef] [PubMed]
27. Liu, C.; Peng, M.; Yu, A.; Liu, J.; Ming, S.; Yang, Z.; Zhai, J. Interface Engineering on p-CuI/n-ZnO Heterojunction for Enhancing Piezoelectric and Piezo-Phototronic Performance. *Nano Energy* **2016**, *26*, 417–424. [CrossRef]
28. Zhang, Y.; Zhai, J.; Wang, Z.L. Piezo-Phototronic Matrix via A Nanowire Array. *Small* **2017**, *13*, 1702377. [CrossRef] [PubMed]
29. Miculescu, F.; Maidaniuc, A.; Voicu, S.I.; Thakur, V.K.; Stan, G.E.; Ciocan, L.T. Progress in Hydroxyapatite-Starch Based Sustainable Biomaterials for Biomedical Bone Substitution Applications. *ACS Sustain. Chem. Eng.* **2017**, *5*, 8491–8512. [CrossRef]
30. Zhao, G.; Zhang, X.; Lu, T.J.; Xu, F. Recent Advances in Electrospun Nanofibrous Scaffolds for Cardiac Tissue Engineering. *Adv. Funct. Mater.* **2015**, *25*, 5726–5738. [CrossRef]
31. Fuh, Y.K.; Wang, B.S. Near Field Sequentially Electrospun Three-Dimensional Piezoelectric Fibers Arrays for Self-Powered Sensors of Human Gesture Recognition. *Nano Energy* **2016**, *30*, 677–683. [CrossRef]
32. Lee, S.; Yun, S.; Park, K.I.; Jang, J.H. Sliding Fibers: Slidable, Injectable, and Gel-Like Electrospun Nanofibers as Versatile Cell Carriers. *ACS Nano* **2016**, *10*, 3282–3294. [CrossRef] [PubMed]
33. Heo, D.N.; Kim, H.J.; Lee, Y.J.; Heo, M.; Lee, S.J.; Lee, D.; Do, S.H.; Lee, S.H.; Kwon, I.K. Flexible and Highly Biocompatible Nanofiber-Based Electrodes for Neural Surface Interfacing. *ACS Nano* **2017**, *11*, 2961–2971. [CrossRef] [PubMed]
34. Kai, D.; Prabhakaran, M.P.; Jin, G.; Ramakrishna, S. Guided Orientation of Cardiomyocytes on Electrospun Aligned Nanofibers for Cardiac Tissue Engineering. *J. Biomed. Mater. Res. B.* **2011**, *98*, 379–386. [CrossRef] [PubMed]
35. Hitscherich, P.; Wu, S.; Gordan, R.; Xie, L.H.; Arinzeh, T.; Lee, E.J. The Effect of PVDF-TrFE Scaffolds on Stem Cell Derived Cardiovascular Cells. *Biotechnol. Bioeng.* **2016**, *113*, 1577–1585. [CrossRef] [PubMed]
36. Genchi, G.G.; Ceseracciu, L.; Marino, A.; Labardi, M.; Marras, S.; Pignatelli, F.; Bruschini, L.; Mattoli, V.; Ciofani, G. P(VDF-TrFE)/BaTiO$_3$ Nanoparticle Composite Films Mediate Piezoelectric Stimulation and Promote Differentiation of SH-SY5Y Neuroblastoma Cells. *Adv. Healthc. Mater.* **2016**, *5*, 1808–1820. [CrossRef] [PubMed]
37. Zhang, X.; Zhang, C.; Lin, Y.; Hu, P.; Shen, Y.; Wang, K.; Meng, S.; Chai, Y.; Dai, X.; Liu, X.; et al. Nanocomposite membranes enhance bone regeneration through restoring physiological electric microenvironment. *ACS Nano* **2016**, *10*, 7279–7286. [CrossRef] [PubMed]
38. Wang, A.; Liu, Z.; Hu, M.; Wang, C.; Zhang, X.; Shi, B.; Fan, Y.; Cui, Y.; Li, Z.; Ren, K. Piezoelectric Nanofibrous Scaffolds as in Vivo Energy Harvesters for Modifying Fibroblast Alignment and Proliferation in Wound Healing. *Nano Energy* **2018**, *43*, 63–71. [CrossRef]
39. Ren, K.; Liu, Y.; Hofmann, H.; Zhang, Q.M. Blottman, An Active Energy Harvesting Scheme with An Electroactive Polymer. *Appl. Phys. Lett.* **2007**, *91*, 132910. [CrossRef]
40. Hodgkin, A.L.; Huxley, A.F. A quantitative description of membrane current and its application to Conduction and Excitation in Nerve. *J. Physiol.* **1952**, *117*, 500–544. [CrossRef] [PubMed]
41. Nelson, M.; Rinzel, J. *The Hodgkin—Huxley Model*; Springer: New York, NY, USA, 1998.
42. Ayres, C.E.; Jha, B.S.; Meredith, H.; Bowman, J.R.; Bowlin, G.L.; Henderson, S.C.; Simpson, D.G. Measuring Fiber Alignment in Electrospun Scaffolds: A User's Guide to The 2D Fast Fourier Transform Approach. *J. Biomater. Sci.-Polym. E* **2008**, *19*, 603–621. [CrossRef] [PubMed]
43. Pérez, J.M.M.; Pascau, J. *Image Processing with ImageJ*; Packt Pub: Birmingham, UK, 2016.
44. Tseng, L.F.; Mather, P.T.; Henderson, J.H. Shape-Memory-Actuated Change in Scaffold Fiber Alignment Directs Stem Cell Morphology. *Acta Biomater.* **2013**, *9*, 8790–8801. [CrossRef] [PubMed]
45. Aqeel, S.M.; Wang, Z.; Than, L.; Sreenivasulu, G.; Zeng, X. Poly (vinylidene fluoride)/Poly (acrylonitrile)-Based Superior Hydrophobic Piezoelectric Solid Derived by Aligned Carbon Nanotube in Electrospinning: Fabrication, the Phase Conversion and Surface Energy. *RSC Adv.* **2015**, *5*, 76383–76391. [CrossRef] [PubMed]
46. Minwei, Y.; Yun, W.; Leonard, T.; Cecilia, C.; Ho, K.S. Improving Nanofiber Membrane Characteristics and Membrane Distillation Performance of Heat-Pressed Membranes via Annealing Post-Treatment. *Appl. Sci.* **2017**, *7*, 78.

47. Papadopoulou, S.K.; Tsioptsias, C.; Pavlou, A.; Kaderides, K.; Sotiriou, S.; Panayiotou, C. Superhydrophobic Surfaces from Hydrophobic or Hydrophilic Polymers via Nanophase Separation Or Electrospinning/Electrospraying. *Colloids Surface A* **2011**, *387*, 71–78. [CrossRef]
48. Persano, L.; Dagdeviren, C.; Su, Y.; Zhang, Y.; Girardo, S.; Pisignano, D.; Huang, Y.; Rogers, J.A. High Performance Piezoelectric Devices Based on Aligned Arrays of Nanofibers of Poly(vinylidenefluoride-co-trifluoroethylene). *Nat. Commun.* **2013**, *4*, 1633. [CrossRef] [PubMed]
49. Koga, K.; Nakano, N.; Hattori, T.; Ohigashi, H. Crystallization, Field-Induced Phase Transformation, Thermally Induced Phase Transition, and Piezoelectric Activity in P(vinylidene fluoride-TrFE) Copolymers with High Molar Content of Vinylidene Fluoride. *J. Appl. Phys.* **1990**, *67*, 965–974. [CrossRef]
50. Bellet-Amalric, E.; Legrand, J.F. Crystalline Structures and Phase Transition of The Ferroelectric P(VDF-TrFE) Copolymers, A Neutron Diffraction Study. *Eur. Phys. J. B* **1998**, *3*, 225–236. [CrossRef]
51. Garcia-Gutierrez, M.C.; Linares, A.; Martin-Fabiani, I.; Hernandez, J.J.; Soccio, M.; Rueda, D.R.; Ezquerra, T.A.; Reynolds, M. Understanding Crystallization Features of P(VDF-TrFE) Copolymers Under Confinement to Optimize Ferroelectricity in Nanostructures. *Nanoscale* **2013**, *5*, 6006–6012. [CrossRef] [PubMed]
52. Salimi, A.; Yousefi, A.A. FTIR Studies of Beta-Phase Crystal Formation in Stretched PVDF Films. *Polym. Test.* **2003**, *22*, 699–704. [CrossRef]
53. Xu, Z.; Baniasadi, M.; Moreno, S.; Cai, J.; Naraghi, M.; Minary-Jolandan, M. Evolution of Electromechanical and Morphological Properties of Piezoelectric Thin Films with Thermomechanical Processing. *Polymer* **2016**, *106*, 62–71. [CrossRef]
54. Murillo, G.; Blanquer, A.; Vargas-Estevez, C.; Barrios, L.; Ibáñez, E.; Nogués, C.; Esteve, J. Electromechanical Nanogenerator-Cell Interaction Modulates Cell Activity. *Adv. Mater.* **2017**, *29*, 1605048. [CrossRef] [PubMed]
55. Kandel, E.R.; Schwartz, J.H.; Jessell, T.M. *Principles of Neural Science*, 4th ed.; Mc Graw Hill Medical: New York, NY, USA, 2013; pp. 139–140.
56. Aishwarya, N.; Jennifer, H. Hodgkin–Huxley Neuron and FPAA Dynamics. *IEEE Trans. Biomed. Circ. Syst.* **2018**, 1–9.
57. Frankenhaeuser, B.; Hodgkin, A.L. The Action of Calcium on The Electrical Properties of Squid Axons. *J. Physiol.* **1957**, *137*, 218–244. [CrossRef] [PubMed]
58. Sydlik, S.A.; Jhunjhunwala, S.; Webber, M.J.; Anderson, D.G.; Langer, R. In Vivo Compatibility of Graphene Oxide with Differing Oxidation States. *ACS Nano* **2015**, *9*, 3866–3874. [CrossRef] [PubMed]
59. Beachley, V.; Katsanevakis, E.; Zhang, N.; Wen, X. *Highly Aligned Polymer Nanofiber Structures: Fabrication and Applications in Tissue Engineering*; Springer: Berlin/Heidelberg, Germany, 2011.
60. Yang, F.; Murugan, R.; Wang, S. Electrospinning of Nano/Micro Scale Poly (L-lactic acid) Aligned Fibers and Their Potential in Neural Tissue Engineering. *Biomaterials* **2005**, *26*, 2603–2610. [CrossRef] [PubMed]
61. Bhattarai, N.; Edmondson, D.; Veiseh, O.; Matsen, F.A.; Zhang, M.Q. Electrospun Chitosan-Based Nanofibers and Their Cellular Compatibility. *Biomaterials* **2005**, *26*, 6176–6184. [CrossRef] [PubMed]
62. Chen, F.; Su, Y.; Mo, X.; He, C.L.; Wang, H.S.; Ikada, Y. Biocompatibility, Alignment Degree and Mechanical Properties of an Electrospun Chitosan–P(LLA-CL) Fibrous Scaffold. *J. Biomater. Sci. Polym. Ed.* **2009**, *20*, 2117–2128. [CrossRef] [PubMed]

© 2019 by the authors. Licensee MDPI, Basel, Switzerland. This article is an open access article distributed under the terms and conditions of the Creative Commons Attribution (CC BY) license (http://creativecommons.org/licenses/by/4.0/).

Article

Poly(vinylidene fluoride) Composite Nanofibers Containing Polyhedral Oligomeric Silsesquioxane–Epigallocatechin Gallate Conjugate for Bone Tissue Regeneration

Hyo-Geun Jeong [1], Yoon-Soo Han [2], Kyung-Hye Jung [2] and Young-Jin Kim [1],*

[1] Department of Biomedical Engineering, Daegu Catholic University, Gyeongsan 38430, Korea; jhg2833@empal.com
[2] Department of Advanced Materials and Chemical Engineering, Daegu Catholic University, Gyeongsan 38430, Korea; yshancu@cu.ac.kr (Y.-S.H.); khjung@cu.ac.kr (K.-H.J.)
* Correspondence: yjkim@cu.ac.kr; Tel.: +82-53-850-2512; Fax: +82-53-359-6752

Received: 14 January 2019; Accepted: 29 January 2019; Published: 1 February 2019

Abstract: To provide adequate conditions for the regeneration of damaged bone, it is necessary to develop piezoelectric porous membranes with antioxidant and anti-inflammatory activities. In this study, composite nanofibers comprising poly(vinylidene fluoride) (PVDF) and a polyhedral oligomeric silsesquioxane–epigallocatechin gallate (POSS–EGCG) conjugate were fabricated by electrospinning methods. The resulting composite nanofibers showed three-dimensionally interconnected porous structures. Their average diameters, ranging from 936 ± 223 nm to 1094 ± 394 nm, were hardly affected by the addition of the POSS–EGCG conjugate. On the other hand, the piezoelectric β-phase increased significantly from 77.4% to 88.1% after adding the POSS–EGCG conjugate. The mechanical strength of the composite nanofibers was ameliorated by the addition of the POSS–EGCG conjugate. The results of in vitro bioactivity tests exhibited that the proliferation and differentiation of osteoblasts (MC3T3-E1) on the nanofibers increased with the content of POSS–EGCG conjugate because of the improved piezoelectricity and antioxidant and anti-inflammatory properties of the nanofibers. All results could suggest that the PVDF composite nanofibers were effective for guided bone regeneration.

Keywords: bone regeneration; poly(vinylidene fluoride); composite nanofiber; piezoelectricity; antioxidant activity

1. Introduction

A variety of techniques have been tried for the regeneration of bone defects, including bone grafting, distraction osteogenesis, and guided bone regeneration (GBR) [1–3]. Among them, GBR has been used as an efficient method for both the reconstruction of the structure of, and reestablishment of the function in, bone defects and damaged tissues. The GBR approach principally uses a membrane acting as a barrier membrane in between soft tissue and bone tissue defect. GBR membranes prevent the migration of faster-growing fibrous tissue into the defect and secure osteoconduction (bone growth) to promote bone regeneration [3]. Thus, GBR membranes need to have biocompatibility to support bone tissue reconstruction and mechanical stability to maintain a protective space during bone restoration process.

Nanofibrous membranes have structural similarity to the extracellular matrix (ECM) and high specific surface area [4]. GBR membranes composed of ultrafine nanofibers are more suitable for cell attachment and proliferation than conventional membranes with large scale structures. Various strategies—such as self-assembly, wet-spinning, and electrospinning—have been used to fabricate GBR membranes [4–10]. Of these strategies, electrospinning has attracted a great deal of attention because this method is very simple and versatile, and can produce various polymer nanofibers. Electrospun polymer nanofibers have unique advantages in the biomedical fields such as drug delivery and tissue engineering, as they can form porous structures that simulate the features of the ECM [2,4,11,12].

Poly(vinylidene fluoride) (PVDF) is an attractive semicrystalline polymers for use in biomedical applications because of its efficient piezoelectricity and biocompatibility [13,14]. The strong piezoelectric effect observed in PVDF is directly linked to the crystalline phase arrangement. PVDF exits in one of five polymorphic crystalline phases (α, β, γ, δ, and ε) dependent on its distinct chain conformation [15]. The nonpolar α-phase does not exhibit an efficient conformation for piezoelectric application, but it does present good mechanical properties and thermal stability. On the other hand, β- and γ-phases exhibit electroactive and polar properties. Among them, the β-phase has a higher density net dipole moment and exhibits more promising piezoelectric properties. Therefore, increasing the piezoelectric β-phase content in PVDF is very important for its biomedical application because piezoelectric materials can provide electrical stimulation to cells to promote tissue formation and can use as a scaffold for tissue engineering [14,16]. This phenomenon is very similar that electrically charged ECM stimulates the cell growth during tissue regeneration [14]. In particular, bone tissue and its constituent collagen fibers possess intrinsic piezoelectricity to play a significant role in regeneration of the damaged bone, and thus a variety of piezoelectric materials have been applied to promote bone regeneration both in vitro and in vivo [16,17].

Polyhedral oligomeric silsesquioxane (POSS) molecules have a unique nanostructure comprised of an inorganic silsesquioxane cage core and an organic functional group shell [18]. A variety of POSS molecules contain reactive functionalities, making them suitable precursors for grafting. Therefore, they have been used in various biomedical applications including tissue engineering due to their distinctive structure and excellent biocompatibility [19,20]. Epigallocatechin gallate (EGCG) is a polyphenolic flavonoid derived from a variety of plants and, in particular, is a major bioactive component in green tea, which has been shown to possess a variety of pharmacological functions; for instance, it exhibits antioxidant, anticancer, and anti-inflammatory properties [21–23]. EGCG inhibits lipopolysaccharide (LPS)-induced osteoclastic bone resorption and attenuates inflammatory bone loss in bone metabolism [24,25]. Moreover, EGCG increases osteoblast proliferation and alkaline phosphatase (ALP) activity, which leads to increased osteoblastic bone formation [26].

Piezoelectric materials can change their surface charge without external energy power source [14]. Several research papers have reported an observed enhancement of the piezoelectric β-phase content in PVDF fabricated by electrospinning with the incorporation of fillers, such as metal nanoparticles, ceramics, and inorganic salts [27–30]. In addition, EGCG-incorporated membranes have been fabricated and have been observed to promote osteoblastic proliferation and prevent inflammatory responses [26]. Therefore, in this study, we fabricated PVDF/POSS–EGCG conjugate composite nanofibers, which showed improved piezoelectric, antioxidant, and anti-inflammatory properties, to support bone tissue regeneration using an electrospinning method. We systematically examined the effect of the POSS–EGCG conjugate content on the structure and physicochemical properties of the composite nanofibers obtained. Furthermore, the in vitro bioactivity of the composite nanofibers was investigated through the cell proliferation, ALP activity, and bone mineralization assays.

2. Materials and Methods

2.1. Materials

PVDF (M_w = 275000 g/mol), EGCG, lipopolysaccharide (LPS), horseradish peroxidase (HRP), hydrogen peroxide (H_2O_2, 30 wt% in H_2O), dimethyl sulfoxide (DMSO), N,N-dimethylformamide (DMF), tetrahydrofuran (THF), and isopropyl alcohol (IPA) were obtained from Sigma-Aldrich Co. (St. Louis, MO, USA). Xanthine oxidase (XO from buttermilk, EC 1.1.3.22), xanthine, and nitro blue tetrazolium (NBT) were purchased from Wako Pure Chemical Industries (Osaka, Japan). Aminopropylisobutyl polyhedral oligomeric silsesquioxane (API-POSS) was purchased from Hybrid Plastics (Hattiesburg, MS, USA). Osteoblast-like cell line MC3T3-E1 derived from mouse calvaria was acquired from the American Type Culture Collection (Manassas, VA, UAS) and a murine macrophage cell line RAW 264.7 was purchased from the Korean Cell Line Bank (Seoul, South Korea). All materials were used as received without further purification.

2.2. Synthesis of the POSS–EGCG Conjugate

The POSS–EGCG conjugate was synthesized as follows. API-POSS (174.9 mg, 0.2 mmol) and EGCG (137.5 mg, 0.3 mmol) were first dissolved in 30 mL of a mixed solvent of THF/IPA/deionized water (DW) at a ratio of 2:1:1 before adding a solution of HRP (5 mg, 750 units) dissolved in 3 mL of Dulbecco's phosphate-buffered saline (DPBS, pH 7.4). To this resultant solution, 5 mL of 5 wt% H_2O_2 was added five times every 10 min. After 4 h, the resultant POSS–EGCG conjugate was dialyzed in DW for 48 h. Then, the precipitated conjugate was isolated by centrifugation and washing with DW in four replicates, followed by drying in vacuo. The amount of EGCG conjugated to API-POSS was determined by elemental analysis using a Flash 2000 elemental analyzer (Thermo Fisher Scientific, USA). The structures of EGCG, API-POSS, and the POSS–EGCG conjugate were determined by ^{1}H NMR spectroscopy (AVANCE III 400, Bruker BioSpin, USA).

^{1}H NMR of EGCG (400 MHz, D_2O), δ (ppm) = 6.91 (C$\underline{H}$ of D-ring), 6.52 (C$\underline{H}$ of B-ring), 6.10, 6.07 (C$\underline{H}$ of A-ring), 5.48, 4.97 (C$\underline{H}$ of C-ring), 2.88 (C$\underline{H}_2$ of C-ring).

^{1}H NMR of API-POSS (400 MHz, $CDCl_3$), δ (ppm) = 2.83 (C$\underline{H}_2$NH), 1.84 (C$\underline{H}$CH$_3$), 1.58 (CH$_2$C$\underline{H}_2$CH$_2$), 0.95 (CHC$\underline{H}_3$), 0.56 (SiC$\underline{H}_2$).

^{1}H NMR of POSS–EGCG (400 MHz, DMSO-d_6), δ (ppm) = 6.81 (C$\underline{H}$ of D-ring), 6.41 (C$\underline{H}$ of B-ring), 5.95, 5.84 (C$\underline{H}$ of A-ring), 5.36, 4.98 (C$\underline{H}$ of C-ring), 2.77 (C$\underline{H}_2$ of C-ring) for EGCG, 2.69 (C$\underline{H}_2$NH), 1.82 (C$\underline{H}$CH$_3$), 1.58 (CH$_2$C$\underline{H}_2$CH$_2$), 0.95 (CHC$\underline{H}_3$), 0.54 (SiC$\underline{H}_2$) for API-POSS.

2.3. Fabrication of PVDF Composite Nanofibers

Before electrospinning the PVDF composite nanofibers, the PVDF (1.8 g) was dissolved in 10 mL of a mixed solvent of DMF/THF in a ratio of 1:1 to obtain an 18 wt% solution. After stirring 12 h, the POSS–EGCG conjugate was added to the PVDF solution with a different weight ratio at 60 °C for 8 h. The contents of POSS–EGCG conjugate in solutions were 0, 2, 4, and 6 wt% with respect to the weight of the PVDF. The mixed solution of PVDF and the POSS–EGCG conjugate was filled into a 20 mL standard syringe with a 22 G stainless steel needle (internal diameter = 0.41 mm). Then, this solution was electrospun into nanofibers at room temperature with a feeding rate of 2 mL/h feed rate, a voltage of 18 kV, and a working distance of 15 cm. After electrospinning, the PVDF composite nanofibers were peeled off from the stainless steel plate and vacuum dried for 12 h.

2.4. Characterization of the PVDF Composite Nanofibers

The morphology of the PVDF composite nanofibers was visualized by scanning electron microscopy (SEM, S-4300, Hitachi, Japan) at an acceleration voltage of 5 kV after sputter coating of samples with gold. Image-Pro Plus software (Media Cybernetics Inc., Rockville, MD, USA) was used to determine the average fiber diameters for the composite nanofibers. Energy dispersive spectroscopy (EDS) spectra were obtained for the determination of the distribution profile of the POSS–EGCG conjugate on the nanofiber surface.

Fourier transform infrared spectroscopy (FTIR) measurement was performed using an ALPHA spectrometer (Bruker OpticsBillerica, MA, USA) with a resolution of 4 cm^{-1}. Attenuated total reflectance (ATR) mode was used to analysis the characteristic bands over a range from 400 to

1600 cm^{-1}. The relative β-phase content (F(β)) in the samples was determined from the following equation [29].

$$F(\beta) = \frac{A_\beta}{\left(\frac{K_\beta}{K_\alpha}\right) A_\alpha + A_\beta}$$

where A_α and A_β represent the absorbance of peaks related to the α- and β-phases at 760 and 840 cm^{-1}, respectively, and K_α and K_β are the absorption coefficients at the respective wavenumbers, which are 7.7×10^4 cm^2/mol and 6.1×10^4 cm^2/mol.

X-ray diffraction (XRD) patterns were also used to evaluate the crystalline phases of the PVDF composite nanofibers and thus XRD measurements were performed using a Rigaku D/MAX-2500V/PC X-ray diffractometer (Japan) with high intensity Cu Kα radiation at 40 kV/100 mA. The diffractograms were scanned in a 2θ range from 10° to 50°. The composite nanofiber samples were analyzed using X-ray photoelectron spectroscopy (XPS) to obtain the surface elemental composition. XPS measurements were performed using a Quantera SXM (ULVAC-PHI Inc., Japan) equipped with a monochromatic Al Kα X-ray source (1486 eV). The photoelectron take-off angle was fixed at 45° relative to the sample surface. Besides XPS measurements, static contact angles were measured with a DSA 100 contact angle meter (KRÜSS, Germany) to examine the hydrophobicity of the composite nanofiber. Each contact angle of the sample is an average of five measurements. The mechanical properties of the composite nanofibers were investigated with a TO-101 universal testing machine (Testone Co., Siheung, South Korea) with 2 kN load capacity at a rate of 10 mm/min. All samples were cut into rectangular specimens with a size of 3 cm × 1 cm and tested five parallel measurements.

2.5. Cell Proliferation

MC3T3-E1 cells were used to evaluate the in vitro bioactivity of the PVDF composite nanofibers by a cell proliferation assay in alpha minimum essential medium (α-MEM) containing 10% of fetal bovine serum (FBS) and 1% of penicillin–streptomycin at 37 °C with 5% CO_2. Before assaying, the composite nanofibers were cut into circular shape with a diameter of 15 mm, followed by putting into 24-well tissue culture plate and fixing with glass ring (inner diameter = 11 mm). Then, they were sterilized with 70% ethanol and rinsed with DPBS and α-MEM. After drying, the composite nanofibers were once again sterilized under UV radiation for 3 h.

The proliferation of viable cells on the composite nanofibers was determined using the 3-(4,5-dimethylthiazol-2-yl)-2,5-diphenyltetrazoliumbromide (MTT) assay. One milliliter of cell suspension (2×10^4 cells/well) was placed onto the sterilized samples and cells were cultured for different periods of time. After culturing for 1, 3, 5, 7, and 14 days, culture media were replaced with 0.2 mL of the MTT solution and further incubation of the cells was maintained for another 4 h. Next, we removed the remaining media and added 1 mL of DMSO to solubilize the precipitated formazan crystals. Finally, the resulting supernatant was transferred to 96-well plate with 0.2 mL per well. The absorbance was determined at 570 nm using a spectrophotometric plate reader (OPSYS-MR, Dynex Technology Inc., USA). Furthermore, after days 3 and 7 of cell culture, the cell-nanofibers were fixed in 4% glutaraldehyde for 1 h, and then dehydrated using different concentrations of ethanol (25, 50, 70, and 100%), followed by vacuum-drying. The morphologies of dried samples were observed using SEM after sputter coating with gold.

2.6. Superoxide Anion Radical Scavenging Capacity Assay

Superoxide anion radicals were generated with xanthine/XO and determined as described in a previous paper of ours [31]. EGCG and the POSS–EGCG conjugate were first dissolved in the mixed solvent of DMSO and DPBS (pH 7.4) and diluted with DPBS. To measure the superoxide anion radical scavenging capacity, xanthine (30 μg/mL) was added to DPBS containing ethylenediaminetetraacetic acid (EDTA, 15 μg/mL), XO (40 milliunits/mL), NBT (130 μg/mL), and various concentrations of test sample. Changes in the absorbance at 560 nm over 10 min were recorded as a measure of the changing number of superoxide anion radicals.

The superoxide anion radical scavenging capacity of samples was determined according to the following formula.

$$\text{Superoxide anion radical scavenging capacity (\%)} = \frac{\text{Abs}_{\text{control}} - \text{Abs}_{\text{sample}}}{\text{Abs}_{\text{control}}} \times 100$$

where $\text{Abs}_{\text{control}}$ is the absorbance of control in the absence of sample and $\text{Abs}_{\text{sample}}$ is the absorbance in the presence of samples.

2.7. ALP Activity and Bone Mineralization Assay

ALP activity can provide a useful index of the osteoblastic phenotype. Thus, the osteoblastic differentiation of MC3T3-E1 cells was assessed by measuring the ALP activity with an ALP assay kit (DALP-250, BioAssay Systems, USA) at periods of 3, 5, 7, and 14 days after cell seeding. The composite nanofibers seeded with cells washed with DPBS and incubated in 0.5 mL of DW containing 0.2% Triton X-100. The cell lysates were mixed with p-nitrophenyl phosphate solution. In the presence of ALP, p-nitrophenyl phosphate can be hydrolyzed to p-nitrophenol and the rate of p-nitrophenol production is proportional to the ALP activity. Therefore, the level of p-nitrophenol production was determined by measuring the absorbance at 405 nm and normalized to 1×10^4 cells.

Bone mineralization capability of the composite nanofibers was assayed by the alizarin red S (ARS) staining method. ARS can selectively bind to calcium ions and thus calcium deposition is easily measured by the use of ARS. After 3, 5, 7, and 14 days of cell culture on the different nanofibers, the cells were fixed with 4% glutaraldehyde for 30 min and then stained with 1 mL of 40 mM ARS (pH 4.1). After incubation for 20 min, the specimens were washed thrice with DW to remove unreacted ARS and dissolved in 1 mL of 10% cetylpyridinium chloride. The absorbance of the supernatant at 540 nm was obtained using microplate reader and was normalized to 1×10^4 cells.

2.8. Quantification of Inflammatory Cytokine

The effect of the PVDF composite nanofibers on the expression of the inflammatory cytokine interleukin-6 (IL-6) was measured using the RAW 264.7 cells. For the quantification of IL-6 production, the cells (2×10^4 cells/well) were precultured in Dulbecco's modified Eagle's medium (DMEM) containing 10% heat-inactivated FBS, 100 U/mL of penicillin, and 100 μg/mL of streptomycin for 24 h at 37 °C before adding 1 μg/mL of LPS into the cell culture plate. After activation of the cells for 24 h, the amount of IL-6 released into the media was measured by mouse IL-6 enzyme-linked immunosorbent assay kit (R&D Systems, USA) according to the manufacturer's protocol.

2.9. Statistical Analysis

All data were represented as mean value ± standard deviation. Differences between two groups were analyzed with a one-way ANOVA followed by a Turkey test using SigmaPlot 13.0 (Systat Software, CA). A significant difference was defined for values of $p^* < 0.05$.

3. Results and Discussion

3.1. Synthesis of the POSS–EGCG Conjugate

To prepare the POSS–EGCG conjugate, EGCG was conjugated to API-POSS via a one-step reaction using HRP as a catalyst (Figure 1). The structure of the resultant POSS–EGCG conjugate was characterized by ^{1}H NMR spectroscopy (Figure S1). The characteristic signals belonging to EGCG units were observed at 6.81, 6.41, 5.95, 5.84, 5.36, 4.98, and 2.77 ppm. Moreover, the peaks due to API-POSS units in the POSS–EGCG conjugate were shown at 2.69, 1.82, 1.58, 0.95, and 0.54 ppm [20,32]. The semiquinone radicals or active oxygen species can easily oxidize either the gallyl moiety (B-ring) or the gallate moiety (D-ring) of EGCG, resulting to transform EGCG into reactive species with a quinone structure [33]. However, gallyl structures are more susceptible to oxidation than gallate structures and they only form catechol quinones through intermediate semiquinones [34]. In addition, the presence of a reactive amino group on API-POSS can provide a site for enzymatic conjugation of catechol through electrophilic addition [35]. The conjugation ratio of EGCG that was introduced to API-POSS was calculated from the result of elemental analysis to be 0.96.

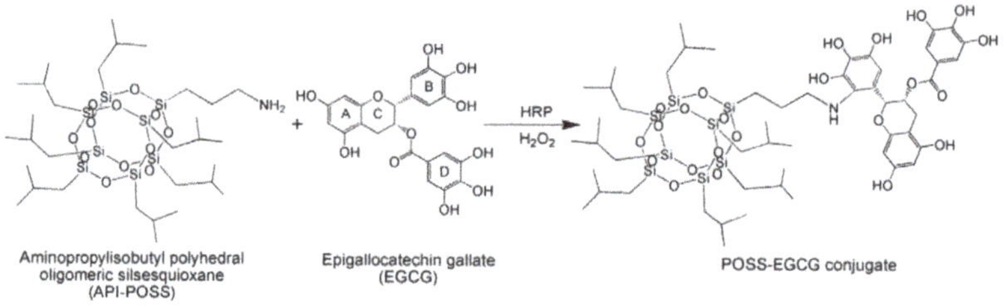

Figure 1. Schematic diagram of enzymatic synthesis of the polyhedral oligomeric silsesquioxane–epigallocatechin gallate (POSS–EGCG) conjugate.

3.2. Fabrication of the PVDF Composite Nanofibers

Composite nanofibers have been proven to act as effective mechanical supports and to promote osteoconduction in bone tissue regeneration [2,4]. In the present study, composite nanofibers were fabricated by electrospinning PVDF solutions containing different amounts of POSS–EGCG conjugate. The contents of POSS–EGCG conjugate in the mixed solutions were 0 (PVDF), 2 (PE02), 4 (PE04), and 6 wt% (PE06) with respect to the weight of the PVDF. In addition, the PVDF nanofiber containing 6 wt% of pure API-POSS (PO06) was also prepared as a control. The SEM observations revealed that thoroughly interconnected porous structures formed between the composite nanofibers (Figure 2). The average fiber diameter of the composite nanofibers, which was 1033 ± 270 nm for PVDF, 971 ± 262 nm for PE02, 936 ± 223 nm for PE04, 1094 ± 394 nm for PE06, and 1131 ± 281 nm for PO06, was hardly affected by adding the POSS–EGCG conjugate. In addition, the distribution of the POSS–EGCG conjugate in the PVDF composite nanofibers was observed using EDS. The EDS Si-mapping analyses of the composite nanofibers represented that the POSS–EGCG conjugate was homogenously distributed over the nanofibers and that a greater density of Si was detected on the PE06 nanofiber (Figure S2).

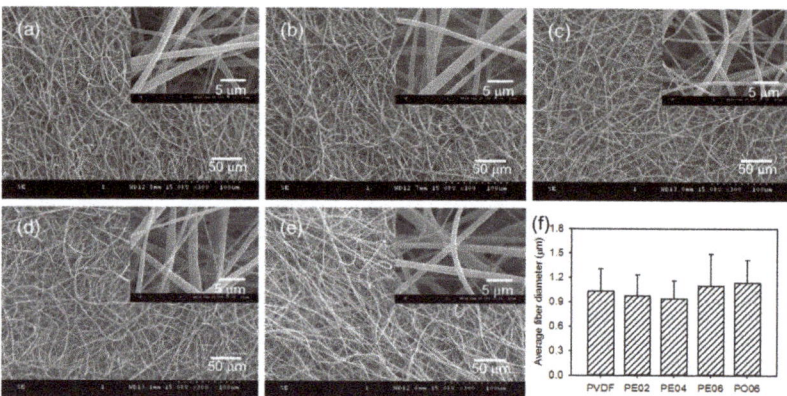

Figure 2. SEM images of (**a**) PVDF, (**b**) PE02, (**c**) PE04, (**d**) PE06, and (**e**) PO06 composite nanofibers. (**f**) Average fiber diameters of the PVDF composite nanofibers analyzed from the SEM images ($n = 4$).

3.3. Physicochemical Properties of the PVDF Composite Nanofibers

The FTIR was used to analyze the chemical structure and phase composition of the PVDF composite nanofibers. Characteristic bands appeared at 1402 and 840 cm^{-1} in the spectrum of the pure PVDF nanofiber corresponded to the stretching and rocking vibration of the CH$_2$ groups, respectively (Figure 3a) [28,29,36]. The four bands observed at 1275, 1178, 760, and 482 cm^{-1} corresponded to the deformation, stretching, bending, and wagging modes of the CF$_2$ groups in PVDF. An absorption peak due to the skeletal vibration of C–C bonds in PVDF was also observed at 880 cm^{-1}. After adding the POSS–EGCG conjugate or pure API-POSS, the PVDF composite nanofibers exhibited a new absorption peak at 1110 cm^{-1} associated with the stretching mode of Si–O–Si bonds, which is the typical absorption peak of the POSS inorganic framework [37]. In particular, the absorption peaks at 1402, 1275, 880, 840, and 482 cm^{-1} represent characteristic vibration modes of the piezoelectric β-phase of PVDF [28,29].

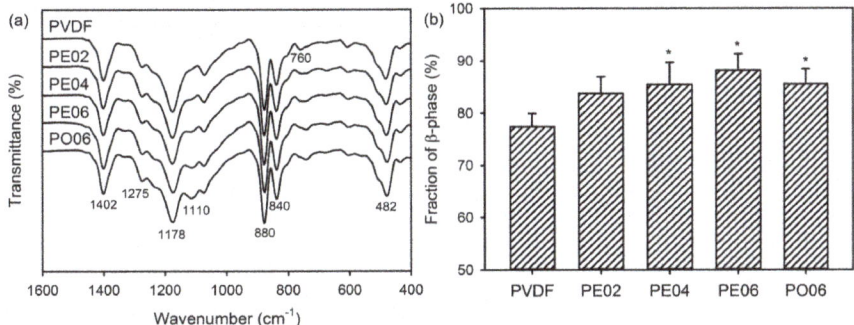

Figure 3. (a) FTIR spectra of the PVDF composite nanofibers and (b) changes in β-phase content in the nanofibers calculated from the FTIR spectra (n = 5). Significant difference from the pure PVDF nanofiber was denoted as $p^* < 0.05$.

The FTIR data was used to calculate the relative β-phase content (F(β)) in the PVDF composite nanofibers. As presented in Figure 3b, the F(β) values of all samples were higher than 75% because the crystalline phase change of the PVDF from α-phase to β-phase was occurred by the electrostatic forces applied to the polymer droplet during the electrospinning [30]. In addition, the piezoelectric β-phase increased from 77.4 ± 2.5% to 88.1 ± 3.2% after adding the POSS–EGCG conjugate. According to previous reports, converting α-phase to piezoelectric β-phase in PVDF/ceramic composites was ascribed to the interaction between the negatively charged ceramic nanoparticles and the positively charged CH$_2$ groups of PVDF [27,28]. Moreover, hydroxylated ceramic nanoparticles interacted more strongly with PVDF through the hydrogen bonds formed by the fluorine atoms on the PVDF with the hydroxyl groups on the surface of the ceramic nanoparticles. Therefore, we may deduce that, in our experiments, PE06 exhibited a higher F(β) value than those of the other composite nanofibers because of strong dipole and hydrogen bonding interactions between the PVDF and POSS–EGCG conjugates.

The surface elemental composition of the PVDF composite nanofibers was assessed using XPS. The XPS survey scan spectrum of the pure PVDF nanofiber exhibited two separate peaks which corresponded to the F 1s orbital (688 eV) and the C 1s orbital (287 eV) (Figure 4a). After adding the POSS–EGCG conjugate or pure API-POSS, the XPS spectra exhibited six separate peaks assigned to PVDF, the POSS–EGCG conjugate, and API-POSS: F 1s (688 eV), O 1s (535 eV), N 1s (400 eV), C 1s (287 eV), Si 2s (155 eV), and Si 2p (105 eV). The intensities of the peaks assigned to the POSS–EGCG conjugate increased with increasing contents of the POSS–EGCG conjugate. In addition, the crystalline phases of the PVDF composite nanofibers were examined by means of XRD as shown in Figure 5. All samples exhibited two diffraction peaks mainly attributed to the piezoelectric β-phase of PVDF at 20.4° and 35.9°, which were indexed to the β(110)/(200) and β(020)/(101) planes, respectively [27,28].

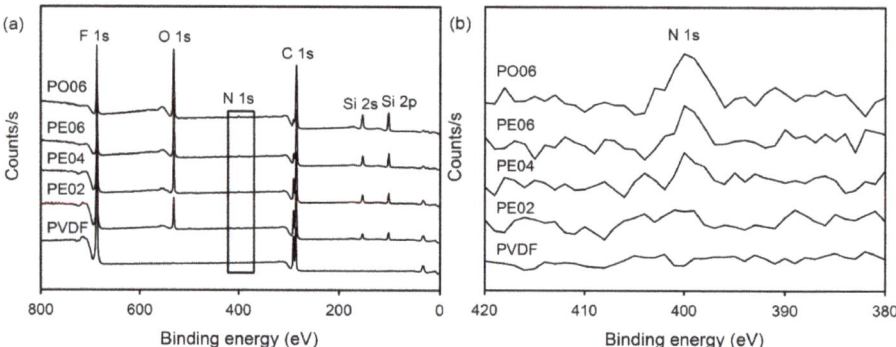

Figure 4. (a) X-ray photoelectron (XPS) spectra of the PVDF composite nanofibers and (b) the magnified XPS spectra of the areas in (a) marked by rectangle.

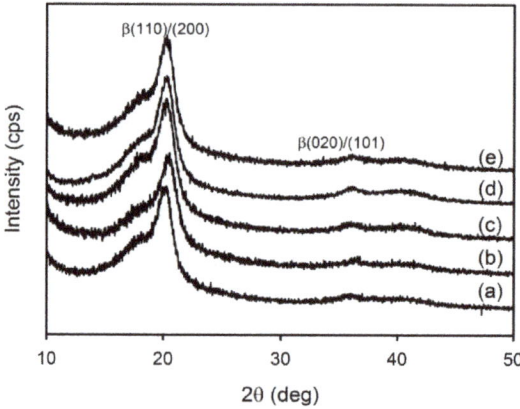

Figure 5. XRD patterns of (a) PVDF, (b) PE02, (c) PE04, (d) PE06, and (e) PO06 composite nanofibers.

The investigation on the surface hydrophilicity of biomedical membranes is very important to confirm the affinity between and the membranes and the cell. Suitable hydrophilicity can influence the attachment and proliferation of the cells on the surface of the membranes. Therefore, the performance of membranes in biomedical applications may depend on the hydrophilicity [38]. Here we measured the water contact angle to evaluate the hydrophilicity of the composite nanofibers. Figure 6 exhibited the water contact angle data measured for the composite nanofibers; the PVDF nanofiber revealed a high water contact angle (127.7 ± 9.5°) due to its hydrophobic property. In addition, the water contact angles were hardly changed by the addition of the POSS–EGCG conjugate, i.e., 129.0 ± 8.7° for PE02, 128.9 ± 6.8° for PE04, and 130.4 ± 8.6° for PE06.

GBR membrane should be strong enough to withstand the force during new bone tissue regeneration. Thus, the effect of the addition of the POSS–EGCG conjugate on the mechanical properties of the composite nanofibers was assayed. As shown in Table 1, the resulting data obtained on the basis of stress–strain measurements exhibited that the tensile strength and Young's modulus increased with increasing contents of the POSS–EGCG conjugate. This means that the POSS–EGCG conjugate can cause temporary intermolecular cross-linking of polymer chains and thus can provide enhanced mechanical strength [39]. However, the addition of POSS–EGCG conjugate induced a slight decrease in elongation at break of the composite nanofibers.

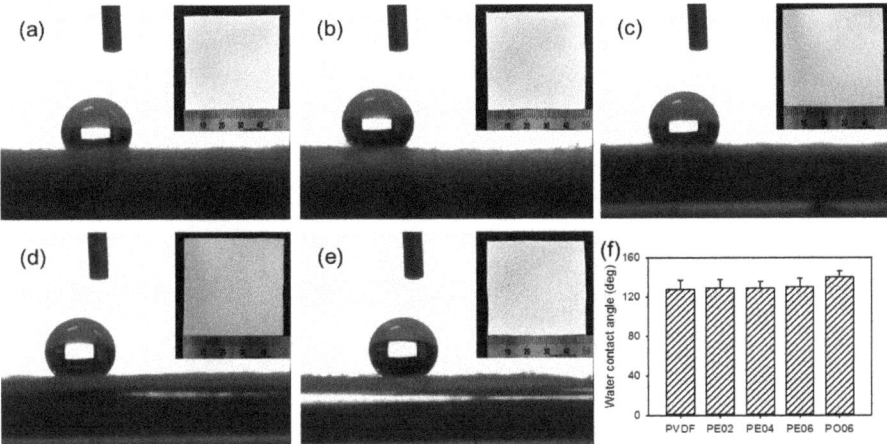

Figure 6. Contact angle images for water droplet on (**a**) PVDF, (**b**) PE02, (**c**) PE04, (**d**) PE06, and (**e**) PO06 composite nanofibers. (**f**) Average contact angles on PVDF composite nanofibers determined by contact angle meter (*n* = 5). Insets are photographs of PVDF composite nanofibers.

Table 1. Mechanical properties of the PVDF composite nanofibers with different contents of the POSS–EGCG conjugate.

Sample	Content of POSS–EGCG Conjugate (wt%)	Young's Modulus (MPa)	Tensile Strength (MPa)	Elongation at Break (%)
PVDF	0	3.5 ± 0.3	1.1 ± 0.1	72.6 ± 6.0
PE02	2	3.6 ± 0.2	1.1 ± 0.3	71.3 ± 6.5
PE04	4	4.0 ± 0.3	1.4 ± 0.2	69.1 ± 7.7
PE06	6	4.5 ± 0.5	1.6 ± 0.3	65.6 ± 7.2
PO06	6[a]	5.2 ± 0.4	1.7 ± 0.3	63.4 ± 8.4

[a] Content of API-POSS (wt%).

3.4. Cell Proliferation

The in vitro bioactivity of the PVDF composite nanofibers was evaluated to assess the potential of the nanofibers as GBR membranes. The cell proliferation on the PVDF composite nanofibers was studied by the MTT assay, resulting that the relative cell viability of the MC3T3-E1 cells cultured on the composite nanofibers indicated good cytocompatibility of the nanofibers (Figure 7). For all the tested nanofibers, time-dependent actions were observed in cell proliferation. Changes in the content of the POSS–EGCG conjugate in the composite nanofibers affected the cellular activities. All the composite nanofibers accelerated more rapid cell proliferation as compared with the pure PVDF nanofiber. Among the composite nanofibers, cell proliferation was significantly enhanced on PE06 in comparison with on the other nanofibers. These results were associated with the contents of piezoelectric β-phase and POSS–EGCG conjugate in the PVDF composite nanofibers.

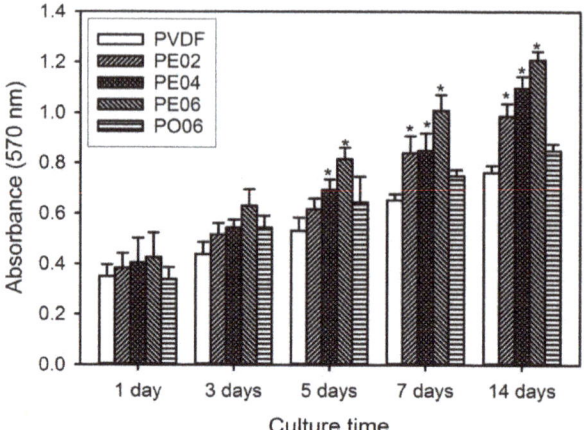

Figure 7. Proliferation behavior of MC3T3-E1 cells cultured on the PVDF composite nanofibers ($n = 6$). Significant difference from the pure PVDF nanofiber at each time point was denoted as $p^* < 0.05$.

During tissue regeneration, piezoelectricity can induce an increase in the surface charge density of ECM materials for the delivery of an electrical stimulus without external energy sources and consequently causes higher levels of cell proliferation and differentiation [14,16]. In addition, EGCG causes a significant elevation of osteoblastic survival as well as a decrease of osteoblastic apoptosis caused by reactive oxygen species, resulting that the proliferation and differentiation of osteoblasts can be stimulated because of antioxidant and free-radical scavenging activities of EGCG [26,40]. As shown in Figure 8, the POSS–EGCG conjugate exhibited excellent superoxide anion radical scavenging activity similar to that of pure EGCG, which could have a positive effect on cell viability. Therefore, the piezoelectric property and antioxidant activity of PVDF composite nanofibers may play an important role in osteoblast cell proliferation.

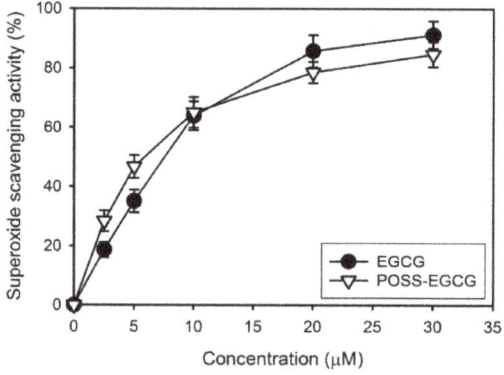

Figure 8. Superoxide anion scavenging activities of EGCG and the POSS–EGCG conjugate ($n = 4$).

To confirm cell growth on the PVDF composite nanofibers, the interactions between the cells and nanofibers was observed using SEM on periods of 3 and 7 days after cell seeding. SEM images exhibited that the cells adhered and spread on the surface of the composite nanofibers, and cell proliferation increased time-dependently in all the tested nanofibers (Figure 9). This indicates excellent biocompatibility of the composite nanofibers. Moreover, almost the whole surface of the PE06 was covered by the cells at day 7 of cell culture.

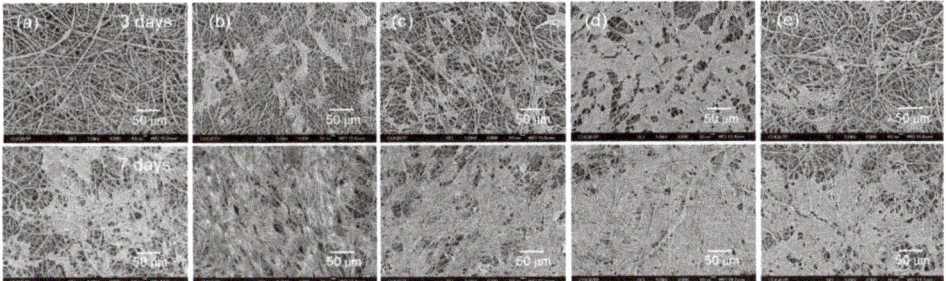

Figure 9. Morphologies of MC3T3-E1 cells on (**a**) PVDF, (**b**) PE02, (**c**) PE04, (**d**) PE06, and (**e**) PO06 composite nanofibers after culturing for 3 and 7 days.

3.5. ALP Activity and Bone Mineralization

ALP catalyzes the cleavage of organic phosphate esters and plays an important role during the formation process of bone nodule [1,2]. ALP activity is considered as an early marker of osteoblastic differentiation because it significantly increases during the differentiation step. Therefore, we examined ALP activity to evaluate the capability of the PVDF composite nanofibers for promoting osteoblastic differentiation. As a result, the cells cultured on PE06 exhibited higher ALP activity than those cultured on the other composite nanofibers at all time points (Figure 10a). Similar results have been previously reported, in that various flavonoids—such as green tea catechins, quercetin, and kaempferol—significantly increased ALP activity by the activation of the extracellular signal-regulated kinase pathway and endoplasmic reticulum binding [40–42].

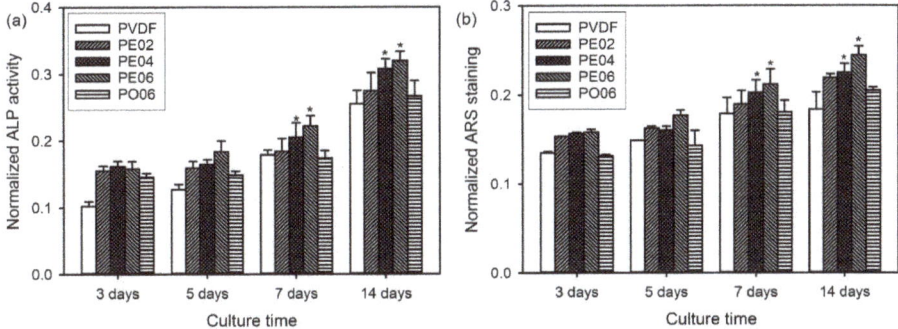

Figure 10. (**a**) ALP activity and (**b**) quantitative calcium deposition of MC3T3-E1 cells on the PVDF composite nanofibers after culturing for 3, 5, 7, and 14 days ($n = 6$). The values of ALP activity and quantitative calcium deposition were normalized by the cell number (1×10^4 cells). Significant difference from the pure PVDF nanofiber at each time point was denoted as $p^* < 0.05$.

The mineralized bone nodules are usually formed via three main stages such as cell proliferation, ECM development, and mineralization. Therefore, the osteogenesis capability of the cells as a late-stage marker for osteogenic differentiation can be estimated by the determination of bone-like calcium deposits produced on different PVDF composite nanofibers using ARS staining [1,4]. After the incubation for 3, 5, 7, and 14 days, the absorbance changes of ARS in the red-stained composite nanofibers were used to quantify the calcium mineralization. Figure 10b exhibits the absorbance of ARS extracted from the stained calcium deposits in different composite nanofibers. The increase of calcium deposition on the composite nanofibers with time was observed in all samples. Among the samples, PE06 exhibited the highest calcium deposition compared to the other samples at all time points caused by increased osteogenesis capability [42]. These results indicated that the amount of

POSS–EGCG conjugate present in the composite nanofibers significantly influenced the bone-forming ability of the MC3T3-E1 cells.

3.6. Quantification of Inflammatory Cytokine

Osteoblasts can produce bone-active cytokines such as tumor necrosis factor-α and IL-6 [21,38]. In bone metabolism, these cytokines act as potent factors for osteoclast formation and bone resorption because they can mediate the effects of many stimulators of bone resorption, i.e., parathyroid hormone and IL-1 [22,23,40]. Among these cytokines, IL-6—the most potent osteoclastogenic factor—promotes the activation of osteoclast precursors and their subsequent differentiation into mature osteoclasts, leading to the bone resorption. In addition, a significant release of IL-6 can suppress cell viability. Thus, we determined the expression of IL-6 in cell cultures to evaluate the bone-forming ability and cytocompatibility of the PVDF composite nanofibers.

It was found that PVDF and PO06 promoted the expression of the IL-6 markedly due to the contamination by LPS and subsequent macrophage activation (Figure 11). However, all composite nanofibers containing the POSS–EGCG conjugate exhibited considerably decreased levels of IL-6 expression compared with the pure PVDF nanofiber. In particular, PE06 exhibited only a very small amount of IL-6 secretion, suggesting that the POSS–EGCG conjugate can downregulate the production of bone-active cytokines and improve the cytocompatibility of the composite nanofibers via the anti-inflammatory effect of EGCG [22,23,40]. Therefore, the content of the POSS–EGCG conjugate present was an essential factor in decreasing IL-6 production on the PVDF composite nanofibers.

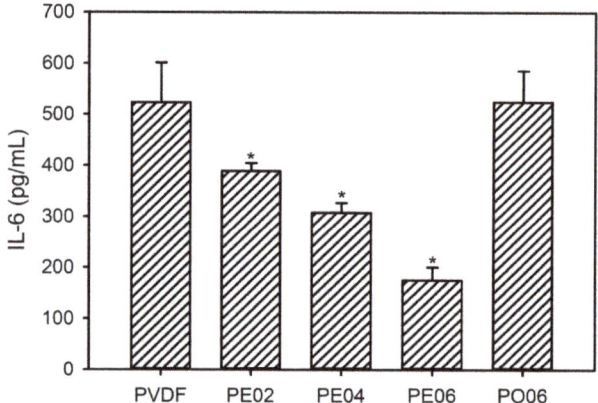

Figure 11. Levels of IL-6 expression by RAW 264.7 cells on the PVDF composite nanofibers ($n = 5$). Significant difference from the pure PVDF nanofiber at each time point was denoted as $p* < 0.05$.

4. Conclusions

In this study, we fabricated PVDF composite nanofibers as potential membranes for bone tissue regeneration using an electrospinning process. The composite nanofibers obtained had fully interconnected porous structures adequate to promote the cell proliferation and the efficient transport of nutrients and metabolic waste. The content of POSS–EGCG conjugate present affected the piezoelectric, antioxidant, and anti-inflammatory properties of the nanofibers, which finally influenced the in vitro bioactivity of the PVDF composite nanofibers. PE06 facilitated the proliferation and differentiation of the MC3T3-E1 cells more effectively compared to the other nanofibers. Based on these results, our simple method is very useful to prepare PVDF composite nanofibers which may contribute to the development of new scaffolding materials for bone tissue regeneration.

Supplementary Materials: The following are available online at http://www.mdpi.com/2079-4991/9/2/184/s1, Figure S1: ^{1}H NMR spectrum of the POSS–EGCG conjugate. Figure S2: EDS Si-mapping images of the PVDF composite nanofibers.

Author Contributions: Y.-J.K. designed the experiments, analyzed the data, and wrote the manuscript; H.-G.J. and Y.-S.H. performed the fabrication and characterization of materials; K.-H.J. carried out in vitro experiments. All authors have read and approved the final manuscript.

Funding: This work was supported by research grants from Daegu Catholic University in 2017.

Conflicts of Interest: The authors declare no conflicts of interest.

References

1. Lee, H.; Hwang, H.; Kim, Y.; Jeon, H.; Kim, G. Physical and bioactive properties of multi-layer PCL/silica composite scaffolds for bone tissue regeneration. *Chem. Eng. J.* **2014**, *250*, 399–408. [CrossRef]
2. Rad, M.M.; Khorasani, S.N.; Ghasemi-Mobarakeh, L.; Prabhakaran, M.P.; Foroughi, M.R.; Kharaziha, M.; Saadatkish, N.; Ramakrishna, S. Fabrication and characterization of two-layered nanofibrous membrane for guided bone and tissue engineering application. *Mater. Sci. Eng. C* **2017**, *80*, 75–87.
3. Cai, Y.; Guo, J.; Chen, C.; Yao, C.; Chung, S.M.; Yao, J.; Lee, I.S.; Kong, X. Silk fibroin membrane used for guided bone tissue regeneration. *Mater. Sci. Eng. C* **2017**, *70*, 148–154. [CrossRef] [PubMed]
4. Ke, R.; Yi, W.; Tao, S.; Wen, Y.; Hongyu, Z. Electrospun PCL/gelatin composite nanofiber structures for effective guided bone regeneration membranes. *Mater. Sci. Eng. C* **2017**, *78*, 324–332.
5. Zhang, S. Fabrication of novel biomaterials through molecular self-assembly. *Nat. Biotechnol.* **2003**, *21*, 1171–1178. [CrossRef] [PubMed]
6. Maeda, H.; Kasuga, T.; Hench, L.L. Preparation of poly(L-lactic acid)-polysiloxane-calcium carbonate hybrid membranes for guided bone regeneration. *Biomaterials* **2006**, *27*, 1216–1222. [CrossRef] [PubMed]
7. Meier, C.; Welland, M.E. Wet-spinning of amyloid protein nanofibers into multifunctional high-performance biofibers. *Biomacromolecules* **2011**, *12*, 3453–3459. [CrossRef]
8. Obata, A.; Ito, S.; Iwanaga, N.; Mizuno, T.; Jones, J.R.; Kasuga, T. Poly(γ-glutamic acid)-silica hybrids with fibrous structure: Effect of cation and silica concentration on molecular structure, degradation rate and tensile properties. *RSC Adv.* **2014**, *4*, 52491–52499. [CrossRef]
9. Poologasundarampillai, G.; Wang, D.; Li, S.; Nakamura, J.; Bradley, R.; Lee, P.D.; Stevens, M.M.; McPhail, D.S.; Kasuga, T.; Jones, J.R. Cotton-wool-like bioactive galsses for bone regeneration. *Acta Biomater.* **2014**, *10*, 3733–3746. [CrossRef]
10. Wang, D.; Nakamura, J.; Poologasundarampillai, G.; Kasuga, T.; Jones, J.R.; McPhail, D.S. ToF-SIMS evaluation of calcium-containing silica/γ-PGA hybrid systems for bone regeneration. *Appl. Surf. Sci.* **2014**, *309*, 231–239. [CrossRef]
11. Okuda, T.; Tominaga, K.; Kidoaki, S. Time-programmed dual release formulation by multilayered drug-loaded nanofiber meshes. *J. Control. Release* **2010**, *143*, 258–264. [CrossRef] [PubMed]
12. Moydeen, A.M.; Padusha, M.S.A.; Aboelfetoh, E.; Al-Deyab, S.S.; El-Newehy, M.H. Fabrication of electrospun poly(vinyl alcohol)/dextran nanofibers via emulsion process as drug delivery system: Kinetics and in vitro release study. *Int. J. Biol. Macromol.* **2018**, *116*, 1250–1259. [CrossRef]
13. Paul, N.E.; Skazik, C.; Harwardt, M.; Bartneck, M.; Denecke, B.; Klee, D.; Salber, J.; Zwadlo-Klarwasser, G. Topographical control of human macrophages by a regularly microstructured polyvinylidene fluoride surface. *Biomaterials* **2008**, *29*, 4056–4064. [CrossRef] [PubMed]
14. Lee, Y.S.; Collins, G.; Arinzeh, T.L. Neurite extension of primary neurons on electrospun piezoelectric scaffolds. *Acta Biomater.* **2011**, *7*, 3877–3886. [CrossRef] [PubMed]
15. Lopes, A.C.; Gutiérrez, J.; Barandiarán, J.M. Direct fabrication of a 3D-shape film of polyvinylidende fluoride (PVDF) in the piezoelectric β-phase for sensor and actuator applications. *Eur. Polym. J.* **2018**, *99*, 111–116. [CrossRef]
16. Rajabi, A.H.; Jaffe, M.; Arinzeh, T.L. Piezoelectric materials for tissue engineering: A review. *Acta Biomater.* **2015**, *24*, 12–23. [CrossRef] [PubMed]
17. Genchi, G.G.; Sinibaldi, E.; Ceseracciu, L.; Labardi, M.; Marino, A.; Marras, S.; De Simoni, G.; Mattoli, V.; Ciofani, G. Ultrasound-activated piezoelectric P(VDF-TrFE)/boron nitride nanotube composite films promote differentiation of human SaOS-2 osteoblast-like cells. *Nanomedicine* **2018**, *14*, 2421–2432. [CrossRef] [PubMed]

18. Lu, L.; Zhang, C.; Li, L.; Zhou, C. Reversible pH-responsive aggregates based on the self-assembly of functionalized POSS and hyaluronic acid. *Carbohydr. Polym.* **2013**, *94*, 444–448. [CrossRef]
19. Ha, Y.M.; Amna, T.; Kim, M.H.; Kim, H.C.; Hassan, M.S.; Khil, M.S. Novel silicificated PVAc/POSS composite nanofibrous mat via facile electrospinning technique: Potential scaffold for hard tissue engineering. *Colloid Surf. B Biointerfaces* **2013**, *102*, 795–802. [CrossRef]
20. Rozga-Wijas, K.; Michalski, A. An efficient synthetic route for a soluble silsesquioxane-daunorubicin conjugate. *Eur. Polym. J.* **2016**, *84*, 490–501. [CrossRef]
21. Johnson, J.J.; Bailey, H.H.; Mukhtar, H. Green tea polyphenols for prostate cancer chemoprevention: A translational perspective. *Phytomedicine* **2010**, *17*, 3–13. [CrossRef] [PubMed]
22. Singh, R.; Akhtar, N.; Haqqi, T.M. Green tea polyphenol epigallocatechin-3-gallate: Inflammation and arthritis. *Life Sci.* **2010**, *86*, 907–918. [CrossRef] [PubMed]
23. Singh, B.N.; Shankar, S.; Srivastava, R.K. Green tea catechin, epigallocatechin-3-gallate (EGCG): Mechanisms, perspectives and clinical applications. *Biochem. Pharmacol.* **2011**, *82*, 1807–1821. [CrossRef] [PubMed]
24. Tominari, T.; Matsumoto, C.; Watanabe, K.; Hirata, M.; Grundler, F.M.W.; Miyaura, C.; Inada, M. Epigallocatechin gallate (EGCG) suppresses lipopolysaccharide-induced inflammatory bone resorption, and protects against alveolar bone loss in mice. *FEBS Open Bio* **2015**, *5*, 522–527. [CrossRef] [PubMed]
25. Kuroyanagi, G.; Tokuda, H.; Yamamoto, N.; Kainuma, S.; Fujita, K.; Ohguchi, R.; Kawabata, T.; Sakai, G.; Matsushima-Nishiwaki, R.; Harada, A.; et al. (−)-Epigallocatechin gallate synergistically potentiates prostaglandin E2-stimulated osteoprotegerin synthesis in osteoblast. *Prostagladins Other Lipid Mediat.* **2017**, *128–129*, 27–33. [CrossRef] [PubMed]
26. Chu, C.; Deng, J.; Hou, Y.; Xiang, L.; Wu, Y.; Qu, Y.; Man, Y. Application of PEG and EGCG modified collagen-base membrane to promote osteoblasts proliferation. *Mater. Sci. Eng. C* **2017**, *76*, 31–36. [CrossRef] [PubMed]
27. Zhou, T.; Zha, J.W.; Cui, R.Y.; Fan, B.H.; Yuan, J.K.; Dang, Z.M. Improving dielectric properties of BaTiO3/ferroelectric polymer composites by employing surface hydroxylated BaTiO3 nanoparticles. *ACS Appl. Mater. Interfaces* **2011**, *3*, 2184–2188. [CrossRef]
28. Martins, P.; Lopes, A.C.; Lanceros-Mendrez, S. Electroactive phases of poly(vinylidene fluoride): Determination, processing and applications. *Prog. Polym. Sci.* **2014**, *39*, 683–706. [CrossRef]
29. Thakur, P.; Kool, A.; Bagchi, B.; Hoque, N.A.; Das, S.; Nandy, P. The role of cerium(III)/yttrium(III) nitrate hexahydrate salts on electroactive β phase nucleation and dielectric properties of poly(vinylidene fluoride) thin films. *RSC Adv.* **2015**, *5*, 28487–28496. [CrossRef]
30. Mokhtari, F.; Shamshirsaz, M.; Latifi, M. Investigation of β phase formation in piezoelectric response of electrospun polyvinylidene fluoride nanofibers: LiCl additive and increasing fibers tension. *Polym. Eng. Sci.* **2016**, *56*, 61–70. [CrossRef]
31. Kim, Y.E.; Kim, Y.J. Effects of nanofibrous membranes containing low molecular weight β-glucan on normal and cancer cells. *J. Nanosci. Nanotechnol.* **2017**, *17*, 3597–3605. [CrossRef]
32. Peres, I.; Rocha, S.; Pereira, M.C.; Coelho, M.; Rangel, M.; Ivanova, G. NMR structural analysis of epigallocatechin gallate loaded polysaccharide nanoparticles. *Carbohydr. Polym.* **2010**, *82*, 861–866. [CrossRef]
33. Chen, R.; Wang, J.B.; Zhang, X.Q.; Ren, J.; Zeng, C.M. Green tea polyphenol epigallocatechin-3-gallate (EGCG) induced intermolecular cross-linking of membrane proteins. *Arch. Biochem. Biophys.* **2011**, *507*, 343–349. [CrossRef] [PubMed]
34. Bors, W.; Michel, C.; Stettmaier, K. Electron paramagnetic resonance studies of radical species of proanthocyanidins and gallate esters. *Arch. Biochem. Biophys.* **2000**, *374*, 347–355. [CrossRef] [PubMed]
35. Brzonova, I.; Steiner, W.; Znakel, A.; Nyanhongo, G.S.; Guebitz, G.M. Enzymatic synthesis of catechol and hydroxyl-carboxic acid functionalized chitosan microspheres for iron overload therapy. *Eur. J. Pharm. Biopharm.* **2011**, *79*, 294–303. [CrossRef] [PubMed]
36. Yi, X.S.; Shi, W.X.; Yu, S.L.; Sun, N.; Jin, L.M.; Wang, S.; Zhang, B.; Ma, C.; Sun, L.P. Comparative study of anion polyacrylamide (APAM) adsorption-related fouling of a PVDF UF membrane and a modified PVDF UF membrane. *Desalination* **2012**, *286*, 254–262. [CrossRef]
37. Valentini, L.; Bon, S.B.; Cardinali, M.; Monticelli, O.; Kenny, J.M. POSS vapor grafting on graphene oxide film. *Chem. Phys. Lett.* **2012**, *537*, 84–87. [CrossRef]
38. Qiao, T.; Jiang, S.; Song, P.; Song, X.; Liu, Q.; Wang, L.; Chen, X. Effect of HA-g-PLLA on xanthohumol-loaded PLGA fiber membrane. *Colloid Surf. B Biointerfaces* **2016**, *146*, 221–227. [CrossRef]

39. Peng, Q.Y.; Cong, P.H.; Liu, X.J.; Liu, T.X.; Huang, S.; Li, T.S. The preparation of PVDF/clay nanocomposites and the investigation of their tribological properties. *Wear* **2009**, *266*, 713–720. [CrossRef]
40. Shen, C.L.; Yeh, J.K.; Cao, J.J.; Wang, J.S. Green tea and bone metabolism. *Nutr. Res.* **2009**, *29*, 437–456. [CrossRef]
41. Prouillet, C.; Mazière, J.C.; Mazière, C.; Wattel, A.; Brazier, M.; Kamel, S. Stimulatory effect of naturally occurring flavonols quercetin and kaempferol on alkaline phosphatase activity in MG-63 human osteoblasts through ERK and estrogen receptor pathway. *Biochem. Pharmacol.* **2004**, *67*, 1307–1313. [CrossRef] [PubMed]
42. Vali, B.; Rao, L.G.; El-Sohemy, A. Epigallocatechin-3-gallate increases the formation of mineralized bone nodules by human osteoblast-like cells. *J. Nutr. Biochem.* **2007**, *18*, 341–347. [CrossRef] [PubMed]

© 2019 by the authors. Licensee MDPI, Basel, Switzerland. This article is an open access article distributed under the terms and conditions of the Creative Commons Attribution (CC BY) license (http://creativecommons.org/licenses/by/4.0/).

Article

Electrospun Filaments Embedding Bioactive Glass Particles with Ion Release and Enhanced Mineralization

Francesca Serio [1,2], Marta Miola [3], Enrica Vernè [3], Dario Pisignano [4,5], Aldo R. Boccaccini [6] and Liliana Liverani [6,*]

1. Dipartimento di Matematica e Fisica "Ennio De Giorgi," Università del Salento, Via Arnesano, I-73100 Lecce, Italy; francescaserio@unisalento.it
2. Istituto Microelettronica e Microsistemi-CNR, Via Monteroni, Campus Unisalento, I-73100 Lecce, Italy
3. Applied Science and Technology Department, Politecnico di Torino, Corso Duca degli Abruzzi 24, I-10129 Torino, Italy; marta.miola@polito.it (M.M.); enrica.verne@polito.it (E.V.)
4. Dipartimento di Fisica, Università di Pisa, Largo B. Pontecorvo 3, I-56127 Pisa, Italy; dario.pisignano@unipi.it
5. NEST, Istituto Nanoscienze-CNR, Piazza S. Silvestro 12, I-56127 Pisa, Italy
6. Institute of Biomaterials, Department of Materials Science and Engineering, University of Erlangen-Nuremberg, Cauerstr. 6, D-91058 Erlangen, Germany; aldo.boccaccini@fau.de
* Correspondence: liliana.liverani@fau.de; Tel.: +49-(0)-9131-85-28618

Received: 15 December 2018; Accepted: 28 January 2019; Published: 1 February 2019

Abstract: Efforts in tissue engineering aim at creating scaffolds that mimic the physiological environment with its structural, topographical and mechanical properties for restoring the function of damaged tissue. In this study we introduce composite fibres made by a biodegradable poly(lactic acid) (PLLA) matrix embedding bioactive silica-based glass particles (SBA2). Electrospinning is performed to achieve porous PLLA filaments with uniform dispersion of bioactive glass powder. The obtained composite fibres show in aligned arrays significantly increased elastic modulus compared with that of neat polymer fibres during uniaxial tensile stress. Additionally, the SBA2 bioactivity is preserved upon encapsulation as highlighted by the promoted deposition of hydroxycarbonate apatite (HCA) upon immersion in simulated body fluid solutions. HCA formation is sequential to earlier processes of polymer erosion and ion release leading to acidification of the surrounding solution environment. These findings suggest PLLA-SBA2 fibres as a composite, multifunctional system which might be appealing for both bone and soft tissue engineering applications.

Keywords: poly(lactic acid) (PLLA); bioactive glass; scaffolds; electrospinning; composite fibres

1. Introduction

Bone tissue continuously undergoes shape remodelling and repair at the microscale through processes of local regeneration, which are regulated by growth factors, hormones and the action of mechanical stresses. In reconstructive surgeries, the regeneration of bone and cartilage by autologous cell transplantation after an injury or tumour removal is one of the most promising strategies in order to reduce issues related to immunocompatibility and consequent immune rejection, as well as to avoid potential pathogen transfer [1]. However, since autologous grafts are often poorly available, an appealing approach is represented by the development of scaffold-based tissue engineering approaches based on bioactive materials for restoring bone morphology and function [2].

Therefore, enormous efforts are being made to create engineered constructs that mimic the physiological environment with its structural, topographical and mechanical properties. To this aim, various biomaterials, metals, natural or synthetic polymers and ceramics, have been investigated but no single one was proved to show all the crucial features required for an optimal scaffold [3,4]. In this framework, a promising approach relies on composite biomaterials with osteoconductive and osteoinductive capabilities, which might allow for osteogenesis stimulation while mimicking the extracellular matrix (ECM) morphology [4,5]. Several studies have been focused on the addition of inorganic and bioactive fillers, such as bioactive glasses, bioceramics or hydroxyapatite [6] in polymeric constructs, with the aim to promote chemical links to bone tissue by forming hydroxycarbonate apatite (HCA) layers as a result of ion leaching, in case of bioactive glass components, into the surrounding physiological fluids. In this process, the precipitation of microcrystalline HCA onto the scaffold surface [7] is due to a well-defined ion exchange mechanism between modifier ions (Na^+ and Ca^{2+}) in the glass and hydronium ions (H_3O^+) in the surrounding fluid, thereby causing dissolution of the glass network [8,9]. Compared to other bioactive materials, silica-based bioactive glasses are available in different compositions, which exhibit remarkable osteoinductive behaviour since they feature ionic dissolution products (Si^{4+}, Mg^{2+}, Ca^{2+}) able to stimulate osteogenesis and angiogenesis [8,9]. In fact, the potential angiogenic effects of silica-based bioactive glass has been recently highlighted, through increased secretion of vascular endothelial growth factor involved in vascularization processes [10]. For these reasons, these materials are not only useful for bone tissue engineering but they might also promote the regeneration of soft tissues as needed in wound healing [11].

In fact, 45S5 Bioglass®has been largely used as inorganic phase in polymer foams and matrices to realize porous composites [12–16]. Silicate [17], borate [18] and phosphate-based glasses [19,20] have been recently tested in bulky bioresorbable polymeric sponges or polymer-coated scaffolds. A number of nano- and micro-fabrication technologies allow these systems to be processed as biocompatible and biodegradable fillers in polymers retaining higher surface-to-volume ratio and interconnected porous networks to better support tissue ingrowth and vascularization. In the last years, the electrospinning technology has been largely developed and notable progress has been made to realize biomimetic porous scaffolds designed for tissue engineering and for drug delivery [21–23]. Electrospinning is a versatile technique, which allows for the fabrication of polymer, ceramic or nanocomposite fibres with diameter ranging from a few tens of nanometres to a few micrometres, which strongly resemble the morphology of the native ECM and provide a networked architecture suitable for cell attachment [22–25]. Various different structures, morphologies and compositions can be achieved in fibres, to make them suitable for different tissue applications including vascular, bone, neural and tendon or ligament [21,25,26]. Particularly, fibres with nanocomposite materials and complex internal [27] or surface [28,29] nanostructures can be electrospun by blends of polymers or from colloidal solutions [30,31].

In this work we introduce nanocomposite electrospun fibres embedding silica-based bioactive glass (SBA2). SBA2 belongs to the SiO_2–Na_2O–CaO–P_2O_5–B_2O_3–Al_2O_3 class of systems, previously investigated as component of antibacterial and bioactive, bulky bone cements [7,32,33]. The FDA-approved polymer, poly(lactic acid) (PLLA), is chosen as matrix because of its excellent biocompatibility and biodegradability, already assessed in clinical treatments [1], as well as for its excellent processability with electrospinning [34]. The obtained PLLA-SBA2 fibres are characterized in their morphology and in their chemical and mechanical properties. The addition of inorganic particles in a polymeric matrix [35–38] leads to composites with varied mechanical properties, depending on the filler size and on their dispersion in the organic phase [35,36,38], as well as on the fabrication parameters, including the solvent used for electrospinning [38]. Additionally, acellular, in vitro bioactivity and cell viability are investigated, evidencing the biocompatibility of the PLLA-SBA2 fibrous composites. Overall, dispersed silica-based bioactive glass in resorbable polymeric composites with microscale texturing are highly promising systems for supporting cell cultures as well as for the development of biomedical applications. The novelty of this work is represented by the successful incorporation of SBA2 in electrospun PLLA fibres, not reported previously in literature and also on the correlation of the nanopores on the fibres surface with fibres degradation and bioactive glass particles release.

2. Materials and Methods

2.1. Bioactive Glass Synthesis and Characterization

The SBA2 glass has the following nominal composition (mol %): 48% SiO_2, 18% Na_2O, 30% CaO, 3% P_2O_5, 0.43% B_2O_3, 0.57% Al_2O_3 and was synthesised by melt-quenching route, as detailed elsewhere [7,32,33]. Briefly, reagent-grade reactants [SiO_2, Na_2CO_3, $CaCO_3$, $Ca_3(PO_4)_2$, H_3BO_3 and Al_2O_3] were melted in a platinum crucible at 1450 °C for 1 h (Carbolite HTF 1800, CARBOLITE GERO, Neuhausen, Germany), the melt was then quenched in water to obtain a frit. The frit was ball milled in aqueous medium. The grain size distribution of milled SBA2 was estimated using a particle size analyser (Sympatec Helos H0621, kindly performed at CERICOL Research Centre, Sovigliana, Vinci (Firenze), Italy); the specific surface area (SSA) of milled SBA2 was assessed by using the Brunauer-Emmet-Teller (BET, ASAP2020Plus-Micromeritics, Aachen, Germany) method [39]. The obtained glass powder was analysed morphologically and compositionally by scanning electron microscopy (SEM, FEI QUANTA INSPECT 200, Eindhoven, The Netherlands) and energy dispersive X-ray spectrometry (EDS) (EDAX PV 9900, Weiterstadt, Germany). Thermal properties were determined by differential thermal analysis (DTA–404 PC instrument, Netzsch, Selb, Germany), in a temperature range of 20–1300 °C, using a heating rate of 10 °C/min and high-purity alumina as reference.

2.2. Electrospinning

PLLA (molecular weight 85−160 kg mol^{-1}, Sigma-Aldrich, Munich, Germany) was dissolved in a mixture of dichloromethane and acetone (80:20 v/v) at a concentration of 20% (w/v) at room temperature. The solution was stirred overnight, then SBA2 was added with a concentration of 7% w/v. The resulting suspension was vigorously mixed and then stirred again overnight to obtain a homogeneous dispersion of SBA2 in the polymer solution. The solution was put in an ultrasound bath for 1 hour prior the electrospinning with the aim to reduce SBA2 clustering and then transferred in a 1 mL syringe. The spinning process was performed by a 21G stainless steel needle and a syringe pump (Harvard Apparatus, Holliston, MA, United States) with feeding rate 0.8 mL h^{-1} and by applying a positive high-voltage of 12 kV (EL60R0.6-22, Glassman High Voltage, XP Power, Milano, Italy) between the needle and a metal collector. Random fibre mats were collected on a static grounded 10×10 cm^2 collector meanwhile aligned fibre mats were collected on a grounded disk (8 cm diameter, 1 cm thickness) rotating at a speed of 5000 rpm, maintaining constant all the other processing parameters. The air relative humidity and temperature during the electrospinning process were about 40% and 20°C, respectively. The needle-collector distance was adjusted to 15 cm. Neat PLLA fibres were fabricated as reference material by using identical set-up parameters.

2.3. Morphology and Mechanical Properties of PLLA-SBA2 Fibres

PLLA-SBA2 electrospun fibres were sputtered with gold by using a Sputter Coater (Q150T, Quorum Technologies, Darmstadt, Germany) and then inspected by SEM and energy dispersive X-ray spectrometry (EDS) (Auriga 0750, ZEISS, Jena, Germany). Average fibre diameters were calculated by analysing a total of at least 100 fibres for each sample, using the software ImageJ [40]. Regarding mechanical properties, as reported by Ricotti et al. [41], mechanical differences in uniaxial aligned arrays are more pronounced. Thus, aligned fibre mats were used in this paper to investigate more specifically the mechanical properties of PLLA-SBA2 composite fibre mats. Mechanical properties of random and aligned composite fibres mats at room temperature were investigated by uniaxial tensile strength tests using a universal testing machine (K. Frank GmbH, Mannheim, Germany). Each sample was cut into a rectangular shape (with cross-section area of 5×4 mm^2) using a paper square framework and its thickness is measured by using a digital micrometre (0.02–0.06 mm). Then measurements were carried out at a crosshead speed of 10 mm/min using a 50 N load cell, according to a previous study [42] and the resulted stress-strain curves are used to obtain Young's modulus, elongation at break and tensile strength.

2.4. Degradation studies

The degradation behaviour of composite fibres was investigated by immersion in Dulbecco's phosphate buffered saline (DPBS, Sigma-Aldrich, Munich, Germany) medium at pH 7.4. Resulting pH values were measured instantly after the immersion of the sample and for different time points, up to 21 days, at 37 °C by using a pH meter (HD8705, Delta OHM, Padova, Italy). The pH measurements of different solutions were correlated to the dissolution of ions in the incubation media. In addition, the degradation of PLLA-SBA2 fibres in physiological solutions was assessed by measuring the weight loss [23] for dried fibrous mats after 21 days of DPBS incubation. The percentage of weight loss, W_L %, was computed as $100 \times (W_0 - W_r)/W_0$, where W_0 and W_r are the initial and the residual weight of the sample, respectively.

2.5. Acellular Bioactivity

The acellular bioactivity of PLLA-SBA2 fibres, related to the surface deposition of HCA layers, was evaluated by immersing the samples inserted in scaffold holders (CellCrownTM 24, Scaffdex, Sigma Aldrich, Munich, Germany) in simulated body fluid (SBF) solution [43] for 1, 7, 14 and 21 days at 37 °C, on an oscillating tray in an incubator. A falcon tube with only SBF was used as control. After each time point, samples were washed with distilled water, dried at room temperature before SEM-EDS characterization and Fourier Transform Infrared Spectroscopy (FTIR) analysis in attenuated total reflectance (ATR) mode. PLLA neat fibres were considered as control. FTIR spectra were recorded by a spectrometer (IRAffinity-1S, Shimadzu, Kyoto, Japan), repeating 32 scans over the wavenumber range 4000–500 cm^{-1}, with a resolution of 4 cm^{-1}.

2.6. Cell Cultures

PLLA-SBA2 fibres were disinfected under an ultraviolet lamp for 1 h. Murine-derived stromal cells ST-2 (obtained from Leibniz-Institut DSMZ-Deutsche Sammlung von Mikroorganismen und Zellkulturen GmbH, Braunschweig, Germany), were cultured to confluence in 75 cm^2 culture flasks in Roswell Park Memorial Institute medium (RPMI 1640) (GibcoTM, Thermo Fisher Scientific, Schwerte, Germany) containing 10% foetal bovine serum (FBS; Lonza) and 1% penicillin/streptomycin (Lonza) at 37 °C and 5 % CO$_2$. Before seeding, cells were detached using Trypsin in DPBS (Sigma Aldrich, Munich, Germany), stained with 0.4% (v/v) trypan blue solution and counted using a Neubauer chamber (VWR). Then, ST-2 cells were seeded onto the electrospun scaffolds (including neat PLLA fibres as control) with a density of 2×10^4 cells/cm^2. All samples were assayed in triplicate and each sample was incubated in the same RPMI medium described above. Cells were cultured for 7 days renewing the culture medium once after 3 days of culture. Viability analyses of ST-2 cells on composite fibres was assessed after a cultivation period of 1 day and 7 days by using a WST-8 assay (Cell Counting Kit-8, Sigma Aldrich, Munich, Germany), which is based on the conversion of tetrazolium salt to highly water-soluble formazan by mitochondrial enzymes of viable cells. At each time point, the culture medium was removed from each sample and each well with samples and cells was washed with DPBS and added with a solution of 10% WST-8 reagent in colourless medium. After an incubation period of 3 hours at 37 °C and 5% CO$_2$, the solution was transferred into a 96 well plate to measure absorbance at 450 nm by using a microplate Elisa reader (PHOmo Elisa reader, Autobio Diagnostics Co. Ltd., Zhengzhou, China).

To investigate cell morphology, a preliminary evaluation was provided by SEM analysis after 1 day and 7 days of culture. Samples were fixed by using a solution containing paraformaldehyde, glutaraldehyde, sodium cacodylate trihydrate and sucrose (Sigma Aldrich, Munich, Germany). Subsequently, samples dehydration was achieved by using a series of aqueous ethanol solutions. The samples were then dried in a critical point drier (Leica EM CPD 300, Leica, Wetzlar, Germany) and sputtered with gold. The cytoskeleton organization and nucleus morphology of cells on PLLA-SBA2 fibres were studied after 1 day and 7 days following seeding by staining with rhodamine phalloidin and DAPI (ThermoFisher Scientific, Schwerte, Germany). Briefly, samples were fixed by using a fixation solution containing 1,4-piperazinediethanesulfonic acid buffer, ethylene glycol tetraacetic acid, polyethylene glycol, paraformaldehyde, DPBS and sodium hydroxide (Sigma Aldrich, Munich, Germany), washed with DPBS and immersed in a permeabilization buffer for intracellular staining. 400 µL of a 8 µL/mL DPBS solution of rhodamine phalloidin was added in each well containing samples, then kept for 1 hour at 37°C. After the removal of the dye, samples were vigorously washed with DPBS and 400 µL of a 1 µL/mL DPBS solution of DAPI was added to each well. Then, samples were washed in DPBS and analysed with a fluorescent microscope (Axio Scope A1, Zeiss, Jena, Germany).

2.7. Statistical Analysis

Each experiment was repeated three times. All results of cell viability and average fibre diameter were expressed as (mean ± standard deviation) and a one-way analysis of variance (ANOVA) was used for statistical analysis. A P value < 0.05 was considered statistically significant.

3. Results and Discussion

The BET analysis evidenced a SSA of 11.6 m^2/g, the particle size analysis showed a non-symmetric distribution with a mode grain size of 2 µm; characteristic particle sizes are d_{50} = 2.0 µm and d_{90} = 4.4 µm, with 94% of grain sizes below 5 µm and the residual 4% below 9 µm.

Figure 1 shows the morphology (a, b) and the compositional analysis (c) of SBA2 glass powders. The powders display the typical sharp-cornered morphology of ball-milled glass. Moreover, as previously mentioned, the majority of SBA2 powders showed a grain size <5 µm. In addition, the performed DTA analysis evidenced a glass transition temperature of about 550 °C, the crystallization onset at 600 °C, a crystallization peak at 655 °C and a melting temperature of 1220 °C.

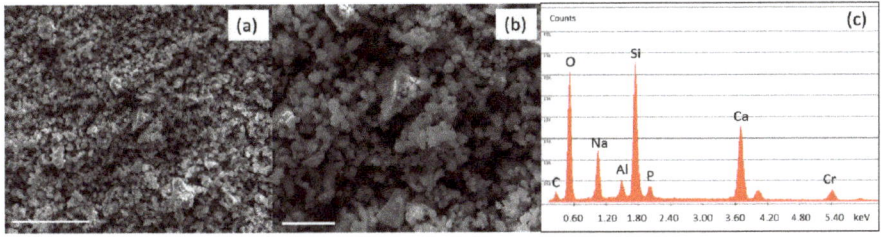

Figure 1. Scanning electron microscopy (SEM) (**a,b**) and energy dispersive x-ray spectrometry (EDS) (**c**) analysis of SBA2 powders after milling process. Scale bars: 40 µm (**a**), 10 µm (**b**).

The electrospinning process of PLLA-SBA2 is carried out by adjusting the voltage bias as well as the solution flow rate to obtain almost bead-free fibres. The morphology of neat PLLA and composite fibres is shown in SEM micrographs in Figure 2. Composite fibres exhibit a quite homogeneous distribution of embedded SBA2 particles (arrows in Figure 2b).

The SBA2 incorporation also leads to an increase of the overall roughness in the electrospun mats. Interestingly, the average diameter of PLLA-SBA2 filaments is around one half of fibres without SBA2 (Table 1), similarly to previous findings on composite fibres made by PCL and commercially available Bioglass®(45S5) particles [42]. This effect might be explained due to the change of intrinsic solution properties, including rheology and conductivity, of the polymeric solutions after the addition of SBA2 particles [44]. In addition, the physical effect of particles in the electrified solution is associated to the onset of enhanced whipping and varicose instabilities, which affect the morphology of the ultimately deposited fibres [45], as shown by the local increase of filament radius in the composite fibres close to the particles. The resulting excess mass due to local polymer accumulation leads to the here found decrease of diameter along the rest of the fibre length due to the overall polymer mass conservation. Furthermore, we find that both the types of fibres display cylindrical shape and high surface porosity, with pore size around 100 nm (Figure 2c,d), which is characteristic of PLLA electrospun with highly volatile solvents [28] and expected to favour cell attachment. Indeed, it is known that both micro- and nanoporosity play an important role in protein adhesion and cell function [46]. In addition, the incorporation of the bioactive glass particles in PLLA-SBA2 fibres is confirmed by SEM-EDS analysis as shown in Figure 3 and Figure S1.

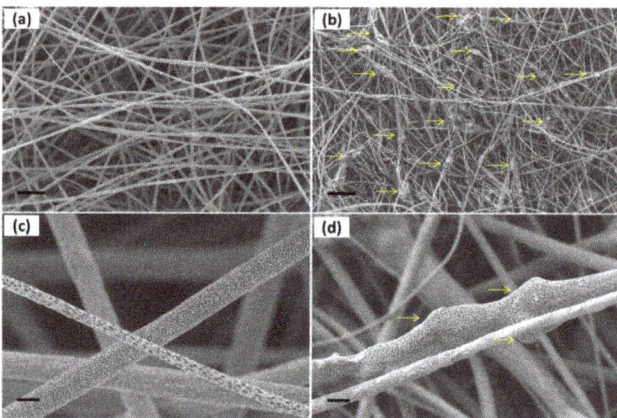

Figure 2. SEM micrographs of (**a**) neat polylactic acid (PLLA) fibres and (**b**) PLLA-SBA2 fibres (scale bar = 25 µm). Neat PLLA fibres (**c**) and composite fibres (**d**) at high magnification, evidencing their highly porous surface. Scale bar: 2 µm.

Table 1. Average diameters, minimum and maximum values measured for the transversal size of neat polymeric and composite fibres.

Sample	Average Fibre Diameter (µm)	Minimum Fibre Transversal Size (µm)	Maximum Fibre Transversal Size (µm)
PLLA	2.0 ± 0.2	1.0	3.9
PLLA-SBA2	1.0 ± 0.2	0.3	2.5

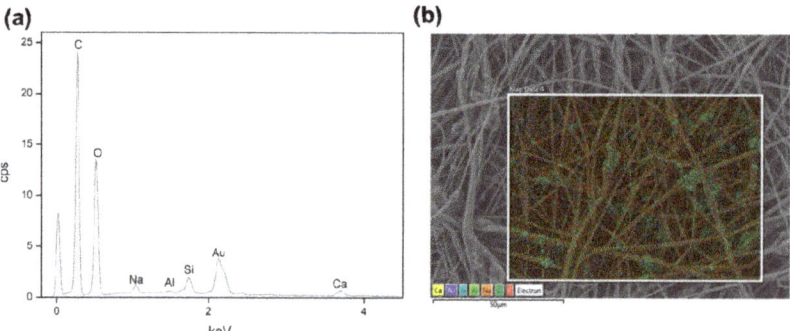

Figure 3. EDS spectrum of electrospun PLLA-SBA2 fibres (**a**) and its comparative SEM-EDS layered image (**b**) showing peaks of SBA2 constituents such as Ca (yellow), Na (orange) and Si (cyan). O (bright green) and C (red) are mainly related to PLLA.

The mechanical properties of the polymeric and composite fibres mats, determined from their stress–strain curves, are summarized in Figure 4, in terms of Young's modulus (Figure 4a), elongation at break (Figure 4b) and tensile strength (Figure 4c) for either randomly oriented and aligned electrospun samples. Particularly, in uniaxially aligned arrays, in which the fibres and to some extent polymeric chains [47–51] are oriented along the traction axis, mechanical differences between composite and neat PLLA are more pronounced [49]. The elastic modulus of PLLA-SBA2 fibres is found to be (34 ± 5) MPa, significantly higher than the value measured for pristine PLLA fibres, (20 ± 2) MPa, just for the aligned fibres. Correspondingly, PLLA-SBA2 exhibits a relatively lower elongation at break (15 ± 2%) compared to the value of PLLA samples in the same alignment conditions (20 ± 2%), as well as higher tensile strength (11 ± 1 MPa), compared to (7 ± 1) MPa. The mechanical behaviour of the composites tightly depends on the interaction at the polymer-bioactive glass interface and on the homogeneity of the dispersion of glass particles in the fibres, which if poor would lead to agglomeration in clusters [51]. The present findings in terms of increased elasticity and concomitant reduction of the elongation at break and tensile strength are indicative of a reinforced fibre system where inorganic fillers lead to embrittlement of the fibres [39,52].

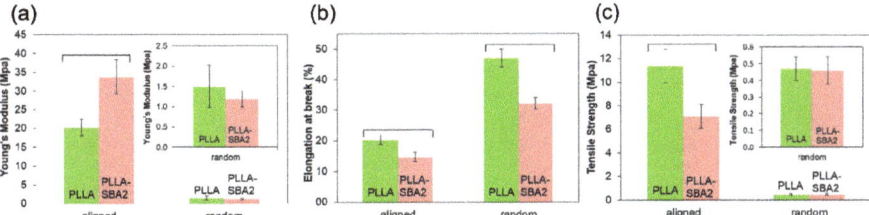

Figure 4. Young's modulus (**a**), elongation at break (**b**) and tensile strength (**c**) of random and aligned fibres. Results are expressed as (mean ± standard deviation). Bars show statistically significant differences ($p < 0.05$). In the inset of (**a**) and (**c**) a zoom view of the properties of randomly oriented fibres is reported.

The analysis of both PLLA-SBA2 and PLLA fibres, performed by collecting SEM micrographs of samples soaked in SBF solution, evidences polymer degradation starting from the 7th day following incubation, resulting in cracks and points of break in the surface of fibres. As highlighted in SEM micrographs in the top of Figure 5, in PLLA-SBA2 fibres this degradation leads to a progressively increasing exposure of glass particles to the surrounding microenvironment. In order to investigate the resulting bioactive glass dissolution from PLLA cracks propagating along the nanopores on the fibre surface, we perform further in vitro degradation studies, incubating the composite scaffolds in DPBS in physiological conditions (at pH = 7.4 and 37 °C), as showed in Figure S2c. During the immersion in DPBS for 21 days, the corresponding changes of the solution pH was monitored, as displayed in Figure 5 together with corresponding SEM micrographs at each time point. PLLA fibres clearly undergo acid hydrolysis in the solution which leads to a decrease of pH, whereas for the composite fibres this trend is overcome by the release of SBA2 dissolution products into the solution. The pH increase related to bioactive glass dissolution products have been already investigated in several medium by Cerutti et al. [52]. This increased pH associated with the dissolution-precipitation of bioactive glass is known to affect various cellular processes, being correlated with increased metabolic activity and proliferation rate in mammalian cells [53]. Our data on pH variations are also supported by the corresponding measurements of sample weight loss, performed after 21 days, which is found to be 2.2% of the initial weight for neat PLLA fibres and 9.7% for PLLA-SBA2 fibres.

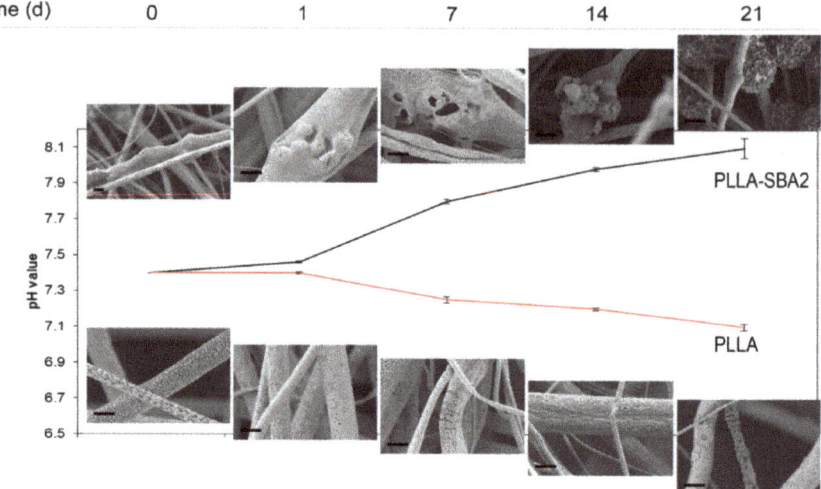

Figure 5. In vitro degradation studies and dissolution of PLLA and PLLA-SBA2 electrospun fibres. Change in the pH of DPBS solution for PLLA and PLLA-SBA2 fibres, which is associated to the acid hydrolysis of the polymer components and to the release of SBA2 into the solutions at 0, 1, 7, 14, 21 days from incubation and the relatives SEM micrographs at high magnification of composite and polymeric fibres along soaking experiments in SBF at the same time point (scale bar = 2 μm).

The acellular bioactivity of the electrospun fibres is evaluated by exploring the formation of HCA on their surfaces upon soaking in SBF, at 37 °C for different time periods. The bioactivity of PLLA-SBA2 fibres can be clearly appreciated when a large quantity of HCA with typical cauliflower-like morphology is found nearby the electrospun filaments (rightmost SEM micrographs in the top of Figure 5). Neat fibres, instead, do not lead to HCA formation. These findings are supported by EDS, detecting high Ca and P peaks belonging to HCA (Figure 6) with Ca/P ratio of 1.9, as well as by FTIR analysis (Figure 7). Indeed, FTIR spectra of PLLA-SBA2 fibres following incubation feature two new peaks at 600 cm^{-1} and 560 cm^{-1} corresponding to HCA-associated P–O groups asymmetric bending [54] and the increase in intensity of the peak centred at 960 cm^{-1} ascribable to the contribution of Si-OH symmetric stretching characteristic of the bioactive glass after immersion in SBF [54,55].

The formation of HCA following immersion in SBF solution confirms that the incorporation of the bioactive glass particles in the PLLA filaments is not preventing the characteristic bioactivity of this specific composition of bioactive glass to be highlighted [32]. In addition, the formation of HCA after 21 days suggests this class of scaffolds not only as useful systems for bone tissue engineering but also for soft tissue repair applications, because release of ions is detected since the earliest time point, as evidenced by the pH variation in the degradation studies. In particular, these fibrous architectures would be highly suitable for the release of therapeutic ions embedded in the bioactive glass structure, occurring without simultaneous damage in the fibrous structure of the scaffolds. Indeed, the double-scale temporal dynamics exhibited by the two mechanisms might be highly advantageous for wound healing processes, where the composite can still provide mechanical support while serving as reservoir of therapeutic ions [56]. The ability to modulate ion release through bioactive glass composition engineering, in relation to the addressed cell type, might be especially important in this respect. This relevance is also suggested by previous works that emphasized the influence of ion release in stimulating vascularization and triggering the production of angiogenic growth factors during soft tissue repair [57,58]. Other application fields potentially benefiting from interplaying ionic release and gradual morphological changes in composite scaffolds include the development of biomedical materials able to modulate the inflammatory response [59], to control cell proliferation [11], to support the regeneration of the peripheral nerve and the treatment of relevant pathological conditions such as chronic osteomyelitis [60–62]. The smart combination of biologically active ion release and nanotopography has been highlighted for tissue engineering [63].

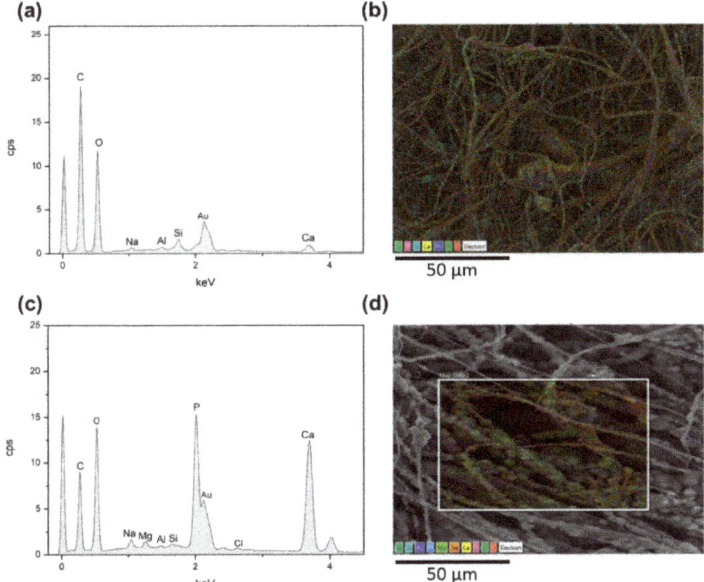

Figure 6. EDS analysis of PLLA-SBA2 samples after 7 days (**a,b**) and 21 days (**c,d**) of soaking in SBF. EDS spectra of PLLA-SBA2 (**a,c**) and comparative EDS layered images (**b,d**) reveal an increment of Ca (yellow) and P (purple) content. SBA2 constituents: Na (orange), Al (green) and Si (cyan). O (bright green) and C (red) mainly relate to PLLA.

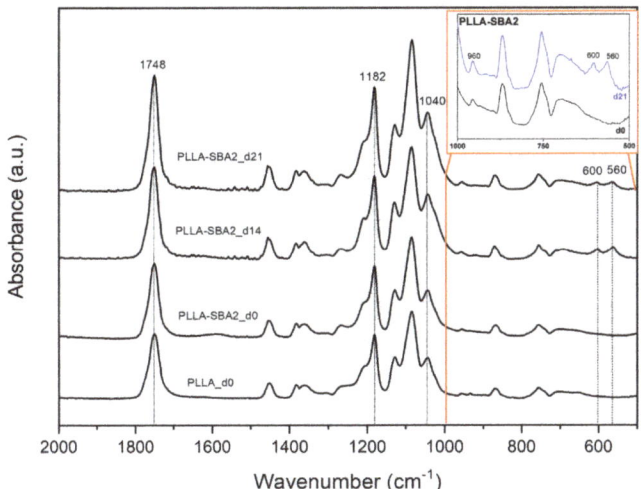

Figure 7. Fourier transform infrared (FTIR) spectra of neat PLLA (PLLA_d0) and composite scaffolds (PLLA-SBA2_d0) before incubation in simulated body fluid (SBF) and of composite fibres after 14 (PLLA-SBA2_d14) and 21 days (PLLA-SBA2_d21) of incubation in SBF. The inset better highlights the new peaks ascribable to hydroxycarbonate apatite (HCA) formation mentioned in the text.

Finally, we investigated the biocompatibility of PLLA-SBA2 fibres using a murine-derived ST2 stromal cell line. The morphology of cells, inspected by SEM, was found to be affected by the fibrous structure of the scaffold, with significant cells elongation along the directions of supporting fibres as shown in Figure 8. The cytoskeleton actin and the nuclei of cultured cells were also analysed by immunofluorescence microscopy (Figure 9). Overall, cells adhesion was efficient on the nanocomposite filaments embedding bioactive glass particles. In previous works, in vitro cytotoxicity tests of SBA2 in PMMA-based cement were accomplished using the indirect contact method [33]. Here, the cell viability is directly evaluated at 1 and 7 days after seeding. WST-8 assays highlight, for absorbance detected at 450 nm at day 1 for cultures on PLLA fibres, an almost double value compared to those measured for cultures on PLLA-SBA2, results shown in Figure 10. This result correlates well with the increase of the pH value found during degradation studies, namely with the release of ions from the bioactive glass particles. Indeed, it is known that the alkalinisation of the medium can influence significantly cell metabolism [64] and can consequently induce distinct phases in cell cycling on composite and neat fibres, respectively, as also suggested by previous reports focused on the behaviour of human osteoblasts exposed to the ionic dissolution products of bioactive glass [65,66].

At day 7 after seeding the values of absorbance measured on PLLA fibres and on PLLA-SBA2 fibres become comparable, as shown in Figure 10, evidencing that the initial conditions of pH changes do not affect cell viability at longer timescales. ST2 cells cultured on composite fibres generally exhibit a relatively lower cell density following seeding, then reaching cell densities comparable to those on PLLA fibres after 7 days. Indeed, despite a significant difference in cell density immediately after cell seeding, these results evidence the ability of PLLA-SBA2 fibres to providing an effective and viable environment for subsequent scaffold colonization.

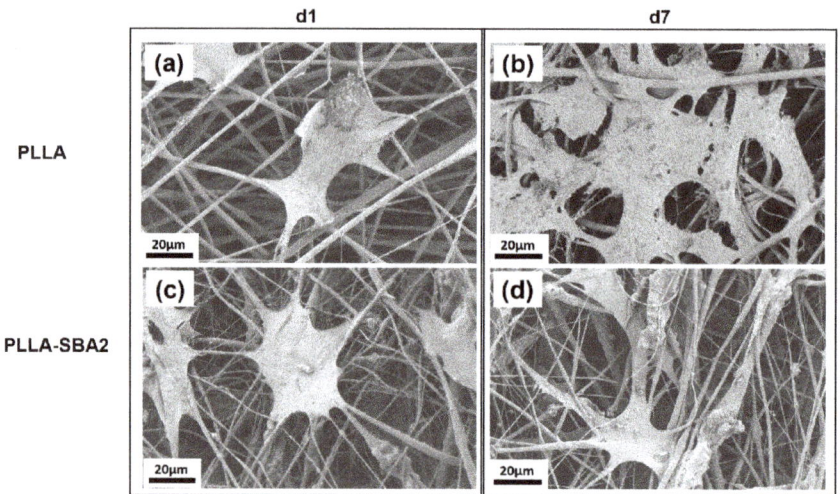

Figure 8. SEM micrographs of dehydrated ST2 cells, cultured on neat (**a**,**b**) and composite (**c**,**d**) fibres for 1 (**a**,**c**) and 7 (**b**,**d**) days. Scale bar: 20 μm.

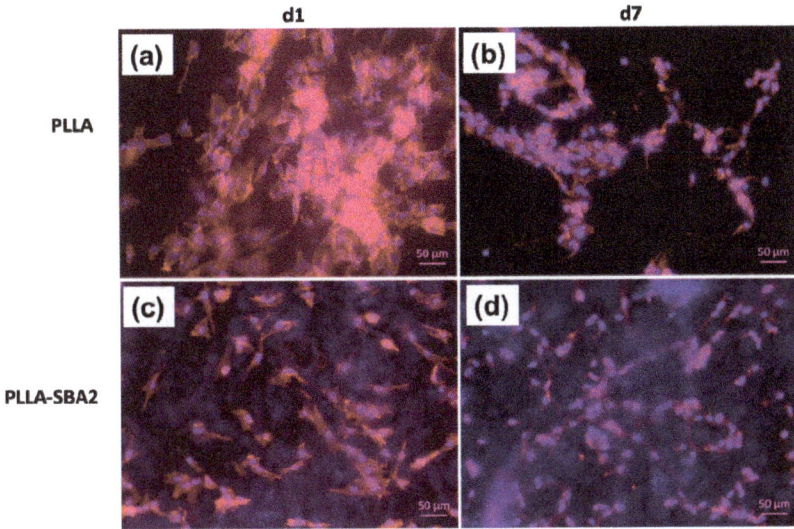

Figure 9. Fluorescent labelled actin filaments (red) and nuclei (blue) in ST2 cells cultured on neat PLLA (**a**,**b**) and PLLA-SBA2 (**c**,**d**) fibres, after 1 day (**a**,**c**) and 7 days from cell seeding. Scale bar: 50 μm.

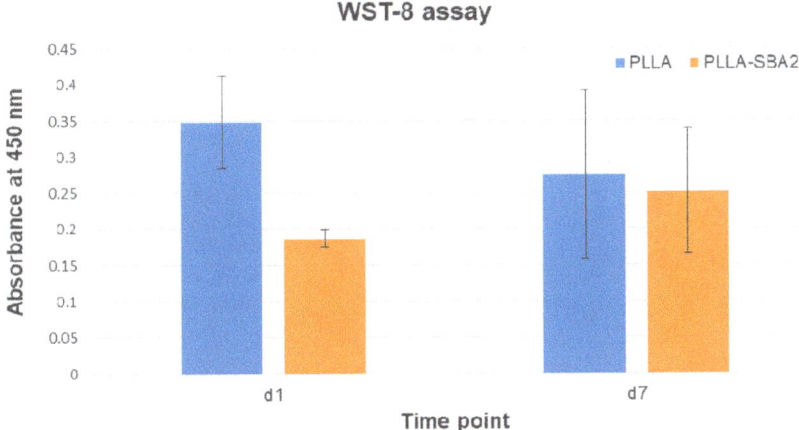

Figure 10. WST-8 assay: histograms of absorbance at 450 nm for PLLA and PLLA-SBA2, one (d1) and seven (d7) days after the seeding.

4. Conclusions

This study investigated the potentiality of electrospun PLLA-SBA2 fibres as potential scaffold material for tissue engineering. The bioactive glass (SBA2) micro-sized powder was effectively incorporated in electrospun polymer filaments and the obtained composite system was characterized in terms of morphology and mechanical properties. The acellular bioactivity and biocompatibility of the PLLA-SBA2 fibres was assessed and HCA deposition after 21 days of immersion in SBF was found to be promoted by the embedded bioactive glass particles at a later stage compared with ion release, suggesting this system as multifunctional scaffold appealing for both bone and soft tissue engineering applications.

Supplementary Materials: The following are available online at http://www.mdpi.com/2079-4991/9/2/182/s1, Figure S1: EDS analysis performed on PLLA-SBA2 sample before (top) and after 21 days of immersion in SBF solution (bottom), Figure S2: SEM micrographs of neat PLLA fibers (a), composite PLLA-SBA2 fibers (b), composite fibers after 7 days in DPBS (c) and 21 days in SBF (d).

Author Contributions: Conceptualization, E.V., A.R.B., D.P. and L.L.; Formal analysis, F.S., M.M. and L.L.; Investigation, F.S., M.M. and L.L.; Methodology, F.S. and L.L.; Resources, E.V., A.R.B. and D.P.; Supervision, A.R.B., D.P. and L.L.; Writing—original draft, F.S.; Writing—review & editing, M.M., E.V., A.R.B., D.P. and L.L.

Acknowledgments: The authors wish to thank G. Baldi (CERICOL Research Centre, Italy) for ball milling and particle size analysis BET facilities.

Conflicts of Interest: The authors declare no conflict of interest.

References

1. Hutmacher, D.W. Scaffolds in tissue engineering bone and cartilage. *Biomaterials* **2000**, *21*, 2529–2543. [CrossRef]
2. Henkel, J.; Woodruff, M.A.; Epari, D.R.; Steck, R.; Glatt, V.; Dickinson, I.C.; Choong, P.F.M.; Schuetz, M.A.; Hutmacher, D.W. Bone Regeneration Based on Tissue Engineering Conceptions—A 21st Century Perspective. *Bone Res.* **2013**, *1*, 216–248. [CrossRef] [PubMed]
3. Wang, M. Developing bioactive composite materials for tissue replacement. *Biomaterials* **2003**, *24*, 2133–2151. [CrossRef]
4. Rezwan, K.; Chen, Q.Z.; Blaker, J.J.; Boccaccini, A.R. Biodegradable and bioactive porous polymer/inorganic composite scaffolds for bone tissue engineering. *Biomaterials* **2006**, *27*, 3413–3431. [CrossRef] [PubMed]
5. Tamayol, A.; Akbari, M.; Annabi, N.; Paul, A.; Khademhosseini, A.; Juncker, D. Fiber-based tissue engineering: Progress, challenges, and opportunities. *Biotechnol. Adv.* **2013**, *31*, 669–687. [CrossRef]
6. Shin, H.; Jo, S.; Mikos, A.G. Biomimetic materials for tissue engineering. *Biomaterials* **2003**, *24*, 4353–4364. [CrossRef]
7. Miola, M.; Fucale, G.; Maina, G.; Verné, E. Antibacterial and bioactive composite bone cements containing surface silver-doped glass particles. *Biomed. Mater.* **2015**, *10*, 055014. [CrossRef]
8. Hoppe, A.; Güldal, N.S.; Boccaccini, A.R. A review of the biological response to ionic dissolution products from bioactive glasses and glass-ceramics. *Biomaterials* **2011**, *32*, 2757–2774. [CrossRef]
9. Xynos, I.D.; Edgar, A.J.; Buttery, L.D.K.; Hench, L.L.; Polak, J.M. Gene-expression profiling of human osteoblasts following treatment with the ionic products of Bioglass® 45S5 dissolution. *J. Biomed. Mater. Res.* **2001**, *55*, 151–157. [CrossRef]
10. Day, R.M.; Boccaccini, A.R.; Shurey, S.; Roether, J.A.; Forbes, A.; Hench, L.L.; Gabe, S.M. Assessment of polyglycolic acid mesh and bioactive glass for soft-tissue engineering scaffolds. *Biomaterials* **2004**, *25*, 5857–5866. [CrossRef]
11. Miguez-Pacheco, V.; Hench, L.L.; Boccaccini, A.R. Bioactive glasses beyond bone and teeth: Emerging applications in contact with soft tissues. *Acta Biomater.* **2015**, *13*, 1–15. [CrossRef] [PubMed]
12. Blaker, J.J.; Gough, J.E.; Maquet, V.; Notingher, I.; Boccaccini, A.R. In vitro evaluation of novel bioactive composites based on Bioglass®-filled polylactide foams for bone tissue engineering scaffolds. *J. Biomed. Mater. Res. Part A* **2003**, *67A*, 1401–1411. [CrossRef] [PubMed]
13. Maquet, V.; Boccaccini, A.R.; Pravata, L.; Notingher, I.; Jérôme, R. Porous poly(α-hydroxyacid)/Bioglass®composite scaffolds for bone tissue engineering. I: Preparation and in vitro characterisation. *Biomaterials* **2004**, *25*, 4185–4194. [CrossRef] [PubMed]
14. Zhang, K.; Wang, Y.; Hillmyer, M.A.; Francis, L.F. Processing and properties of porous poly(l-lactide)/bioactive glass composites. *Biomaterials* **2004**, *25*, 2489–2500. [CrossRef] [PubMed]
15. Allo, B.A.; Rizkalla, A.S.; Mequanint, K. Synthesis and Electrospinning of ε-Polycaprolactone-Bioactive Glass Hybrid Biomaterials via a Sol–Gel Process. *Langmuir* **2010**, *26*, 18340–18348. [CrossRef] [PubMed]
16. Gönen, S.Ö.; Taygun, M.E.; Küçükbayrak, S. Fabrication of Bioactive Glass Containing Nanocomposite Fiber Mats For Bone Tissue Engineering Applications. *Compos. Struct.* **2016**, *138*, 96–106. [CrossRef]
17. Baino, F.; Verné, E.; Vitale-Brovarone, C. Feasibility, tailoring and properties of polyurethane/bioactive glass composite scaffolds for tissue engineering. *J. Mater. Sci. Mater. Med.* **2009**, *20*, 2189–2195. [CrossRef]

18. Jia, W.-T.; Zhang, X.; Luo, S.-H.; Liu, X.; Huang, W.-H.; Rahaman, M.N.; Day, D.E.; Zhang, C.-Q.; Xie, Z.-P.; Wang, J.-Q. Novel borate glass/chitosan composite as a delivery vehicle for teicoplanin in the treatment of chronic osteomyelitis. *Acta Biomater.* **2010**, *6*, 812–819. [CrossRef]
19. Navarro, M.; Ginebra, M.P.; Planell, J.A.; Zeppetelli, S.; Ambrosio, L. Development and cell response of a new biodegradable composite scaffold for guided bone regeneration. *J. Mater. Sci. Mater. Med.* **2004**, *15*, 419–422. [CrossRef]
20. Navarro, M.; Aparicio, C.; Charles-Harris, M.; Ginebra, M.P.; Engel, E.; Planell, J.A. Development of a Biodegradable Composite Scaffold for Bone Tissue Engineering: Physicochemical, Topographical, Mechanical, Degradation, and Biological Properties. In *Ordered Polymeric Nanostructures at Surfaces*; Springer: Berlin/Heidelberg, Germany, 2006; pp. 209–231.
21. Xu, C.; Inai, R.; Kotaki, M.; Ramakrishna, S. Aligned biodegradable nanofibrous structure: A potential scaffold for blood vessel engineering. *Biomaterials* **2004**, *25*, 877–886. [CrossRef]
22. Yang, F.; Murugan, R.; Wang, S.; Ramakrishna, S. Electrospinning of nano/micro scale poly(L-lactic acid) aligned fibers and their potential in neural tissue engineering. *Biomaterials* **2005**, *26*, 2603–2610. [CrossRef] [PubMed]
23. Liu, W.; Thomopoulos, S.; Xia, Y. Electrospun Nanofibers for Regenerative Medicine. *Adv. Healthc. Mater.* **2012**, *1*, 10–25. [CrossRef] [PubMed]
24. Yoshimoto, H.; Shin, Y.M.; Terai, H.; Vacanti, J.P. A biodegradable nanofiber scaffold by electrospinning and its potential for bone tissue engineering. *Biomaterials* **2003**, *24*, 2077–2082. [CrossRef]
25. Hidalgo Pitaluga, L.; Trevelin Souza, M.; Dutra Zanotto, E.; Santocildes Romero, M.E.; Hatton, P.V. Electrospun F18 Bioactive Glass/PCL—Poly (ε-caprolactone)—Membrane for Guided Tissue Regeneration. *Materials* **2018**, *11*, 400. [CrossRef] [PubMed]
26. Sill, T.J.; von Recum, H.A. Electrospinning: Applications in drug delivery and tissue engineering. *Biomaterials* **2008**, *29*, 1989–2006. [CrossRef] [PubMed]
27. Camposeo, A.; Greenfeld, I.; Tantussi, F.; Moffa, M.; Fuso, F.; Allegrini, M.; Zussman, E.; Pisignano, D. Conformational Evolution of Elongated Polymer Solutions Tailors the Polarization of Light-Emission from Organic Nanofibers. *Macromolecules* **2014**, *47*, 4704–4710. [CrossRef] [PubMed]
28. Bognitzki, M.; Frese, T.; Steinhart, M.; Greiner, A.; Wendorff, J.H.; Schaper, A.; Hellwig, M. Preparation of fibers with nanoscaled morphologies: Electrospinning of polymer blends. *Polym. Eng. Sci.* **2001**, *41*, 982–989. [CrossRef]
29. Megelski, S.; Stephens, J.S.; Chase, D.B.; Rabolt, J.F. Micro-and Nanostructured Surface Morphology on Electrospum Polymer Fibers. *Macromolecules* **2002**, *35*, 8456–8466. [CrossRef]
30. Ramakrishna, S.; Fujihara, K.; Teo, W.-E.; Yong, T.; Ma, Z.; Ramaseshan, R. Electrospun nanofibers: Solving global issues. *Mater. Today* **2006**, *9*, 40–50. [CrossRef]
31. Liverani, L.; Roether, J.A.; Boccaccini, A.R. Nanofiber composites in bone tissue engineering. In *Nanofiber Composite Materials for Biomedical Applications*; Ramalingam, M., Ramakrishna, S., Eds.; Elsevier: Amsterdam, The Netherlands, 2017; pp. 301–323.
32. Miola, M.; Bruno, M.; Maina, G.; Fucale, G.; Lucchetta, G.; Vernè, E. Antibiotic-free composite bone cements with antibacterial and bioactive properties. A preliminary study. *Mater. Sci. Eng. C* **2014**, *43*, 65–75. [CrossRef]
33. Miola, M.; Fucale, G.; Maina, G.; Verné, E. Composites bone cements with different viscosities loaded with a bioactive and antibacterial glass. *J. Mater. Sci.* **2017**, *52*, 5133–5146. [CrossRef]
34. Huang, Z.M.; Zhang, Y.Z.; Kotaki, M.; Ramakrishna, S. A review on polymer nanofibers by electrospinning and their applications in nanocomposites. *Compos. Sci. Technol.* **2003**, *63*, 2223–2253. [CrossRef]
35. Misra, S.K.; Mohn, D.; Brunner, T.J.; Stark, W.J.; Philip, S.E.; Roy, I.; Salih, V.; Knowles, J.C.; Boccaccini, A.R. Comparison of nanoscale and microscale bioactive glass on the properties of P(3HB)/Bioglass® composites. *Biomaterials* **2008**, *29*, 1750–1761. [CrossRef] [PubMed]
36. Jo, J.H.; Lee, E.J.; Shin, D.S.; Kim, H.E.; Kim, H.W.; Koh, Y.H.; Jang, J.H. In vitro/in vivo biocompatibility and mechanical properties of bioactive glass nanofiber and poly(ε-caprolactone) composite materials. *J. Biomed. Mater. Res. Part B Appl. Biomater.* **2009**, *91*, 213–220. [CrossRef] [PubMed]
37. Venugopal, J.R.; Low, S.; Choon, A.T.; Bharath Kumar, A.; Ramakrishna, S. Nanobioengineered electrospun composite nanofibers and osteoblasts for bone regeneration. *Artif. Organs* **2008**, *32*, 388–397. [CrossRef] [PubMed]

38. Patlolla, A.; Collins, G.; Livingston Arinzeh, T. Solvent-dependent properties of electrospun fibrous composites for bone tissue regeneration. *Acta Biomater.* **2010**, *6*, 90–101. [CrossRef]
39. Brunauer, S.; Emmett, P.H.; Teller, E. Adsorption of Gases in Multimolecular Layers. *J. Am. Chem. Soc.* **1938**, *60*, 309–319. [CrossRef]
40. Schneider, C.A.; Rasband, W.S.; Eliceiri, K.W. NIH Image to ImageJ: 25 years of image analysis. *Nat Meth* **2012**, *9*, 671–675. [CrossRef]
41. Ricotti, L.; Polini, A.; Genchi, G.G.; Ciofani, G.; Iandolo, D.; Vazão, H.; Mattoli, V.; Ferreira, L.; Menciassi, A.; Pisignano, D. Proliferation and Skeletal Myotube Formation Capability of C2C12 and H9c2 Cells on Isotropic and Anisotropic Electrospun Nanofibrous PHB Scaffolds. *Biomed. Mater.* **2012**, *7*, 035010. [CrossRef]
42. Liverani, L.; Boccaccini, A.R. Versatile production of poly(Epsilon-caprolactone) fibers by electrospinning using benign solvents. *Nanomaterials* **2016**, *6*, 75. [CrossRef]
43. Kokubo, T.; Takadama, H. How useful is SBF in predicting in vivo bone bioactivity? *Biomaterials* **2006**, *27*, 2907–2915. [CrossRef] [PubMed]
44. Theron, S.A.; Zussman, E.; Yarin, A.L. Experimental Investigation of the Governing Parameters in the Electrospinning of Polymer Solutions. *Polymer* **2004**, *45*, 2017–2030. [CrossRef]
45. Lauricella, M.; Cipolletta, F.; Pontrelli, G.; Pisignano, D.; Succi, S. Effects of Orthogonal Rotating Electric Fields on Electrospinning Process. *Phys. Fluids* **2017**, *29*, 082003. [CrossRef]
46. Karageorgiou, V.; Kaplan, D. Porosity of 3D Biomaterial Scaffolds and Osteogenesis. *Biomaterials* **2005**, *26*, 5474–5491. [CrossRef] [PubMed]
47. Kakade, M.V.; Givens, S.; Gardner, K.; Lee, K.H.; Chase, D.B.; Rabolt, J.F. Electric Field Induced Orientation of Polymer Chains in Macroscopically Aligned Electrospun Polymer Nanofibers. *J. Am. Chem. Soc.* **2007**, *129*, 2777–2782. [CrossRef] [PubMed]
48. Arinstein, A.; Burman, M.; Gendelman, O.; Zussman, E. Effect of Supramolecular Structure on Polymer Nanofibre Elasticity. *Nat. Nanotechnol.* **2007**, *2*, 59–62. [CrossRef]
49. Pagliara, S.; Vitiello, M.S.; Camposeo, A.; Polini, A.; Cingolani, R.; Scamarcio, G.; Pisignano, D. Optical Anisotropy in Single Light-Emitting Polymer Nanofibers. *J. Phys. Chem. C* **2011**, *115*, 20399–20405. [CrossRef]
50. Richard-Lacroix, M.; Pellerin, C. Molecular Orientation in Electrospun Fibers: From Mats to Single Fibers. *Macromolecules* **2013**, *46*, 9473–9493. [CrossRef]
51. Hao, J.; Yuan, M.; Deng, X. Biodegradable and Biocompatible Nanocomposites of Poly(ε-Caprolactone) with Hydroxyapatite Nanocrystals: Thermal and Mechanical Properties. *J. Appl. Polym. Sci.* **2002**, *86*, 676–683. [CrossRef]
52. Cerruti, M.; Greenspan, D.; Powers, K. Effect of pH and ionic strength on the reactivity of Bioglass® 45S5. *Biomaterials* **2005**, *26*, 1665–1674. [CrossRef]
53. Schreiber, R. Ca2+ Signaling, Intracellular pH and Cell Volume in Cell Proliferation. *J. Membr. Biol.* **2005**, *205*, 129–137. [CrossRef] [PubMed]
54. Zheng, K.; Solodovnyk, A.; Li, W.; Goudouri, O.-M.; Stähli, C.; Nazhat, S.N.; Boccaccini, A.R. Aging Time and Temperature Effects on the Structure and Bioactivity of Gel-Derived 45S5 Glass-Ceramics. *J. Am. Ceram. Soc.* **2015**, *98*, 30–38. [CrossRef]
55. Aguiar, H.; Serra, J.; González, P.; León, B. Structural Study of Sol–gel Silicate Glasses by IR and Raman Spectroscopies. *J. Non-Cryst. Solids* **2009**, *355*, 475–480. [CrossRef]
56. Zhao, S.; Li, L.; Wang, H.; Zhang, Y.; Cheng, X.; Zhou, N.; Rahaman, M.N.; Liu, Z.; Huang, W.; Zhang, C. Wound Dressings Composed of Copper-Doped Borate Bioactive Glass Microfibers Stimulate Angiogenesis and Heal Full-Thickness Skin Defects in a Rodent Model. *Biomaterials* **2015**, *53*, 379–391. [CrossRef] [PubMed]
57. Day, R.M. Bioactive Glass Stimulates the Secretion of Angiogenic Growth Factors and Angiogenesis in Vitro. *Tissue Eng.* **2005**, *11*, 768–777. [CrossRef] [PubMed]
58. Rahaman, M.N.; Day, D.E.; Bal, B.S.; Fu, Q.; Jung, S.B.; Bonewald, L.F.; Tomsia, A.P. Bioactive Glass in Tissue Engineering. *Acta Biomater.* **2011**, *7*, 2355–2373. [CrossRef] [PubMed]
59. Day, R.M.; Boccaccini, A.R. Effect of Particulate Bioactive Glasses on Human Macrophages and Monocytesin Vitro. *J. Biomed. Mater. Res. Part A* **2005**, *73A*, 73–79. [CrossRef]
60. Xynos, I.D.; Edgar, A.J.; Buttery, L.D.K.; Hench, L.L.; Polak, J.M. Ionic Products of Bioactive Glass Dissolution Increase Proliferation of Human Osteoblasts and Induce Insulin-like Growth Factor II MRNA Expression and Protein Synthesis. *Biochem. Biophys. Res. Commun.* **2000**, *276*, 461–465. [CrossRef]

61. Romanò, C.L.; Logoluso, N.; Meani, E.; Romanò, D.; De Vecchi, E.; Vassena, C.; Drago, L. A Comparative Study of the Use of Bioactive Glass S53P4 and Antibiotic-Loaded Calcium-Based Bone Substitutes in the Treatment of Chronic Osteomyelitis. *Bone Joint J.* **2014**, *96*, 845–850. [CrossRef]
62. Baino, F.; Novajra, G.; Miguez-Pacheco, V.; Vitale-Brovarone, C. Bioactive Glasses: Special Applications Outside the Skeletal System. *J. Non-Cryst. Solids* **2016**, *432*, 15–30. [CrossRef]
63. Xu, Y.; Peng, J.; Dong, X.; Xu, Y.; Li, H.; Chang, J. Combined Chemical and Structural Signals of Biomaterials Synergistically Activate Cell-Cell Communications for Improving Tissue Regeneration. *Acta Biomater.* **2017**, *55*, 249–261. [CrossRef] [PubMed]
64. Busa, W.B.; Nuccitelli, R. Metabolic Regulation via Intracellular PH. *Am. J. Physiol.* **1984**, *246 Pt 2*, R409–R438. [CrossRef] [PubMed]
65. Xynos, I.D.; Hukkanen, M.V.J.; Batten, J.J.; Buttery, L.D.; Hench, L.L.; Polak, J.M. Bioglass®45S5 Stimulates Osteoblast Turnover and Enhances Bone Formation In Vitro: Implications and Applications for Bone Tissue Engineering. *Calcif. Tissue Int.* **2000**, *67*, 321–329. [CrossRef] [PubMed]
66. Bosetti, M.; Zanardi, L.; Hench, L.; Cannas, M. Type I Collagen Production by Osteoblast-like Cells Cultured in Contact with Different Bioactive Glasses. *J. Biomed. Mater. Res.* **2003**, *64A*, 189–195. [CrossRef] [PubMed]

© 2019 by the authors. Licensee MDPI, Basel, Switzerland. This article is an open access article distributed under the terms and conditions of the Creative Commons Attribution (CC BY) license (http://creativecommons.org/licenses/by/4.0/).

Article

In Vitro and In Vivo Studies of Hydrophilic Electrospun PLA95/β-TCP Membranes for Guided Tissue Regeneration (GTR) Applications

Chien-Chung Chen [1,2,†], Sheng-Yang Lee [3,4,†], Nai-Chia Teng [3], Hsin-Tai Hu [3], Pei-Chi Huang [4] and Jen-Chang Yang [5,6,*]

1. Graduate Institute of Biomedical Materials & Engineering, College of Biomedical Engineering, Taipei Medical University, Taipei 110, Taiwan; polyjack@tmu.edu.tw
2. Ph.D. Program in Biotechnology Research and Development, College of Pharmacy, Taipei Medical University, Taipei 110, Taiwan
3. School of Dentistry, College of Oral Medicine, Taipei Medical University, Taipei 110, Taiwan; seanlee@tmu.edu.tw (S.-Y.L.); tengnaichia@hotmail.com (N.-C.T.); hsintaihu1011@hotmail.com (H.-T.H.)
4. Department of Dentistry, Wan-Fang Medical Center, Taipei Medical University, Taipei 116, Taiwan; littlepeichi@yahoo.com.tw
5. Graduate Institute of Nanomedicine and Medical Engineering, College of Biomedical Engineering, Taipei Medical University, Taipei 110, Taiwan
6. International Ph.D. Program in Biomedical Engineering, College of Biomedical Engineering, Taipei Medical University, Taipei 110, Taiwan
* Correspondence: yang820065@tmu.edu.tw; Tel.: +886-2-2736-1661 (ext. 5124)
† Chien-Chung Chen and Sheng-Yang Lee contributed equally to this work.

Received: 17 February 2019; Accepted: 27 March 2019; Published: 11 April 2019

Abstract: The guided tissue regeneration (GTR) membrane is a barrier intended to maintain a space for alveolar bone and periodontal ligament tissue regeneration but prevent the migration of fast-growing soft tissue into the defect sites. This study evaluated the physical properties, in vivo animal study, and clinical efficacy of hydrophilic PLA95/β-TCP GTR membranes prepared by electrospinning (ES). The morphology and cytotoxicity of ES PLA95/β-TCP membranes were evaluated by SEM and 3-(4,5-cimethylthiazol-2-yl)-2,5-diphenyl tetrazolium bromide (MTT) respectively. The cementum and bone height were measured by an animal study at 8 and 16 weeks after surgery. Fifteen periodontal patients were selected for the clinical trial by using a commercial product and the ES PLA95/β-TCP membrane. Radiographs and various indexes were measured six months before and after surgery. The average fiber diameter for this ES PLA95/β-TCP membrane was 2.37 ± 0.86 µm. The MTT result for the ES PLA95/β-TCP membrane showed negative for cytotoxicity. The significant differences in the cementum and bone height were observed between empty control and the ES PLA95/β-TCP membrane in the animal model ($p < 0.05$). Clinical trial results showed clinical attachment level (CAL) of both control and ES PLA95/β-TCP groups, with a significant difference from the pre-surgery results after six months. This study demonstrated that the ES PLA95/β-TCP membrane can be used as an alternative GTR membrane for clinical applications.

Keywords: PLA95; biocompatibility; guided tissue regeneration (GTR); electrospinning

1. Introduction

Periodontitis is one of the most destructive diseases that destroys the tooth-supporting tissues, including the alveolar bone, periodontal ligament, and cementum, ultimately leading to tooth loss [1–3]. For patients with severe periodontitis, it is critical to remove dental calculus and plaque by scaling and root planning [4]. Guided tissue regeneration (GTR) membranes are typically used to block the migration of fast-growing connective tissue into the bony defect and to create space for the regeneration of slow-growing alveolar bone and periodontal ligament [5]. Over the years, the materials of the GTR barrier matrix from non-resorbable polytetrafluoroethylene-like expanded e-PTFE or dense d-PTFE [6] and titanium mesh [7], evolved to resorbable polymer to dispense with the operation of secondary GTR removal [8,9]. Most commercial resorbable synthetic polymer membranes are based on aliphatic polyesters, such as poly(lactic acid) (PLA), poly(glycolic acid) (PGA), poly(ε-caprolactone) (PCL), or their copolymers, to match the resorption time period for various clinical needs [10]. Among these polymers, a novel copolymer composed of poly-5D/95L-lactide (PLA95) has been successfully used in distal radius fractures [11] and craniomaxillofacial applications like skull flap fixation [12] and facial fracture fixation [13] due to its relatively strong mechanical properties. The feasibility of using PLA95 resorbable GTR membranes is worth exploring.

Due to the potential inflammation risk caused by acid release from the monomer or crystalline debris during degradation which could result in a foreign body immune response [14], bio-ceramics such as β-tricalcium phosphate (β-TCP) and hydroxyapatite (HAp) were used for their pH buffering effects [15] and bone cell response enhancement [16]. In recent years, various osteo-conductive membranes such as polylactic acid (PLA)/HAp [17,18], gelatin/HAp [19], three-layered HAp/collagen/PLGA [20], and nano-apatite/polycaprolactone (PCL) [21] have been fabricated using the electrospinning (ES) technique. These resorbable hybrid membranes help bone reconstruction with calcium ions releasing and good pH buffering properties.

In this study, we prepared the PLA95/β-TCP GTR membranes by ES and dip-coating techniques [22], then the safety and effectiveness of this GTR membrane were assayed by cytotoxicity testing, in vivo, and clinical studies.

2. Materials and Methods

2.1. ES PLA95/β-TCP Fibrous Membranes

Poly-5D/95L-lactide (PLA95) was provided by BioTech One Inc. (Taipei, Taiwan) with inherent viscosity (I.V.) of 0.6 dL/g. The solution dope was prepared by mixing 20 w/v% PLA95 used a mixed dichloromethane/dimethylformamide (DCM/DMF: 7/3 (v/v)) solvent and 3% (w/v) β-TCP powders (<23 μm) under an ultrasonic vibrator to prevent the β-TCP powders from agglomerating, then spun it via electrospinning (ES) technique using the setup shown in Figure 1.

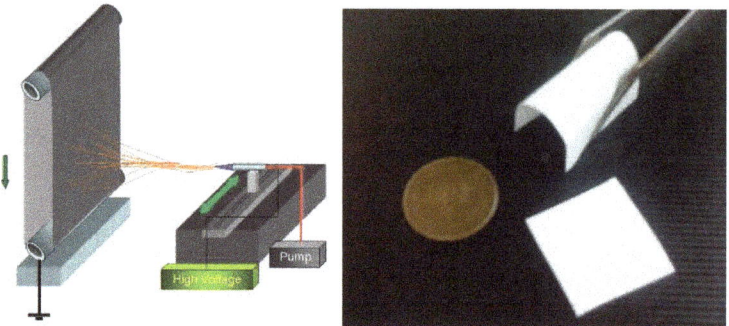

Figure 1. Electrospinning setup and ES PLA95/β-TCP membranes.

The ES PLA95/β-TCP membranes were sterilized by gamma-radiation. The membranes were gold-coated, and their morphology was examined by scanning electron microscopy (SEM, S-2400; Hitachi, Tokyo, Japan), followed by characterization with Image J analytical software (NIH, MD, USA)

2.2. Cytotoxicity Testing

Cytotoxicity testing was performed using the method described in the ISO-10993-5 guideline. Accordingly, the specimens were divided into the following four groups: test (ES PLA95/β-TCP), reagent blank control (medium), negative control (HDPE material), and positive control (zinc sulfate). The samples were extracted with Eagle's minimal essential medium (α-MEM; GIBCO BRL, OK, USA) containing 10% fetal bovine serum (GIBCO BRL, USA) at 37 ± 1 °C for 24 ± 2 h. Cell line (NCTC clone 929; ATCC) was cultured in each of the extraction medium, with 5% CO_2 at 37 °C for 48 h ($N = 3$). A light microscope was used for qualitative morphological grading of the cytotoxicity test findings.

2.3. In Vivo Test (Animal Model)

Four healthy LanYu pigs (weight: 20–25 kg) were used for animal studies. The protocol was approved by Taipei Medical University (No. LAC-99-0087). Buccal mucoperiosteal flaps were reflected in the bilateral mandibular premolar and molar areas. Buccal alveolar bone was reduced to a level 5-mm apical to the cement–enamel junction (CEJ). The root surface was denuded of the periodontal ligament (PDL) and cementum (CE), and notches were placed at the bone level of each root as in Figure 2. The ES PLA95/β-TCP and control membranes were placed on individual bone defect areas without bone grafting. Flaps were positioned and sutured. All LanYu pigs were sacrificed at the designated times after surgery. Histological and histometric evaluation at 8 and 16 weeks were performed after surgery respectively, to determine the healing response of each treatment modality. Hematoxylin and eosin stain (H&E) staining of the demineralized animal sections were evaluated under a light microscope (40×), and the CE and bone height were measured using the Image J software (NIH, MD, USA).

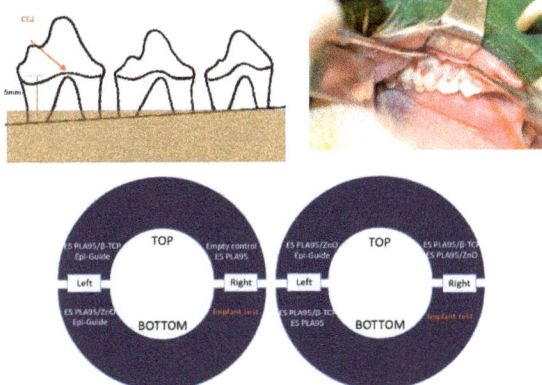

Figure 2. From left to right: Animal model for guided tissue regeneration (GTR) membrane; representative surgery photo and overview of the animal study: the GTR membranes of ES PLA95 and ES PLA95/β-TCP were experimental groups, while the Epi-Guide and empty defect site were used as control groups. Histological and histometric evaluation at week 8 and 16 were performed to determine the healing response of each modality.

2.4. Clinical Trial

A commercially available PLA dental membrane (PLA; Epi-Guide®, Kensey Nash Corp., Exton, PA, USA) was purchased as a control group; while the hydrophilic ES PLA95/β-TCP membranes were used as the experimental group. The protocol was approved by Taipei Medical University Joint Institutional Review Board (TMU-IRB) (No. 201105011). Fifteen periodontal patients with 20 defects were enrolled in the study. The exclusion criteria were patients with unstable vital signs, pregnant women, infection with oral ulcer, participation in other clinical trial, and systemic disease such as leukemia, aplastic anemia, and diabetes. The inclusion criteria were class II or class III furcation defect or intrabony defects with probing depth ≥4 mm in the vertical direction in need of periodontal surgery at the Dental Department of Wan-Fang Hospital. All patients selected for inclusion in the study received a comprehensive periodontal examination and radiographs. The patients were assigned randomly to the control and test groups by the single-blind method. The procedure involved open-flap surgery, scaling and root planning, and additive bone-grafting (particle size 250–500 μm; BonaGraft™, BioTech One Inc., New Taipei City, Taiwan). The GTR membranes were used on the defect sites, the flap was sutured, and patients were instructed on oral health. After 1 week of the surgery, the patients were recalled for adjustment and evaluation and scheduled for follow-up every 4 weeks. Clinical indices such as probing depth (PD), plaque index (PI), gingival index (GI), bleeding on probing (BOP), gingival recession (GR), mobility (MOB), and clinical attachment level (CAL) were assessed 6 months after the surgery.

2.5. Statistical Analysis

All data are expressed as mean ± standard deviation (SD). For the in vivo test, the Student's t-test was used. In the clinical trial, the clinical indices (PD, PI, GI, BOP, GR, MOB, and CAL) were compared by Wilcoxon signed-rank test. Statistical differences between the control and test groups was analyzed by Mann–Whitney U test, and differences were considered statistically significant when the *p*-value was <0.05.

3. Results and Discussion

3.1. Morphology

The morphology of ES PLA95/β-TCP fibrous membranes are shown in Figure 3. The average fiber diameter for this ES PLA95/β-TCP membrane was 2.37 ± 0.86 μm. The ES PLA95/β-TCP membranes were prepared using dip-coating technique with a dimension of 3.0 × 4.0 cm^2 (width × length), thickness of 0.3–0.4 mm, suture pull-out force of >200 gf, average porosity of 53.0 ± 4.5%, and average pore size of 25.0 ± 1.0 μm. Unlike the ES PLA95/β-TCP-N with contact angle of 122.6 ± 0.1°, the PEO (polyethylene oxide) dip-coated ES PLA95/β-TCP-T revealed the contact angle of 50.7 ± 0.2° [22]. The hydrophilic surface would help cell adhesion to prevent the membrane expose and avoid infection during healing process.

The typical criteria for ideal GTR membranes are known as cell-occlusive, space making, tissue integrative, clinically manageable and biocompatible [23]. Among the fabrication processes electrospinning, a versatile physical processing technology that does not affect the inherent material properties, has substantially more advantages to manufacture membranes for biomedical and tissue-engineering applications due to their high surface area-to-volume ratio, porosity, and three-dimensional (3-D) structure to mimic an extracellular matrix for enhancing cell-surface interactions [24]. The fiber morphology and diameter of electro-spun poly lactic acid (PLA) fibers were mainly affected by solution properties and process parameters [25,26]. In recent years, growth factors [27–29], doxycycline [30,31], and bio-ceramic materials [32,33] have been incorporated into the GTR membranes to improve bioactivity and antibacterial properties. Hydroxyapatite (HAp), β-tricalcium phosphate (β-TCP), and calcium sulfate (CaSO$_4$), are osteo-conductive bio-ceramics additives that are widely used in orthopedic and dental applications.

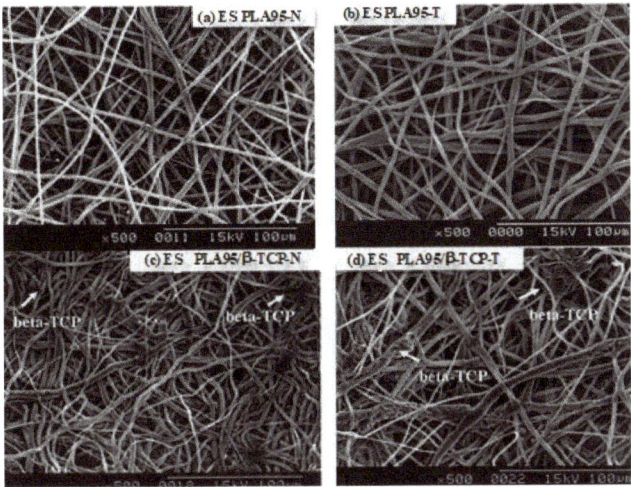

Figure 3. Scanning electron micrographs of the four electro-spun fibrous membranes: (**a**) ES PLA95-N, (**b**) ES PLA95-T, (**c**) ES PLA95/β-TCP-N and (**d**) ES PLA95/β-TCP-T. (N: Without polyethylene oxide (PEO) dip-coating treatment, T: PEO dip-coating treatment). Images courtesy of [22].

The control group chosen in this study is Epi-Guide®, a porous three-layer self-supporting poly-D,L-lactic acid (DL-PLA) barrier for up to 20 weeks with complete bio-resorption between 6–12 months. The Epi-Guide® barrier is claimed to be a hydrophilic membrane that quickly absorbs blood fluid and supports healthy clot formation to maintain gingival flap viability and coverage.

3.2. Cytotoxicity Test

As a surgical implant, it is important to verify the biocompatible properties of the ES PLA95/β-TCP membrane. In a previous study, the results indicated that the MC-3T3-E1 cells could adhere and proliferate on the surface of ES PLA95/β-TCP membrane [22]. Therefore, we alternatively evaluated the cytotoxicity of the ES PLA95/β-TCP membrane to ensure the safety of the manufacturing process as well as the PLA95 material. In this study, the MEM elution assay was used to verify the cytotoxicity of the ES PLA95/β-TCP membrane in accordance with the ISO-10993-5 guideline. The results revealed no cell lysis in the membrane extract, reagent blank, and the negative control extracts under light microscope. Thus, it was inferred that the ES PLA95/β-TCP membrane did not exhibit cytotoxic reactivity (Table 1).

Table 1. Cytotoxicity assay of ES PLA95/β-TCP membrane by indirect method.

Extracts of Test Item and Controls	Observation	Grade	Reactivity
Membrane extract, reagent blank, and negative control extract ($N = 3$)	Discrete intracytoplasmic granules; no cell lysis; no reduction of cell growth	0	None
Positive control ($N = 3$)	Complete destruction of the cell layers	4	Severe

Note: (Grade 0: no cell lysis or reduction of cell growth, Grade 1: not more than 20% of the cells are round, Grade 2: not more than 50% of the cells are round and devoid of intra-cytoplasmic granules, Grade 3: not more than 70% of the cell layers contain rounded cells or are lysed, Grade 4: nearly complete or complete destruction of the cell layers).

3.3. In Vivo Test (Animal Model)

There were no severe inflammations and swellings at the flaps in the defect. The periodontal tissues were healthy on the day of sacrifice. The GTR animal model for cementum and bone height were H&E stained and observed under a light microscope (Figure 4).

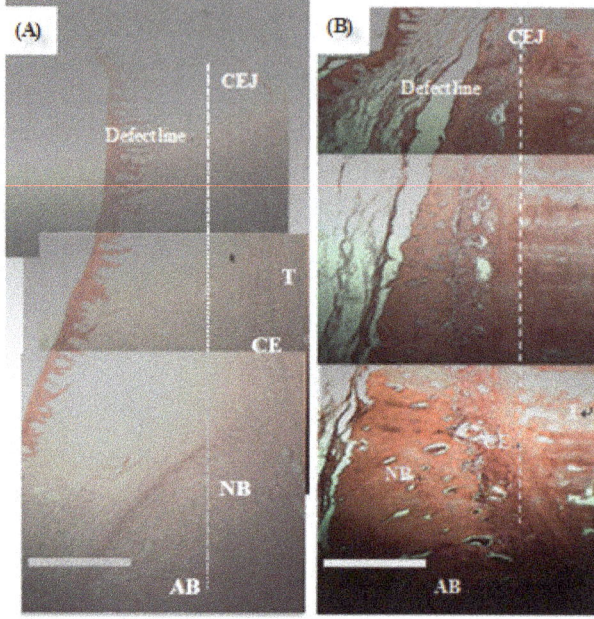

Figure 4. The GTR animal model for cementum and bone height evaluation of (**A**) empty control group and (**B**) representative experimental group. NB: New bone, T: Tooth, AB: Alveolar bone, CE: Cementum, CEJ: Cementoenamel junction. White scale bar noted as 1 mm. Time of implantation: 8 weeks.

The histological results of various 8-week ES PLA95/β–TCP GTR membranes were shown in Figure 5. The histometric evaluation were carried out and listed in Table 2 and Figure 6. The cementum height of the test and control membranes was significantly different between the empty defect after 8 and 16 weeks ($p < 0.05$). The results of bone height showed difference only at 16 weeks ($p > 0.05$) with respective test and control values of 2.67 (±0.33) mm and 2.58 (±0.15) mm; these values were significantly different from the corresponding value of the empty group at 16 weeks ($p < 0.05$). The ES PLA95/β-TCP membrane was effective to block the migration of fast-growing connective tissue into the defect area and in creating some space for the regeneration of new tissues.

In a typical GTR animal study, the usage of bone grafts usually interferes with the efficacy of the GTR membrane [34,35]. Therefore, we intentionally adopted the similar procedure but without bone grafts to confirm the blocking function of the proposed GTR membrane in this study. The results, however, still showed significant differences between the experimental and empty control groups. In addition, few previous studies have shown more bone formation after GTR as compared to that in the empty control without membrane [36,37]. In this study, the cementum height of the experimental group was larger than that of the empty control group without bone grafts at eight weeks. From these results, we conjectured that the addition of bone graft would affect the growth of new bone tissues. The ES PLA95/β-TCP membrane and Epi-Guide® groups were all with similar PLA materials and mechanical properties, with a limited amount of β-TCP content for buffering properties. Therefore, the osteoconductive effect on tissue growth was not significantly different from ES PLA95/β-TCP membrane and Epi-Guide® groups.

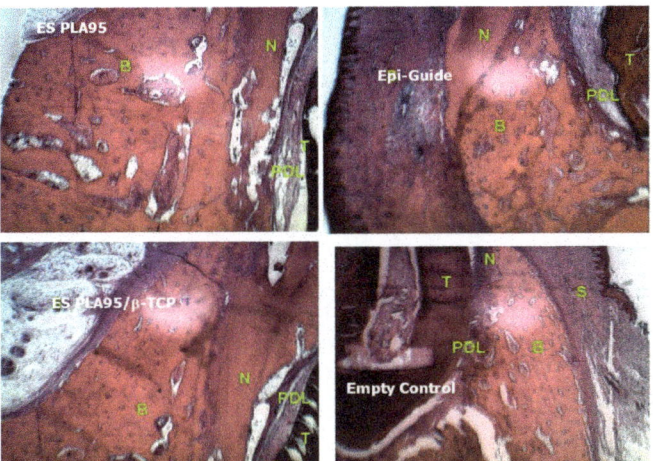

Figure 5. The 8-week H&E stain of various ES PLA95/β-TCP GTR barriers (H&E Stain, 40X) T: Tooth, B: Cementum, N: New bone, PDL: Periodontal ligament.

Table 2. The GTR model results of cementum and bone height measurements with empty control and ES PLA95/β-TCP membranes in LanYu pigs after 8 and 16 weeks.

Time	Membranes	Cementum Height (Mean ± SD) mm	Bone Height (Mean ± SD) mm
8 weeks	Empty control	0.82 ± 0.15	0.98 ± 0.12
	ES PLA95/β-TCP	1.82 ±0.38	1.42 ± 0.37
	Epi-Guide® membrane	1.65 ± 0.28	1.25 ± 0.36
16 weeks	Empty control	1.18 ± 0.22 *,#	1.28 ± 0.33 *,#
	ES PLA95/β-TCP	3.67 ± 0.42 *	2.67 ± 0.33 *
	Epi-Guide® membrane	3.42 ±0.33 #	2.58 ± 0.36 #

*,#: Statistical significant differences when $p < 0.05$. Values are reported as mean (SD) ($N = 3$).

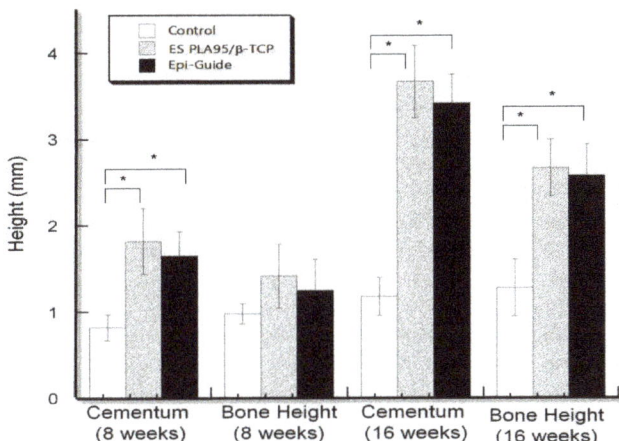

Figure 6. The cementum and bone height of ES PLA95/β-TCP and commercial membranes were measured in LanYu pigs after 8 and 16 weeks. *Differences were considered statistically significant when $p < 0.05$.

3.4. Clinical Trial

All patients selected for inclusion in the study received a comprehensive periodontal examination and radiographs. The study patients were comprised of seven men and eight women with an age range of 35-65 years (mean age: 53 years). Among the seven patients who used Epi-Guide®, two patients collected more than one site due to their severity in periodontal disease (7 + 2 = 9 results). Among the eight patients who used ES PLA95/β-TCP GTR membrane, three patients collected more than one site due to their severity in periodontal disease (8 + 3 = 11 results). In patient response, four patients of the control group had a sore tooth at the surgical sites six months after the surgery, while the ES PLA95/β-TCP membrane group did not. In clinical observation, ES PLA95/β-TCP membrane did not show early exposure, implying that the hydrophilic membrane might help gingival tissue adhesion. Six months after the surgery, the clinical indices of each site were measured and re-recorded, such as PI, GI, PI, GI, BOP, PD, GR, MOB, and CAL. Furthermore, pre-surgery and post-surgery radiographies were observed, and reconstruction of bony defect was compared (Figure 7). Several new bone tissues were detected in the defect, indicating that the surgery had good bone-material compatibility outcome.

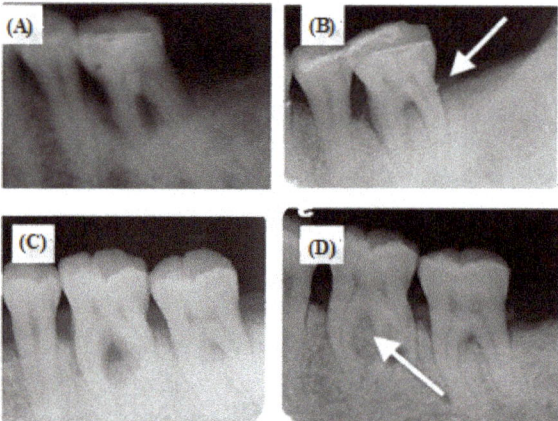

Figure 7. The patient A is with (**A**) pre-surgical radiograph (Tooth No. 36), (**B**) post-surgical radiograph with Epi-Guide® membrane after 6 months. The patent B is with (**C**) pre-surgical radiograph (Tooth No. 37), (**D**) post-surgical radiograph with ES PLA95/β-TCP membrane after 6 months. Arrow (white) indicates the bone regeneration.

The clinical indices for the commercial Epi-Guide® group in PD, GI, BOP, and CAL, showed statistically significant differences ($p < 0.05$; Table 3) after treatment, while the experimental group of the ES PLA95/β-TCP membrane showed statistically significant differences for PD, GI, GR, and CAL indices ($p < 0.05$; Table 4) after treatment. The results showed significantly more attachment gain (Epi-Guide®, 2 mm; PLA95/β-TCP GTR group, 3 mm; $p = 0.28$) and shallower probing depths (Epi-Guide®, 3.3 mm; PLA95/β-TCP GTR group, 2.25 mm; $p = 0.85$) than the empty control group. The change of clinical indices indicated direct improvement of periodontal inflammation and the efficacy for both the Epi-Guide® and the ES PLA95/β-TCP GTR membrane groups. However, these results did not show statistically significant differences between them (Table 5).

Table 3. Clinical indices before and after treatment for Epi-Guide® group by Wilcoxon signed-rank test.

Clinical Index	N	Before	After	p-Value
		Median (Q1–Q3)		
Probing depth (PD) mm	9	6.3 (6–7)	3 (2.7–4.5)	0.0039 *
Plaque index (PI)%	9	16 (10–16)	16 (16–16)	0.62
Gingival index (GI)	9	1 (1–1)	0.5 (0.5–0.5)	0.015 *
Bleeding on probing (BOP)%	9	17 (16–33)	0 (0–16)	0.031 *
Gingival recession (GR) mm	9	0 (0–0)	2 (0–2)	0.25
Mobility (MOB)	9	0 (0–1)	0 (0–1)	1.00
Clinical attachment level (CAL = PD + GR) mm	9	7 (7–8)	5 (4–5)	0.0039 *

*: Differences were considered statistically significant when $p < 0.05$.

Table 4. Clinical indices before and after treatment for ES PLA95/β-TCP group by Wilcoxon signed-rank test.

Clinical Index	N	Before	After	p-Value
		Median (Q1–Q3)		
Probing depth (PD) mm	11	5.75 (5–8)	3.5 (3–5)	0.001*
Plaque index (PI)%	11	16 (10–16)	10 (0–16)	0.125
Gingival index (GI)	11	1 (1–1)	0.5 (0–0.5)	0.0098*
Bleeding on probing (BOP)%	11	17 (16–50)	16 (0–16)	0.1250
Gingival recession (GR) mm	11	0 (0–2)	2 (1–3)	0.0078*
Mobility (MOB)	11	0 (0–1)	0 (0–1)	0.7500
Clinical attachment level (CAL = PD + GR) mm	11	8 (6–9)	5 (5–8)	0.0020*

*: Differences were considered statistically significant when $p < 0.05$.

Table 5. Itemized clinical index differences between the Epi-Guide® and ES PLA95/β-TCP groups after treatment.

Clinical Index	Epi-Guide® Group				ES PLA95/β-TCP Group				p-Value
	N	Median	Q1	Q3	N	Median	Q1	Q3	
Probing depth (PD) mm	9	2	3.7	2	11	2.5	3	1.5	0.85
Plaque index (PI)%	9	0	0	0	11	0	16	0	0.33
Gingival index (GI)	9	0.5	0.5	0.5	11	0.5	0.5	0.5	0.24
Bleeding on probing(BOP) %	9	16	17	0	11	0	34	0	0.60
Gingival recession (GR) mm	9	0	0	2	11	1	0	2	0.31
Mobility (MOB)	9	0	0	0	11	0	0	0	1.00
Clinical attachment level (CAL = PD + GR) mm	9	2	4	2	11	2	3	1	0.28

Mann–Whitney U test, *: Differences were considered statistically significant when $p < 0.05$.

In the clinical study, a few patients of the control group showed soreness at the surgical sites 6 months after the surgery, while the ES PLA95/β-TCP membrane did not. We conjectured that the small amount of β-TCP in this membrane acts as a buffer to reduce the acid releasing during the hydrolysis of ES PLA95 membrane. Therefore, the ES technique is suitable for manufacturing the ES PLA95/β-TCP GTR membrane.

4. Conclusions

In this study, ES PLA95/β-TCP membranes were prepared by ES technology. Their effectiveness and safety with regards to cytotoxicity, in vivo animal, and clinical studies were investigated. The ES PLA95/β-TCP membrane did not show cytotoxicity, nor did it result in any inflammation. Significant difference was observed in cementum and bone height before and after surgery using the ES

PLA95/β-TCP membrane in animal study. Furthermore, the ES PLA95/β-TCP membrane have a hydrophilic property would prevent early exposure and healing efficacy in this study. In intrabony defects, the use of Epi-Guide® or ES PLA95/β-TCP membranes in GTR procedures yielded comparable clinical results in reducing the probing depth and increasing attachment gain for periodontal patients. The results extended the data bank of resorbable polymer for medical applications where contradictory use of current commercial solution due to clinical condition/preexisting condition of patients

Author Contributions: The contributions for each author are as follows: Conceptualization by S.-Y.L. and J.-C.Y.; methodology by N.-C.T.; validation by C.-C.C. and J.-C.Y.; investigation by H.-T.H.; clinical study and data curation by P.-C.H.; writing—original draft preparation by C.-C.C. and H.-T.H.; writing—review and editing by J.-C.Y.; project administration by S.-Y.L.

Funding: This research was funded by National Science Council (NSC) under grant number of NSC99-2321-B038-002.

Acknowledgments: The bone grafts (BonaGraft™) used in this clinical study were kindly donated by BioTech One Inc., Taiwan.

Conflicts of Interest: The authors declare no conflicts of interest.

References

1. Slots, J.; MacDonald, E.S.; Nowzari, H. Infectious aspects of periodontal regeneration. *Periodontol 2000* **1999**, *19*, 164–172. [CrossRef] [PubMed]
2. Haffajee, A.D.; Socransky, S.S. Microbial etiological agents of destructive periodontal diseases. *Periodontol 2000* **1994**, *5*, 78–111. [CrossRef] [PubMed]
3. Pihlstrom, B.L.; Michalowicz, B.S.; Johnson, N.W. Periodontal diseases. *Lancet* **2005**, *366*, 1809–1820. [CrossRef]
4. Stefanac, S.J. The Systemic Phase of Treatment. In *Treatment Planning in Dentistry (Second Edition)*; Stefanac, S.J., Nesbit, S.P., Eds.; Mosby: Saint Louis, MO, USA, 2007; Chapter 5; pp. 91–111. [CrossRef]
5. Buser, D.; Dahlin, C.; Schenk, R.K. *Guided Bone Regeneration in Implant Dentistry*; Quintessence Books: Batavia, IL, USA, 1994.
6. Ronda, M.; Rebaudi, A.; Torelli, L.; Stacchi, C. Expanded vs. dense polytetrafluoroethylene membranes in vertical ridge augmentation around dental implants: A prospective randomized controlled clinical trial. *Clin. Oral Implant. Res.* **2014**, *25*, 859–866. [CrossRef] [PubMed]
7. Buser, D.; Ruskin, J.; Higginbottom, F.; Hardwick, R.; Dahlin, C. Osseointegration of titanium implants in bone regenerated in membrane-protected defects: A histologic study in the canine mandible. *Implant Dent.* **1996**, *5*, 212. [CrossRef]
8. Zhang, Y.; Zhang, X.; Shi, B.; Miron, R.J. Membranes for guided tissue and bone regeneration. *Ann. Oral Maxillofac. Surg.* **2013**, *1*, 10. [CrossRef]
9. Jo, Y.-Y.; Oh, J.-H. New Resorbable Membrane Materials for Guided Bone Regeneration. *Appl. Sci.* **2018**, *8*, 2157. [CrossRef]
10. Wang, J.; Wang, L.; Zhou, Z.; Lai, H.; Xu, P.; Liao, L.; Wei, J. Biodegradable Polymer Membranes Applied in Guided Bone/Tissue Regeneration: A Review. *Polymers* **2016**, *8*, 115. [CrossRef]
11. Chang, I.-L.; Yang, J.-C.; Lee, S.-Y.; Wang, L.-F.; Hu, H.-T.; Yang, S.-Y.; Lin, H.-C. Early clinical experience with resorbable poly-5D/95L-lactide (PLA95) plate system for treating distal radius fractures. *J. Dent. Sci.* **2013**, *8*, 44–52. [CrossRef]
12. Lin, J.W.; Lin, C.-M.; Chiu, W.T. Clinical Experience with Bioresorbable Plates for Skull Flap Fixation. *J. Dent. Sci.* **2006**, *1*, 187–194.
13. Lee, S.-Y.; Wang, L.-F.; Wang, Y.-C.; Chiou, S.-Y.; Lee, S.-A.; Yang, J.-C. Treatment of facial fractures using poly-5D/95L-lactide (PLA95) copolymer plates and screws. *J. Dent.* **2008**, *3*, 150–158.
14. Ignatius, A.A.; Claes, L.E. In vitro biocompatibility of bioresorbable polymers: Poly(L, DL-lactide) and poly(L-lactide-co-glycolide). *Biomaterials* **1996**, *17*, 831–839. [CrossRef]
15. Kikuchi, M.; Koyama, Y.; Takakuda, K.; Miyairi, H.; Shirahama, N.; Tanaka, J. In vitro change in mechanical strength of β-tricalcium phosphate/copolymerized poly-L-lactide composites and their application for guided bone regeneration. *J. Biomed. Mater. Res.* **2002**, *62*, 265–272. [CrossRef]

16. Kikuchi, M.; Koyama, Y.; Yamada, T.; Imamura, Y.; Okada, T.; Shirahama, N.; Akita, K.; Takakuda, K.; Tanaka, J. Development of guided bone regeneration membrane composed of β-tricalcium phosphate and poly (l-lactide-co-glycolide-co-ε-caprolactone) composites. *Biomaterials* **2004**, *25*, 5979–5986. [CrossRef]
17. Sui, G.; Yang, X.; Mei, F.; Hu, X.; Chen, G.; Deng, X.; Ryu, S. Poly-L-lactic acid/hydroxyapatite hybrid membrane for bone tissue regeneration. *J. Biomed. Mater. Res. A* **2007**, *82*, 445–454. [CrossRef]
18. Jeong, S.I.; Ko, E.K.; Yum, J.; Jung, C.H.; Lee, Y.M.; Shin, H. Nanofibrous Poly(lactic acid)/Hydroxyapatite Composite Scaffolds for Guided Tissue Regeneration. *Macromol. Biosci.* **2008**, *8*, 328–338. [CrossRef]
19. Kim, H.-W.; Song, J.-H.; Kim, H.-E. Nanofiber Generation of Gelatin–Hydroxyapatite Biomimetics for Guided Tissue Regeneration. *Adv. Funct. Mater.* **2005**, *15*, 1988–1994. [CrossRef]
20. Wu, S.; Xiao, Z.; Song, J.; Li, M.; Li, W. Evaluation of BMP-2 Enhances the Osteoblast Differentiation of Human Amnion Mesenchymal Stem Cells Seeded on Nano-Hydroxyapatite/Collagen/Poly(l-Lactide). *Int. J. Mol. Sci.* **2018**, *19*, 2171. [CrossRef]
21. Yang, F.; Both, S.K.; Yang, X.; Walboomers, X.F.; Jansen, J.A. Development of an electrospun nano-apatite/PCL composite membrane for GTR/GBR application. *Acta Biomater.* **2009**, *5*, 3295–3304. [CrossRef]
22. Hu, H.-T.; Lee, S.-Y.; Chen, C.-C.; Yang, Y.-C.; Yang, J.-C. Processing and properties of hydrophilic electrospun polylactic acid/beta-tricalcium phosphate membrane for dental applications. *Polym. Eng. Sci.* **2013**, *53*, 833–842. [CrossRef]
23. Gorgieva, S.; Tomaž, V.; Vanja, K. Processing of Biopolymers-Based, Efficiently Integrated Bilayer Membranes for Use in Guided Tissue Regeneration Procedures. *IOP Conf. Ser. Mater. Sci. Eng.* **2018**, *460*, 012036.
24. Senthamizhan, A.; Balusamy, B.; Uyar, T. 1—Electrospinning: A versatile processing technology for producing nanofibrous materials for biomedical and tissue-engineering applications. In *Electrospun Materials for Tissue Engineering and Biomedical Applications*; Uyar, T., Kny, E., Eds.; Woodhead Publishing: Witney, Oxford, UK, 2017; pp. 3–41.
25. Theron, S.A.; Zussman, E.; Yarin, A.L. Experimental investigation of the governing parameters in the electrospinning of polymer solutions. *Polymer* **2004**, *45*, 2017–2030. [CrossRef]
26. Ramakrishna, S.; Lim, Te.; Fujihara, K.; Teo, W.E.; Ma, Z. *An Introduction to Electrospinning and Nanofibers*; World Scientific: Singapore, 2005.
27. Jung, R.E.; Glauser, R.; Scharer, P.; Hammerle, C.H.; Sailer, H.F.; Weber, F.E. Effect of rhBMP-2 on guided bone regeneration in humans. *Clin. Oral Implant. Res.* **2003**, *14*, 556–568. [CrossRef]
28. Wikesjo, U.M.; Qahash, M.; Thomson, R.C.; Cook, A.D.; Rohrer, M.D.; Wozney, J.M.; Hardwick, W.R. rhBMP-2 significantly enhances guided bone regeneration. *Clin. Oral Implant. Res.* **2004**, *15*, 194–204. [CrossRef]
29. Yin, L.; Yang, S.; He, M.; Chang, Y.; Wang, K.; Zhu, Y.; Liu, Y.; Chang, Y.; Yu, Z. Physicochemical and biological characteristics of BMP-2/IGF-1-loaded three-dimensional coaxial electrospun fibrous membranes for bone defect repair. *J. Mater. Sci. Mater. Med.* **2017**, *28*, 94. [CrossRef]
30. Chaturvedi, R.; Gill, A.S.; Sikri, P. Evaluation of the regenerative potential of 25% doxycycline-loaded biodegradable membrane vs biodegradable membrane alone in the treatment of human periodontal infrabony defects: A clinical and radiological study. *Indian J. Dent. Res.* **2008**, *19*, 116–123. [CrossRef]
31. Lyons, L.C.; Weltman, R.L.; Moretti, A.J.; Trejo, P.M. Regeneration of degree ii furcation defects with a 4% doxycycline hyclate bioabsorbable barrier. *J. Periodontol.* **2008**, *79*, 72–79. [CrossRef]
32. Zafiropoulos, G.K.; Hoffmann, O.; Kasaj, A.; Willershausen, B.; Weiss, O.; Van Dyke, T.E. Treatment of Intrabony Defects Using Guided Tissue Regeneration and Autogenous Spongiosa Alone or Combined With Hydroxyapatite/beta-Tricalcium Phosphate Bone Substitute or Bovine-Derived Xenograft. *J. Periodontol.* **2007**, *78*, 2216–2225. [CrossRef]
33. Schwarz, F.; Herten, M.; Ferrari, D.; Wieland, M.; Schmitz, L.; Engelhardt, E.; Becker, J. Guided bone regeneration at dehiscence-type defects using biphasic hydroxyapatite + beta tricalcium phosphate (Bone Ceramic) or a collagen-coated natural bone mineral (BioOss Collagen): An immunohistochemical study in dogs. *Int. J. Oral Maxillofac. Surg.* **2007**, *36*, 1198–1206. [CrossRef]
34. Coonts, B.A.; Whitman, S.L.; O'Donnell, M.; Polson, A.M.; Bogle, G.; Garrett, S.; Swanbom, D.D.; Fulfs, J.C.; Rodgers, P.W.; Southard, G.L.; et al. Biodegradation and biocompatibility of a guided tissue regeneration barrier membrane formed from a liquid polymer material. *J. Biomed. Mater. Res.* **1998**, *42*, 303–311. [CrossRef]
35. Da Silva Pereira, S.L.; Sallum, A.W.; Casati, M.Z.; Caffesse, R.G.; Weng, D.; Nociti, F.H., Jr.; Sallum, E.A. Comparison of bioabsorbable and non-resorbable membranes in the treatment of dehiscence-type defects. A histomorphometric study in dogs. *J. Periodontol.* **2000**, *71*, 1306–1314. [CrossRef]

36. Magnusson, I.; Batich, C.; Collins, B.R. New Attachment Formation Following Controlled Tissue Regeneration Using Biodegradable Membranes. *J. Periodontol.* **1988**, *59*, 1–6. [CrossRef]
37. Pitaru, S.; Tal, H.; Soldinger, M.; Grosskopf, A.; Noff, M. Partial Regeneration of Periodontal Tissues Using Collagen Barriers. *J. Periodontol.* **1988**, *59*, 380–386. [CrossRef]

© 2019 by the authors. Licensee MDPI, Basel, Switzerland. This article is an open access article distributed under the terms and conditions of the Creative Commons Attribution (CC BY) license (http://creativecommons.org/licenses/by/4.0/).

Article

Biofunctional Nanofibrous Substrate for Local TNF-Capturing as a Strategy to Control Inflammation in Arthritic Joints

Elisa Bacelo [1,2], Marta Alves da Silva [1,2], Cristina Cunha [2,3], Susana Faria [4], Agostinho Carvalho [2,3], Rui L. Reis [1,2,5], Albino Martins [1,2,*] and Nuno M. Neves [1,2,5]

1. 3B's Research Group, I3Bs—Research Institute of Biomaterials, Biodegradables and Biomimetics, Headquarters of the European Institute of Excellence on Tissue Engineering and Regenerative Medicine, AvePark—Parque de Ciência e Tecnologia, Zona Industrial da Gandra, University of Minho, Barco, 4805-017 Guimarães, Portugal; elisabacelo@live.com.pt (E.B.); estranged28@gmail.com (M.A.d.S.); rgreis@i3bs.uminho.pt (R.L.R.); nuno@i3bs.uminho.pt (N.M.N.)
2. ICVS/3B's—PT Government Associate Laboratory, Barco, 4805-017 Guimarães, Portugal; cristinacunha@med.uminho.pt (C.C.); agostinhocarvalho@med.uminho.pt (A.C.)
3. Life and Health Sciences Research Institute, Scholl of Medicine, Campus of Gualtar, University of Minho, 4710-057 Braga, Portugal
4. Department of Mathematics for Science and Technology Research CMAT, Campus of Azurém, University of Minho, 4800-058 Guimarães, Portugal; sfaria@math.uminho.pt
5. The Discoveries Centre for Regenerative and Precision Medicine, Headquarters at University of Minho, Avepark, Barco, 4805-017 Guimarães, Portugal
* Correspondence: amartins@i3bs.uminho.pt; Tel.: +351-253-510-900

Received: 19 February 2019; Accepted: 28 March 2019; Published: 8 April 2019

Abstract: Rheumatoid arthritis (RA) is an autoimmune disease that affects the synovial cavity of joints, and its pathogenesis is associated with an increased expression of pro-inflammatory cytokines, namely tumour necrosis factor-alpha (TNF-α). It has been clinically shown to have an adequate response to systemic administration of TNF-α inhibitors, although with many shortcomings. To overcome such limitations, the immobilization of a TNF-α antibody on a nanofibrous substrate to promote a localized action is herein proposed. By using this approach, the antibody has its maximum therapeutic efficacy and a prolonged therapeutic benefit, avoiding the systemic side-effects associated with conventional biological agents' therapies. To technically achieve such a purpose, the surface of electrospun nanofibers is initially activated and functionalized, allowing TNF-α antibody immobilization at a maximum concentration of 6 µg/mL. Experimental results evidence that the biofunctionalized nanofibrous substrate is effective in achieving a sustained capture of soluble TNF-α over time. Moreover, cell biology assays demonstrate that this system has no deleterious effect over human articular chondrocytes metabolism and activity. Therefore, the developed TNF-capturing system may represent a potential therapeutic approach for the local management of severely affected joints.

Keywords: antibody immobilization; electrospun nanofibers; TNF-α capture; human articular chondrocytes; rheumatoid arthritis

1. Introduction

Rheumatoid arthritis (RA) is an autoimmune disease that affects primarily the synovial cavity tissues on the small diarthrodial joints of the hands and feet [1], leading to persistent synovial inflammation [2] and progressive erosion of the articular structures [3,4]. The triggers for the disease susceptibility and the pathological cascade of events encompass environmental, genetic and stochastic factors [3,4]. For example, RA has special incidence on females [5], which is related with genetic [5–7] and hormonal factors [5,8–10], but also presents a heterogeneous geographic distribution, being less common in developing countries [11,12], which confirms the involvement of environmental and socio-cultural factors [12].

The disease pathophysiology involves several interconnected mechanisms. Specific antibodies for immunoglobulin G (IgG) mediate the autoimmune process, as well as the imbalanced expression of the pro-inflammatory cytokine's profile and its functionality [1,12–14]. Consequently, this leads to inflammatory processes, autoimmunity enhancement, long-term inflammatory synovitis and joint damage [13,15]. Additionally, locally expressed degradative enzymes digest the cartilaginous matrix and destroy the articular surfaces [1]. Infiltration of B cells, CD4+ T cells and macrophages into the synovium, which in normal conditions is relatively acellular [1,14,16], leads to soft tissue oedema and stiffness [2]. Moreover, other inflammatory cells such as neutrophils, natural killer cells and mast cells play key roles in the disease's progression [13].

From the autoimmune disease (AD) subsets that entail the multiple inflammatory cascades, one of the most significant players is the TNF-α, particularly the over-expression of the TNF-α [11,13,17]. TNF-α, a 233 amino acid protein [18], is a key signaling cytokine in the immune system mainly produced by monocytes, macrophages [19], and B and T cells [2,13,17]. TNF-α stimulates the production of other inflammatory mediators, namely IL's, as well as the recruitment of immune and inflammatory cells into the joint [20,21]. As a regulatory cytokine that manages communication between immune cells and controls many of their functions when deregulated, TNF-α plays a key role in the pathogenesis of chronic inflammatory diseases, such as RA [22,23]. TNF deregulation in RA is linked to TNF-α converting enzyme (TACE or ADAM17), a metalloproteinase that cleaves trans-membrane TNF, releasing the soluble segment [18,24]; and to TNF receptors (mostly TNFr-I) [18,25,26], which are likely to be related with proinflammatory, cytotoxic and apoptotic responses [2].

The increasing knowledge on RA pathogenesis stimulated the development of different therapeutic modalities aiming to avoid joint destruction, minimize the symptomatic profile, and enhance physical function [27]. These options comprise analgesics, symptomatic management and inflammatory drugs [11,28]. Analgesics enclose the non-steroidal anti-inflammatory drugs (NSAIDs) that are commonly used to treat RA. The inflammatory suppressive drugs include glucocorticoids and disease-modifying anti-rheumatic drugs (DMARDs) both non-biologic and biologic. Due to the disease heterogeneity, therapeutic strategies must be tailored to the individual patient in order to achieve a low level of disease activity within a limited period of time [29]. Thus, the combination of two treatment modalities has been proposed to achieve such a purpose [30].

Considering the significant involvement of TNF-α in RA, as well as the rising evidences that this cytokine heads the pro-inflammatory cytokine cascade, it becomes a significant therapeutic target [31,32]. Specifically, these evidences led to the development of TNF inhibitors for the treatment of ADs [33–37], as they became the first class of biologic agents to be used in RA treatment [38]. A specific high affinity monoclonal antibody is used to recognize and neutralize selectively its antigen, i.e., TNF-α [18,27,39]. Five different types of TNF-α inhibitors are currently licensed for human clinical use in RA treatment, namely Infliximab, Entarnecept, Adalimumab, Certalizumab Pegol and Golimumab. With the exception of Infliximab, which is administrated by intravenous (IV) infusion, all other medicines are administered subcutaneously [13]. For long-term control of RA, a continuous treatment is required because of disease flares' risk when the therapy is discontinued [1]. Although these therapeutic modalities have been used for quite some time, they present many shortcomings such as lack of specificity, limited antibody half-life, high cost, response variability, or even lack of response to treatments [11,40]. Due to the systemic character of these treatments, not only target tissues but also healthy tissues are exposed to a significant dose of drug, leading to adverse side effects [41]. These effects are common to many tissues, organs and systems, such as cardiovascular, renal, dermatological, and neurological, or risk of severe infection [11,27,28,40].

Furthermore, some attention has been given to drug carriers, in order to maintain effective concentration levels in plasma for extended periods [41]. Examples of these systems are the encapsulation of Infliximab in polylactide-co-glycolide microspheres [42] or porous silicon 3D structures [43], as well as long-term release of Etanercept by polyelectrolyte complex formulated particles [44]. Despite the promising preliminary results, these systems are still poorly developed.

Considering the shortcomings of these treatments, there is a need for innovative approaches that circumvent the aforementioned adverse side effects. Therefore, the main objective of this work was to develop an implantable system capable of capturing excessive TNF-alpha present in intra-articular cavities of an RA patient. For that, we immobilized a neutralizing TNF-α antibody at the surface of a polymeric substrate (i.e., electrospun nanofibers). This strategy takes advantage of specific interactions between the antibody and the antigen, where the antibody (TNF-α antibody) binds to the antigen (TNF-α) avoiding/limiting its harmful pro-inflammatory effects. After the system's assembly and proved ability to capture TNF-α secreted by activated monocyte-derived macrophages, the cytotoxicity of the biofunctionalized system was tested with human articular chondrocytes. Indeed, the cytocompatibility and chondrogenic differentiation potential of bare electrospun nanofibers were previously reported by us [45–47], corroborating their potential use in cartilage regeneration approaches.

2. Materials and Methods

2.1. Materials

Polycaprolactone (PCL; Mn = 70,000–90,000 determined by Gas Permeation Chromatography), chloroform, N,N-dimethylformamide (DMF), 1,6-hexamethylenediamine (HMD), Ellmans reagent (DTNB), 2-iminiothiolane (2IT), 2-(N-morpholino)-ethanesulfonic acid (MES hydrate), phosphate buffered saline (PBS), N-hydroxysuccinimide (NHS), and 1-[3-(dimethylamino)propyl]-3-ethylcarbodiimide hydrochloride (EDC) were purchased from Sigma-Aldrich (Saint Louis, MO, USA) and kept at room temperature (RT), with the exception of 2IT which was kept at 4 °C and EDC which was kept at −20 °C. 4-(dimethylamino) pyridine (DMAP) was purchased from Millipore (Darmstadt, Germany) and kept at RT. Anti-TNF-α antibody [B-C7] was purchased from Abcam (Cambridge, UK) and kept at 4 °C until further use. Alexa Fluor® 488 goat anti-mouse rabbit IgG (H + L) and TNF-α human ELISA kit were purchased from Life Technologies (Carlsbad, CA, USA) and kept at 4 °C until further use.

2.2. Production and Functionalization of Nanofiber Meshes

2.2.1. Production of Nanofiber Meshes

Nanofiber meshes (NFMs) were prepared as described elsewhere [48,49]. Briefly, the NFMs were produced by electrospinning of a polymeric solution of 15% (w/v) PCL dissolved in an organic solvent mixture of chloroform and dimethylformamide (7:3 ratio). This PCL solution was electrospun by applying a voltage of 12 kV, a needle tip-to-ground collector distance of 20 cm and a flow rate of 1 mL/h. After the complete processing of 1 mL of PCL solution, the NFMs were left to dry for 1 day at RT inside a chemical hood, in order to completely evaporate the solvents. All further tests described were performed in samples of 10-mm squares.

2.2.2. Surface Functionalization of Electrospun NFMs

The surface functionalization of electrospun NFMs was performed as described elsewhere [48,49]. Briefly, the NFMs' surface was activated by 4 min of UV-Ozone irradiation (UV-O-Cleaner®, ProCleaner 220, Bioforce Nanoscience, Salt Lake City, UT, USA). Afterwards, the amine groups (–NH$_2$) were inserted by immersion of the irradiated NFM in a 1 M hexanediamine (HMD) solution during 1 hour at 37 °C.

2.2.3. Quantification of Amine Groups Present at the Functionalized NFMs

Quantification of amine groups present in each NFM was performed as described elsewhere [50,51]. For that, SH groups were inserted at the surface of the aminolysis-treated NFMs through the reaction of the amine groups with a 20 mM 2IT solution (0.1 mM PBS at pH8) and with a 20 mM DMAP solution during 1 h at 37 °C. Ellman's reagent method was used to quantify the amino groups. For that, samples were immersed in 0.1 mM DTNB solution (in PBS 0.1 mM at pH 7.27) and incubated during 1 h at 37 °C. The absorbance of supernatants was measured in triplicate at 412 nm in a quartz plate, using the DTNB solution as blank. For the calculation of the 2-nitro-5-thiobenzoate (NTB^{-2}) molar absorption coefficient, the value of 14,151 M^{-1} cm^{-1} was used [50,51].

2.3. Antibody Immobilization

The TNF-α antibody was immobilized at the surface of activated and functionalized electrospun NFMs. To determine the substrate's maximum immobilization capacity, a wide range of primary antibody concentrations were considered (from 0 to 12 µg/mL). The TNF-α antibody was firstly activated by a 15-minute incubation with a solution of EDC/NHS (50 mM EDC and 200 mM NHS), dissolved in a 0.1 M 2-(N-morpholino)-ethanesulfonic acid (MES) buffer with 0.9% (w/w) NaCl, followed by pH adjustment to 4.7 [48].

The functionalized NFMs were incubated with 200 µL of the activated TNF-α antibody overnight at 4 °C. Biofunctionalized NFMs were washed three times with PBS. A 3% bovine serum albumin (BSA) solution was used as a blocking step. To determine the degree of TNF-α antibody immobilization, biofunctionalized NFMs were incubated with an Alexa Fluor® 488 solution diluted in PBS (1:200 ratio). As a negative control, NFMs with no antibody immobilized (plotted as 0 µg/mL) were used. The fluorescence of the supernatant was determined in triplicate, using an excitation of 495/20 nm and an emission of 519/20 nm, and used to quantify the amount of antibody effectively immobilized.

2.4. Characterization Biofunctionalized NFMs

2.4.1. Scanning Electron Microscopy (SEM)

SEM was used to analyse the morphology of biofunctionalized NFMs. Briefly, the biofunctionalized NFMs were sputter-coated with a thin layer (9–12 nm) of gold/palladium (Cressington 208 HR) and then analysed by SEM (NanoSEM, Nova 200, FEI company, Hillsboro, OR, USA). Micrographs were recorded at 15 kV with magnifications ranging from 100 to 2000 times.

2.4.2. Fluorescence Microscopy

A fluorescence microscope (Axio Imager Z1m, Zeiss, Gottingen, Germany) was used to analyze the spatial distribution of the TNF-α antibody at the surface of electrospun NFMs. The TNF-α antibody was linked by a secondary antibody with green fluorescence (Alexa Fluor™ 488). Photographs were recorded at magnifications of 50, 200 and 400 times.

2.5. Capturing of TNF-α Present in Conditioned Medium of Macrophage Culture

A human monocytic cell line THP-1 was maintained in RPMI 1640 media supplemented with 2 mM L-glutamine, 100 µg/mL of penicillin, 100 µg/mL of streptomycin, 10 mM HEPES, and 10% fetal bovine serum (complete RPMI, cRPMI) (Life Technologies, Carlsbad, CA, USA). For the induction of cell differentiation, cells (106 per mL) were seeded in cRPMI with 100 nM phorbol 12-myristate-13-acetate (PMA) for 24 h. After incubation, non-attached cells were removed by aspiration, and the adherent cells were washed three times with cRPMI. To ensure reversion of cells to a resting macrophage phenotype before stimulation, cells were incubated for an additional 48 h in cRPMI without PMA. For stimulation and retrieval of conditioned media, cells were further incubated for 4 h with 100 ng/mL of lipopolysaccharide (LPS) in fresh cRPMI and the supernatants were collected and stored at −80 °C. Production of TNF-α was assessed in the conditioned media by commercial ELISA (Life Technologies, Carlsbad, CA, USA). Four systems were tested at least in quadruplicates, in two independent assays: (1) the biofunctionalized NFM (with immobilized TNF-α antibody); (2) a positive control with the soluble TNF-α antibody; (3) a negative control with the UV-O activated NFM; and (4) the conditioned cell culture medium alone. Each system was placed in sterile Eppendorf tubes and 2 mL of monocyte-derived macrophage conditioned medium was added individually, and kept at 37 °C in agitation in an orbital shaker at 120 rpm. Initially, a 3 days assay was performed with a 1 ng/mL TNF-α concentration from monocyte-derived macrophage conditioned medium. A specimen was collected at the following time points (2, 4, 6, 8, 10, 12, 24, 32, 48, and 72 h of incubation) without conditioned medium replacement. Afterwards, a 15-day assay was conducted, where a sample was collected daily and recharged with the same volume of recovered cell culture medium each day, which had a TNF-α concentration of 500 pg/mL. This concentration was chosen due to being closer to TNF-α levels in active RA patients (76.1 ± 103.2 pg/mL) [19], considering a margin of error affected by the TNF-α degradation.

Enzyme-Linked Immunosorbent Assay (ELISA)

For the quantification of TNF-α in the conditioned medium at each time point, sandwich TNF-α ELISA (KHC3012, TermoFisher Scientific, Carlsbad, CA, USA) was performed according to the manufacturer. Absorbance was read at 450 nm on a microplate reader (Synergy HT, BioTek, Winooski, VT, USA). A standard curve was built with concentrations ranging from 0 to 1000 pg/ml. The final concentrations were calculated by subtracting the value of each sample (i.e., testing conditions 1–3) to the conditioned culture medium baseline (i.e., testing condition 4), and the average was plotted for each testing condition along time.

2.6. Biological Assays

2.6.1. Isolation and Cell Culture

Cartilage tissue consists of only one cell type, the chondrocyte, embedded in a dense extracellular matrix (ECM). To evaluate the cytotoxicity of the developed biofunctionalized nanofibrous substrates, human articular chondrocytes (hAC) isolated from knee cartilage samples, collected from arthroplasties surgeries biopsies, were used. These cells were chosen as they were isolated from diseased joints cartilage. Samples were collected under the cooperation agreement between Centro Hospitalar do Alto Ave, Guimarães, Portugal, and the 3B's Research Group, after informed donor consent. Briefly, cells were isolated by enzymatic digestion, according to the protocol described previously [46].

Dulbecco's modified Eagle's medium (DMEM, Sigma D5671), containing 10 mM Hepes buffer (Life Technologies, Paisley, UK), L-alanyl-L-glutamine (Sigma), Non Essential Aminoacids (Sigma) 10,000 units/ml penicillin, 10,000 µg/mL streptomycin, and 10% foetal calf serum was the basis for the expansion and differentiation medium used herein (Basic medium). Basic medium was supplemented with 10 ng/mL of bFGF for hAC expansion. Basic medium was further supplemented with 1 mg/mL of insulin and 1 mg/mL of ascorbic acid, when hAC were seeded onto the PCL NFM (differentiation medium). These supplements enhanced extracellular matrix (ECM) deposition by the cultured hACs.

Human articular chondrocytes where used at passage 4. Expansion medium was changed every 2 days until the cells reached a confluence of 90%. The cells were harvested and seeded onto both the activated NFM, as well the NFM with the TNF-α antibody immobilized using the differentiation medium.

2.6.2. Seeding onto NFMs

Electrospun PCL NFMs were sterilized by UV-O treatment. All the surface functionalization and TNF-α antibody immobilization steps were performed in sterile conditions. For the biofunctionalized NFMs, antibody immobilization was performed overnight, and after the BSA blocking step, cell seeding was performed. Activated and biofunctionalized electrospun NFMs were used as controls.

Confluent hACs were detached from the culture flask using trypsin, counted on a hemocytometer and seeded at a density of 200.000 cells/NFM. Seeding was performed using the droplet method, using a 50 µL drop of cell suspension on all sample groups in 24-well plates. The plates were placed at 37 °C and 5% CO_2 over 4 hours. After cell attachment to the NFMs, 1 mL of expansion medium was added to each well. Afterwards, the medium was changed into differentiation medium. The seeding was performed in three independent experiments. For each independent experiment, constructs were cultured for 1, 14, 21 and 28 days under static conditions and collected at each time point for quantification of cell viability, DNA, total protein analysis, as well as GAGs deposition. Cell morphology was evaluated by SEM.

2.6.3. DNA Quantification

DNA quantification was assessed using Quant-iT™ PicoGreen® dsDNA Reagent and Kit (Life Technologies, Eugene, OR, USA), according to the manufacturer's instructions. Triplicates of each condition collected at 1, 14, 21 and 28 days were evaluated. The specimens were collected, 1 mL of sterile distilled water was added, and then samples were stored at −80 °C until further analysis. Prior quantification, the samples were defrosted and sonicated for 15 min. To extrapolate the DNA values for each sample, a set of standards were prepared with concentrations ranging from 0 to 1.5 µg/mL. The fluorescence of each sample was measured on an opaque 96-well plate using an excitation of 485/20 nm and an emission of 528/20 nm, using a microplate reader (Synergy HT, BioTek, Winooski, VT, USA).

2.6.4. Total Protein Synthesis Quantification

Total protein was performed using Micro BCA™ Protein Assay Kit (ThermoFisher Scientific, Rockford, IL, USA) according to the manufacturer's instructions. Triplicates of each condition collected at 1, 14, 21, and 28 days were evaluated. The samples were collected as described in Section 2.6.3. A standard curve was prepared ranging from 0 to 200 µg/mL. The absorbance of each sample was measured at 562 nm using a microplate reader (Synergy HT, BioTek, Winooski, VT, USA).

2.6.5. Glycosaminoglycan (GAG) Quantification

GAG quantification was performed using papain digestion. Triplicates of each condition collected at each previously established time point were tested. The samples were collected and stored in eppendorf tubes at −80 °C until further analysis. Digestion solution was prepared by adding papain (Sigma Aldrich, Saint Louis, MO, USA) and N-acetyl cysteine (Sigma Aldrich) at concentrations of 0.05% and 0.096%, respectively, to 50 mL of digestion buffer (200 mM of phosphate buffer containing 1 mM EDTA (Sigma Aldrich), pH 6.8). Each specimen was incubated with 600 µL of digestion buffer, overnight at 60 °C. After a 10-min centrifugation at 1300 rpm, the supernatant was collected. Dimethymethylene Blue (DMB) stock solution was prepared dissolving 16 mg of DMB powder in 900 mL of distilled water containing 3.04 g of glycine and 2.73 g of NaCl. pH was adjusted to 3.0 with HCl and the volume adjusted to 1 L. The solution was stored at RT covered by aluminum foil. Chondroitin sulphate (Sigma, C8529) solution was prepared in water in a 5 mg/mL stock solution and kept refrigerated. Standards were prepared from serial dilutions of this solution. Three samples per condition were considered per time point, and the absorbance of each sample was measured at 530 nm using a microplate reader.

2.6.6. Histological Analysis

Samples were collected at the end of the experiment and processed for histology. Samples were then fixed in 10% neutral-buffered formalin and then kept at 4 °C until the staining procedures. Alcian Blue staining was performed by rinsing the samples in 3% acetic acid, keeping them in 1% alcian blue solution (Sigma, A-3157) for 1 h. After that, the stain was poured off and sections were washed with water, let to dry, and then rinsed in absolute alcohol, cleared in xylene, and mounted in Entellan rapid (Merck Millipore, Darmstadt, Germany, 1.07960.0500).

2.7. Statistical Analysis

Statistical analysis of values related with antibody immobilization, TNF-α capturing profiles and biological studies was performed using IBM SPSS software (version 21; SPSS Inc., Armonk, NY, USA). Firstly, a Shapiro-Wilk test was used to ascertain the assumption of data normality. For antibody immobilization, protein capture and biological assays, the results showed that the data was not following a normal distribution. P values lower than 0.01 were considered statistically significant in the analysis of the results. A Kruskal-Wallis test was also performed for the comparison of more than two independent groups of samples for one variable.

3. Results and Discussion

3.1. Antibody Immobilization Efficiency

The primary antibody (anti-TNF-α antibody) immobilization at the surface of electrospun nanofiber meshes (NFMs) was performed in a wide range of concentrations (0–12 µg/mL) to determine the maximum antibody immobilization capacity. The fluorescence of the unbound secondary antibody was measured and the values were plotted against the various primary antibody concentrations (Figure 1). As this is an indirect quantification method, higher fluorescence values of the unbound secondary antibody correspond to lower concentrations of anti-TNF-α primary antibody immobilization. Moreover, when the substrate is saturated, the fluorescence values of unbound secondary antibody reach a plateau, despite the increment on antibody concentration.

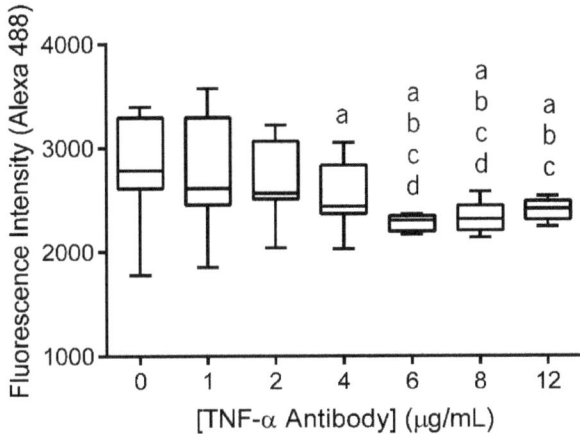

Figure 1. Box plot of anti-TNF-α antibody immobilization at concentrations ranging from 0 to 12 µg/mL. Data were analysed by nonparametric way of a Kruskal-Wallis test followed by Tukey's HSD test: (**a**) denotes significant differences compared to concentration 0, (**b**) denotes significant differences compared to concentration 1, (**c**) denotes significant differences compared to concentration 2, and (**d**) denotes significant differences compared to concentration 4.

The statistical analysis showed that the immobilized primary antibody, with concentrations above 4 µg/mL, displayed significantly lower values of fluorescence intensity than 0 µg/mL ($p < 0.01$). Concentrations of 6 µg/mL and 8 µg/mL exhibited significantly lower values of fluorescence intensity when compared to all lower concentrations ($p < 0.001$). Moreover, these two concentration values did not display differences between themselves. Finally, when the antibody was immobilized at a concentration of 12 µg/mL, the fluorescence intensity of secondary antibody did not present significant differences when compared to the fluorescence intensity of NFMs with primary antibody immobilized at concentrations above 2 µg/mL ($p < 0.01$). Therefore, primary antibody immobilization at the surface of the nanofibrous substrate reached a maximum concentration of 6 µg/mL, since no statistically significant differences were found between the concentrations above. This value is within the range of concentrations systemically administrated in clinical practice using Infliximab (3–10 mg/kg or 3–10 µL/mL, considering water as body mass) [52].

3.2. Spatial Distribution of the Antibody at the Surface of Electrospun Nanofibers

After the optimization of the anti-TNF-α antibody immobilization, its spatial distribution at the surface of electrospun NFMs was evaluated by fluorescence microscopy (Figure 2). The fluorescence images display a random mesh-like arrangement comparable to the typical morphology of the electrospun NFM (Figure 2A), showing that the antibody immobilization was successfully performed and uniformly distributed along the surface of the nanofibers (Figure 2B). To ensure that the secondary antibody was only bound to the immobilized primary antibody, a control experiment was defined, in which all biofunctionalization steps were performed, except for incubation with the primary antibody. The absence of a fluorescent signal proves that Alexa Fluor 488® secondary antibody was not immobilized at the surface of activated and functionalized NFMs (data not shown), confirming the specific bond between the primary and secondary antibodies.

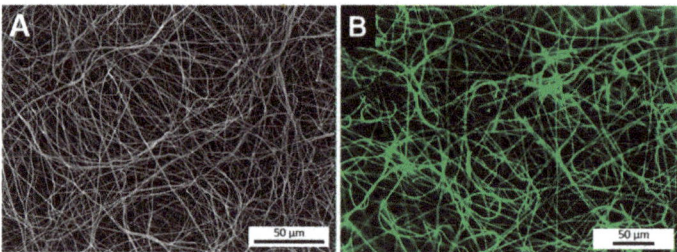

Figure 2. SEM micrograph of a biofunctionalized NFM (**A**). Fluorescence micrograph of a NFM with the TNF-α antibody immobilized at the maximum concentration (6 µg/mL) (**B**).

3.3. Quantification of Available Amine Groups

Antibody immobilization was achieved by the amine groups inserted at the nanofibers' surface, which provides binding sites. Consequently, the availability of this functional group influences directly the primary antibody immobilization. To confirm the success of aminolysis functionalization, as well as to determine if the binding points were enough for primary antibody immobilization, quantification of available amine groups was performed. For this, SH groups were inserted at the surface of the aminolysis-treated NFMs and determined by the Ellman's reagent method [49]. Table 1 presents SH groups quantification at the surface of bare, UV-O irradiated, aminolysis-treated, and fully biofunctionalized (with primary antibody immobilized) NFMs. The concentration of free SH groups at the surface of electrospun NFMs is at its maximum value for the aminolysis-treated NFM, whereas the bare NFM presents the minimum value, as expected. These values are in agreement with the ones described in the literature [49] and confirm that the insertion of amine groups was successful. Moreover, it was also proved that the amine groups inserted at the surface of activated NFM were enough for primary antibody immobilization, since free amine groups were still present. Indeed, only around 48% of the inserted amine groups were used in the immobilization of the TNF-α antibody at 6 µg/mL.

Table 1. Quantification of the free SH groups at the surface of untreated, activated, aminolys-treated, and biofunctionalized NFMs.

Condition	SH/NH$_2$ Groups [mol/cm]
Untreated	$(3.54 \pm 5.61) \times 10^{-9}$
UV-O activated	$(5.86 \pm 1.12) \times 10^{-9}$
Aminolysis-treatment	$(17.9 \pm 6.06) \times 10^{-9}$
Immobilized Antibody	$(8.65 \pm 3.41) \times 10^{-9}$

3.4. Quantification of Captured TNF-α

The ultimate goal of this work was to engineer a system that could capture TNF-α. Accordingly, after the primary antibody immobilization optimization, the TNF-α capturing capacity of the biofunctionalized substrate was assessed. A TNF-α rich conditioned culture medium from stimulated monocyte-derived macrophages was used over two distinct time intervals: 3 days representing the half-life of the anti-TNF-α antibody and 15 days representing the time interval of anti-TNF-α antibody administration in the clinic. Four conditions were tested and compared: (1) the biofunctionalized NFM (with primary antibody); (2) a positive control with the soluble form of the anti-TNF-α; (3) a negative control with UV-O activated NFM; and (4) the conditioned culture medium alone. The quantification of TNF-α levels from the conditioned culture medium (condition 4) was used as a baseline, eliminating the quantification of degraded TNF-α but not captured. Therefore, to guarantee the values' accuracy and reproducibility between experiments, the values herein presented do not correspond to the ones quantified by ELISA, but just to the concentration of the TNF-α captured.

3.4.1. TNF-α Capturing during 3 days

The proposed system comprising anti-TNF-α immobilized at the surface of NFMs (NFM + Ab) was found to capture soluble TNF-α in significant amounts after 8 hours of incubation when compared to the UV-O activated NFM (NFM) ($p < 0.01$) (Figure 3). This observation was also valid when the biofunctionalized substrate was compared to the soluble antibody (sAb), as the NFM + Ab exhibited significantly higher cytokine clearance for time points above the 10th hour of incubation ($p < 0.01$). In addition, no significant differences were found between the sAb and the control condition UV-O activated NFM (NFM).

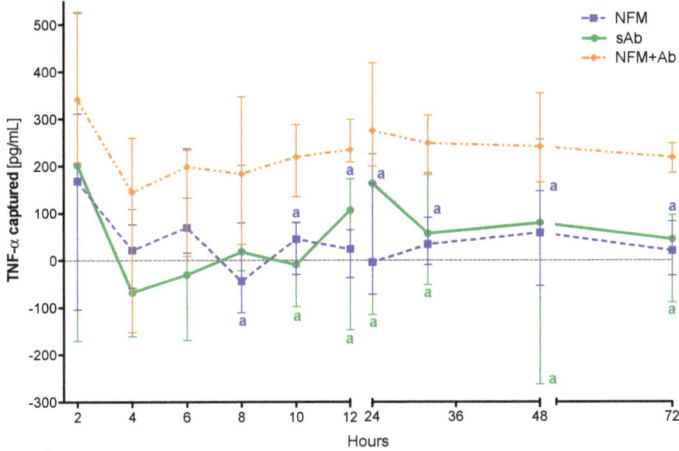

Figure 3. TNF-α captured (median and interquartile range) over 72 h. Data were analysed by nonparametric way of a Kruskal-Wallis test, followed by Tukey's HSD test: (**a**) denotes significant differences compared to immobilized anti-TNF-α at the surface of electrospun NFM (condition NFM + Ab).

The initial efficiency of the activated NFM over the TNF-α levels at early time points might be justified by the presence of free binding sites (such as the available amine groups quantified in Section 2.4) that unspecifically immobilize/adsorb proteins present in the conditioned medium, including TNF-α. Nevertheless, the capturing levels in this condition stabilize around zero for time points above 6 hours, unlike the biofunctionalized NFM + Ab, which maintains the capturing ability until the last time point. Likewise, the positive control sAb has a similar biological effect to the NFM + Ab testing condition during the first 10 hours. The sAb capturing ability decreases after this time point and becomes comparable to the NFM negative control condition. These results show that the immobilized form of anti-TNF-α is more stable and has longer action times, when compared to the non-immobilized one (i.e., sAb). However, this decrease might be partially justified by the gradual clearance of the soluble anti-TNF-α from the conditioned medium along the samples collection. It is envisioned that this decrease of sAb mimics its clearance in a living system, since the circulating antibody has higher propensity to be cleared by biological means [41]. Even if the sAb was locally administered, part of it would escape the articular cavity and enter into circulation, suffering also the clearance phenomena.

3.4.2. TNF-α Capturing during 15 Days

To confirm the efficiency of the primary antibody immobilized at the NFMs' surface to clear TNF-α from the conditioned medium over time, as well as to evaluate the action time of the immobilized antibody, an extended assay (15 days) was performed. Experimental results show that the sAb only has a biological effect for the first 3 days, since from the fourth time point on, it does not present significant differences when compared to the UV-O activated NFM (Figure 4). The NFM + Ab has a significantly higher antigen-neutralizing activity ranging from the 2nd day until the 11th day when compared to the sAb condition. Moreover, the biofunctionalized NFM + Ab does not have a significant biological effect from the 11th day on, as no significant differences were found when compared to the control NFM.

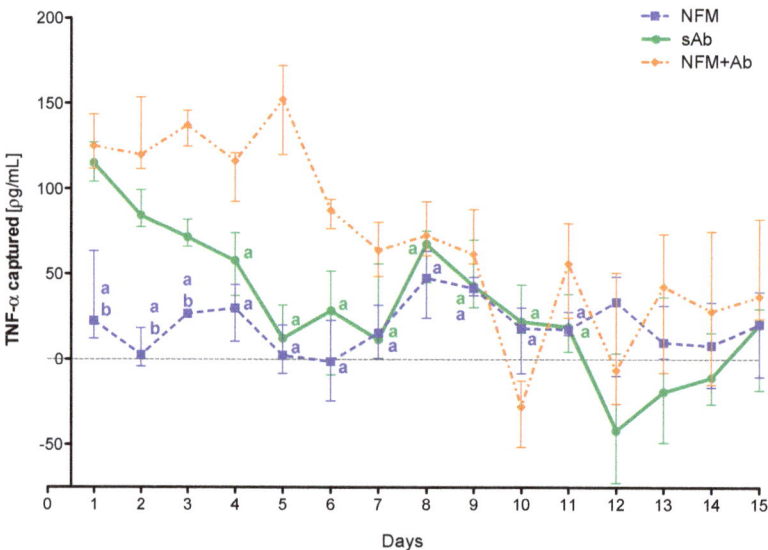

Figure 4. TNF-α captured (median and interquartile range) over 15 days. Data were analyzed by nonparametric way of a Kruskal-Wallis test, followed by Tukey's HSD test: (**a**) denotes significant differences compared to immobilized anti-TNF at the surface of electrospun NFM (condition NFM + Ab); and (**b**) denotes significant differences compared to soluble anti-TNF (condition sAb).

These results confirm the initial hypothesis that the immobilization of the primary antibody at the NFMs' surface had a longer effect than the soluble antibody. This longer effect might be supported by the increased stability that the antibody immobilization provides [53]. Moreover, the physical attachment prevents the antibody clearance from the system by biological means. As each NFM has a high specific surface area, it allows the immobilization of high concentrations of primary antibody, requiring a small amount of material for the development of an effective cytokine capture system. Moreover, as the NFMs are made of polycaprolactone (PCL), a well-known biodegradable polymer, these membranes will be completely degraded within a maximum period of 2 years. Thus, the degradation time ensures that the captured TNF-α, as well as the attached antibody, may be slowly degraded/metabolized by biological processes and that there is no abrupt release of these molecules to the exterior of the synovial cavity.

3.5. Biologic Assays

After the successful assembly of the TNF-capturing system, we then tested it in contact with arthritic joint cells. Therefore, the toxicity of the biofunctionalized NFMs was assessed by culturing them with human articular chondrocytes (hACs). These cells were isolated from diseased knee arthroplasties, thus presenting a phenotype that is associated with RA and osteoarthritis pathologies, where inflammation was present. Different biological assays were conducted to assess the hACs morphology, proliferation, total protein, and glycosaminoglycans (GAGs) synthesis. hACs were cultured at the surface of three different substrates: (1) anti-TNF-α antibody immobilized at the surface of NFMs (NFM + Ab), (2) UV-O activated NFM (NFM) and (3) tissue culture polystyrene (TCPs) as positive control. The influence of the anti-TNF-α immobilization at the surface of NFMs over hACs morphology was assessed by SEM. Micrographs (Figure 5) showed that hACs maintained their round-shape morphology, typical of these chondrocytic cells.

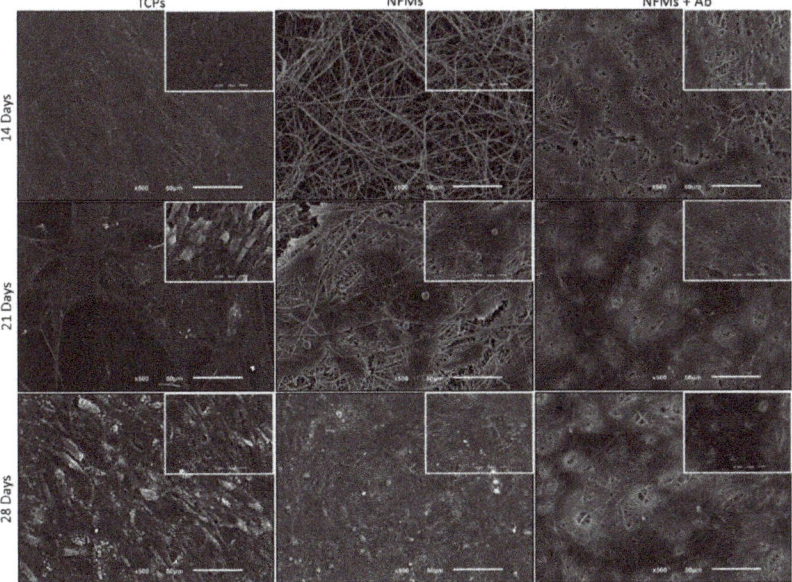

Figure 5. SEM micrographs of hACs cultures on tissue culture polystyrene (TCPs), UV-O treated electrospun PCL NFM (NFMs) and anti-TNF-α immobilized at the NFMs' surface (NFM + Ab) for 14, 21 and 28 days.

Chondrocytes proliferation was also evaluated by quantifying the dsDNA concentration. Experimental results showed that the TCP's condition displayed a significantly higher DNA concentration, for 14 and 28 days of culture, when compared to all other culture conditions ($p < 0.001$) (Figure 6A). Furthermore, on the 1st and 21st days of hACs culture, the control condition (TCPs) present similar proliferation to the NFM + Ab testing condition. Finally, the NFM condition presented a significantly lower cell proliferation at 21 and 28 days of culture, when compared to the NFM + AB condition ($p < 0.001$). Taken together, these results indicate that the immobilized anti-TNF-α may inhibit chondrocytes growth by mimicking an inflammatory environment, where cells would respond by limiting their growth and proliferation. This effect is not negative, since our primary goal was not to promote an extensive cell proliferation, but merely to assess the device's potential cell toxicity. Nevertheless, hACs keep proliferating, even if at a lower rate when compared to the controls. Indeed, IL1-β and TNF-α have been shown to inhibit the migratory potential of chondrogenic progenitor cells in osteoarthritic cartilage [54]. Furthermore, OA-derived hACs cultures have been reported to excrete around 3000 pm/µL of TNF-α [55], thus these cells have mechanisms to respond to high concentrations of this cytokine in the surrounding environment.

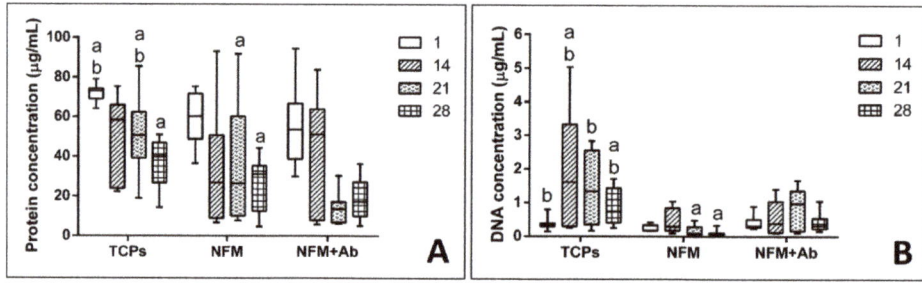

Figure 6. Box plot of human articular chondrocytes proliferation (**A**) and protein synthesis (**B**) when cultured on tissue culture polystyrene (TCPs), UV-O treated electrospun PCL NFM (NFM) and anti-TNF-α immobilized at the NFMs' surface (NFM + Ab). Data were analysed by nonparametric way of a Kruskal-Wallis test, followed by Tukey's HSD test: (**a**) denotes significant differences compared to NFM + Ab and (**b**) denotes significant differences compared to NFM.

Quantification of total protein synthesis is presented in Figure 6B. Experimental data showed that TCP's control condition displayed significantly higher concentration values than all the other testing conditions for the 1st and 21st days of culture ($p < 0.01$). On the other hand, the NFM + Ab testing condition displayed a significantly lower protein concentration when compared to all other culture conditions for 28 days of culture ($p < 0.01$). At 14 days of hACs culture, no significant differences between conditions were found ($p = 0.148$). The protein synthesis in the NFM + Ab condition was lower for longer culture periods, which is possibly due to the extracellular accumulation of proteins by hACs, rather than the intracellular synthesizing rate.

Furthermore, the impact of the anti-TNF-α antibody immobilization at the surface of NFMs on GAG production and accumulation (Figure 7A) by hACs was also assessed. No significant differences were found between all culture conditions for the 1st ($p = 0.186$) and 14th days ($p = 0.088$) of the experiment. On the 21st day, the NFM + Ab testing condition displayed significantly lower GAG concentrations than control condition TCPs ($p < 0.01$). Moreover, on the 28th day, TCPs presented significant higher GAGs concentration than all other culture conditions ($p < 0.01$). Concerning the GAGs production by hACs cultured onto the NFM + Ab, the obtained results confirm the evidences found in the intracellular protein synthesis. There is a decrease in the total protein concentration values for longer time points, with a concomitant increase of GAGs production for these same periods of time, indicating that protein synthesis was being directed by the GAGs production. This increase of GAGs concentration for late time points of the experiment are in accordance to our hypothesis, as they will protect chondrocytes against the inflammatory environment.

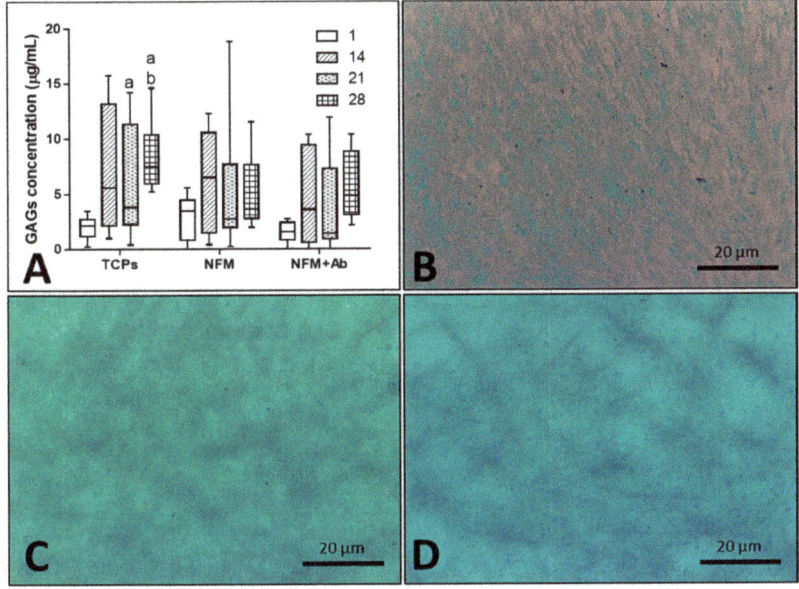

Figure 7. Blox plot of human articular chondrocytes GAGs accumulation when cultured on tissue culture polystyrene (TCPs), UV-O treated electrospun PCL NFM (NFM) and anti-TNF-α immobilized at the NFMs' surface (NFM + Ab) (**A**). Data were analysed by nonparametric way of a Kruskal-Wallis test, followed by Tukey's HSD test: (**a**) denotes significant differences compared to NFM + Ab and (**b**) denotes significant differences compared to NFM. Alcian blue staining evidencing sulphated proteoglycans deposition in samples from TCPs (**B**), NFMs (**C**), and NFM + Ab (**D**) on the 28th day of the experiment.

As it can be observed in Figure 7B–D, sulphated proteoglycans, an important component of articular cartilage ECM, were detected on all culture conditions by alcian blue staining. Furthermore, these observations are reassured by the quantification of GAGs presented in the box plot of Figure 7A. Altogether, these results on intra- and extra-cellular protein synthesis confirmed that hACs are not only able to proliferate and maintain their typical round morphology, but are also capable of performing their metabolic functions properly, in the presence of the NFM + Ab testing system. With this study, we validated the use of this TNF-α capturing system nearby cartilaginous structures. There is, indeed, an influence on these parameters when cells are cultured onto NFM + Ab nanofibers, but it is not enough to restrain human articular chondrocytes functions. In fact, it has been described that the blockade of TNF-α by Infliximab treatment, during fracture healing, leads to an increased callus size. This increase was accompanied by an increase in cartilage tissue [56].

4. Conclusions

The present work proposes the use of nanofibrous meshes (NFM) functionalized with a neutralizing anti-TNF-α antibody, for the selective clearance of TNF-α. From a spectrum of primary antibody concentrations, 6 μg/mL was determined as the maximum immobilization capacity of a UV-O activated and aminolysis-functionalized NFM. Experimental results demonstrated the bioactivity of the developed biofunctional substrate on clearing TNF-α from conditioned culture medium of macrophages and a longer time of action when compared to the circulating/soluble antibody.

Furthermore, cell biology data showed that the immobilized anti-TNF-α antibody exhibited no cytotoxicity over human articular chondrocytes and that these cells were able to maintain their metabolic functions in the presence of the TNF-capturing system. In conclusion, we herein describe an effective system for local clearance of TNF-α, which is particularly relevant in a RA scenario, but adaptable to other autoimmune inflammatory conditions.

5. Patents

Martins A, Oliveira C, Reis RL, Neves NM. Polymeric substrates with immobilized antibodies and method of production thereof. WO/2015/166414 A1, EP 3137906 A1, US 2017/0051050 A1. Priority date 28/04/2014.

Author Contributions: Conceptualization, E.B., A.M. and N.M.N.; methodology, E.B., M.A.d.S., A.C., A.M. and N.M.N.; validation, A.M. and N.M.N.; formal analysis, E.B., S.F. and A.M.; investigation, E.B., M.A.d.S. and C.C.; resources, A.C., R.L.R., A.M. and N.M.N.; data curation, E.B., M.A.d.S. and C.C.; writing—Original draft preparation, E.B., M.A.d.S. and A.M.; writing—Review and editing, A.M. and N.M.N.; visualization, E.B., A.M. and N.M.N.; supervision, A.C., R.L.R., A.M. and N.M.N.; project administration, R.L.R., A.M. and N.M.N.; funding acquisition, R.L.R., A.M. and N.M.N.

Funding: This research was funded by the Portuguese Foundation for Science and Technology (FCT), namely the Post-doc fellowships of Marta Alves da Silva and Cristina Cunha (SFRH/BPD/73322/2010 and SFRH/BPD/96176/2013, respectively) and the Starting Investigator Grant of Agostinho Carvalho and Albino Martins (IF/00735/2014 and IF/00376/2014, respectively). It was also acknowledged the funded projects SPARTAN (PTDC/CTM-BIO/4388/2014) and FROnTHERA (NORTE-01-0145-FEDER-000023).

Conflicts of Interest: The authors declare no conflict of interest. The funders had no role in the design of the study; in the collection, analyses, or interpretation of data; in the writing of the manuscript, or in the decision to publish the results.

References

1. Firestein, G.S. Evolving concepts of rheumatoid arthritis. *Nature* **2003**, *423*, 356–361. [CrossRef]
2. Apostolaki, M.; Armaka, M.; Victoratos, P.; Kollias, G. Cellular mechanisms of TNF function in models of inflammation and autoimmunity. *Curr. Dir. Autoimmun.* **2010**, *11*, 1–26. [CrossRef]
3. Korczowska, I. Rheumatoid arthritis susceptibility genes: An overview. *World J. Orthop.* **2014**, *5*, 544–549. [CrossRef] [PubMed]
4. Klareskog, L.; Catrina, A.I.; Paget, S. Rheumatoid arthritis. *Lancet* **2009**, *373*, 659–672. [CrossRef]
5. Quintero, O.L.; Amador-Patarroyo, M.J.; Montoya-Ortiz, G.; Rojas-Villarraga, A.; Anaya, J.-M. Autoimmune disease and gender: Plausible mechanisms for the female predominance of autoimmunity. *J. Autoimmun.* **2012**, *38*, J109–J119. [CrossRef] [PubMed]
6. Ngo, S.T.; Steyn, F.J.; McCombe, P.A. Gender differences in autoimmune disease. *Front. Neuroendocrinol.* **2014**, *35*, 347–369. [CrossRef] [PubMed]
7. Moroni, L.; Bianchi, I.; Lleo, A. Geoepidemiology, gender and autoimmune disease. *Autoimmun. Rev.* **2012**, *11*, A386–A392. [CrossRef] [PubMed]
8. Amur, S.; Parekh, A.; Mummaneni, P. Sex differences and genomics in autoimmune diseases. *J. Autoimmun.* **2012**, *38*, J254–J265. [CrossRef] [PubMed]
9. Nussinovitch, U.; Shoenfeld, Y. The role of gender and organ specific autoimmunity. *Autoimmun. Rev.* **2012**, *11*, A377–A385. [CrossRef] [PubMed]
10. Pennell, L.M.; Galligan, C.L.; Fish, E.N. Sex affects immunity. *J. Autoimmun.* **2012**, *38*, J282–J291. [CrossRef] [PubMed]
11. Scott, D.L.; Wolfe, F.; Huizinga, T.W.J. Rheumatoid arthritis. *Lancet* **2010**, *376*, 1094–1108. [CrossRef]
12. Perricone, C.; Ceccarelli, F.; Valesini, G. An overview on the genetic of rheumatoid arthritis: A never-ending story. *Autoimmun. Rev.* **2011**, *10*, 599–608. [CrossRef] [PubMed]
13. Meroni, P.L.; Valesini, G. Tumour necrosis factor α antagonists in the treatment of rheumatoid arthritis: An immunological perspective. *BioDrugs* **2014**, *28* (Suppl. 1), 5–13. [CrossRef] [PubMed]
14. McInnes, I.B.; Schett, G. Cytokines in the pathogenesis of rheumatoid arthritis. *Nat. Rev. Immunol.* **2007**, *7*, 429–442. [CrossRef]

15. Mcinnes, I.B.; Schett, G. The Pathogenesis of Rheumatoid Arthritis. *N. Engl. J. Med.* **2012**, 2205–2219. [CrossRef]
16. Brennan, F.M.; Foey, A.D. Cytokine regulation in RA synovial tissue: Role of T cell/macrophage contact-dependent interactions. *Arthritis Res.* **2002**, *4* (Suppl. 3), S177–S182. [CrossRef]
17. Caplan, M.S.; Jilling, T. The pathophysiology of necrotizing enterocolitis. *NeoReviews* **2001**, *2*, e103–e109. [CrossRef]
18. Taylor, P.C. Anti-TNF therapy for rheumatoid arthritis and other inflammatory diseases. *Mol. Biotechnol.* **2001**, *19*, 153–168. [CrossRef]
19. Edrees, A.F.; Misra, S.N.; Abdou, N.I. Anti-tumor necrosis factor (TNF) therapy in rheumatoid arthritis: Correlation of TNF-alpha serum level with clinical response and benefit from changing dose or frequency of infliximab infusions. *Clin. Exp. Rheumatol.* **2005**, *23*, 469–474.
20. Rubbert-Roth, A. Assessing the safety of biologic agents in patients with rheumatoid arthritis. *Rheumatology* **2012**, *51*, v38–v47. [CrossRef] [PubMed]
21. Brennan, F.M. Inhibitory effect of TNF alpha antibodies on synovial cell interleukin-1 production in rheumatoid arthritis. *Lancet* **1989**, *334*, 244–247. [CrossRef]
22. Horiuchi, T.; Mitoma, H.; Harashima, S.; Tsukamoto, H.; Shimoda, T. Transmembrane TNF-α: Structure, function and interaction with anti-TNF agents. *Rheumatology* **2010**, *49*, 1215–1228. [CrossRef]
23. Faustman, D.; Davis, M. TNF receptor 2 pathway: Drug target for autoimmune diseases. *Nat. Rev. Drug Discov.* **2010**, *9*, 482–493. [CrossRef]
24. Jones, E.Y.; Stuart, D.I.; Walker, N.P.C. Structure of tumour necrosis factor. *Nature* **1989**, *338*, 225–228. [CrossRef]
25. Bazzoni, F.; Beutler, B. The Tumor Necrosis Factor Ligand and receptor families. *N. Engl. J. Med.* **1996**, *334*, 1717–1725. [CrossRef]
26. Vandenabeele, P.; Declercq, W.; Beyaert, R.; Fiers, W. Two tumour necrosis factor receptors: Structure and function. *Trends Cell Biol.* **1995**, *5*, 392–399. [CrossRef]
27. Willrich, M.A.V.; Murray, D.L.; Snyder, M.R. Tumor necrosis factor inhibitors: Clinical utility in autoimmune diseases. *Transl. Res.* **2015**, *165*, 270–282. [CrossRef]
28. Croft, M.; Duan, W.; Choi, H.; Eun, S.-Y.; Madireddi, S.; Mehta, A. TNF superfamily in inflammatory disease: Translating basic insights. *Trends Immunol.* **2012**, *33*, 144–152. [CrossRef]
29. Hoes, J.N.; Jacobs, J.W.G.; Buttgereit, F.; Bijlsma, J.W.J. Current view of glucocorticoid co-therapy with DMARDs in rheumatoid arthritis. *Nat. Rev. Rheumatol.* **2010**, *6*, 693–702. [CrossRef]
30. Walsh, D.A.; McWilliams, D.F. Mechanisms, impact and management of pain in rheumatoid arthritis. *Nat. Rev. Rheumatol.* **2014**, *10*, 581–592. [CrossRef]
31. Rossi, D.; Modena, V.; Sciascia, S.; Roccatello, D. Rheumatoid arthritis: Biological therapy other than anti-TNF. *Int. Immunopharmacol.* **2015**, *27*, 185–188. [CrossRef]
32. Alves da Silva, M.; Martins, A.; Teixeira, A.A.; Reis, R.L.; Neves, N.M. Impact of biological agents and tissue engineering approaches on the treatment of rheumatic diseases. *Tissue Eng. Part B Rev.* **2010**, *16*, 331–339. [CrossRef]
33. Elliott, M.J.; Maini, R.N.; Feldmann, M.; Long-Fox, A.; Charles, P.; Katsikis, P.; Brennan, F.M.; Walker, J.; Bijl, H.; Ghrayeb, J.; et al. Treatment of rheumatoid arthritis with chimeric monoclonal antibodies to tumor necrosis factor alpha. *Arthritis Rheum.* **1993**, *36*, 1681–1690. [CrossRef]
34. Buchan, G.; Barrett, K.; Turner, M.; Chantry, D.; Maini, R.N.; Feldmann, M. Interleukin-1 and tumour necrosis factor mRNA expression in rheumatoid arthritis: Prolonged production of IL-1 alpha. *Clin. Exp. Immunol.* **1988**, *73*, 449–455.
35. Keffer, J.; Probert, L.; Caziaris, H.; Georgopoulos, S.; Kaslaris, E.; Kioussis, D.; Kollias, G. Transgenic mice expressing human tumour necrosis factor: A predictive genetic model of arthritis. *EMBO J.* **1991**, *10*, 4025–4031. [CrossRef] [PubMed]
36. Haworth, C.; Brennan, F.M.; Chantry, D.; Turner, M.; Maini, R.N.; Feldmann, M. Expression of granulocyte-macrophage colony-stimulating factor in rheumatoid arthritis: Regulation by tumor necrosis factor-alpha. *Eur. J. Immunol.* **1991**, *21*, 2575–2579. [CrossRef] [PubMed]
37. Williams, R.O.; Feldmann, M.; Maini, R.N. Anti-tumor necrosis factor ameliorates joint disease in murine collagen-induced arthritis. *Proc. Natl. Acad. Sci. USA* **1992**, *89*, 9784–9788. [CrossRef]

38. Dimitroulas, T.; Nikas, S.N.; Trontzas, P.; Kitas, G.D. Biologic therapies and systemic bone loss in rheumatoid arthritis. *Autoimmun. Rev.* **2013**, *12*, 958–966. [CrossRef] [PubMed]
39. Astrakhantseva, I.V.; Efimov, G.A.; Drutskaya, M.S.; Kruglov, A.A.; Nedospasov, S.A. Modern anti-cytokine therapy of autoimmune diseases. *Biochemistry* **2014**, *79*, 1308–1321. [CrossRef]
40. Quan, L.-D.; Thiele, G.M.; Tian, J.; Wang, D. The development of novel therapies for rheumatoid arthritis. *Expert Opin. Ther. Pat.* **2008**, *18*, 723–738. [CrossRef]
41. Tarner, I.H.; Müller-Ladner, U. Drug delivery systems for the treatment of rheumatoid arthritis. *Expert Opin. Drug Deliv.* **2008**, *5*, 1027–1037. [CrossRef]
42. Gokhale, K.S.; Jonnalagadda, S. Preparation and evaluation of sustained release infliximab microspheres. *PDA J. Pharm. Sci. Technol.* **2013**, *67*, 255–266. [CrossRef]
43. McInnes, S.J.P.; Turner, C.T.; Al-Bataineh, S.A.; Airaghi Leccardi, M.J.I.; Irani, Y.; Williams, K.A.; Cowin, A.J.; Voelcker, N.H. Surface engineering of porous silicon to optimise therapeutic antibody loading and release. *J. Mater. Chem. B* **2015**, *3*, 4123–4133. [CrossRef]
44. Jung, Y.S.; Park, W.; Na, K. Temperature-modulated noncovalent interaction controllable complex for the long-term delivery of etanercept to treat rheumatoid arthritis. *J. Control. Release* **2013**, *171*, 143–151. [CrossRef]
45. Alves da Silva, M.L.; Martins, A.; Costa-Pinto, A.R.; Costa, P.; Faria, S.; Gomes, M.; Reis, R.L.; Neves, N.M. Cartilage Tissue Engineering Using Electrospun PCL Nanofiber Meshes and MSCs. *Biomacromolecules* **2010**, *11*, 3228–3236. [CrossRef]
46. Alves da Silva, M.L.; Costa-Pinto, A.R.; Martins, A.; Correlo, V.M.; Sol, P.; Bhattacharya, M.; Faria, S.; Reis, R.L.; Neves, N.M. Conditioned medium as a strategy for human stem cells chondrogenic differentiation. *J. Tissue Eng. Regen. Med.* **2015**, *9*, 714–723. [CrossRef]
47. Piai, J.F.; Alves da Silva, M.L.; Martins, A.; Torres, A.; Faria, S.; Reis, R.L.; Muniz, E.C.; Neves, N.M. Chondroitin sulfate immobilization at the surface of electrospun nanofiber meshes for cartilage tissue regeneration approaches. *Appl. Surf. Sci.* **2017**, *403*, 112–125. [CrossRef]
48. Oliveira, C.; Costa-Pinto, A.R.; Reis, R.L.; Martins, A.; Neves, N.M. Biofunctional nanofibrous substrate comprising immobilized antibodies and selective binding of autologous growth factors. *Biomacromolecules* **2014**, *15*, 2196–2205. [CrossRef]
49. Monteiro, N.; Martins, A.; Pires, R.; Faria, S.; Fonseca, N.A.; Moreira, J.N.; Reis, R.L.; Neves, N.M. Immobilization of bioactive factor-loaded liposomes on the surface of electrospun nanofibers targeting tissue engineering. *Biomater. Sci.* **2014**, *2*, 1195–1209. [CrossRef]
50. Kakabakos, S.E.; Tyllianakis, P.E.; Evangelatos, G.P.; Ithakissios, D.S. Colorimetric determination of reactive solid-supported primary and secondary amino groups. *Biomaterials* **1994**, *15*, 289–297. [CrossRef]
51. Riddles, P.W.; Blakeley, R.L.; Zerner, B. Ellman's reagent: 5,5'-dithiobis(2-nitrobenzoic acid)—A reexamination. *Anal. Biochem.* **1979**, *94*, 75–81. [CrossRef]
52. Caporali, R.; Pallavicini, F.B.; Filippini, M.; Gorla, R.; Marchesoni, A.; Favalli, F.B.; Sarzi-Puttini, P.; Atzeni, F.; Montecucco, C. Treatment of rheumatoid arthritis with anti-TNF-alpha agents: A reappraisal. *Autoimmun. Rev.* **2009**, *8*, 274–280. [CrossRef]
53. Turková, J. Oriented immobilization of biologically active proteins as a tool for revealing protein interactions and function. *J. Chromatogr. B Biomed. Sci. Appl.* **1999**, *722*, 11–31. [CrossRef]
54. Joos, H.; Wildner, A.; Hogrefe, C.; Reichel, H.; Brenner, R.E. Interleukin-1 beta and tumor necrosis factor alpha inhibit migration activity of chondrogenic progenitor cells from non-fibrillated osteoarthritic cartilage. *Arthritis Res. Ther.* **2013**, *15*, R119. [CrossRef] [PubMed]
55. Tsuchida, A.I.; Beekhuizen, M.; 't Hart, M.C.; Radstake, T.; Dhert, W.; Saris, D.; van Osch, G.; Creemers, L.B. Cytokine profiles in the joint depend on pathology, but are different between synovial fluid, cartilage tissue and cultured chondrocytes. *Arthritis Res. Ther.* **2014**, *16*, 441. [CrossRef] [PubMed]
56. Timmen, M.; Hidding, H.; Wieskötter, B.; Baum, W.; Pap, T.; Raschke, M.J.; Schett, G.; Zwerina, J.; Stange, R. Influence of antiTNF-alpha antibody treatment on fracture healing under chronic inflammation. *BMC Musculoskelet. Disord.* **2014**, *15*, 184. [CrossRef]

© 2019 by the authors. Licensee MDPI, Basel, Switzerland. This article is an open access article distributed under the terms and conditions of the Creative Commons Attribution (CC BY) license (http://creativecommons.org/licenses/by/4.0/).

Article

Antibacterial and Bioactive Surface Modifications of Titanium Implants by PCL/TiO$_2$ Nanocomposite Coatings

A. Sandeep Kranthi Kiran [1,2,3], T.S. Sampath Kumar [1,*], Rutvi Sanghavi [2], Mukesh Doble [2] and Seeram Ramakrishna [3,*]

[1] Medical Materials Laboratory, Department of Metallurgical and Materials Engineering, Indian Institute of Technology Madras, Chennai 600036, India; urskranthi.kiran@gmail.com
[2] Department of Biotechnology, Bhupat and Jyoti Mehta School of Biosciences, Indian Institute of Technology Madras, Chennai 600036, India; rutvirs93@gmail.com (R.S.); mukeshd@iitm.ac.in (M.D.)
[3] NUS Centre for Nanofibers and Nanotechnology, Department of Mechanical Engineering, National University of Singapore, 2 Engineering Drive 3, Singapore 117581, Singapore
* Correspondences: tssk@iitm.ac.in (T.S.S.K.); seeram@nus.edu.sg (S.R.)

Received: 5 October 2018; Accepted: 18 October 2018; Published: 20 October 2018

Abstract: Surface modification of biomedical implants is an established strategy to improve tissue regeneration, osseointegration and also to minimize the bacterial accumulation. In the present study, electrospun poly(ε-caprolactone)/titania (PCL/TiO$_2$) nanocomposite coatings were developed on commercially pure titanium (cpTi) substrates for an improved biological and antibacterial properties for bone tissue engineering. TiO$_2$ nanoparticles in various amounts (2, 5, and 7 wt %) were incorporated into a biodegradable PCL matrix to form a homogeneous solution. Further, PCL/TiO$_2$ coatings on cpTi were obtained by electrospinning of PCL/TiO$_2$ solution onto the substrate. The resulted coatings were structurally characterized and inspected by employing scanning electron microscope (SEM), X-ray diffraction (XRD), and Fourier transform infrared (FTIR) spectroscopy. Given the potential biological applications of PCL/TiO$_2$ coated cpTi substrates, the apatite-forming capacity was examined by immersing in simulated body fluid (SBF) for upto 21 days. Biocompatibility has been evaluated through adhesion/proliferation of hFOB osteoblast cell lines and cytotoxicity by MTT assay. Antimicrobial activity of PCL/TiO$_2$ nanocomposites has been tested using UV light against gram-positive Staphylococcus aureus (*S.aureus*). The resulting surface displays good bioactive properties against osteoblast cell lines with increased viability of 40% at day 3 and superior antibacterial property against *S.aureus* with a significant reduction of bacteria to almost 76%. Surface modification by PCL/TiO$_2$ nanocomposites makes a viable approach for improving dual properties, i.e., biological and antibacterial properties on titanium implants which might be used to prevent implant-associated infections and promoting cell attachment of orthopedic devices at the same time.

Keywords: titanium; antibacterial coatings; electrospinning; nanocomposite coatings; TiO$_2$ photocatalytic; orthopedic infections

1. Introduction

Biocompatible titanium (Ti) and its alloys are broadly accepted metallic materials for hard tissue repair (orthopedic and dental) for its exceptional combination of biomedical and mechanical properties [1]. Even though Ti and its alloys are used as an implant material for more than three decades, there are still some inadequacies that need to be addressed. Especially, bacterial associated diseases/infections during surgery always carry serious hazards leading to a severe clinical economic consequence such as re-hospitalization, complex re-surgeries, implant loosening, high economic associated costs and sometimes even death. Recent studies estimated the current incidence of bacterial infection had incurred a total financial cost of $10 billion with close to 100,000 infections and 8000 reported deaths in the United States alone [2,3]. The reason being, when compared to bioactivity enhancement modifications, relatively very few efforts have been made to address antibacterial activity on the surface before the implantation. It is a known fact that treating an infected orthopedic implant materials post-surgery is hugely complicated, primarily due to the inherent difficulties of treating an established biofilm formed by microorganisms on the surface [4].

Nevertheless, many preventative strategies have been proposed by academics to improve the antibacterial ability of the material before the implantation/surgery [5,6]. But, most of the solutions proposed for obtaining antibacterial surfaces without losing its bioactivity require a complex coatings technique. In this view, several novel strategies such as topographical modifications (nanotubes), incorporating antibacterial agents (Ag, Cu) and various surface treatments [7–10] were suggested/developed to disinfect the bacterial colonization on biomedical implants before the implantation. However, still traces of evidence of bacterial invasion can be still found even after the post-surgery. Nevertheless, these findings highlight the crucial need modifications to the material to prevent bacterial implant-associated infections at early stages. Among these, surface modification either by treatment or coating on the implant material has been well recognized as the best substitute to design and alter the biological performance of the Ti and its alloys [11].

Titanium dioxide (titania, TiO_2), a bioceramic material have become a focus of significant research due to its versatile characteristics [12,13]. TiO_2 nanoparticles are well-known for its stability, non-toxicity, UV resistance and found its application in cosmetics, electronics, biomedical, optics and also as a cleaning reagent [14]. Ever since first reported by Matsunaga et al., TiO_2 photocatalytic properties have drawn more attention in the biomedical field for its specific ability to a kill wide variety of microorganisms under a strong UV radiation [15,16]. In brief, when the surface of TiO_2 is exposed to a strong UV light, electron-hole pairs (e^--h^+) are generated in the valence band and reacts with oxygen and atmospheric water (OH^-) thereby yielding to reactive oxygen species (ROS). The generated ROS acts a powerful oxidizing agent capable of decomposing organic molecules and inactivating micro-organisms through a series of chemical reactions, leading to the powerful antibacterial agent [17–19]. Also, for enhancing the composite cell attachment and proliferation properties, TiO_2 nanoparticles are projected as a secondary phase material for biodegradable polymer matrices [20–25].

Over the past few years, polymer/ceramic nanocomposites as scaffold materials have attracted more attention for bone tissue engineering. Many works have been described in the literature which explains the enhancement of Ti-based implants with organic or organic-inorganic substitute surface coatings [26–28]. In particular, because of the high porosity and large specific surface area, nanocomposite fiber scaffolds have been successfully explored in tissue engineering for orthopedic implants. Among the various techniques for nanocomposite scaffold fabrication, the electrospinning process is described as the most reliable process for producing long and continuous fibers. Electrospinning is a simple and economical fiber fabrication technique that utilizes electrical forces to produce ultrafine micro and nanofibers templets with a wide range of polymers for a variety of applications [29,30]. In most typical tissue engineering strategies, the engineered 3D porous scaffolds serve as a pattern for cell adhesion, expansion, and proliferation of cells ingrowth. Electrospun nanofibers are favorably proficient of imitating microarchitecture of native ECMs owing to their high surface area to volume ratio and relatively large internal porosity. This technique also enables to entrap inorganic ceramic nanoparticles into the organic polymer in a very convenient way to enhance physical, chemical and mechanical properties.

Polycaprolactone (PCL), a semi-crystalline biodegradable polymer, known for its superior mechanical properties, excellent biocompatibility, and slower degradation rate. It is a widely accepted polymer for drug release carriers, biodegradable packaging materials, and more importantly for the development of 3D scaffolds for bone tissue engineering applications. Numerous techniques have been developed to fabricate PCL-based scaffold to a simple two-dimensional structure (casting) to complex three-dimensional (3D printing) objects [31,32]. However, PCL in current form is hydrophobic, which results in lack of wettability and poor cell attachment. Successful blending with bioceramics haven been reported elsewhere [9,23,27,33] for improved biological properties.

The current study aims to incorporate TiO_2 nanoparticles into the PCL scaffolds for improving mechanical properties, biological properties (bioactivity, anti-bacterial, cell adhesion, and cell proliferation) and physiochemical properties (hydrophilicity) for orthopedic applications. To achieve that, PCL/TiO_2-nanocomposite scaffolds were synthesized by electrospinning and coated on cpTi substrates.

2. Materials and Methods

2.1. Sample Preparation

Commercial pure titanium (cpTi), biomedical grade 2 (0.015% carbon, 0.1156% oxygen, 0.0095% nitrogen, 0.0013% hydrogen, 0.04% iron) plates with dimensions of 10 mm × 10 mm × 2 mm (MIDHANI, Hyderabad, India) were used as the base material. The substrates were polished using 300, 600, 800 and 1200 emery sheets followed by disc polishing to mirror finish using diamond paste. The polished samples were further degreased by cleanser, ultrasonically washed with ethanol and sonicated in acetone several times before drying in air overnight.

2.2. Synthesis of PCL/TiO_2 Nanocomposites

PCL/TiO_2 polymer-ceramic hybrid composites containing 2, 5 and 7 wt % of the ceramic content was synthesized by means of the electrospinning process. In brief, poly(ε-caprolactone) (PCL, 80 kDa; Sigma-Aldrich, Singapore) pellets were dissolved in chloroform and methanol (Sigma-Aldrich, Singapore) at 3:1 ratio to make an 8% (w/v) solution and kept under stirring conditions for 4 h. TiO_2 was synthesized as reported earlier *via* electrospraying method by using titanium isopropoxide as the main precursor [34]. To prepare PCL/TiO_2 nanocomposites with a certain PCL to TiO_2 ratio, a known amount of TiO_2 (2, 5 and 7 wt %) powders were added into the solution batch wise under string conditions. Because of the nature of TiO_2 nanoparticles, it was very critical to homogeneously disperse TiO_2 nanoparticles in the PCL matrix to acquire satisfactory dispersion without agglomeration. The mixed solution was under stirring conditions for 72 h using a magnetic stirrer with sonication for every 6 h. Upon attaining satisfactory homogeneous conditions, PCL/TiO_2 solution was taken in a 5 mL syringe equipped with a 22-gauge rounded metal needle. A constant flow rate was set to optimized 500 µL/h through a syringe pump (NE1000, U.S.A). A constant high voltage of 12.0 kV was applied to the needle tip which was positioned at a distance of 10 cm from the grounded collector. Previously cut cpTi substrates of dimensions 10 mm × 10 mm × 2 mm were fixed on the aluminum foil as a collector. Prior to the coating procedure, the substrates were washed, sonicated in dry ethanol for 1 h and left overnight for drying. After the coating procedure, all the samples were dried under vacuum for 1 day to completely remove any solvent residues present.

2.3. Characterization

The morphology, phase analysis, and chemical composition of the obtained PCL/TiO_2 coatings were characterized by scanning electron microscopy (SEM, Hitachi, S-4300, Tokyo, Japan) attached with energy dispersive X-ray spectroscopy (EDX, Thermo Noran, Sonora, CA, USA), X-ray diffractometer (XRD, D8 DISCOVER, Bruker, Billerica, MA, USA) and Fourier transform infrared spectroscope (FT-IR, Perkin-Elmer Spectrum Two spectroscopy) respectively. Morphological analysis of the surface before and after coatings were examined by SEM, operated at an accelerating voltage ranging from 10–15 kV. To avoid any 'charging' and increase conductivity, the dry substrates were sputter-coated for 30 s with a thin deposits of gold (Au) before observing in SEM. XRD (at $2\theta = 10°–90°$) is used for analyzing the phases present in the synthesized nanocomposites coatings with Cu-Kα radiation (λ = 1.540 Å) at a scanning rate of 0.1 step/s. To analyse the chemical composition, FT-IR measurements were performed over a range of 4000–500 cm^{-1}. To identify functional groups of the synthesized samples, KBr powder is mixed and made of pellets. In addition, surface wettability was calculated by measuring static contact angles of deionized water with a contact angle system (Easy DROP, KRUSS, Stuttgart, Germany) at the ambient temperature. For each set of samples, a 10 µL drop of water was deposited on each substrate and allowed to rest for 5 s.

2.4. Biological Studies

2.4.1. Biominerilization Studies

To evaluate the apatite-forming ability (bioactivity), the substrates were immersed in 30 mL of simulated body fluid (SBF) with material composition and composition nearly equal to human blood plasma for 14 days. The SBF solution is prepared in the laboratory by mixing laboratory-grade chemicals (NaCl, NaHCO$_3$, KCl, K$_2$HPO$_4$·3H$_2$O, MgCl$_2$·6H$_2$O, CaCl$_2$, and Na$_2$SO$_4$) in deionized water and maintained at pH 7.4 with tris (hydroxy-methyl) aminomethane and 1 M HCl. The detailed information on ion concentrations of SBF proposed by Kokubo et al. without organic species and its correlation with human blood plasma can be found elsewhere [35]. The falcon tubes containing SBF and substrates were immersed in water bath, maintained at static conditions at 37 °C. After immersing for 14 days, the substrates were carefully taken out from the SBF solution, washed with deionized water and freeze dried at room temperature for further morphological and elemental analysis.

2.4.2. Antibacterial Assay

Prior to inoculation, PCL/TiO$_2$ samples with different TiO$_2$ concentration were UV-irradiated for 5 h to stimulate the photocatalytic reaction in the TiO$_2$ material. Other samples (pure PCL and cpTi substrate) were UV-irradiated and sterilized for the same time interval. These specimens were used for carrying out the antibacterial assay against bacterium Staphylococcus aureus (NCIM 5021). In brief, the samples were placed in 12 welled plates containing 2 mL nutrient broth. A bacterial inoculum of 1×10^7 CFU/mL was added to each well. The plates were then allowed to incubate for 24 h at 37 °C.

After 24 h the samples were carefully removed and rinsed. The samples were sonicated for one minute with an interval of one minute. Five cycles were repeated to ensure complete extraction of bacteria. These samples were centrifuged, and the bacterial pellet obtained, was resuspended in PBS. The bacterial solution was then spread onto pre-cooled agar plates and incubated for 24 h at 37 °C. The resulting colonies were then counted, and the log values were calculated for them. The reduction in bacterial growth was estimated as a reduction in log CFU/mL values

2.4.3. Cell Culture Studies

The cytotoxicity, cell adhesion and proliferation of surface coated PCL/TiO$_2$ nanocomposites were determined by using human fetal osteoblastic cell lines (hFOB 1.19, ATTC CRL 11372) for day 1 and day 3. Five substrates (cpTi, PCL with 2, 5 and 7 wt % of TiO$_2$ content) were adopted to evaluate the potential of using them for biomedical applications. In brief, cells were cultured in base medium of hFOB cell lines i.e., Dulbecco's Modified Eagle's Medium/Ham's Nutrient Mixture F12 (1:1 DMEM/F12) and complemented with 10% fetal bovine serum (FBS), 1% of non-essential amino acids (MEM) and antibiotics (penicillin G, and streptomycin). For the entire duration of experiment, the culture medium was replaced every alternative day and was preserved in a humidified incubator at 37 °C under an atmosphere of 5% CO$_2$. After attaining about 80% confluence, the cells were trypsinated and digested at a final concentration of 5×10^4 cells/cm^2 onto the substrates in 24-well plates. Prior to cell seeding, the substrates with PCL/TiO$_2$ coatings were sterilized by immersing in 70% ethanol for 1 h followed by washing 3 times with sterilized phosphate buffered saline solution (pH = 7.4, PBS).

MTT assay for toxicity connected with cell viability and proliferation were also observed on surface coated cpTi substrates for day 1 and day 3. Pure cpTi without any coating is used as the control.

After removing the culture medium, hFOB cells were quantitatively assessed by seeding 4 mg mL^{-1} MTT 3-(4,5-dimethylazol-2-yl)-2,5-diphenyltetrazolium bromide (yellow) reagent on to the substrates and determined at day 1 and day 3. In brief, both days of incubation, culture medium was removed and washed with 400 µL of prewarmed PBS. Then 400 µL of culture medium accompanied with 60 µL of MTT solution was added to each well-plate containing samples and incubated at 37 °C in a 5% CO_2 humidified atmosphere. After an incubation period of 4 h, 100 µL of the resulting supernatant was added to each well of 96-well ELISA plate. The plates were gently agitated for 3 min to establish complete crystal dissolution. Percentage cell viability was determined by recording optical absorbance at 570 nm with reference to 690 nm using a microplate reader (Bio TEK Instrument, Winooski, VT, USA, EL307C). To ensure reproducibility, tests were carried out by performing triplicates of samples.

2.4.4. Cell Morphology

Samples for SEM analysis were withdrawn from culture after 3 days incubation. For cell adhesion and proliferation studies, before fixing with 3% glutaraldehyde solution, cell-seeded substrates were washed with PBS thrice to confiscate any separated cells. The samples were cleaned again with PBS after 30 min, then kept at 4 °C. After cell fixation, the substrates were dehydrated in a series of ethanol solutions at varying concentrations from 30% to 100% for 15 min each. 1 mL of hexamethyldisilane (HMDS) was added on each sample and left to dry for 2 days. Prior to SEM image analysis, all the samples were gold coated for 20 s.

2.5. Statistical Analysis

All data values are presented a mean ± standard deviation for each group of samples. A paired sample t-test and one-way analysis of variance (ANOVA) were performed to determine statistically significant differences between groups. All tests were conducted with 95% confidence intervals (p-value < 0.05).

3. Results and Discussion

3.1. Surface Characterization

SEM images of unmodified cpTi without any surface coating (Figure 1a) and PCL/TiO_2 nanocomposites with and without TiO_2 particulate additions were presented in Figure 1b–e respectively. Due to an exact balance between the solution viscosity and electrical conductivity, continuous uniformity of the fibrous structures without any beads were observed in all the samples. Figure 1b shows the pure PCL nanofibers (0 wt % TiO_2) in uniform diameter with a smooth surface with an average diameter of 540 ± 40 nm. But, the morphology of the fibers was adversely affected by the addition of TiO_2 nanoparticles. 8 wt % of PCL was used as the constant polymeric solution for electrospinning and a gradual growth in fiber diameter was observed when TiO_2 content increased from 2 wt % (Figure 1c) to 5 wt % (Figure 1d) and to 7 wt % (Figure 1e). With the increase of TiO_2 particle concentration, the size of nanofibers tends to become more significant and visible agglomeration to some extent were observed inside the fibers. For example, the average fiber diameter increases from 640 ± 60 nm to 710 ± 20 nm and 900 ± 89 nm for as-spun PCL/2TiO_2, PCL/5TiO_2 and PCL/7TiO_2 composite nanofibers respectively. The results were in agreement with another study where the addition of nanoparticles increased the diameter of fiber [33].

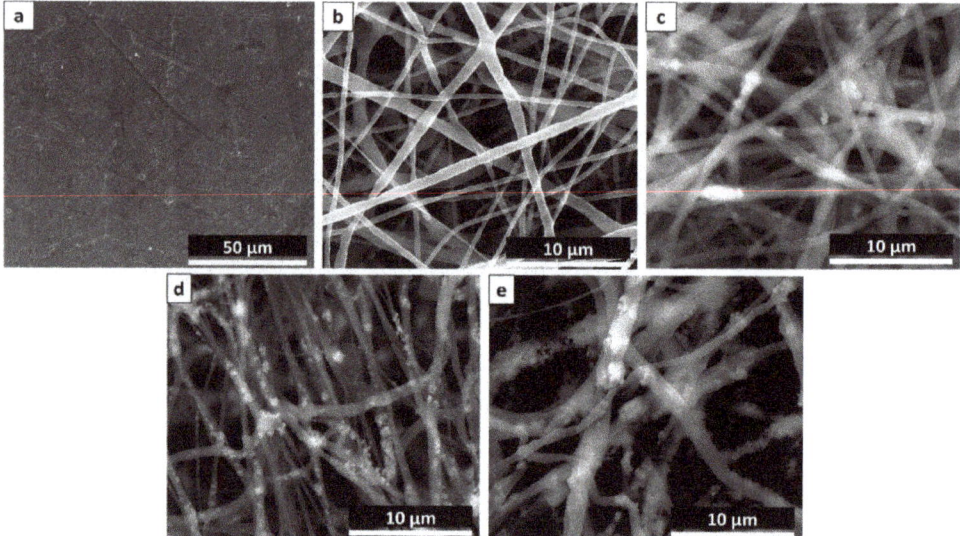

Figure 1. SEM Images displaying (**a**) cpTi substrate; PCL mat containing (**b**) 0 wt % TiO$_2$, (**c**) 3 wt % TiO$_2$, (**d**) 5 wt % TiO$_2$, and (**e**) 7 wt % TiO$_2$.

Also, based on the SEM analysis of Figure 1c,d, the results also indicate that TiO$_2$ nanoparticles are directly embedded inside the PCL nanofiber matrix rather than exposed on top of the fibers. The concentration of 7 wt % of ceramic TiO$_2$ (Figure 1e) causes some of the nanoparticles surfaced on the as-spun nanofiber mats, indicating inhomogeneous mixing of the particles. In view of this, no further experiments above 7 wt % TiO$_2$ have been conducted as it gives the notion that *"outer"* particles on fibers would be leached away easily as no chemical interaction with the PCL nanofibrous mat is made. Due to this, there may also be a negative effect on the mechanical properties (tensile strength and Young's modulus) of the composite. Also, the morphologies of fibers became more irregular when the increasing of TiO$_2$ content which it might because of the influence of ceramic particles on the solution viscosity, surface tension, and concentration. So, optimization of nanoparticle concentration in connection to the mechanical properties of the nanocomposite is sometimes essential.

3.2. Phase Analysis

The XRD patterns of PCL mat and PCL/TiO$_2$ coatings with 2, 5 and 7 wt % of TiO$_2$ content is shown in Figure 2a along with electrosprayed TiO$_2$ nanoparticles for reference. PCL is a semi-crystalline polymer, which can be spotted by two characteristic peaks in the region of 20°–25° (2θ = 22.1° and 24.5°). The TiO$_2$ anatase nanoparticles exhibit characteristic peaks at 2θ values of 25.6°, 35.9°, 37.9°, 38.9°, 48.4°, 53.9° and 56.1°, corresponding to the diffraction patterns of (101), (103), (004), (112), (200), (105), and (211) crystalline planes respectively (JCPDS data No. 36–1451). From the Figure 2a, it can be noted that the relative intensities of TiO$_2$ are increased with increasing TiO$_2$ content. Also, incorporation of TiO$_2$ nanoparticles into the PCL fibrous structure resulted in the widening of peak widths and reduction in its peak height implying the decrease in the crystallinity of PCL structure.

FTIR spectra of the PCL/TiO$_2$ coatings with 2, 5 and 7 wt % of TiO$_2$ content and pure PCL mat for comparison are shown in Figure 2b. The analysis shows the presence of H-bonds between organic (PCL) and inorganic (TiO$_2$) components of the hybrid materials. The spectrum of the TiO$_2$ coatings are visible at wavenumber lower than 998 cm^{-1} are due to Ti-O and Ti-O-Ti vibration bands in the lattice [36]. The peaks located at 1725 cm^{-1}, 1182 cm^{-1}, 1054 cm^{-1}, and 2865 cm^{-1} corresponds to carbonyl groups of C=O, C–O–C, C–O, and alkyl group of C=H stretching vibrations of PCL polymer respectively.

The bands present at 564–647 cm^{-1} (PO$_4^{3-}$ bending vibration), 878 cm^{-1} (P–OH stretching) and 999–1102 cm^{-1} (PO$_4^{3-}$ asymmetric stretching) [37]. Most corresponding bands of TiO$_2$ were detected in the PCL/TiO$_2$ spectrum recommends that TiO$_2$ is effectively incorporated into the PCL nanofibrous mat to form nanocomposites. The reduced intensities of PCL in PCL-TiO$_2$ nanocomposites is because of the presence of physical interaction between PCL and TiO$_2$ nanoparticles [38].

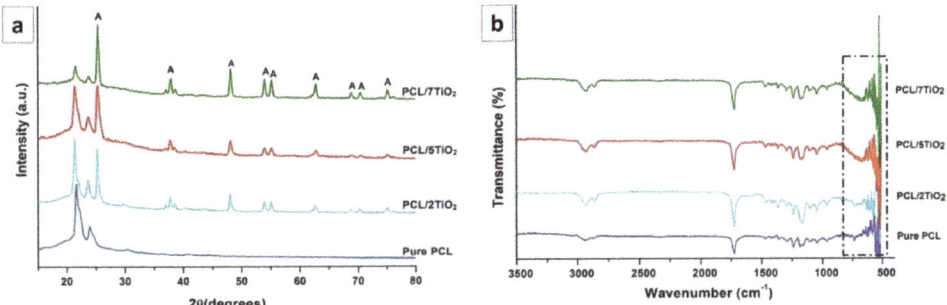

Figure 2. (a) XRD and (b) FTIR patterns of PCL/TiO$_2$ (2, 5 and 7 wt %) nanocomposites in comparison with pure PCL.

3.3. Wettability Studies

To evaluate the effect of TiO$_2$ nanoparticles on the hydrophilicity of electrospun PCL/TiO$_2$ nanocomposites, water contact angles were measured on PCL/TiO$_2$ mats and compared to those of cpTi and pure PCL. From Figure 3, it can be noted that a considerable reduction in water contact angle is observed in PCL/TiO$_2$ nanocomposite (2, 5, and 7 wt %), indicating improved wetting properties. The water contact angle of the pure PCL sample reached above 140°, demonstrating a hydrophobic surface. Presence of polar surface groups (TiO$_2$ nanoparticles) inside the PCL matrix enhanced its hydrophilicity due to a higher interaction between the composite mat and solvent [39]. It is a well-known fact that enhancement in hydrophilicity of a scaffold used in tissue engineering is desirable for initial cell adhesion and cell migration.

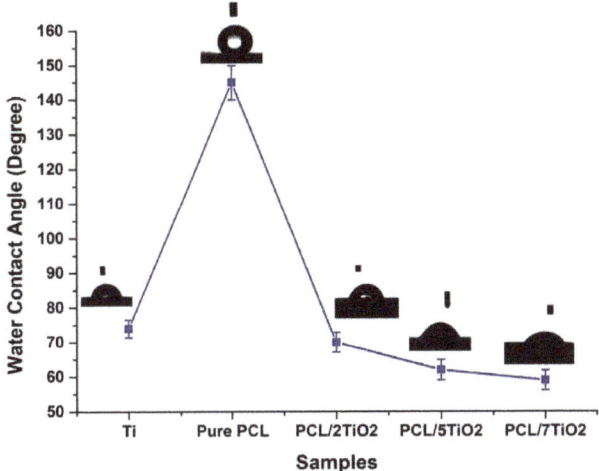

Figure 3. Water contact measurements on various surfaces obtained as a function of TiO$_2$ wt % with distinctive water droplet images after 5 s from droplet.

3.4. Bioactivity Studies

The bioactivity features of PCL/TiO$_2$ nanocomposites are well standardized before. Figure 4 presents the SEM images of the cpTi, pure PCL coated and PCL/TiO$_2$ nanocomposites with 2, 5 and 7 wt % TiO$_2$ content after incubating in SBF for 21 days. The control substrates, i.e., cpTi (Figure 4a), and pure PCL (Figure 4b) exhibit very insignificant mineralization while substantial mineralization takes place in PCL/TiO$_2$ nanocomposite samples. It was observed that PCL bio-nanocomposite with higher TiO$_2$ concentration i.e., PCL/5TiO$_2$ (Figure 4d) and PCL/7TiO$_2$ sample (Figure 4e) have the highest nucleation rate of apatite formation after 21 days. It is worth mentioning that even after immersing in SBF for 21 days, the scaffold preserved its microstructure and showed better interconnectivity of the pores.

Development and growth of apatite layer on the substrate is a dynamic process where the material surfaces dissolve and the new bundle of layers precipitates on the surface [40]. When soaked in SBF, the high content of TiO$_2$ in PCL/5TiO$_2$ and PCL/7TiO$_2$ composites leads to an increase of Ti–OH groups on the surface. The unique development of new Ti–OH groups can stimulus the formation of apatite nucleation. Once the apatite nuclei formed, the growth occurs spontaneously by consuming the positive ions of calcium (Ca^{2+}) and negative ions of phosphate (PO^{3-}) from the SBF fluid to form an amorphous phosphate, which impulsively transforms into hydroxyapatite [Ca$_{10}$(PO$_4$)$_6$(OH)$_2$] [HA] [41]. Based on the EDS analysis for PCL/5TiO$_2$ (Figure 4f), the Ca/P molar ratio was estimated to be approximately 1.5 which is in close agreement with the chemical formulation of the biomineral HA [35].

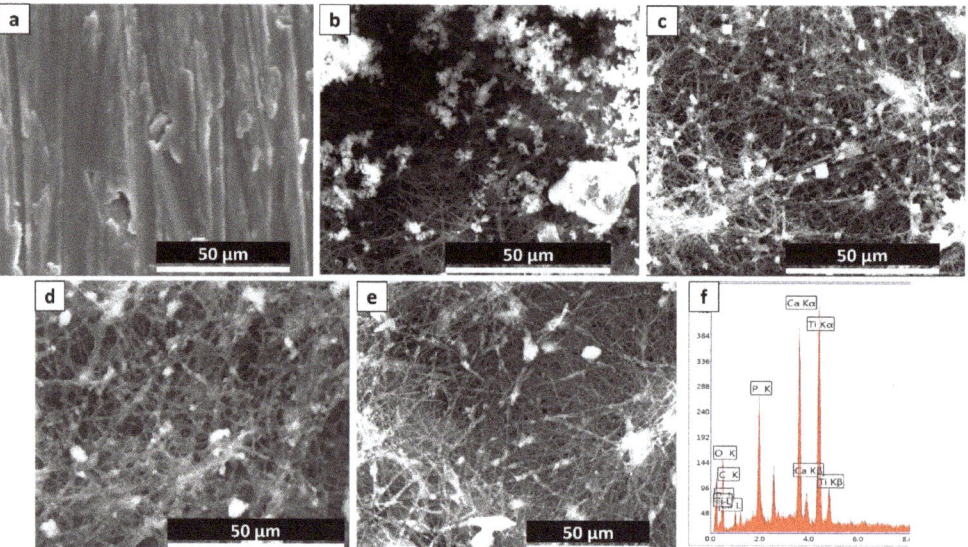

Figure 4. SEM images of (**a**) An, (**b**) PCL (**c**) PCL with 2 wt % TiO$_2$ (**d**) PCL with 5 wt % TiO$_2$ and (**e**) PCL with 7 wt % TiO$_2$ after immersing in SBF for 21 days. (**f**) EDS analysis results for the newly formed calcium and phosphate of PCL with 5 wt % TiO$_2$.

3.5. Cytotoxicity-MTT Assay

With the intention of using the PCL/TiO$_2$ scaffolds in bone tissue engineering application, the influence of the prepared composites on the growth and viability with hFOB cell lines were investigated for day 1 and day 3 and presented in Figure 5. PCL has widely demonstrated polymer for bone tissue engineering application for its slow degradation kinetics and biocompatibility [42]. In vitro cell culture experiments in the current research manifested the importance of PCL as a function of the TiO$_2$ content. In fact, the presence of PCL has enhanced cell proliferation as a function of the TiO$_2$ amount. The addition of TiO$_2$ nanoparticles at low concentration (2 wt % and 5 wt %) to polymeric PCL scaffolds has favoured the proliferation of hFOB cells. Compared to cpTi and PCL fiber mat, initial attachment and proliferation of PCL/2TiO$_2$ and PCL/5TiO$_2$ nanocomposites supported the growth of the cells and mediated their proliferation by approximately 20% and 38% respectively. The results demonstrate the substantial and time-dependent growth in cell viability of TiO$_2$ at lower concentrations.

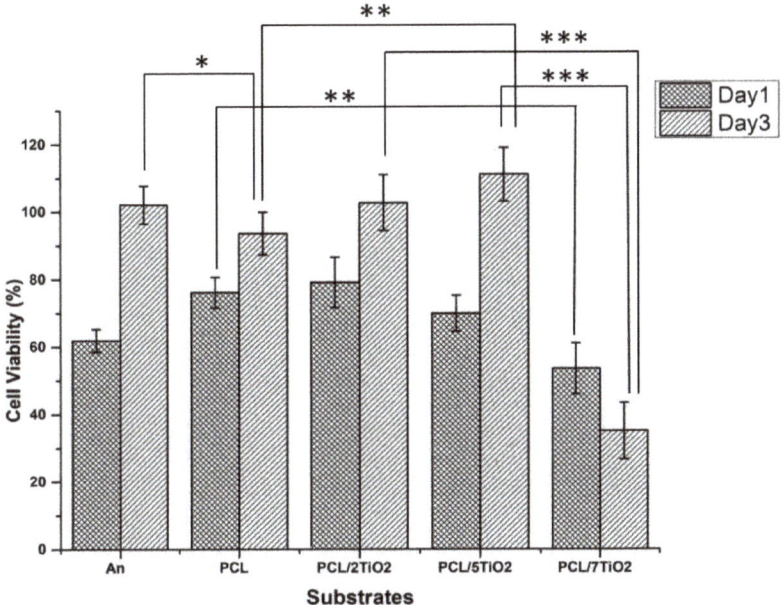

Figure 5. hFOB Cell viability on An, Pure PCL coated, PCL/2TiO$_2$ coated, PCL/5TiO$_2$ coated and PCL/7TiO$_2$ coated cpTi samples cultured for day 1 and day 3 ($p < 0.05$).

However, a noticeable reduction in the cell viabilities is being observed in the PCL/7TiO$_2$ sample for both day 1 and day 3. At a higher content of TiO$_2$, the TiO$_2$ nanoparticles have fetched a significant cytotoxic effect with viabilities of 53 ± 4% and 35 ± 5% as observed on day 1 and day 3 respectively. As we know, the cellular behaviour particle cytotoxicity of TiO$_2$ nanoparticles relay on numerous aspects which includes particle shape/size, chemical stability, and mechanical stimulation. Many studies conclusively showed that the effect of TiO$_2$ nanoparticles concentrations on cellular behaviour at higher concentrations [25,43]. From the above analysis, it can be implied that the incorporation of TiO$_2$ nanoparticles have a positive impact and suggests that PCL/5TiO$_2$ can be considered as best suitable biocompatible material that can be employed as a tissue scaffold for the orthopedic application.

3.6. Cell Morphology (Cell Adhesion and Proliferation) Studies

Cell morphology (cell adhesion and proliferation) of An, PCL and PCL/TiO$_2$ (2, 5 and 7 wt %) scaffolds were carried out using hFOB, a human osteoblast-like cell lines were presented in Figure 6a–j. Integration of TiO$_2$ nanoparticles into PCL, thereby surface coatings on cpTi has significantly increased cell adhesion on day 1 as hFOB cells spread well over the surface and showed the normal morphology of phenotype. Form the SEM analysis, when compared to An (Figure 6a–b) and PCL (Figure 6c–d), it is evident that higher cell proliferation is been observed on the scaffolds containing TiO$_2$ nanoparticles. This early interaction of biomolecules and cells with the material is strongly dependent on the

PCL/TiO$_2$ surface properties, among which hydrophilicity is a key factor. After just 3 days of culture, hFOB cells on PCL/5TiO$_2$ nanocomposites showed an excellent homogeneous structure with a clear evidence of cells penetrating into the scaffolds through their pores (Figure 6g–h). The results are well in connection with the wettability studies (Section 3.3). It is evident that the added TiO$_2$ nanoparticles increase the surface area and surface hydrophilicity, thus by favoring cell adhesion on day 1 and cell proliferation on day 3.

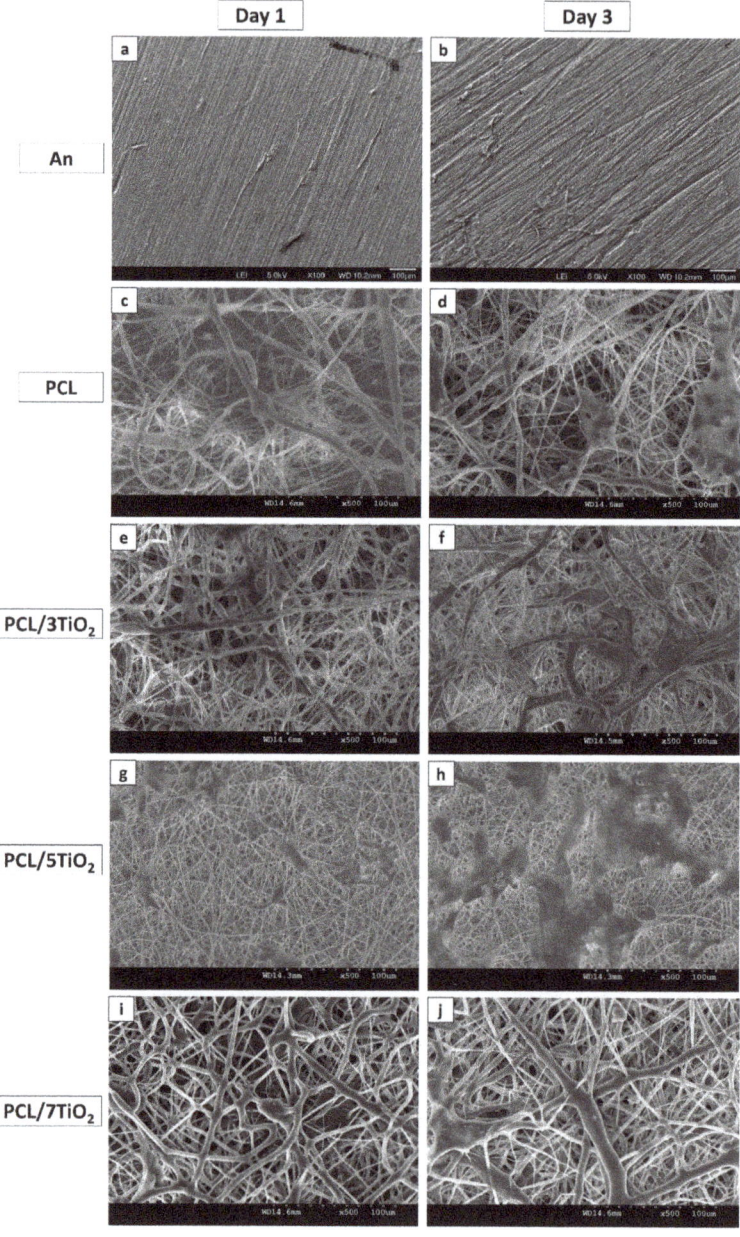

Figure 6. SEM images of hFOB cells seeded on various substrates after day 1, and day 3.

Interestingly, when compared to other scaffolds, cell behavior to higher TiO$_2$ content i.e., with 7 wt % demonstrated little attachment and proliferation (Figure 6i–j). This is probably due to some leachables and an inhomogeneous mixture of TiO$_2$ nanoparticles (Figure 1e). Also, it has been stated in many reports that higher TiO$_2$ content leads to toxicity into the substrates [25,43,44]. As found in this study, for a given PCL concentration, depending on type of cells used, seeding densities and cell viability test method, it must be definite and established that at given polymeric concentration, the amount of TiO$_2$ nanoparticles in scaffolds should not surpass a higher activation level, which should be of the order of 5 wt % in the present case. However, optimized TiO$_2$ nanoparticle concentration (5 wt %) will help to engineer the PCL/TiO$_2$ based scaffold favoring new tissue formation with an antimicrobial property.

3.7. Antibacterial Activity

Due to the significant toxicity and minimal cell proliferation, the sample with 7% TiO$_2$ was eliminated as a candidate from further studies. Amongst the remaining samples, annealed sample with no coating was taken as the control. The electrospun PCL fiber mat shows a relative increase in the bacterial count compared to the control (Figure 7). This is attributed to the fibrous nature of the interface which provides ample opportunities for cellular attachment. This attachment promoting behaviour is also seen in previous studies where the increased surface area contributes to the increase in hFOB proliferation and attachment to electrospun PCL mat and PCL mat containing 2% TiO$_2$. This factor proves to be advantageous for bone growth and repair as observed from the study with mammalian cells. However, it may also cause problems due to increased adhesion of bacterial cells. So, it is advisable to use an antimicrobial agent blended with PCL to retard and offset the bacterial growth. TiO$_2$ has proven to be such an agent. The antimicrobial properties of TiO$_2$ are well known.

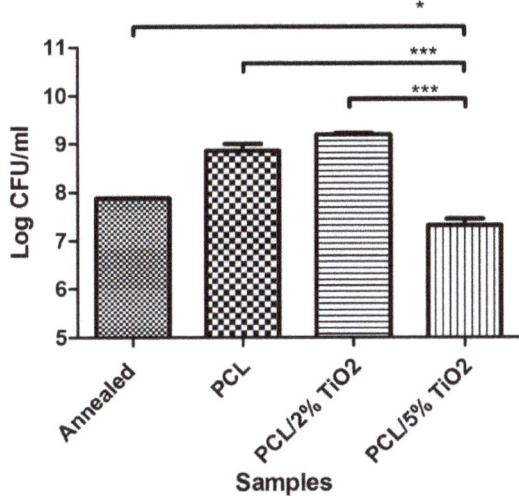

Figure 7. Antibacterial activity on different substrates against *S.aureus*.

As presented in Figure 8, several fundamental mechanisms for cell killing and bacterial growth inhibition by the TiO$_2$ photocatalytic processes were presented in the literature [45,46]. As a result the blend of PCL and 2% TiO$_2$ was initially tested against *S.aureus*. However, there is no significant difference between the PCL fiber mat and the PCL fibers containing 2% TiO$_2$. This implies that this concentration of TiO$_2$ is not enough to bring about a photocatalytic effect that reduces the bacterial count. On increasing the TiO$_2$ concentration to 5%, significant reduction is seen in the bacterial colony count compared to plain annealed sample (control). Thus 5% TiO$_2$ also seems to be the minimal concentration necessary to be incorporated with PCL to achieve antibacterial activity. It also appears to be the optimal concentration of TiO_2 necessary for to give an antibacterial action without significant toxicity. This nanocomposite thus proves to be an efficient approach to achieving bioactivity and improving integration of the titanium implants.

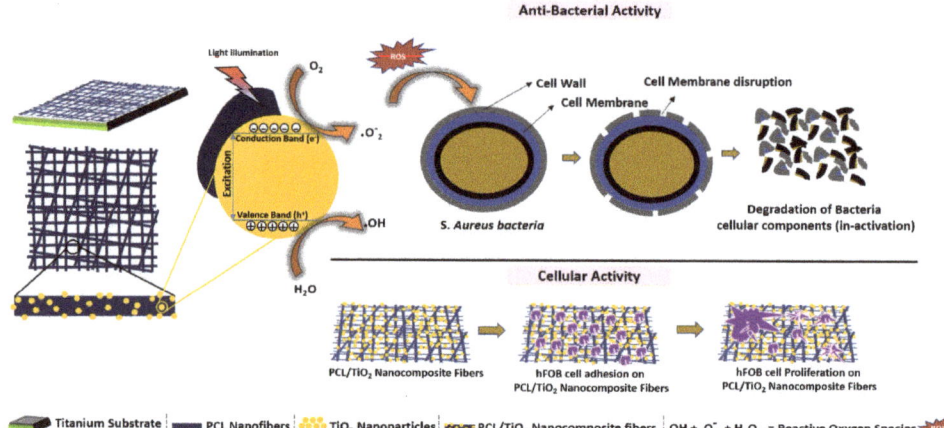

Figure 8. Schematic illustration for the mechanism of degradation bacteria by TiO_2 nanoparticles under UV radiation and cellular activity of PCL/TiO_2 nanocomposites.

4. Conclusions

We report the fabrication of functional PCL-TiO_2 coatings by means of the electrospinning technique and coated on cpTi substrates for bone tissue engineering. A range of electrospun nanocomposite scaffolds were fabricated with a novel and simple material consisting of biodegradable polymer PCL (8 wt %) and TiO_2 (2, 5 and 7 wt %). When subjected to SBF, the enrichment of a PCL scaffolds was observed when TiO_2 nanoparticles were incorporated into it. The TiO_2 nanoparticles serves as a basis of inorganic phosphate to enhance bone cell mineralization, demonstrating the bioactive features of the hybrid scaffolds. The results obtained by hybrid PCL/TiO_2 coatings have a cell viability more than the uncoated one and pure PCL suggesting the effect of nanoparticles. On the contrary, uncoated cpTi and PCL/7TiO_2 appears to be less biocompatible mainly due to minimum contact between the cells-substrate and toxic effect of higher TiO_2 content respectively. The synthesized nanocomposite PCL/TiO_2 fibers possess the capacity to replicate surface chemical properties of bone and also acts as an antibacterial surface with greater efficacy and performance under strong UV radiation. In particular, PCL/5TiO_2 coatings mimic the excellent biological response with respectable antibacterial activity. Because of the effective antibacterial property, no additional covering of the wound for infection prevention is required. The obtained coatings thus can be successfully used to change the surface of cpTi implants to improve their biological properties and antibacterial performance.

Author Contributions: Conceptualization, A.S.K.K.; Data curation, A.S.K.K.; Formal analysis, A.S.K.K.; Investigation, A.S.K.K.; Methodology, A.S.K.K.; Supervision, T.S.S.K., M.D. and S.R.; Validation, A.S.K.K. and R.S.; Writing—original draft, A.S.K.K.; Writing—review & editing, A.S.K.K. and R.S.

Funding: This research received no external funding.

Conflicts of Interest: The authors declare no conflict of interest.

References

1. Niinomi, M.; Nakai, M.; Hieda, J. Development of new metallic alloys for biomedical applications. *Acta Biomater.* **2012**, *8*, 3888–3903. [CrossRef] [PubMed]
2. Darouiche, R.O. Treatment of infections associated with surgical implants. *N. Engl. J. Med.* **2004**, *350*, 1422–1429. [CrossRef] [PubMed]
3. Douglas Scott II, R. *The Direct Medical Costs of Healthcare-Associated Infections in U.S. Hospitals and the Benefits of Prevention*; Centers for Disease Control and Prevention: Atlanta, GA, USA, 2009.
4. Gristina, A. Biomaterial-centered infection: Microbial adhesion versus tissue integration. *Science* **1987**, *237*, 1588–1595. [CrossRef] [PubMed]
5. Romano, C.L.; Scarponi, S.; Gallazzi, E.; Romano, D.; Drago, L. Antibacterial coating of implants in orthopaedics and trauma: A classification proposal in an evolving panorama. *J. Orthop. Surg. Res.* **2015**, *10*, 157. [CrossRef] [PubMed]
6. Kazemzadeh-Narbat, M.; Lai, B.F.; Ding, C.; Kizhakkedathu, J.N.; Hancock, R.E.; Wang, R. Multilayered coating on titanium for controlled release of antimicrobial peptides for the prevention of implant-associated infections. *Biomaterials* **2013**, *34*, 5969–5977. [CrossRef] [PubMed]
7. Zhao, L.; Wang, H.; Huo, K.; Cui, L.; Zhang, W.; Ni, H.; Zhang, Y.; Wu, Z.; Chu, P.K. Antibacterial nano-structured titania coating incorporated with silver nanoparticles. *Biomaterials* **2011**, *32*, 5706–5716. [CrossRef] [PubMed]
8. Wan, Y.Z.; Raman, S.; He, F.; Huang, Y. Surface modification of medical metals by ion implantation of silver and copper. *Vacuum* **2007**, *81*, 1114–1118. [CrossRef]
9. Kiran, A.S.K.; Sampath Kumar, T.S.; Perumal, G.; Sanghavi, R.; Doble, M.; Ramakrishna, S. Dual nanofibrous bioactive coating and antimicrobial surface treatment for infection resistant titanium implants. *Prog. Org. Coat.* **2018**, *121*, 112–119. [CrossRef]
10. Jasinski, J.; Kurpaska, L.; Lubas, M.; Jasinski, J.; Sitarz, M. Duplex titanium oxide layers for biomedical applications. *Mater. Perform. Charact.* **2016**, *5*, 461–471. [CrossRef]
11. Jasinski, J.J.; Kurpaska, L.; Lubas, M.; Lesniak, M.; Jasinski, J.; Sitarz, M. Effect of hybrid oxidation on the titanium oxide layer's properties investigated by spectroscopic methods. *J. Mol. Struct.* **2016**, *1126*, 165–171. [CrossRef]
12. Gupta, S.M.; Tripathi, M. A review of TiO_2 nanoparticles. *Chin. Sci. Bull.* **2011**, *56*, 1639. [CrossRef]
13. Nohynek, G.J.; Lademann, J.; Ribaud, C.; Roberts, M.S. Grey goo on the skin? Nanotechnology, cosmetic and sunscreen safety. *Crit. Rev. Toxicol.* **2007**, *37*, 251–277. [CrossRef] [PubMed]
14. Chen, X.; Mao, S.S. Titanium dioxide nanomaterials: Synthesis, properties, modifications, and applications. *Chem. Rev.* **2007**, *107*, 2891–2959. [CrossRef] [PubMed]
15. Matsunaga, T.; Tomoda, R.; Nakajima, T.; Wake, H. Photoelectrochemical sterilization of microbial cells by semiconductor powders. *FEMS Microbiol. Lett.* **1985**, *29*, 211–214. [CrossRef]
16. Maness, P.-C.; Smolinski, S.; Blake, D.M.; Huang, Z.; Wolfrum, E.J.; Jacoby, W.A. Bactericidal activity of photocatalytic TiO_2 reaction: Toward an understanding of its killing mechanism. *Appl. Environ. Microbiol.* **1999**, *65*, 4094–4098. [PubMed]
17. Nakata, K.; Fujishima, A. TiO_2 photocatalysis: Design and applications. *J. Photochem. Photobiol. C Photochem. Rev.* **2012**, *13*, 169–189. [CrossRef]
18. Wang, W.; Huang, G.; Yu, J.C.; Wong, P.K. Advances in photocatalytic disinfection of bacteria: Development of photocatalysts and mechanisms. *J. Environ. Sci.* **2015**, *34*, 232–247. [CrossRef] [PubMed]
19. Luttrell, T.; Halpegamage, S.; Tao, J.; Kramer, A.; Sutter, E.; Batzill, M. Why is anatase a better photocatalyst than rutile?—Model studies on epitaxial TiO_2 films. *Sci. Rep.* **2014**, *4*, 4043. [CrossRef] [PubMed]
20. Huinan, L.; Elliott, B.S.; Thomas, J.W. Increased osteoblast functions on nanophase titania dispersed in poly-lactic-*co*-glycolic acid composites. *Nanotechnology* **2005**, *16*, S601.
21. Santis, R.D.; Catauro, M.; Silvio, L.D.; Manto, L.; Raucci, M.G.; Ambrosio, L.; Nicolais, L. Effects of polymer amount and processing conditions on the in vitro behaviour of hybrid titanium dioxide/polycaprolactone composites. *Biomaterials* **2007**, *28*, 2801–2809. [CrossRef] [PubMed]
22. Boccaccini, A.R.; Blaker, J.J. Bioactive composite materials for tissue engineering scaffolds. *Expert Rev. Med. Devices* **2005**, *2*, 303–317. [CrossRef] [PubMed]

23. Catauro, M.; Raucci, M.G.; Marco, D.D.; Ambrosio, L. Release kinetics of ampicillin, characterization and bioactivity of TiO$_2$/PCL hybrid materials synthesized by sol–gel processing. *J. Biomed. Mater. Res. Part A* **2006**, *77A*, 340–350. [CrossRef] [PubMed]
24. Boccaccini, A.R.; Blaker, J.J.; Maquet, V.; Chung, W.; Jérôme, R.; Nazhat, S.N. Poly(D,L-lactide) (PDLLA) foams with TiO$_2$ nanoparticles and PDLLA/TiO$_2$-Bioglass® foam composites for tissue engineering scaffolds. *J. Mater. Sci.* **2006**, *41*, 3999–4008. [CrossRef]
25. Gerhardt, L.C.; Jell, G.M.; Boccaccini, A.R. Titanium dioxide TiO$_2$ nanoparticles filled poly(D,L lactid acid) (PDLLA) matrix composites for bone tissue engineering. *J. Mater. Sci. Mater. Med.* **2007**, *18*, 1287–1298. [CrossRef] [PubMed]
26. Simchi, A.; Tamjid, E.; Pishbin, F.; Boccaccini, A.R. Recent progress in inorganic and composite coatings with bactericidal capability for orthopaedic applications. *Nanomedicine* **2011**, *7*, 22–39. [CrossRef] [PubMed]
27. Liao, L.; Zhang, C.; Gong, S. Preparation of poly(ε-caprolactone)/clay nanocomposites by microwave-assisted in situ ring-opening polymerization. *Macromol. Rapid Commun.* **2007**, *28*, 1148–1154. [CrossRef]
28. Gawęda, M.; Jeleń, P.; Długoń, E.; Wajda, A.; Leśniak, M.; Simka, W.; Sowa, M.; Detsch, R.; Boccaccini, A.R.; Sitarz, M. Bioactive layers based on black glasses on titanium substrates. *J. Am. Ceram. Soc.* **2018**, *101*, 590–601. [CrossRef]
29. Zhang, B.G.; Myers, D.E.; Wallace, G.G.; Brandt, M.; Choong, P.F. Bioactive coatings for orthopaedic implants-recent trends in development of implant coatings. *Int. J. Mol. Sci.* **2014**, *15*, 11878–11921. [CrossRef] [PubMed]
30. Kumar, P.S.; Jayaraman, S.; Singh, G. Polymer and composite nanofiber: Electrospinning parameters and rheology properties. In *Rheology and Processing of Polymer Nanocomposites*; John Wiley & Sons: Hoboken, NJ, USA, 2016.
31. Woodruff, M.A.; Hutmacher, D.W. The return of a forgotten polymer-polycaprolactone in the 21st century. *Progress Polym. Sci.* **2010**, *35*, 1217–1256. [CrossRef]
32. Hutmacher, D.W. Scaffolds in tissue engineering bone and cartilage. *Biomaterials* **2000**, *21*, 2529–2543. [CrossRef]
33. Wutticharoenmongkol, P.; Sanchavanakit, N.; Pavasant, P.; Supaphol, P. Preparation and characterization of novel bone scaffolds based on electrospun polycaprolactone fibers filled with nanoparticles. *Macromol. Biosci.* **2006**, *6*, 70–77. [CrossRef] [PubMed]
34. Sandeep Kranthi Kiran, A.; Madhumathi, K.; Sampath Kumar, T.S. Electrosprayed titania nanocups for protein delivery. *Colloid Interface Sci. Commun.* **2016**, *12*, 17–20. [CrossRef]
35. Kokubo, T.; Takadama, H. How useful is SBF in predicting in vivo bone bioactivity? *Biomaterials* **2006**, *27*, 2907–2915. [CrossRef] [PubMed]
36. Ashiri, R. Detailed FT-IR spectroscopy characterization and thermal analysis of synthesis of barium titanate nanoscale particles through a newly developed process. *Vib. Spectrosc.* **2013**, *66*, 24–29. [CrossRef]
37. Jastrzębski, W.; Sitarz, M.; Rokita, M.; Bułat, K. Infrared spectroscopy of different phosphates structures. *Spectrochim. Acta Part A Mol. Biomol. Spectrosc.* **2011**, *79*, 722–727. [CrossRef] [PubMed]
38. Coates, J. Interpretation of infrared spectra, a practical approach. In *Encyclopedia of Analytical Chemistry*; John Wiley & Sons: Hoboken, NJ, USA, 2006.
39. Watson, C.L.; Letey, J. Indices for characterizing soil-water repellency based upon contact angle-surface tension relationships1. *Soil Sci. Soc. Am. J.* **1970**, *34*, 841–844. [CrossRef]
40. Zhang, J.; Dai, C.; Wei, J.; Wen, Z.; Zhang, S.; Lin, L. Calcium phosphate/chitosan composite coating: Effect of different concentrations of Mg^{2+} in the m-SBF on its bioactivity. *Appl. Surf. Sci.* **2013**, *280*, 256–262. [CrossRef]
41. Satoshi, H.; Kanji, T.; Chikara, O.; Akiyoshi, O. Mechanism of apatite formation on a sodium silicate glass in a simulated body fluid. *J. Am. Ceram. Soc.* **1999**, *82*, 2155–2160.
42. Sung, H.-J.; Meredith, C.; Johnson, C.; Galis, Z.S. The effect of scaffold degradation rate on three-dimensional cell growth and angiogenesis. *Biomaterials* **2004**, *25*, 5735–5742. [CrossRef] [PubMed]
43. Shi, H.; Magaye, R.; Castranova, V.; Zhao, J. Titanium dioxide nanoparticles: A review of current toxicological data. *Part. Fibre Toxicol.* **2013**, *10*, 15. [CrossRef] [PubMed]

44. Ghosal, K.; Manakhov, A.; Zajíčková, L.; Thomas, S. Structural and surface compatibility study of modified electrospun poly(ε-caprolactone) (PCL) composites for skin tissue engineering. *AAPS PharmSciTech* **2017**, *18*, 72–81. [CrossRef] [PubMed]
45. Lu, Z.-X.; Zhou, L.; Zhang, Z.-L.; Shi, W.-L.; Xie, Z.-X.; Xie, H.-Y.; Pang, D.-W.; Shen, P. Cell damage induced by photocatalysis of TiO_2 thin films. *Langmuir* **2003**, *19*, 8765–8768. [CrossRef]
46. Sunada, K.; Kikuchi, Y.; Hashimoto, K.; Fujishima, A. Bactericidal and detoxification effects of TiO_2 thin film photocatalysts. *Environ. Sci. Technol.* **1998**, *32*, 726–728. [CrossRef]

© 2018 by the authors. Licensee MDPI, Basel, Switzerland. This article is an open access article distributed under the terms and conditions of the Creative Commons Attribution (CC BY) license (http://creativecommons.org/licenses/by/4.0/).

Article

Electrospun Gelatin Fibers Surface Loaded ZnO Particles as a Potential Biodegradable Antibacterial Wound Dressing

Yu Chen [1,2,3], Weipeng Lu [1,3,*], Yanchuan Guo [1,2,3,*], Yi Zhu [3] and Yeping Song [3]

1. Key Laboratory of Photochemical Conversion and Optoelectronic Material, Technical Institute of Physics and Chemistry, Chinese Academy of Sciences, Beijing 100190, China; chenyubrc@mail.ipc.ac.cn
2. University of Chinese Academy of Sciences, Beijing 100049, China
3. Hangzhou Research Institute of Technical Institute of Physics and Chemistry, Chinese Academy of Sciences, Hangzhou 310018, China; zhuyi@mail.ipc.ac.cn (Y.Z.); songyeping@mail.ipc.ac.cn (Y.S.)
* Correspondence: luweipeng@mail.ipc.ac.cn (W.L.); yanchuanguo@mail.ipc.ac.cn (Y.G.); Tel.: +86-571-8785-3765 (W.L.)

Received: 6 March 2019; Accepted: 25 March 2019; Published: 3 April 2019

Abstract: Traditional wound dressings require frequent replacement, are prone to bacterial growth and cause a lot of environmental pollution. Therefore, biodegradable and antibacterial dressings are eagerly desired. In this paper, gelatin/ZnO fibers were first prepared by side-by-side electrospinning for potential wound dressing materials. The morphology, composition, cytotoxicity and antibacterial activity were characterized by using Fourier transform infrared spectroscopy (FTIR), X-ray diffractometry (XRD), particle size analyzer (DLS), scanning electron microscopy (SEM), energy dispersive X-ray spectroscopy (EDX), thermogravimetry (TGA) and Incucyte™ Zoom system. The results show that ZnO particles are uniformly dispersed on the surface of gelatin fibers and have no cytotoxicity. In addition, the gelatin/ZnO fibers exhibit excellent antibacterial activity against *Staphylococcus aureus* (*S. aureus*) and *Escherichia coli* (*E. coli*) with a significant reduction of bacteria to more than 90%. Therefore, such a biodegradable, nontoxic and antibacterial fiber has excellent application prospects in wound dressing.

Keywords: gelatin fibers; ZnO particles; antibacterial activity

---

1. Introduction

Skin, as the human body's largest organ, exerts a vital role in protecting the human body from external harm. Skin damage can lead to microbial invasion of the human body, resulting in a threat to human health [1–3]. Medical dressing is a kind of medical equipment which can cover damaged skin and form a microenvironment conducive to wound healing, thus playing an effective role in wound care and treatment. At present, most medical dressings still use traditional cotton gauze. Nevertheless, traditional cotton gauze needs to be replaced frequently, and easily adheres to the wound, which can easily lead to secondary tissue trauma and bacterial breeding [4–6]. Moreover, huge dressing wastes cause great harm to the environment. Therefore, it is urgent to design new biodegradable and biocompatible dressing materials with good antimicrobial activity.

Gelatin, as a kind of natural material, is hydrolyzed from collagen in animals, which has many advantages, such as good accessibility, a wide source of raw materials and low cost. Its amino acid composition is similar to collagen and has good biocompatibility, biodegradability and low immunogenicity [7–9]. Especially in the past decades, with the development of electrospinning technology, gelatin fibers can be prepared simply and quickly. Furthermore, there are many small secondary structures on the surface of gelatin fibers, which is similar to the structure of extracellular matrix (ECM) and closer to the structural size of organisms. Therefore, it plays an important part in cell attachment, growth, migration and differentiation, as well as the formation of new tissues [10–12]. Meanwhile, gelatin fibers have strong adsorption, good filtration, barriers, adhesion and hygroscopicity [13–16]. Therefore, the gelatin 3D nanostructures prepared by electrospinning can be widely used in the biomedical materials, which receives more and more attention. In addition, in recent years nano-inorganic ions and nano-metal oxides have been found to have extensive antibacterial properties, which caused widespread concern [17–19]. Among them, ZnO particles were generally recognized as a safe (GRAS) material by the Food and Drug Administration (FDA) [20]. It has excellent antibacterial properties and minimal effect on human cells, which is extensively used in biomedicine and health products [21–23].

Recently, the blending of ZnO particles with gelatin has been used in the research of antimicrobial materials. Liu et al. prepared gelatin/ethyl cellulose/ZnO nanofibers by electrospinning. The results showed that the nanofibers had a good inhibitory effect on *Escherichia coli* (*E. coli*) and *Staphylococcus aureus* (*S. aureus*) [24]. Chhabra et al. synthesized ZnO doped gelatin and poly-methyl vinyl ether-*alt*-maleic anhydride (PMVE/MA) composite electrospun scaffolds, which can inhibit bacterial activity and have no cytotoxicity to mammalian cells [25]. Münchow et al. prepared ZnO loaded gelatin/polycaprolactone (PCL) composite nanofibers by electrospinning technique [26]. At present, the current preparation of ZnO/gelatin fibers is usually carried out in the form of blending spinning. Nevertheless, ZnO particles are hard to fully exert efficient antibacterial activity when encapsulated in gelatin fibers. Consequently, it is necessary to increase the content of ZnO particles to achieve better antibacterial effects [27]. However, with that, excess ZnO particles cause waste of materials, and even more seriously excess ZnO particles may lead to cytotoxic effects and affect the growth of the tissue [28–30].

In this paper, unlike the traditional blending method, ZnO particles were dispersed in ethanol, so that ZnO particles can follow the solvent to be sprayed on gelatin fibers by using the side-by-side spray nozzles in the electrospinning process. With the volatilization of ethanol solvent, ZnO particles can be uniform spread only on the surface of gelatin fibers (as shown in Figure 1), so as to achieve the best antibacterial effect with the minimum content of ZnO particles.

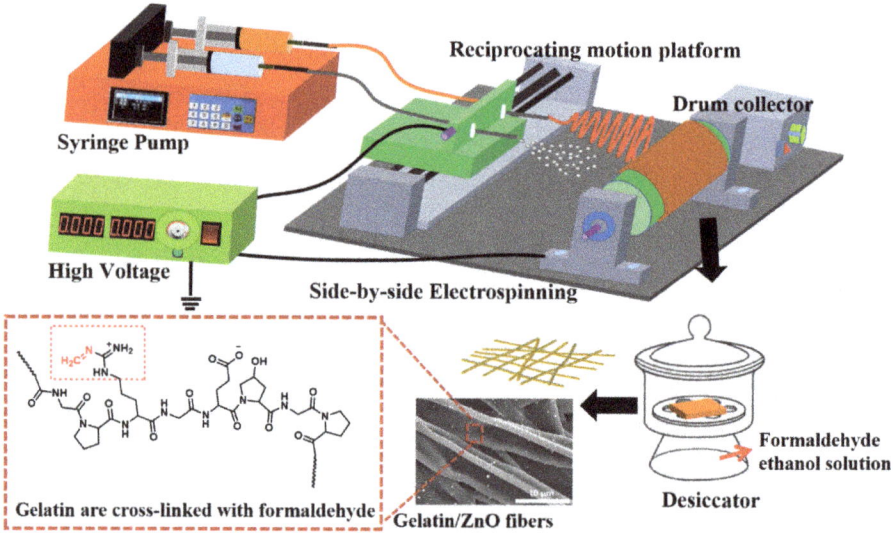

Figure 1. Schematic illustration of the fabrication of gelatin/ZnO fibers by side-by-side electrospinning.

2. Materials and Methods

2.1. Materials

Gelatin (type B, basic-processed, prepared by bones, with a molecular weight of 100,000, viscosity value of 4.9 MPa·s^{-1}) was obtained from Dongbao Bio-Tech Co., Ltd. (Baotou, China). 2,2,2-trifluoroethanol, formaldehyde solution (37–40 wt %), zinc acetate dihydrate ($Zn(CH_3COO)_2 \cdot 2H_2O$), diethylene glycol and ethanol absolute were purchased from Titan Scientific Co., Ltd. (Shanghai, China). The human lung fibroblast cell lines (MRC-5) were purchased from the Cell Bank of Chinese Academy of Sciences (Shanghai, China). *Staphylococcus aureus* (*S. aureus*) and *Escherichia coli* (*E. coli*) were obtained from the China General Microbiological Culture Collection Center (Beijing, China). The deionized water had a resistivity of 18.2 MΩ·cm in the process of the experiments. All chemical reagents were of analytical grade and required no post-treatment.

2.2. Preparation of ZnO Particles

For preparation of ZnO particles [31], 878 mg (0.004 mol) $Zn(CH_3COO)_2 \cdot 2H_2O$ was added into 40 mL diethylene glycol with continued ultrasonic until dissolved completely, followed by transferring into a teflon-lined stainless steel autoclave (45 mL). Then, the autoclave was heated to 160 °C from room temperature in an oven and maintained for 5 h. After the reaction, the white power was obtained by centrifugation. Then, the power was washed with ethanol and water in turn, and finally dried in the oven at 100 °C for 5 h.

2.3. Preparation of Gelatin/ZnO Fibers

For preparation of gelatin/ZnO fibers, gelatin (12.5%, w/v) was added into 2,2,2-trifluoroethanol with constant stirring at 45 °C until it was completely dissolved. Then, ZnO particles (0%, 0.10%, 0.25%, respectively) were evenly dispersed in the ethanol solution by ultrasonic. In the electrospinning process, gelatin solution and ZnO particles solution were added into 50 mL syringes respectively, in which the diameter of the syringe needle containing gelatin was 1.25 mm and that containing ZnO particles was 0.80 mm. The distance from the drum collector to the needle tip was 15 cm, and the applied voltage was 17 kV. The gelatin fibers and ZnO particles were spun with a flow rate of 5 and 2 mL·h^{-1}, respectively. In addition, the speed of reciprocating motion platform and drum collector were 55 and 70 r·min^{-1}, respectively. After electrospinning, the fibers were placed on the porous ceramic shelf of the desiccator, and the ethanol solution containing 1% formaldehyde solution was added at the bottom. Then, the desiccator was put into the oven at 25 °C for 48 h. After that, the cross-linked fibers were put into the oven at 40 °C for 24 h to remove excess formaldehyde. GZ0', GZ1' and GZ2' represent the gelatin/ZnO fibers, in which the concentration of ZnO particles solution is 0%, 0.1% and 0.25% in the electrospinning process, respectively. GZ0, GZ1 and GZ2 represent GZ0', GZ1' and GZ2' after cross-linking.

2.4. Characterization

The microstructure and chemical composition of the gelatin/ZnO fibers were characterized by a S-4300 scanning electron microscope (SEM, Hitachi, Tokyo, Japan) and energy dispersive X-ray spectroscopy (EDX, Hitachi, Tokyo, Japan). The size of ZnO particles was analyzed by the dynamic light scattering (DLS) method (litesizer^{TM} 500, Anton Paar GmbH, Graz, Austria). Samples were dispersed in the ethanol, the concentration of the sample was 1 mg·mL^{-1}, and the pH value was 6.5. The functional groups in the fibers were characterized by Fourier transform infrared spectroscopy (FTIR) (Bruker Tensor II, Karlsruhe, Germany), and the scanning range of the samples was 4000–400 cm^{-1}. The crystal structures of the samples were determined by X-ray diffraction (XRD, ARL XTRA, Zurich, Switzerland). The scanning rate was 0.1 s·step^{-1} and the scanning range was 10–80°. The thermal stability of the samples was analyzed by thermogravimetry (TGA, NETZSCH, Selb, Germany) in nitrogen atmosphere, and the heating rate was 10 °C·min^{-1}. The Incucyte^{TM} Zoom system (EssenBio, Ann Arbor, MI, USA) was used to observe effect of the extraction on MRC-5 cells in real time. The extraction and 10% FBS were placed in 96-well plates. Each well was inoculated with 2000 cells and cultured in an atmosphere of 5% CO_2 at 37 °C for seven days. Phase contrast images were acquired every 3 h.

2.5. Stability of GZ2 in Aqueous Solutions

The stability of cross-linked fibers in aqueous solution was studied by taking GZ2 as an example. Specifically, GZ2 was cut into a $2 \times 0.5 \times 0.2$ cm^3 shape (20 mg) and placed in a bottle containing 10 mL phosphate buffer saline (PBS, 0.01 M, pH = 7.2–7.4). Then, the bottles were placed in the oven at 37 °C and recorded every 24 h. The experiment was done in triplicate.

2.6. Cell Culture and Proliferation

MRC-5 cells were used to assess the cell culture and proliferation. The cells were cultured in a humidified chamber containing 5% CO_2 at 37 °C using Dulbecco's modified eagle's medium/F-12 (DME/F-12) containing 10% fetal bovine serum. Extraction experiment was carried out according to the instruction of ISO 10993-12: 2002. Serum free cell culture medium (DME/F-12) was used to obtain the extraction media of the GZ0, GZ1 and GZ2. The extraction ratio (the ratio of sample quality to extraction medium) was 0.1 g·mL^{-1}. GZ0, GZ1 and GZ2 soaked in the medium were incubated in a humidified atmosphere with 5% CO_2 at 37 °C for 24 h. Cell culture medium (DME/F-12) was used as a negative control.

2.7. Antibacterial Evaluation

E. coli and *S. aureus* were chosen to explore the antibacterial properties of gelatin/ZnO fibers by using the viable colony count method. Tryptic Soy Broth (TSB) and Trypticase Soy Agar (TSA) were used as culturing nutrient sources. Specifically, *E. coli* and *S. aureus* were aseptically inoculated in TSB and then cultured in a shaker at 37 °C for 16 h. Each of the cultured broths were continually cultured in new TBS in a shaker at 37 °C for another 3 h, and then each of the cultured broths were centrifuged and washed twice with sterile PBS (0.01 M, pH = 7.2–7.4). Finally, the bacterial PBS suspension with the concentration of 1×10^6 colony forming units per milliliter (CFU·mL^{-1}) was obtained by the gradient dispersion method. The sterile gelatin/ZnO fibers (GZ0, GZ1 and GZ2) were cut into the shape of 2×2 cm^2 (30 mg) and then placed into tube containing 10 mL sterile PBS. One group of tubes was irradiated for 1 h under ultraviolet light (UV, 365 nm, 50 w), the other group was not irradiated. After the pretreatment, 0.1 mL of each bacterial suspension was added, and incubated in the shaker at 37 °C for 3 h. After that, a 10 μL solution was taken and serially diluted in sterile PBS. Then, 30 μL of each diluent was taken and spread onto a TSA plate, and then all plates were incubated for 16 h at 37 °C. The numbers of the suitable colonies that formed were counted. In addition, medium with only inoculum was used as negative control, and pure ZnO particles (30 mg) were used as positive control. All experiments were performed in triplicate.

2.8. Statistical Analysis

The data were expressed as mean ± standard deviation. Statistically significant differences of the samples were assessed using a Student's *t*-test. $p < 0.05$ was considered to be statistically significant.

3. Results and Discussion

3.1. Morphologies of Gelatin/ZnO Fibers

3.1.1. Characterization of ZnO Particles

The characterization data of the ZnO particles are shown in Figure 2. Specifically, Figure 2a represents the SEM images of ZnO particles. It indicates that the ZnO particles have a regular spherical shape and are uniformly dispersed without agglomeration. The DLS measurements show that average hydrodynamic diameter is about 589.3 nm (Figure 2b). Figure 2c illustrates the FTIR spectrum of the ZnO particles. The single peak at 407 cm^{-1} can be assigned to stretching mode of the Zn-O bond [32]. To further determine the composition of the resulting particles, XRD was carried out. As shown in Figure 2d, the diffraction peaks at 2θ values of 31.8°, 34.5°, 36.3°, 47.8°, 56.3°, 63.2°, 68,0° and 69.2° correspond to (100), (002), (101), (102), (110), (103), (112) and (201) crystal planes of ZnO, respectively [33].

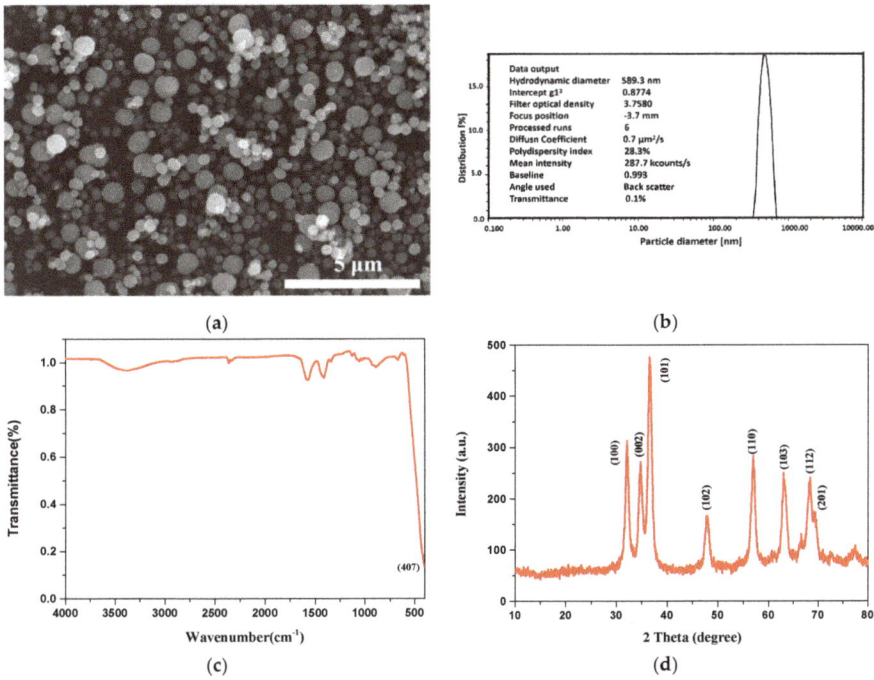

Figure 2. Characterization of ZnO particles (**a**) scanning electron microscopy (SEM) image, (**b**) particle size distribution by intensity, (**c**) Fourier transform infrared spectroscopy (FTIR) spectrum and (**d**) X-ray diffractometry (XRD) spectrum.

3.1.2. Characterization of Gelatin/ZnO Fibers

SEM measurements were performed to study morphology of the fibers before and after crosslinking. As shown in Figure 3a–c, GZ0', GZ1' and GZ2' represent the gelatin/ZnO fibers, in which the concentration of ZnO particles solution is 0%, 0.1% and 0.25% in the electrospinning process, respectively. The average diameters of GZ0', GZ1' and GZ2' are about 6.22, 5.67 and 7.32 μM. The pure gelatin fiber (GZ0') exhibits a uniform smooth surface, and the fiber thickness is consistent. With the addition of ZnO particles in electrospinning process, ZnO particles can be evenly dispersed on the surface of gelatin fibers (Figure 3b,c), but no particles are found inside GZ1' and GZ2' (indicated by blue arrow in Figure 4). The morphology of the fibers remains basically unchanged. Furthermore, as a potential wound dressing, gelatin/ZnO fibers need to have good physical and chemical stability in aqueous solution. Thus, formaldehyde was chosen as the cross-linking agent. GZ0, GZ1 and GZ2 (Figure 3d–f) represent the GZ0', GZ1' and GZ2' after crosslinking, respectively. Figure 5 shows the degradation process of GZ2 in PBS at 37 °C. It can be seen that the degradation of GZ2 is basically completed after five days, which indicates that GZ2 has good water resistance. In addition, Figure 3d–e shows that the average diameter of the nanofibers increases after the cross-linking, and the fibers are curly (indicated by red arrow above). Meanwhile, the fibers become denser and fuse at some intersection points (shown by yellow arrow above).

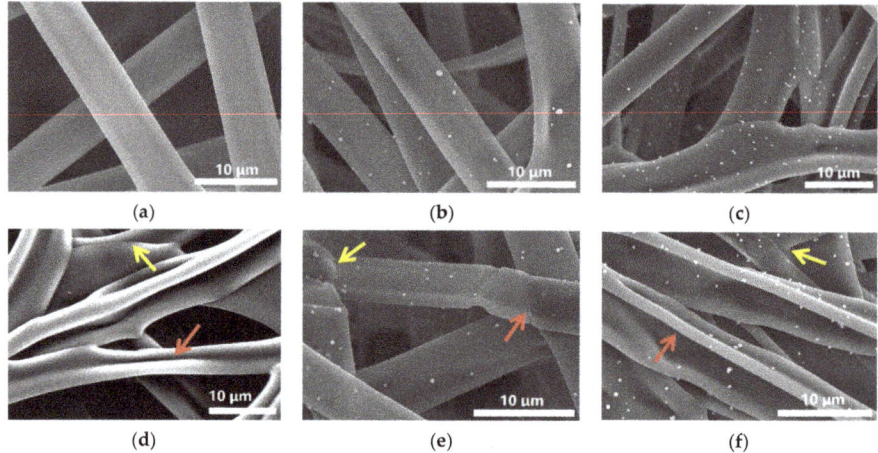

Figure 3. SEM images of (**a**) GZ0′, (**b**) GZ1′, (**c**) GZ2′, (**d**) GZ0, (**e**) GZ1 and (**f**) GZ2.

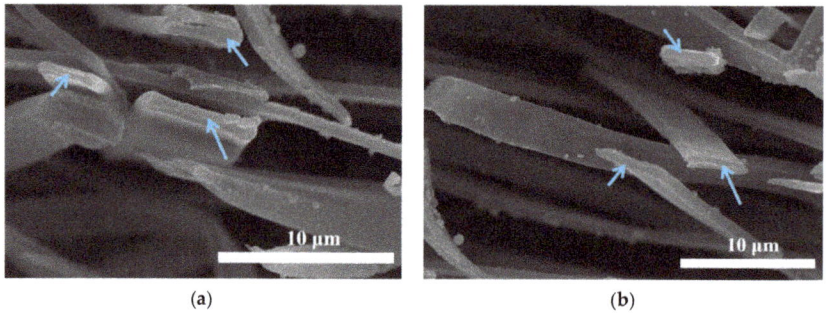

Figure 4. The cross sections SEM images of (**a**) GZ1′, (**b**) GZ2′.

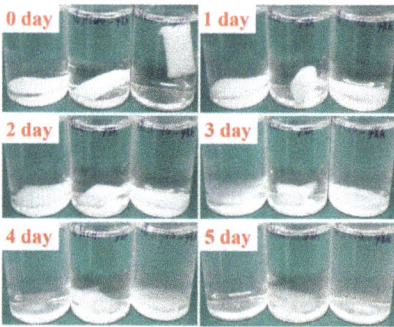

Figure 5. Degradation process diagram of GZ2 in phosphate buffer saline (PBS) at 37 °C.

To ascertain the chemical structures of gelatin/ZnO fibers, FTIR was first tested. As shown in Figure 6a, the spectra of GZ1 and GZ2 represent three characteristic peaks at 1629, 1531 and 1236 cm^{-1} corresponding to amide I, amide II and amide III, respectively. The amide I band is mainly attributed to the tensile vibration of -C=O, and the amide II and III bands are caused by the bending vibration of -NH and the stretching vibration of -C-N, respectively [34]. A weak absorption peak at 407 cm^{-1} in the GZ1 and GZ2 spectra is consistent with the characteristic absorption peak of ZnO (as mentioned in Section 3.1.1), which belongs to the stretching vibration of the Zn-O bond. In addition, due to the reaction between the aldehyde group of formaldehyde and the amino lysine residue of gelatin, the stretching vibration peak of the imide group (-CH=N) appears at 1448 cm^{-1} [35]. Furthermore, Figure 6b shows the XRD spectra of gelatin/Zn fibers as well as the ZnO particles. GZ0 has a broad diffraction peak at 2θ = 20° [36]. Unfortunately, no ZnO peaks are found in the GZ1 and GZ2 spectra, which may be due to the low content of ZnO particles on the surface of the gelatin [24]. In order to determine the types of elements contained in gelatin/ZnO fibers, EDX was carried out. The selected area for mapping is a part of GZ2, in which the ZnO particles are evenly loaded on the surface. Figure 6c,d represent EDX analysis and element mapping of GZ2, respectively. GZ2 shows the presence of C, N, O and Zn elements. The appearance of C and N elements is due to the presence of these two elements in gelatin molecules. The Zn element is caused by the presence of the element in ZnO particles. In addition, the O element exists both in gelatin molecules and ZnO particles. Moreover, Zn mapping of GZ2 verifies that ZnO particles are evenly dispersed on the surface of gelatin fibers.

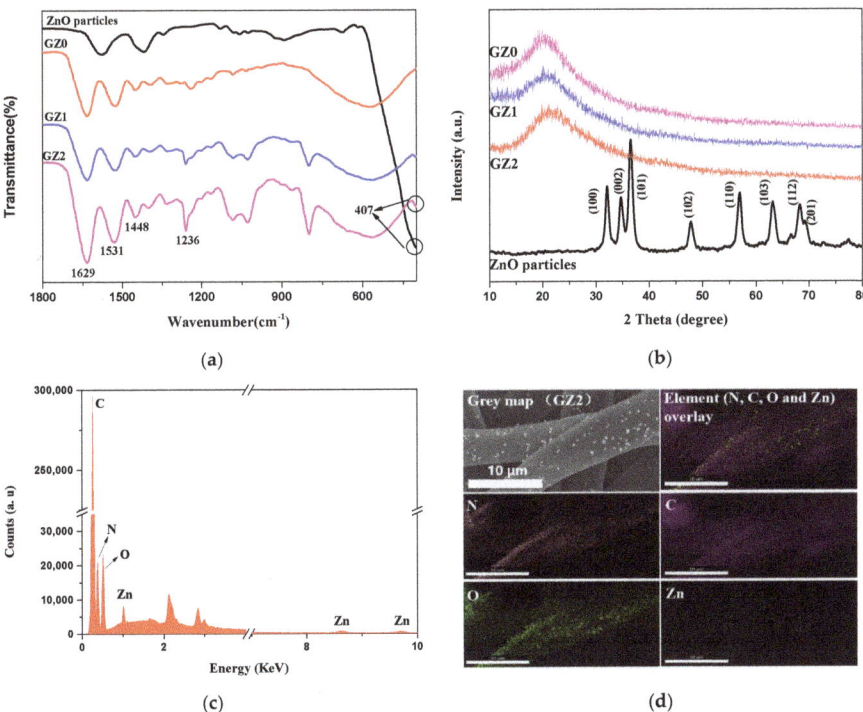

Figure 6. (a) FTIR spectra of ZnO particles, GZ0, GZ1 and GZ2, (b) XRD patterns of ZnO particles, GZ0, GZ1 and GZ2, (c) energy dispersive X-ray spectroscopy (EDX) analysis of GZ2, and (d) EDX mapping of GZ2.

TGA was carried out to study the thermal stability of gelatin/ZnO fibers. As shown in Figure 7, TGA curve of ZnO particles shows 9.02% weight lost at 500 °C, caused by the evaporation of physically adsorbed water and decomposition of residual acetate. Meanwhile, the TGA curves of GZ0, GZ1 and GZ2 are basically consistent, and the thermal decomposition process is divided into two stages [37,38]. In the first stage (30–224 °C), the weight lost is approximately 8.97%, which is caused by the evaporation of physically adsorbed water. The second stage is related to the decomposition of gelatin, and with the increase of ZnO content, the weight loss decreases gradually at the end of the decomposition process (500 °C). Obviously, the presence of ZnO has little effect on the thermal decomposition process of gelatin fibers, and the gelatin/ZnO fibers have good thermal stability below 224 °C.

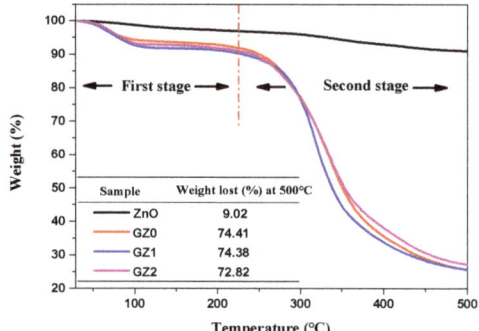

Figure 7. Thermogravimetric analysis (TGA) curve of ZnO, GZ0, GZ1 and GZ2.

3.2. Bioactivity Studies of Gelatin/ZnO Fibers

3.2.1. Cell Activity

The proliferation of GZ0, GZ1 and GZ2 were analyzed by real-time monitoring of MRC-5 cells using an Incucyte^{TM} Zoom microscope. In Figure 8, time-lapse imaging of the control exhibits a standard growth curve up to 55.5% cell confluence level by day 7, and the GZ0, GZ1 and GZ2 reach 45.6, 43.1 and 41.8% cell confluence level by day 7, respectively. The growth curve trends for GZ0, GZ1 and GZ2 are basically the same, and lower than that of the control. The results indicate that the GZ0, GZ1 and GZ2 have certain inhibiting effects on cell proliferation. Relative growth rate (RGR, %) was used to express the cytotoxicity. Table 1 shows that all the relative growth rates are more than 75%, corresponding to the cytotoxicity level of 0 or 1 per the standard in Table 2. This indicates that the gelatin/ZnO fibers (GZ0, GZ1 and GZ2) have no cytotoxicity. In addition, as shown in Figure 9, MRC-5 cells cultured in extraction from GZ0, GZ1 and GZ2 are marked by a green phase object mask via the software of Incucyte^{TM} Zoom. It can be seen that the cells display healthy spindle-like or star-like shape, and the number of cells rose with the extension of culture time, which suggests that GZ1 and GZ2 impose no suppression on the growth of cells.

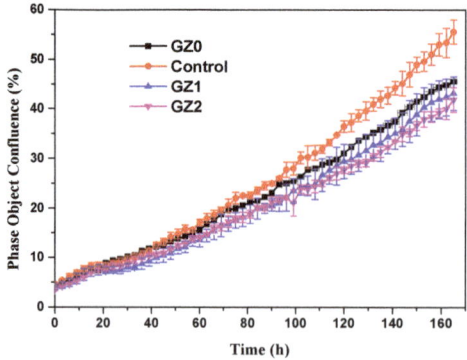

Figure 8. Real-time cell confluence study MRC-5 cells. The cell population was monitored for 168 h using an Incucyte^{TM} Zoom system in an incubator (5% CO_2 and 37 °C).

Table 1. The relative growth rate (RGR) and cytotoxicity level of GZ0, GZ1 and GZ1.

		Day 1	Day 3	Day 5	Day 7
GZ0	RGR (%)	105.8	95.7	85.2	82.1
	Cytotoxicity level	0	1	1	1
GZ1	RGR (%)	84.6	81.2	80.7	77.6
	Cytotoxicity level	1	1	1	1
GZ2	RGR (%)	92.4	84.1	75.8	75.3
	Cytotoxicity level	1	1	1	1

Table 2. The standard of cytotoxicity determined from RGR.

Cytotoxicity Level	0	1	2	3	4	5
RGR (%)	>100	75–99	50–74	25–49	1–24	0

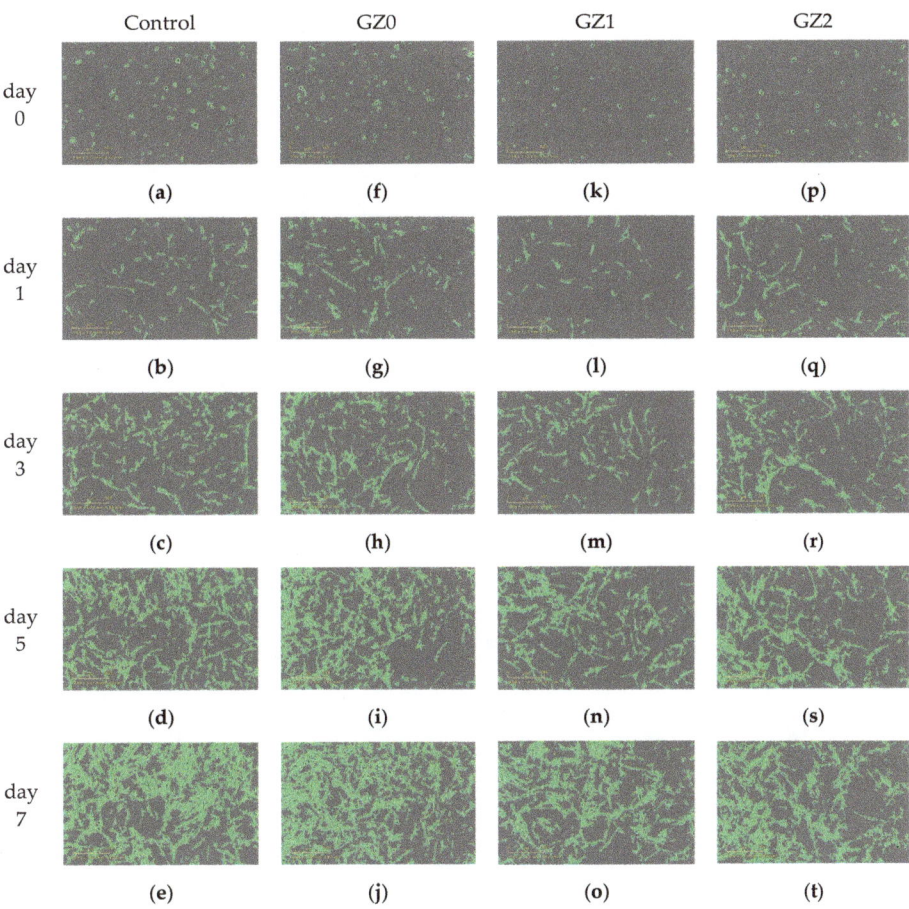

Figure 9. Cell morphology of (**a–e**) Control, (**f–j**) GZ0, (**k–o**) GZ1 and (**p–t**) GZ2 at point of day 0, 1, 3, 5 and 7, respectively. Scale bar: 500 μm.

3.2.2. Antibacterial Activity

Figure 10 illustrates the antibacterial activity of the gelatin/ZnO fibers against *E. coli* and *S. aureus* with and without UV light. The bacteria were determined by using the viable colony count method. Figure 11 represents photographs showing the antibacterial activity of gelatin/ZnO fibers against *S. aureus* and *E. coli*. Obviously, pure gelatin fibers (GZ0) have no toxic effect on the bacterial strains, and pure ZnO, GZ1 and GZ2 have excellent bacteriostatic effects. In addition, GZ2 containing a higher concentration of ZnO has stronger antibacterial activity than GZ1 containing a low concentration of ZnO. According to the reported literature [39–41], the reason for the antibacterial activity of gelatin/ZnO fibers is the superoxide radicals ($\cdot O_2^-$) produced by ZnO particles. The superoxide radicals can attack the bacterial cell wall and lead to cell wall leakage, resulting in the death of bacteria. In addition, under UV irradiation, ZnO particles can produce a large number of superoxide radicals. The obtained data (Figure 10) show that the antibacterial effect of gelatin/ZnO fibers irradiated by UV light is stronger than that of non-irradiated. This also proves the antibacterial mechanism of gelatin/ZnO fibers.

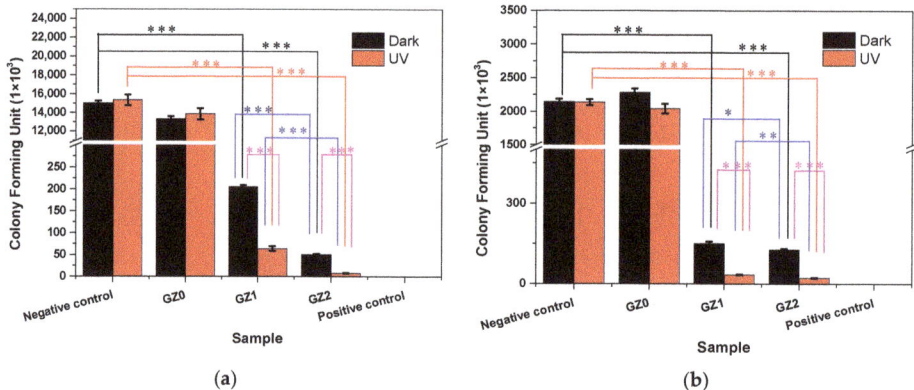

Figure 10. The antibacterial activity of the gelatin/ZnO fibers against (**a**) *Staphylococcus aureus* and (**b**) *Escherichia coli* with and without UV light. (* represents $p < 0.05$, ** represents $p < 0.01$ and *** represents $p < 0.001$.).

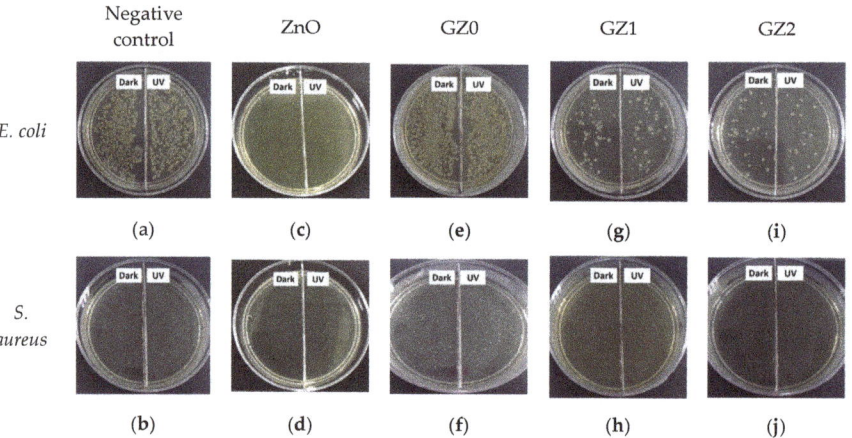

Figure 11. The photographs showing the antibacterial activity of (**a,b**) negative control, (**c,d**) ZnO particles, (**e,f**) GZ0, (**g,h**) GZ1 and (**i,j**) GZ2 against *S. aureus* and *E. coli*, respectively.

4. Conclusions

In summary, we first fabricated gelatin/ZnO fibers via side-by-side electrospinning technique as a potential wound dressing. The data indicate that ZnO particles can be uniformly dispersed on the surface of gelatin fibers. Although gelatin/ZnO fibers have some inhibitory effect on MRC-5 cells, the corresponding cytotoxicity level is 0 or 1 per the standard, which indicates that fibers have no cytotoxicity. Moreover, the gelatin/ZnO fibers show excellent antibacterial activity against *E. coli* and *S. aureus*, which is caused by the superoxide radicals ($\cdot O_2^-$) produced by ZnO particles. In addition, the reductions of bacteria are all more than 90%. The experiments imply that gelatin/ZnO fibers have excellent application prospects for wound dressing.

Author Contributions: W.L., Y.G. and Y.C. Conceived and Designed the Experiments; Y.C. and Y.Z. Performed the Experiments and Analyzed the Data; Y.S. Contributed Reagents/Materials/Analysis Tools; Y.C., W.L. Wrote the Paper.

Funding: This work was supported by the State Natural Sciences Fund, China (Project No. 21506236, 51372276), Research and Application of Gelatin Green Manufacturing 2.0 Technology by Enzymatic method (KFJ-STS-ZDTP-016) and Hangzhou Research Institute of Technical Institute of Physics and Chemistry, CAS Fund (Project No. 2016050201, 2016050202).

Conflicts of Interest: The authors declare no conflict of interest.

References

1. Liang, D.; Lu, Z.; Yang, H.; Gao, J.; Chen, R. A novel asymmetric wettable agnps/chitosan wound dressing: In vitro and in vivo evaluation. *ACS Appl. Mater. Interfaces* **2016**, *8*, 3958–3968. [CrossRef] [PubMed]
2. Pei, Y.; Ye, D.; Zhao, Q.; Wang, X.; Zhang, C.; Huang, W.; Zhang, N.; Liu, S.; Zhang, L. Effectively promoting wound healing with cellulose/gelatin sponges constructed directly from a cellulose solution. *J. Mater. Chem. B* **2015**, *3*, 7518–7528. [CrossRef]
3. Wang, Y.; Dou, C.; He, G.; Ban, L.; Huang, L.; Li, Z.; Gong, J.; Zhang, J.; Yu, P. Biomedical Potential of Ultrafine Ag particles Coated on Poly (Gamma-Glutamic Acid) Hydrogel with Special Reference to Wound Healing. *Nanomaterials* **2018**, *8*, 324. [CrossRef]
4. Guo, Y.; Pan, S.; Jiang, F.; Wang, E.; Miinea, L.; Marchant, N.; Cakmak, M. Anisotropic swelling wound dressings with vertically aligned water absorptive particles. *RSC Adv.* **2018**, *8*, 8173–8180. [CrossRef]
5. Zhao, J.; Qu, Y.; Chen, H.; Xu, R.; Yu, Q.; Yang, P. Self-assembled proteinaceous wound dressings attenuate secondary trauma and improve wound healing in vivo. *J. Mater. Chem B* **2018**, *6*, 4645–4655. [CrossRef]
6. Farokhi, M.; Mottaghitalab, F.; Fatahi, Y.; Khademhosseini, A.; Kaplan, D.L. Overview of Silk Fibroin Use in Wound Dressings. *Trends Biotechnol.* **2018**, *36*, 907–922. [CrossRef] [PubMed]
7. Aldana, A.A.; Malatto, L.; Rehman, M.A.U.; Boccaccini, A.R.; Abraham, G.A. Fabrication of Gelatin Methacrylate (GelMA) Scaffolds with Nano- and Micro-Topographical and Morphological Features. *Nanomaterials* **2019**, *9*, 120. [CrossRef] [PubMed]
8. Chen, Y.; Lu, W.; Guo, Y.; Zhu, Y.; Lu, H.; Wu, Y. Superhydrophobic coatings on gelatin-based films: Fabrication, characterization and cytotoxicity studies. *RSC Adv.* **2018**, *8*, 23712–23719. [CrossRef]
9. Ou, Q.; Miao, Y.; Yang, F.; Lin, X.; Zhang, L.-M.; Wang, Y. Zein/gelatin/nanohydroxyapatite nanofibrous scaffolds are biocompatible and promote osteogenic differentiation of human periodontal ligament stem cells. *Biomater. Sci.* **2019**. [CrossRef] [PubMed]
10. Kuppan, P.; Sethuraman, S.; Krishnan, U.M. Fabrication and investigation of nanofibrous matrices as esophageal tissue scaffolds using human non-keratinized, stratified, squamous epithelial cells. *RSC Adv.* **2016**, *6*, 26461–26473. [CrossRef]
11. Sahay, R.; Kumar, P.S.; Sridhar, R.; Sundaramurthy, J.; Venugopal, J.; Mhaisalkar, S.G.; Ramakrishna, S. Electrospun composite fibers and their multifaceted applications. *J. Mater. Chem.* **2016**, *22*, 12953–12971. [CrossRef]
12. Lei, B.; Shin, K.-H.; Noh, D.-Y.; Jo, I.-H.; Koh, Y.-H.; Choi, W.-Y.; Kim, H.-E. Nanofibrous gelatin-silica hybrid scaffolds mimicking the native extracellular matrix (ECM) using thermally induced phase separation. *J. Mater. Chem.* **2012**, *22*, 14133–14140. [CrossRef]

13. Liu, Y.; Deng, L.; Zhang, C.; Feng, F.; Zhang, H. Tunable Physical Properties of Ethylcellulose/Gelatin Composite fibers by Electrospinning. *J. Agric. Food Chem.* **2018**, *66*, 1907–1915. [CrossRef] [PubMed]
14. Ren, X.; Feng, Y.; Guo, J.; Wang, H.; Li, Q.; Yang, J.; Hao, X.; Lv, J.; Ma, N.; Li, W. Surface modification and endothelialization of biomaterials as potential scaffolds for vascular tissue engineering applications. *Chem. Soc. Rev.* **2015**, *44*, 5680–5742. [CrossRef] [PubMed]
15. Jalaja, K.; Sreehari, V.S.; Kumar, P.R.A.; Nirmala, R.J. Graphene oxide decorated electrospun gelatin fibers: Fabrication, properties and applications. *Mater. Sci. Eng. C* **2016**, *64*, 11–19. [CrossRef]
16. Li, H.; Wang, M.; Williams, G.R.; Wu, J.; Sun, X.; Lv, Y.; Zhu, L.-M. Electrospun gelatin fibers loaded with vitamins A and E as antibacterial wound dressing materials. *RSC Adv.* **2016**, *6*, 50267–50277. [CrossRef]
17. Perelshtein, I.; Applerot, G.; Perkas, N.; Wehrschetz-Sigl, E.; Hasmann, A.; Guebitz, G.M.; Gedanken, A. Antibacterial Properties of an In Situ Generated and Simultaneously Deposited Nanocrystalline ZnO on Fabrics. *ACS Appl. Mater. Interfaces* **2008**, *1*, 361–366. [CrossRef] [PubMed]
18. Manna, J.; Begum, G.; Kumar, K.P.; Misra, S.; Rana, R.K. Enabling Antibacterial Coating via Bioinspired Mineralization of Nanostructured ZnO on Fabrics under Mild Conditions. *ACS Appl. Mater. Interfaces* **2013**, *5*, 4457–4463. [CrossRef]
19. Dutta, R.K.; Sharma, P.K.; Bhargava, R.; Kumar, N.; Pandey, A.C. Differential Susceptibility of Escherichia coli Cells toward Transition Metal-Doped and Matrix-Embedded ZnO particles. *J. Phys. Chem. B* **2010**, *114*, 5594–5599. [CrossRef]
20. Vijayakumar, S.; Vaseeharan, B. Antibiofilm, anti cancer and ecotoxicity properties of collagen based ZnO particles. *Adv. Powder Technol.* **2018**, *29*, 2331–2345. [CrossRef]
21. Huang, Z.; Zheng, X.; Yan, D.; Yin, G.; Liao, X.; Kang, Y.; Yao, Y.; Huang, D.; Hao, B. Toxicological Effect of ZnO particles Based on Bacteria. *Langmuir* **2008**, *24*, 4140–4144. [CrossRef]
22. Brayner, R.; Dahoumane, S.A.; Yéprémian, C.; Djediat, C.; Meyer, M.; Couté, A.; Fiévet, F. ZnO particles: Synthesis, Characterization, and Ecotoxicological Studies. *Langmuir* **2010**, *26*, 6522–6528. [CrossRef]
23. Chen, Y.; Ding, H.; Sun, S. Preparation and Characterization of ZnO particles Supported on Amorphous SiO_2. *Nanomaterials* **2017**, *7*, 217. [CrossRef]
24. Liu, Y.; Li, Y.; Deng, L.; Zou, L.; Feng, F.; Zhang, H. Hydrophobic Ethylcellulose/Gelatin fibers Containing Zinc Oxide particles for Antimicrobial Packaging. *J. Agric. Food Chem.* **2018**, *66*, 9498–9506. [CrossRef]
25. Chhabra, H.; Deshpande, R.; Kanitkar, M.; Jaiswal, A.; Kale, V.P.; Bellare, J.R. A nano zinc oxide doped electrospun scaffold improves wound healing in a rodent model. *RSC Adv.* **2016**, *6*, 1428–1439. [CrossRef]
26. Münchow, E.A.; Albuquerque, M.T.P.; Zero, B.; Kamocki, K.; Piva, E.; Gregory, R.L.; Bottino, M.C. Development and characterization of novel ZnO-loaded electrospun membranes for periodontal regeneration. *Dent. Mater.* **2015**, *31*, 1038–1051. [CrossRef]
27. Sudheesh Kumar, P.T.; Lakshmanan, V.-K.; Anilkumar, T.V.; Ramya, C.; Reshmi, P.; Unnikrishnan, A.G.; Nair, S.V.; Jayakumar, R. Flexible and Microporous Chitosan Hydrogel/Nano ZnO Composite Bandages for Wound Dressing: In Vitro and In Vivo Evaluation. *ACS Appl. Mater. Interfaces* **2012**, *4*, 2618–2629. [CrossRef]
28. Luo, M.; Shen, C.; Feltis, B.N.; Martin, L.L.; Hughes, A.E.; Wright, P.F.A.; Turney, T.W. Reducing ZnO particle cytotoxicity by surface modification. *Nanoscale* **2014**, *6*, 5791. [CrossRef]
29. Hsiao, I.-L.; Huang, Y.-J. Titanium Oxide Shell Coatings Decrease the Cytotoxicity of ZnO particles. *Chem. Res. Toxicol.* **2011**, *24*, 303–313. [CrossRef]
30. Hong, T.-K.; Tripathy, N.; Son, H.-J.; Ha, K.-T.; Jeong, H.-S.; Hahn, Y.-B. A comprehensive in vitro and in vivo study of ZnO particles toxicity. *J. Mater. Chem. B* **2013**, *1*, 2985–2992. [CrossRef]
31. Zhang, Y.; Chung, J.; Lee, J.; Myoung, J.; Lim, S. Synthesis of ZnO nanospheres with uniform nanopores by a hydrothermal process. *J. Phys. Chem. Solids* **2011**, *72*, 1548–1553. [CrossRef]
32. Xiong, G.; Pal, U.; Serrano, J.G.; Ucer, K.B.; Williams, R.T. Photoluminesence and ftir study of ZnO particles: The impurity and defect perspective. *Phys. Status Solidi* **2010**, *3*, 3577–3581. [CrossRef]
33. Ran, J.; He, M.; Li, W.; Cheng, D.; Wang, X. Growing ZnO particles on polydopamine-templated cotton fabrics for durable antimicrobial activity and UV protection. *Polymers* **2018**, *10*, 495. [CrossRef]
34. Chen, Y.; Ma, Y.; Lu, W.; Guo, Y.; Zhu, Y.; Lu, H.; Song, Y. Environmentally Friendly Gelatin/β-Cyclodextrin Composite Fiber Adsorbents for the Efficient Removal of Dyes from Wastewater. *Molecules* **2018**, *23*, 2473. [CrossRef]
35. Gupta, N.; Santhiya, D. In situ mineralization of bioactive glass in gelatin matrix. *Mater. Lett.* **2017**, *188*, 127–129. [CrossRef]

36. Lomakin, S.M.; Rogovina, S.Z.; Grachev, A.V.; Prut, E.V.; Alexanyan, C.V. Thermal Degradation of Biodegradable Blends of Polyethylene with Cellulose and Ethylcellulose. *Thermochim. Acta* **2011**, *521*, 66–73. [CrossRef]
37. Kwak, H.W.; Kim, J.E.; Lee, K.H. Green fabrication of antibacterial gelatin fiber for biomedical application. *React. Funct. Polym.* **2019**, *136*, 86–94. [CrossRef]
38. Lonkar, S.P.; Pillai, V.; Abdala, A. Solvent-free Synthesis of ZnO-Graphene Nanocomposite with Superior Photocatalytic Activity. *Appl. Surf. Sci.* **2019**, *465*, 1107–1113. [CrossRef]
39. Patil, P.P.; Meshram, J.V.; Bohara, R.A.; Nanaware, S.G.; Pawar, S.H. ZnO particle-embedded silk fibroin-polyvinyl alcohol composite film: A potential dressing material for infected wounds. *New J. Chem.* **2018**, *42*, 14620–14629. [CrossRef]
40. Bai, X.; Li, L.; Liu, H.; Tan, L.; Liu, T.; Meng, X. Solvothermal Synthesis of ZnO particles and Anti-Infection Application in Vivo. *ACS Appl. Mater. Interfaces* **2015**, *7*, 1308–1317. [CrossRef]
41. Raghupathi, K.R.; Koodali, R.T.; Manna, A.C. Size-Dependent Bacterial Growth Inhibition and Mechanism of Antibacterial Activity of Zinc Oxide particles. *Langmuir* **2011**, *27*, 4020–4028. [CrossRef]

© 2019 by the authors. Licensee MDPI, Basel, Switzerland. This article is an open access article distributed under the terms and conditions of the Creative Commons Attribution (CC BY) license (http://creativecommons.org/licenses/by/4.0/).

Article

Fabrication and Characteristics of Porous Hydroxyapatite-CaO Composite Nanofibers for Biomedical Applications

Shiao-Wen Tsai [1,2,3], Sheng-Siang Huang [4], Wen-Xin Yu [4], Yu-Wei Hsu [4] and Fu-Yin Hsu [4,*]

1. Graduate Institute of Biomedical Engineering, Chang Gung University, Taoyuan City 33302, Taiwan; swtsai@mail.cgu.edu.tw
2. Department of Orthopedic Surgery, Chang Gung Memorial Hospital, Linkou 33305, Taiwan
3. Department of Periodontics, Chang Gung Memorial Hospital, Taipei 10507, Taiwan
4. Department of Bioscience and Biotechnology, National Taiwan Ocean University, Keelung City 20224, Taiwan; mm0070360@gmail.com (S.-S.H.); andy54861@yahoo.com.tw (W.-X.Y.); qazwest74@gmail.com (Y.W.H.)
* Correspondence: fyhsu@mail.ntou.edu.tw; Tel.: +886-02-2462-2192

Received: 9 July 2018; Accepted: 24 July 2018; Published: 26 July 2018

Abstract: Hydroxyapatite (HAp), a major inorganic and essential component of normal bone and teeth, is a promising biomaterial due to its excellent biocompatibility, bioactivity, and osteoconductivity. Therefore, synthetic HAp has been widely used as a bone substitute, cell carrier, and delivery carrier of therapeutic genes or drugs. Mesoporous materials have attracted considerable attention due to their relatively high surface area, large pore volume, high porosity, and tunable pore size. Recently, mesoporous HAp has also been successfully synthesized by the traditional template-based process and has been demonstrated to possess better drug-loading and release efficiencies than traditional HAp. It is widely accepted that cell adhesion and most cellular activities, including spreading, migration, proliferation, gene expression, surface antigen display, and cytoskeletal functioning, are sensitive to the topography and molecular composition of the matrix. The native extracellular matrix is a porous, nanofibrous structure. The major focus of this study is the fabrication of porous hydroxyapatite-CaO composite nanofibers (p-HApFs) and the investigation of its drug-release property. In this study, nanofibers were prepared by the sol-gel route and an electrospinning technique to mimic the three-dimensional structure of the natural extracellular matrix. We analyzed the components of fibers using X-ray diffraction and determined the morphology of fibers using scanning and transmission electron microscopy. The average diameter of the nanofibers was approximately 461 ± 186 nm. The N_2 adsorption–desorption isotherms were type IV isotherms. Moreover, p-HApFs had better drug-loading efficiency and could retard the burst release of tetracycline and maintain antibacterial activity for a period of 7 days. Hence, p-HApFs have the potential to become a new bone graft material.

Keywords: sol-gel; electrospinning; hydroxyapatite; nanofiber; antibacterial

1. Introduction

Bone is a natural inorganic–organic composite consisting of collagen fibrils containing well-arrayed apatite nanocrystals. Hydroxyapatite [HAp, $Ca_{10}(PO_4)_6(OH)_2$] is chemically similar to the inorganic component of the bone matrix, which has led to extensive research efforts to use this material as a bone substitute in both orthopedic and dental fields. Recently, HAp has been used in range of biomedical applications, such as matrices for drug release and bone tissue engineering scaffold materials.

HAp can be synthesized through a variety of well-developed techniques, such as the sol-gel process [1], the wet-chemical reaction [2], the solid-state reaction [3] and the chemical vapor deposition [4]. The synthesis of HAp using the sol-gel method consists of the molecular-level mixing of calcium and phosphorus precursors under significantly milder conditions. Hence, a sol-gel synthesis of HAp has recently attracted considerable attention.

The sol-gel synthesis technique is also used in conjunction with various spinning techniques to fabricate HAp fibers [5]. Electrospinning is an easy and simple method that utilizes electrical fields to fabricate nano- to microfibers from a polymer solution. The topological structure of the electrospun matrix closely mimics the dimensions of the natural extracellular matrix (ECM). The ECM plays a key role in triggering intracellular signaling cascades for tissue regeneration. Thus, the development of new ECM substitutes has attracted a wide attention as a scaffold for tissue engineering. Franco et al. successfully fabricated HAp nanofibers by combining electrospinning and a non-alkoxide sol-gel system [6]. Pasuri noted that electrospun HAp nanofibers do not activate macrophages in vitro and can be resorbed by human osteoclasts [7].

Recently, much attention has been attracted by HAp as a drug delivery carrier for the loading and delivery of therapeutic agents, such as proteins, growth factors, genes, and drugs [8–10]. Mesoporous HA nanoparticles possess a higher drug-loading capacity and drug-release efficiency than HAp nanoparticles as a result of their large surface areas and high pore volumes [11]. The synthesis of mesoporous HAp nanoparticles using template reagents, such as Pluronic P123 [12], cetyltrimethylammonium bromide (CTAB) [13], and Pluronic F127 [14], is the most well-known method. In the synthesis of mesoporous HAp, the self-assembly of inorganic HAp phases and template reagents followed by template removal can produce the mesoporous structure.

Münchow [15] incorporated CaO nanoparticles into electrospun matrices and demonstrated improved cell viability and osteogenic differentiation. Hence, the aims of this study were to fabricate and characterize a nanofibrous structure of porous hydroxyapatite-CaO (p-HApFs) by utilizing an electrospinning process based on a sol-gel precursor and to evaluate the release profiles and antibacterial activity of tetracycline from p-HApFs.

2. Materials and Methods

2.1. Synthesis and Characterization of Porous Hydroxyapatite Nanofibers

Mesoporous hydroxyapatite nanofibers were synthesized using poly(ethylene glycol)-block-poly (propylene glycol)-block-poly(ethylene glycol) (P123, Sigma-Aldrich, St. Louis, MO, USA) as the structure-directing agent. Briefly, 1.0 g of P123 and 6.172 mL of triethyl phosphite (TEP, Merck, Germany) were mixed in a 10 mL ethanol aqueous solution (50% v/v), which was then continuously stirred to form a clear solution. $Ca(NO_3)_2$ (14 g, Sigma-Aldrich, St. Louis, MO, USA) was dissolved in 10 mL of absolute ethanol under magnetic stirring at room temperature. The calcium nitrate solution was slowly added under stirring to the above P123/TEP solution to form a precursor solution. The precursor solution was tightly capped and placed in an oven at 60 °C for 12 h. Then, 1.5 g poly(vinyl pyrrolidone) (PVP, Sigma-Aldrich, St. Louis, MO, USA), 0.45 g P123 and 7 mL absolute ethanol were prepared and incorporated into the 3 mL precursor solution to obtain a transparent mixture solution. The mixture solution was placed inside a plastic syringe fitted with a stainless needle (18 G, inner diameter = 0.838 mm) and then inserted into a syringe pump to supply a steady flow rate (1.27 mL/h). An electrical field (1.3 kV/cm) was applied between the needle and an aluminum substrate (grounded collector) using a high-voltage power supply (SL 60, Spellman, New York, NY, USA). The polymer solution, which formed a Taylor cone upon exit, was collected on an aluminum substrate in the form of nonwoven nanofiber structures and was then calcined at different temperatures (600 °C, 800 °C and 1000 °C) and environments (air and nitrogen) to obtain the calcined p-HApFs nonwoven structures.

2.2. Characterization of the p-HApF

The p-HApFs morphology was observed using scanning electron microscopy (SEM, Hitachi S-4800, Tokyo, Japan), which operated at an accelerating voltage of 15 kV The average fiber diameter of the p-HApFs was analyzed by image analysis software (Image-Pro Express Version 6.0, Media Cybernetics, Rockville, MD, USA) based on the SEM images. Moreover, the p-HApFs pore structure was observed using transmission electron microscopy (TEM, JEOL JEM-2100, Tokyo, Japan) at an accelerating voltage of 100 kV. Nitrogen adsorption–desorption measurements were used to obtain the Brunauer-Emmett-Teller (BET) specific surface area, and the pore size was also analyzed (Micromeritic ASAP 2020 instrument, Norcross, GA, USA). The phase of p-HApFs was characterized by X-ray diffraction (XRD, Bruker D2-Phaser, Madison, WI, USA) with a copper target. Powder diffraction patterns were acquired over 2-theta, ranging from 20 to 60° with a step size of 0.04°. The functional groups of p-HApFs were analysed with Fourier transitioned infrared spectroscopy (FTIR, Bruker tensor II, Madison, WI, USA). The spectra were recorded from 4000 to 400 cm^{-1} wave number with a resolution of 2 cm^{-1}.

2.3. In Vitro Study of Drug Loading and Release

Tetracycline (TC) was selected as the model drug for the evaluation of drug loading and release. p-HApFs (10 mg) was added to a TC aqueous solution (5 mg/mL, 10 mL) and stirred for 24 h. After the loading procedure, the amount of TC adsorbed onto p-HApFs, which calcined at N_2 atmosphere 800 °C, was calculated by determining the difference in the TC concentration before and after loading. The TC concentration was analyzed at a wavelength of 360 nm using a UV-Vis spectrophotometer (Ultrospec 1100 Pro, Amersham Biosciences, Piscataway, NJ, USA).

The drug contents and loading efficiency were calculated according to the following formulas:

Drug content (w/w) = weight of the TC in the p-HApFs/weight of the p-HApFs

Loading efficiency (%) = (weight of the TC in the p-HApFs/initial weight of the p-HApFs) × 100%

After loading, the p-HApFs were dried using a lyophilization process. TC-loaded p-HApFs (20 mg) were added to a phosphate-buffered saline (PBS) solution (2 mL) and agitated in a horizontally shaking bath at 37 °C. The release medium was withdrawn and replaced with fresh PBS at each measurement. The mechanism of drug release was analyzed by fitting the experimental data to equations describing different kinetic orders. Linear regression analyses were performed for zero-order $[M_t/M_0 = K_0 \cdot t]$ and first-order $[\ln(M_0 - M_t) = K_1 \cdot t]$ kinetics, as well as the Higuchi $[M_t/M_0 = K_H \cdot t^{1/2}]$ model, where K is the kinetic constant and M_t/M_0 is the fraction of TC released at time t. The best-fitted model was assessed on the basis of the correlation coefficient (r^2). The drug release data were further analyzed with the Ritger-Peppas equation $[M_t/M_\infty = k_r \cdot t^n]$, where Mt is the amount of drug released at time t, M_∞ is the amount of drug released at time ∞, and k_r and n represent the release rate constant and release exponent, respectively.

2.4. Antibiotic Activity against Staphylococcus aureus and Pseudomonas aeruginosa

The antibiotic activity of TC in the in vitro release study was assessed against a Gram-positive bacterial strain, *Staphylococcus aureus*, and a Gram-negative bacterial strain, *Pseudomonas aeruginosa*, using the antibiotic elution samples from a turbidity assay. TC (500 μL) eluated during the in vitro release study or from PBS (as the control group) was added to the inoculum containing 500 μL of the 2× nutrient broth solution and 100 μL of bacteria in 5-mL test tubes. After 24 h of incubation, the optical density (OD) of the culture medium was measured using a spectrophotometer at a wavelength of 600 nm. The relative antibacterial activity (R%) was calculated as follows:

$$R\% = [(OD_{growth\ medium} - OD_{released\ tetracycline})/OD_{growth\ medium}] \times 100\%$$

3. Results and Discussions

All of the electrospun nanofibers were calcined at three different temperatures, namely, 600, 800 and 1000 °C, to study the crystal phase structure. Figure 1 shows the XRD pattern for the nanofibers calcined at different temperatures. The electrospun fibers were mostly crystalline calcium carbonate at 600 °C. The amount of CaO increased when the calcination temperature increased, which was due to the breakdown of calcium carbonate that formed calcium oxide, accompanied by the evolution of CO_2. The HAp phase formed at 800 °C.

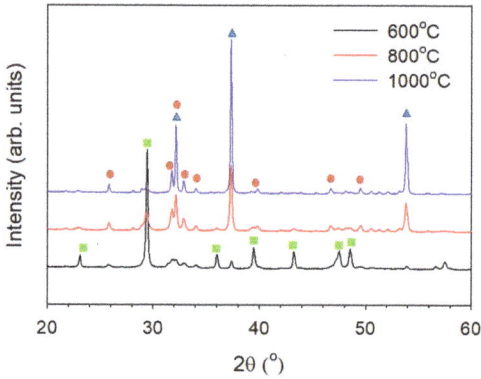

Figure 1. XRD pattern of the fibers calcined at different temperatures under an air atmosphere. (▲: CaO, PDF 70-4068; ●: hydroxyapatite, PDF 84-1998; ■: $CaCO_3$, PDF 85-1108).

Figure 2 shows the XRD pattern for the nanofibers calcined under air and nitrogen atmosphere at 800 °C. The main crystal phase comprised HAp and CaO. However, the amount of HAp was higher under the nitrogen than under the air atmosphere. The amount of dissolved ambient CO_2 can be limited by performing the reaction under a nitrogen atmosphere [16]. Hatzistavrou [17] fabricated hydroxyapatite-CaO composites using the sol-gel method and found that the presence of CaO accelerated the formation of carbonate hydroxyapatite in simulated body fluid. This was due to the dissolution of the CaO phase that rapidly formed carbonate hydroxyapatite. Hence, HAp/CaO composites have been used as scaffolds with tuneable properties by varying the composition.

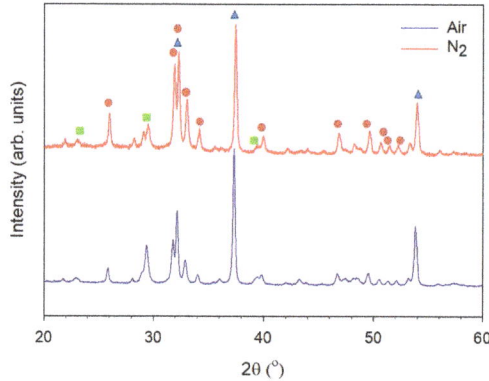

Figure 2. XRD pattern of the fibers calcined at 800 °C under various calcination atmospheres. (▲: CaO, PDF 70-4068; ●: hydroxyapatite, PDF 84-1998; ■: $CaCO_3$, PDF 85-1108).

Figure 3 shows a Fourier transform infrared (FTIR) characteristic spectrum for the nanofibers that were calcined at 800 °C under a nitrogen atmosphere. Peaks at approximately 566 and 609 cm^{-1} are due to the bending vibration of the P-O bond in PO_4^{3-} [18]. The bands at approximately 960 and 1000~1100 cm^{-1} are associated with the stretching modes of the PO_4^{3-} bonds in Hap [19]. The characteristic bands at approximately 873 and 1440~1470 cm^{-1} are attributable to the CO_3^{-2} group. Taherian et al. [20] pointed out CO_3^{2-} ions may form due to the incomplete pyrolysis of organic compounds, or the absorption of CO_2 from the atmosphere. Bilton et al. [21] pointed out that the evaporation of unreacted triethyl phosphite from the sol or gel could constitute to the presence of $CaCO_3$. $CaCO_3$ decomposes to CaO after calcination at 800 °C

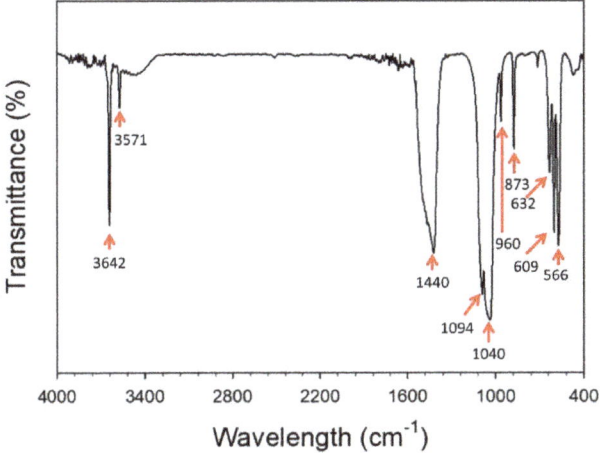

Figure 3. FTIR spectra of the fibers calcined at 800 °C under an N_2 atmosphere.

The dissolution of CO_2 from the atmosphere occurs by the following reaction:

$$CO_2 \text{ (g)} + 2\, OH^- \text{ (aq)} \rightarrow CO_3^{2-} \text{ (aq)} + H_2O \text{ (l)}$$

The characteristic peaks observed at 632 and 3571 cm^{-1} are attributable to the respective hydroxyl functional group (-OH) deformation vibration and stretching vibrations of HAp [22]. Additionally, a sharp peak was observed at 3642 cm^{-1}, which confirms the formation of the CaO phase [23].

The morphology and structure of the p-HApFs were observed under SEM and TEM. Figure 4a shows the SEM image of the p-HApFs. The average diameter of the p-HApFs was 461 ± 186 nm. Figure 4b shows the TEM image of the p-HApFs. The TEM image showed that the nanofiber was composed of a number of nanocrystals and clearly revealed the existence of mesopores within the nanocrystals. The p-HApFs did not exhibit an ordered orientation of the mesopores, which generally had a random arrangement.

The nitrogen adsorption–desorption isotherms of the p-HApFs were type IV hysteresis loops, which are typical for mesoporous materials (shown in Figure 5a). The specific surface areas of the p-HApFs calculated from the BET equation were 7.2 m^2/g. The pore size distribution was plotted according to the BJH nitrogen desorption model as shown in Figure 5b. The average pore size of the p-HApFs calculated from the BJH equation was approximately 28 nm. Figure 5b shows a broad peak ranging from 10 to 130 nm, centered at approximately 50 nm, which suggests that most pores had a size of approximately 50 nm; however, the pore sizes were not uniform.

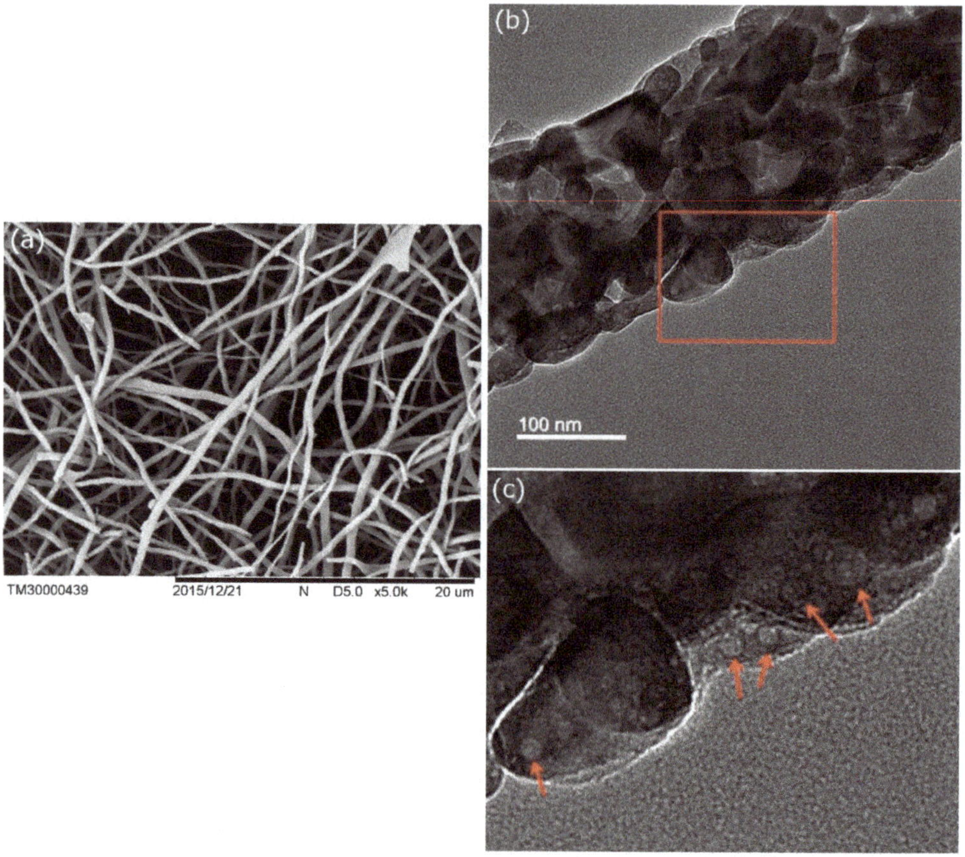

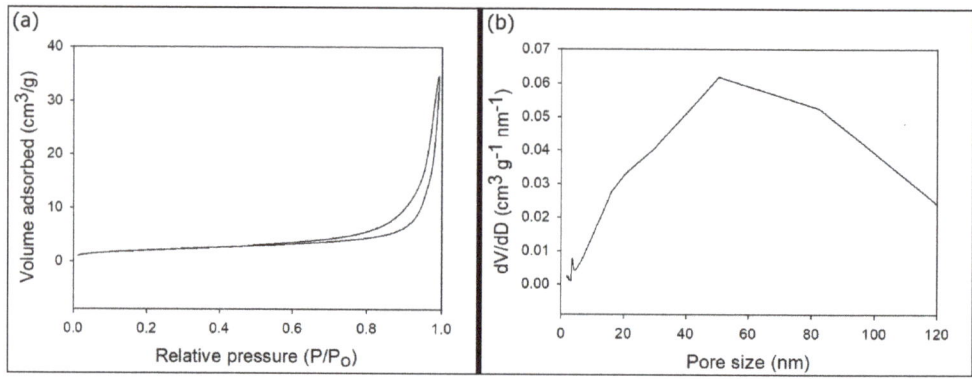

Figure 4. (**a**) SEM and (**b**) TEM micrographs of the nanofibers after heat treatment at 800 °C under an N_2 atmosphere. (**c**) an enlarged graph in red square of (**b**). The red arrows in (**c**) indicates mesopores within the nanocrystals.

Figure 5. (**a**) N_2 adsorption–desorption isotherm and (**b**) pore size distribution curve of p-HApFs.

The amount of TC loaded within the p-HApFs was 8.83 ± 0.09 mg/10 mg (TC/p-HApFs). The loading efficiency of TC was 88.34 ± 0.89% (w/w). The cumulative drug-release curve as a function of time for TC release from the TC-loaded p-HApFs was analyzed in triplicate (shown in Figure 6). The drug-release data indicated that a no-burst release phenomenon occurred at the initial step. The TC release from the TC-loaded p-HApFs was steady and slow over a period of 14 days. The drug-release data were analyzed using different kinetic models: zero-order, first-order, Higuchi, and Korsmeyer-Peppas. The best fit with the highest correlation coefficient (r^2) was obtained using the first-order equation (r^2 = 0.996), followed by the Higuchi model (r^2 = 0.941), and zero-order equation (r^2 = 0.983). The diffusion exponent n-value of the Korsmeyer–Peppas model was 0.75, which indicates anomalous diffusion or both diffusion- and erosion-controlled drug release [24].

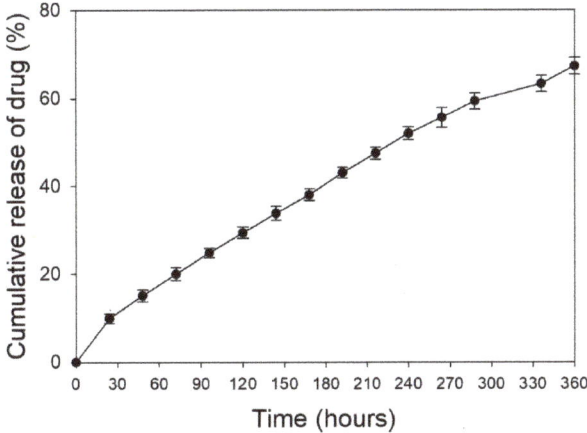

Figure 6. In vitro cumulative tetracycline (TC) release from TC-loaded p-HApFs.

To clarify the antibacterial activity of TC released from the TC-loaded p-HApFs, the release solution was used to cultivate bacterial strains. As shown in Figure 7, the release solution had a strong ability to inhibit bacterial growth, even on day 7.

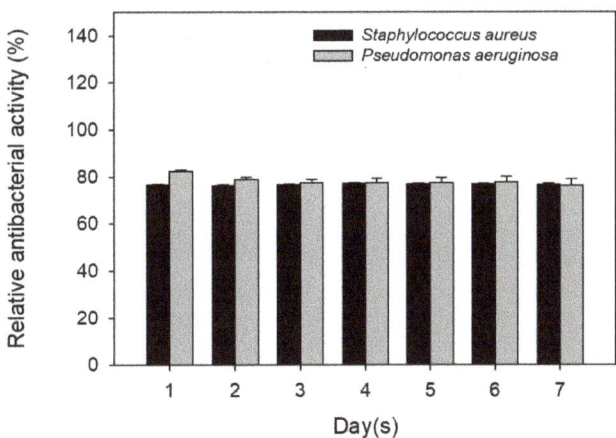

Figure 7. Susceptibility trend of the released tetracycline against *Staphylococcus aureus* and *Pseudomonas aeruginosa*.

4. Conclusions

A mesoporous hydroxyapatite nanofibrous matrix to mimic the three-dimensional structure of the natural extracellular matrix was successfully fabricated by an electrospinning process based on a sol-gel precursor. p-HApFs exhibited a high drug-loading efficiency and could retard the burst release of TC, as well as maintain antibacterial activity over a period of 7 days. Hence, it is anticipated that p-HApFs have the potential to be used as drug delivery carriers and as bone graft substitutes.

Author Contributions: Data curation, S.-W.T. and F.-Y.H.; Formal analysis, S.-W.T., S.-S.H. and F.-Y.H.; Methodology, S.-S.H., W.-X.Y. and Y.-W.H.; Project administration, F.-Y.H.; Writing—original draft, S.-W.T. and F.-Y.H.

Funding: This research was funded by the Ministry of Science and Technology of Taiwan, grant number MOST 102-2221-E-019-003-MY3 and MOST 105-2221-E-019-008.

Conflicts of Interest: The authors declare no conflict of interests.

References

1. Ben-Arfa, B.A.; Salvado, I.M.; Ferreira, J.M.; Pullar, R.C. Novel route for rapid sol-gel synthesis of hydroxyapatite, avoiding ageing and using fast drying with a 50-fold to 200-fold reduction in process time. *Mater. Sci. Eng. C Mater. Biol. Appl.* **2017**, *70*, 796–804. [CrossRef] [PubMed]
2. Cüneyt Taş, A.; Korkusuz, F.; Timuçin, M.; Akkaş, N. An investigation of the chemical synthesis and high-temperature sintering behaviour of calcium hydroxyapatite (HA) and tricalcium phosphate (TCP) bioceramics. *J. Mater. Sci. Mater. Med.* **1997**, *8*, 91–96. [CrossRef] [PubMed]
3. Rao, R.R.; Roopa, H.N.; Kannan, T.S. Solid state synthesis and thermal stability of HAP and HAP—beta-TCP composite ceramic powders. *J. Mater. Sci. Mater. Med.* **1997**, *8*, 511–518. [CrossRef] [PubMed]
4. Li, H.; Zhang, N.; Wang, X.; Geng, X.; Yang, X.; Wang, H.; Li, B.; Liang, C. Synthesis and cytotoxicity of carbon nanotube/hydroxyapatite in situ composite powders prepared by chemical vapour deposition. *Mater. Res. Innov.* **2014**, *18*, 338–343. [CrossRef]
5. Ramanan, S.R.; Venkatesh, R. A study of hydroxyapatite fibers prepared via sol-gel route. *Mater. Lett.* **2004**, *58*, 3320–3323. [CrossRef]
6. Franco, P.Q.; João, C.F.C.; Silva, J.C.; Borges, J.P. Electrospun hydroxyapatite fibers from a simple sol–gel system. *Mater. Lett.* **2012**, *67*, 233–236. [CrossRef]
7. Pasuri, J.; Holopainen, J.; Kokkonen, H.; Persson, M.; Kauppinen, K.; Lehenkari, P.; Santala, E.; Ritala, M.; Tuukkanen, J. Osteoclasts in the interface with electrospun hydroxyapatite. *Colloids Surf. B Biointerfaces* **2015**, *135*, 774–783. [CrossRef] [PubMed]
8. Zhang, W.; Chai, Y.; Xu, X.; Wang, Y.; Cao, N. Rod-shaped hydroxyapatite with mesoporous structure as drug carriers for proteins. *Appl. Surf. Sci.* **2014**, *322*, 71–77. [CrossRef]
9. Li, D.; Huang, X.; Wu, Y.; Li, J.; Cheng, W.; He, J.; Tian, H.; Huang, Y. Preparation of pH-responsive mesoporous hydroxyapatite nanoparticles for intracellular controlled release of an anticancer drug. *Biomater. Sci.* **2016**, *4*, 272–280. [CrossRef] [PubMed]
10. Pistone, A.; Iannazzo, D.; Espro, C.; Galvagno, S.; Tampieri, A.; Montesi, M.; Panseri, S.; Sandri, M. Tethering of Gly-Arg-Gly-Asp-Ser-Pro-Lys Peptides on Mg-Doped Hydroxyapatite. *Engineering* **2017**, *3*, 55–59. [CrossRef]
11. Gu, L.; He, X.; Wu, Z. Mesoporous hydroxyapatite: Preparation, drug adsorption, and release properties. *Mater. Chem. Phys.* **2014**, *148*, 153–158. [CrossRef]
12. Cheah, W.K.; Ooi, C.W.; Othman, R.; Yeoh, F.Y. Synthesis of mesoporous hydroxyapatite through a soft template route using non-ionic surfactant. *ASEAN Eng. J. Part B* **2012**, *1*, 57–69.
13. Yao, J.; Tjandra, W.; Chen, Y.Z.; Tam, K.C.; Ma, J.; Soh, B. Hydroxyapatite nanostructure material derived using cationic surfactant as a template. *J. Mater. Chem.* **2013**, *13*, 3053–3057. [CrossRef]
14. Zhao, Y.F.; Ma, J.; Tan, G. Synthesis of mesoporous hydroxyapatite through neutral templating. *Int. J. Nanosci.* **2006**, *5*, 499–503. [CrossRef]
15. Münchow, E.A.; Pankajakshan, D.; Albuquerque, M.T.; Kamocki, K.; Piva, E.; Gregory, R.L.; Bottino, M.C. Synthesis and characterization of CaO-loaded electrospun matrices for bone tissue engineering. *Clin. Oral Investig.* **2016**, *20*, 1921–1933. [CrossRef] [PubMed]

16. Nazeer, M.A.; Yilgor, E.; Yagci, M.B.; Unal, U.; Yilgor, I. Effect of reaction solvent on hydroxyapatite synthesis in sol-gel process. *R. Soc. Open Sci.* **2017**, *4*, 171098. [CrossRef] [PubMed]
17. Hatzistavrou, E.; Chatzistavrou, X.; Papadopoulou, L.; Kantiranis, N.; Chrissafis, K.; Boccaccini, A.R.; Paraskevopoulos, K.M. Sol-gel hydroxyapatite-CaO composites: Fabrication and bioactivity studies. *Key Eng. Mater.* **2009**, *396–398*, 99–102. [CrossRef]
18. Baddiel, C.B.; Berry, E.E. Spectra structure correlations in hydroxy and fluorapatite. *Spectrochim. Acta* **1966**, *22*, 1407–1416. [CrossRef]
19. Wang, H.; Zhang, P.; Ma, X.; Jiang, S.; Huang, Y.; Zhai, L.; Jiang, S. Preparation, characterization of electrospun meso-hydroxylapatite nanofibers and their sorptions on Co(II). *J. Hazard Mater.* **2014**, *265*, 158–165. [CrossRef] [PubMed]
20. Taherian, M.; Rojaee, R.; Fathi, M.; Tamizifar, M. Effect of different sol–gel synthesis processes on microstructural and morphological characteristics of hydroxyapatite-bioactive glass composite nanopowders. *J. Adv. Ceram.* **2014**, *3*, 207–214. [CrossRef]
21. Bilton, M.; Milne, S.J.; Brown, A.P. Comparison of hydrothermal and sol-gel synthesis of nano-particulate hydroxyapatite by characterisation at the bulk and particle level. *Open J. Inorg. Non-Met. Mater.* **2012**, *2*, 1–10. [CrossRef]
22. Russell, S.W.; Luptak, K.A.; Suchicital, C.T.A.; Alford, T.L.; Pizziconi, V.B. Chemical and structural evolution of sol-gel-derived hydroxyapatite thin films under rapid thermal processing. *J. Am. Ceram. Soc.* **1996**, *79*, 837–842. [CrossRef]
23. Hatzistavrou, E.; Chatzistavrou, X.; Papadopoulou, L.; Kantiranis, N.; Kontonasaki, E.; Boccaccini, A.R.; Paraskevopoulos, K.M. Characterization of the bioactive behaviour of sol-gel hydroxyapatite-CaO and Hydroxyapatite-CaO-bioactive glass composites. *Mater. Sci. Eng. C* **2010**, *30*, 497–502. [CrossRef]
24. Dash, S.; Murthy, P.N.; Nath, L.; Chowdhury, P. Kinetic modeling on drug release from controlled drug delivery systems. *Acta Pol. Pharm.* **2010**, *67*, 217–223. [PubMed]

© 2018 by the authors. Licensee MDPI, Basel, Switzerland. This article is an open access article distributed under the terms and conditions of the Creative Commons Attribution (CC BY) license (http://creativecommons.org/licenses/by/4.0/).

Article

Physical Properties and In Vitro Biocompatible Evaluation of Silicone-Modified Polyurethane Nanofibers and Films

Chuan Yin [1], Sélène Rozet [1], Rino Okamoto [1], Mikihisa Kondo [1], Yasushi Tamada [1], Toshihisa Tanaka [1,*], Hatsuhiko Hattori [2], Masaki Tanaka [2], Hiromasa Sato [3] and Shota Iino [3]

1. Interdisciplinary Graduate School of Science and Technology, Shinshu University, 3-15-1, Tokida, Ueda-shi, Nagano 386-8567, Japan; yinkawa@outlook.com (C.Y.); selene.rozet@gmail.com (S.R.); j3992a1953m@gmail.com (R.O.); Kondo.mikihisa@exc.epson.co.jp (M.K.); ytamada@shinshu-u.ac.jp (Y.T.)
2. Silicone-Electronics Materials Research Center, Shin-Etsu Chemical Co., 1-10, Hitomi, Matsuida-Machi, Annaka-Shi, Gunma 379-0224, Japan; hhattori@shinetsu.jp (H.H.); m.tanaka@shinetsu.jp (M.T.)
3. Dainichiseika Color & Chemicals Mfg. Co., 1-4-3, Ukima, Kita-ku, Tokyo 115-8622, Japan; hrsato@daicolor.co.jp (H.S.); iino@daicolor.co.jp (S.I.)
* Correspondence: tanakat@shinshu-u.ac.jp

Received: 29 January 2019; Accepted: 22 February 2019; Published: 5 March 2019

Abstract: In this study, the physical properties and the biocompatibility of electrospun silicone-modified polyurethane (PUSX) nanofibers were discussed and compared with PUSX films. To investigate the effects of different structures on the physical properties, tensile strength, elongation at break, Young's modulus, water retention, water contact angle (WCA) and thermal conductivity measurements were performed. To prove the in vitro biocompatibility of the materials, cell adhesion, cell proliferation, and cytotoxicity were studied by NIH3T3 mouse embryonic fibroblasts cells following by lactate dehydrogenase (LDH) analysis. As a conclusion, the mechanical properties, water retention, and WCA were proven to be able to be controlled and improved by adjusting the structure of PUSX. A higher hydrophobicity and lower thermal conductivity were found in PUSX nanofibers compared with polyurethane (PU) nanofibers and films. An in vitro biocompatibility evaluation shows that the cell proliferation can be performed on both PUSX nanofibers and films. However, within a short period, cells prefer to attach and entangle on PUSX nanofibers rather than PUSX films. PUSX nanofibers were proven to be a nontoxic alternative for PU nano-membranes or films in the biomedical field, because of the controllable physical properties and the biocompatibility.

Keywords: silicone modified polyurethane nanofibers; physical properties; cell attachment; cell proliferation; cytotoxicity

1. Introduction

Electrospun nanofibers have been used in various fields such as filtration, catalysis, clothing, and biomedical applications, because of their submicron size and high surface area, along with their porous architecture [1,2]. Especially for biomedical applications, electrospun mats provide the lightness in weight, porosity, flexibility in technique, as well as a good support for cells to attach and grow. Their capacity to exchange nutrients and gases makes them suitable for tissue engineering, wound dressing, drug delivery, health care, etc. [3]. A non-woven matrix composed of nanofibers is easily produced via electrospinning, and it is architecturally similar to the nanofibrous structure of extracellular matrix [4]. If necessary, the nanofibers can be further functionalized via the incorporation of bioactive species (e.g. enzymes, DNAs, and growth factors) to better control the proliferation and differentiation of cells seeded on the scaffolds [5]. These attributes make electrospun nanofibers well-suited as scaffolds for tissue engineering.

Among various kinds of nanofibers, polyurethane nanofibers were selected as one of the most suitable choices for biomedical applications, thanks to the unique properties of polyurethane. Electrospun polyurethane nanofibers have been successfully used in wound dressing, due to an excellent oxygen permeability and barrier properties [6]. Water permeability is also important, as it keeps the wound moist and prevents the accumulation of fluid around the wound and on its cover. These covers perform a preventive function against infection with microorganisms, to absorb blood and wound fluids to contribute to the healing process, and in some cases, to apply medical treatment to the wound [7–9].

However, there are still several limitations and disadvantages of polyurethane nanofibers to be applied in the biomedical field, such as poor thermal capability, poor weatherability, and flammability. In order to improve the properties of polyurethane nanofibers, silicone groups were introduced into polyurethane polymer chains to fabricate silicone-modified polyurethane (PUSX) and to optimize the electrospinning parameters. In this study, PUSX nanofibers were evaluated by physical properties and cell culture studies, and they were compared with the films. The advantages of polyurethane, silicone, and nanofibers are very attractive for this work. This new material is expected to be applied in many fields as an improved alternative for polyurethane nanofibers, such as wound dressing and tissue engineering, due to the biocompatibility of silicone. Before going for the in vitro cell attachment and proliferation applications, all of the prepared nanofibers were analyzed in detail by various methods, and compared with films. To investigate the effects of different structure (block and graft type), chain lengths, and silicone concentration, evaluation of the physical properties was performed. Tensile tests were performed to investigate mechanical properties such as tensile strength, elongation at break, and Young's modulus. The water contact angle (WCA) measurement and water retention tests were carried out to determine the hydrophobicity of the PUSX material. The thermal conductivity was analyzed in order to discuss the heat retention ability of the PUSX nanofibers and films. In order to reveal the potential for cell adhesion and proliferation, NIH3T3 mouse embryonic fibroblasts cells were cultured on all the samples, followed by lactate dehydrogenase (LDH) activity. The toxicity of the PUSX nanofibers and films were evaluated by using direct contact based on ISO 10993-5. Therefore, as the purpose of our research, the influence of PUSX structures on the physical properties and biocompatibility is investigated. PUSX nanofibers might be expected as an ideal alternative for PU nanofibrous membranes or films in the biomedical fields.

2. Experimental Section

2.1. Materials

The 12 kinds of PUSX solutions and films were kindly synthesized and provided by Shin-Etsu Chemical Co., Ltd., and Dainichiseika Color & Chemicals Mfg. Co., Ltd. These synthesized PUSX solutions in *N,N*-dimethylformamide/ ethyl methyl ketone (DMF/MEK) standardized a solid content of 30 wt%. The structures of two types of PUSX are shown in Figure 1. PUSX with different silicone concentrations and chain lengths are listed in Table 1. Solvents such as *N,N*-dimethylformamide (DMF) and ethyl methyl ketone (MEK) were purchased from Wako Pure Chemical Industries., Ltd., and used as received.

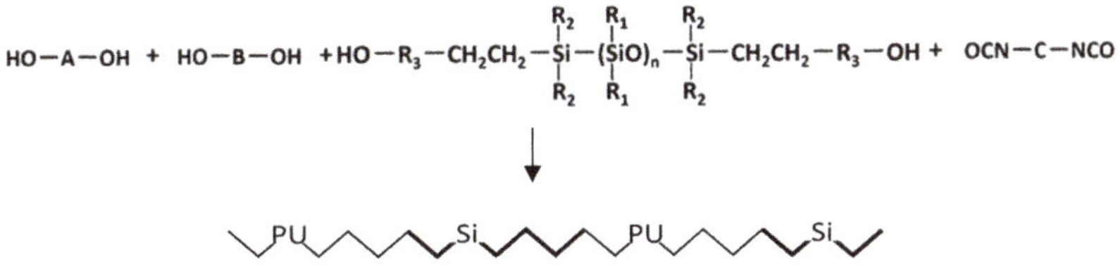

Figure 1. Structures of block-type silicone-modified polyurethane (PUSX) (a) and graft-type PUSX (b).

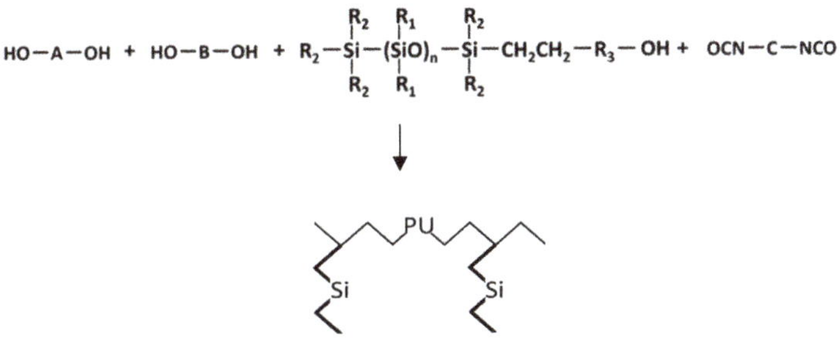

Table 1. Polyurethane (PU) samples and silicone modified polyurethane (PUSX) samples of block type and graft type.

Sample	PU	Si01	Si02	Si03	Si04	Si01-20	Si01-40	Si01-59	Si05	Si06	Si07	Si08
Molecular Structure	×	Block Type							Graft Type			
M_w (×10^5)	1.48	1.69	1.39	1.59	1.66	1.74	2.01	2.33	1.56	1.61	1.57	1.62
M_n (×10^5)	0.75	0.87	0.73	0.79	0.75	0.88	1.02	1.11	0.71	0.70	0.72	0.78
Silicone Concentration (wt%)	0	10	10	10	10	20	40	59	10	10	10	10
Silicone Chain Length (n)	×	20	10	30	50	20	20	20	10	25	30	120

2.2. Electrospinning Method of PUSX Nanofibers

All the electrospinning solutions were prepared by diluting the PUSX solutions (30 wt%) in DMF/MEK mixed solvent (v/v = 64:36), and stirring at room temperature for 48 h in order to obtain homogeneous solutions. All electrospinning experiments were performed at room temperature (22 °C) under the optimized parameters of our previous study [10], and the deposited nanofibers were collected on a moving metallic collector. A 10–20 kV voltage was applied while the needle tip-to-collector distance was 10 cm with an irradiation angle of 30°, and the air flow rate in the spinning environment was 0.1 mL/min.

The surface morphology of the nanofibers was investigated by scanning electron microscope (SEM, JSM-6010LA JEOL, Tokyo, Japan) at an accelerating voltage of 10 kV. Before SEM analysis, the prepared samples were coated by using a platinum sputter coater (Ion sputter JFC-1600 JEOL, Ltd, Tokyo, Japan) under 30 mA for 60 s. The diameters of the nanofibers were measured by ImageJ (National Institutes of Health, Bethesda, MD, USA). The average fiber diameters were calculated from data of at least 50 measurements per sample.

2.3. Physical Properties

Tensile tests were performed by a compact tabletop universal tensile tester (EZTest/EZ-S, Shimadzu Corporation, Kyoto, Japan) for samples 10 mm long and 5 mm wide, at a crosshead speed of 10 mm/min. At least 10 specimens were tested for each sample. To compare the mechanical properties of the PUSX nanofibers with PUSX films, the same tests were performed on the PUSX films.

Thermal conductivities were determined by a KES-F7 Thermal LaboIIB precision rapid thermal property measurement unit (KES-F7 Thermal Labo, Kato Tech Co., Ltd, Kyoto, Japan). The temperature of the water box was set to room temperature. Samples (5 × 5 cm^2) were then placed on the water box, and the heat plate of the bottom temperature box (B. T. box) was placed on the upper surface of the samples. After reaching a constant value, the heat flow loss W (watts) of the B.T. was recorded using a panel meter. Steady heat flow lost was calculated as the following equation:

$$W = K \cdot A \cdot \triangle T / D$$

where D is the thickness of the samples (cm), A is the area of the B.T. heat plate (cm^2), $\triangle T$ is the temperature difference of sample (°C), and K is the thermal conductivity. The thermal conductivity K was calculated by the following equation:

$$K = W \cdot D / A \triangle T \, (W/cm \, °C) = 100 \, W \cdot D / A \triangle T (W/mK).$$

While taking measurements with the B. T. Box, the applied pressure could be adjusted. The standard value was set as 6 g/cm^2. The temperature of the B. T. Box heat plate was controlled with an error of less than 0.1 °C.

Water retention tests were performed based on JIS L 1913 6.9.2. The electrospun PUSX nanofibers, and films of dimensions 100 × 100 mm^2 were incubated in distilled water for a period of 15 min, and then weighed. Water retention capacity was determined as the increase in the weight of the fibers. The percentage of water absorption was calculated as in the following equation:

$$m = (m2 - m1)/m1 \times 100\%$$

where m2 and m1 are the weights of the samples in wet and dry environments, respectively.

WCA is an easy measurement for determining the wettability of the materials by a liquid. The static contact angle of pure water for the surfaces of the PUSX samples was measured by an automated contact angle meter (DM-501Hi, Kyowa Interface Science Co., Ltd, Saitama, Japan) after randomly dripping 2 µL of purified water on the surfaces of the samples. The droplets on the samples were captured after 1000 ms through an image analyzer, and the WCA, θ, was calculated by the software through analyzing the shape of the drop. When depositing a droplet onto the material, the water will form a droplet shape. The point where the surface, the water, and the vapor meet, is called the three-face point, and it determines the contact angle. The relationship between the contact angle, the surface free energy, the liquid surface tension, and the interfacial tension between solid and liquid is defined by the Young equation:

$$\gamma S = \gamma L \cos\theta + \gamma SL,$$

where θ is the contact angle, γL is the surface free energy of the solid, and γSL is the interfacial tension between the solid and liquid.

Usually, when the WCA is less than 90°, the material can be considered to be hydrophilic while the material is hydrophobic. It is worth mentioning that if the WCA is between 150° and 180°, it shows the high water-repellency of the material.

2.4. In Vitro Biocompatible Evaluation

2.4.1. Cell Culture Studies

Before using the samples for in vitro cell culture, it is essential to remove the DMF/MEK mixed solvents because of the possible cytotoxicity. All of the samples were washed with distilled water for 48 h and dried in oven at 80 °C overnight. Then, the nanofibers and films were cut into round shapes with a diameter of 10 mm, with three replicates per sample prepared. Sterilization was performed by deeply soaking the samples in 70% ethanol aqueous solution in a multi-well tissue culture polystyrene (TCPS) dish for 1 h, followed by rinsing three times in phosphate buffer saline (PBS) to remove all traces of ethanol.

NIH3T3 mouse embryonic fibroblasts were used to measure the cell adhesion and proliferation. For the cell adhesion test, 50,000 cells (in 1 mL of medium) were mixed well into the sample. After 3 h, the cells were harvested in 1 mL of 0.5% Triton X-100/PBS solution, and evaluated by LDH assay for adhesion evaluation of the cells.

The cell proliferation test was a quantitative investigation of the capacity for the cells to grow on the electrospun nanofibers and films. At the seeding step of the proliferation culture, 5000 cells were added into the sample. The experiment lasted for a total of seven days, and the results of the first, third, fifth, and seventh days were compared.

The LDH activity was immediately measured by ultraviolet absorption at a wavelength of 340 nm, using the Thermo Scientific Multiskan FC microplate photometer (Thermo Fisher Scientific Inc., Waltham, MA, USA) with a recorder. The enzyme activity of LDH can be measure from chemical reaction of LDH, when it is released into the cells' medium from the damaged or dead cells, due to cell membrane damage. LDH converts lactate using NAD as a coenzyme, and produces pyruvic acid and NADH. The number of cells was calculated from the calibration curve obtained by the relation between the known number of cells, and the absorbance value at 340 nm of NADH in the assay supernatant.

The shape of the cells was observed by SEM to qualitatively investigate the cell reactions when in contact with the electrospun nanofibers. After each incubation period, the sample was fixed with paraformaldehyde (PFA) as a cross-linking fixation agent, to stop the proliferation of cells, and to preserve their shapes. The sample was further dehydrated by serially using ethanol gradient solutions of 50, 70, 95, and 99.5% for 30 min each, over a continuous process. Then, the sample was coated with platinum for SEM analysis.

2.4.2. Toxicity Evaluation

The toxicities of the PUSX nanofibers and films were evaluated by using direct contact based on ISO 10993-5. Briefly, cells were seeded evenly over the surface of each plate and incubated at 37 °C until the cells covered the whole surface. The samples were then placed on the cell layers in the center of the plates, and the culture medium was replaced. In order to determine the toxicity in accordance with Grade 0 (nontoxic) to Grade 4 (severe toxic) evaluation, Trypan blue was added after 24 h in each plate, and observed by morphological changes. This is the most obvious and direct way to reflect the impact of testing the materials on the cell [11].

2.5. Statistical Analysis

Significance in vitro biocompatibility and physical properties were statistically analyzed by a one-way analysis of variance (ANOVA) using R free software. Statistical significance was set at $p < 0.05$ to identify which groups were significantly different from the other groups.

3. Results and Discussions

3.1. Morphology of PUSX Nanofibers Prepared for Physical and Biocompatability Evaluations

In the previous study [10], we successfully prepared 12 kinds of PUSX nanofibers under the optimized conditions, and investigated the effects of solvents and solution concentrations on the as-spun nanofibers, along with the morphological appearance.

Representative SEM images of PUSX nanofibers are shown in Table 2. It can be seen that the surface morphology of the block-type PUSX was smooth and continuous, with fiber diameters ranging from 400 nm to 720 nm. The mean diameters decreased with the increase of both the chain lengths of silicone, and the silicone concentration. The surface morphology of graft-type PUSX nanofibers was also fine and continuous, with the mean diameters ranging from 460 nm to 560 nm. Compared with block-type PUSX, graft-type PUSX nanofibers did not show a clear influence of the chain length on the electrospinning parameters. Compared with the PU nanofibers, the PUSX nanofibers showed a more uniform surface with a smaller diameter, due to the hydrophobicity of the silicone group. In this study, the physical properties and biocompatibilities of the prepared samples are discussed.

Table 2. Scanning electron microscope (SEM) morphologies and average diameters of the PUSX nanofibers under the optimized electrospinning parameters (magnification: 2000×) [10].

Sample	PU	Si01	Si02	Si03
SEM				
Average Diameter (nm)	720	636	690	548
Sample	Si04	Si01-20	Si01-40	Si01-59
SEM				
Average Diameter (nm)	440	531	402	471
Sample	Si05	Si06	Si07	Si08
SEM				
Average Diameter (nm)	564	544	456	456

3.2. Mechanical Properties

3.2.1. Tensile Strength

Figure 2a–c shows the graphs of the trend of tensile strength, and a comparison of all the samples. Referring to all three graphs, the PUSX films had better tensile strength than the nanofibers, because of the porosity and the fiber orientation of the nanofibers. Due to the porosity of the nanofibers, the cross-sectional area was apparent, compared to the cross-sectional area of the films, which made the tensile strength supposedly higher than the results obtained. Pure PU nanofibers had the highest strength because of the high mechanical properties of PU. Figure 2a,b show that for block-type PUSX nanofibers, the tensile strength increased with an increase of silicone chain length, and Si04 samples with the longest silicone chain length (n = 50) showed the highest tensile strength of 5.9 MPa. Samples with longer silicone chain lengths showed smaller diameters and a more improved orientation of the molecular chains while being prepared under the optimized parameters, which contributed to the higher tensile strength. Meanwhile, the tensile strength decreased with the increase of silicone concentration, because of the low cohesive force of the silicone structure, and the decreasing concentration of PU.

When the silicone concentration increased, the low cohesive force changed the tensile strength of the material. On the other hand, the increase of silicone concentration caused the decrease of the ratio of PU in the polymer, and the weight percent of PU as well, which means there was a lack of the higher mechanical structure (PU). The tensile strengths of the films did not show much statistical significance because of the random orientation of the molecular chains.

In the graph of graft-type PUSX nanofibers and films (Figure 2c), Si08 nanofibers showed the highest tensile strength, at 6.8 MPa, with the Si05 sample showing the lowest result of 6.1 MPa. There was almost no difference in the observed trends, because the results were in the same range.

In this case, the silicone groups on the side chain were not able to influence the properties much, because the tensile strength was mainly determined by the high mechanical properties of PU in the main chain.

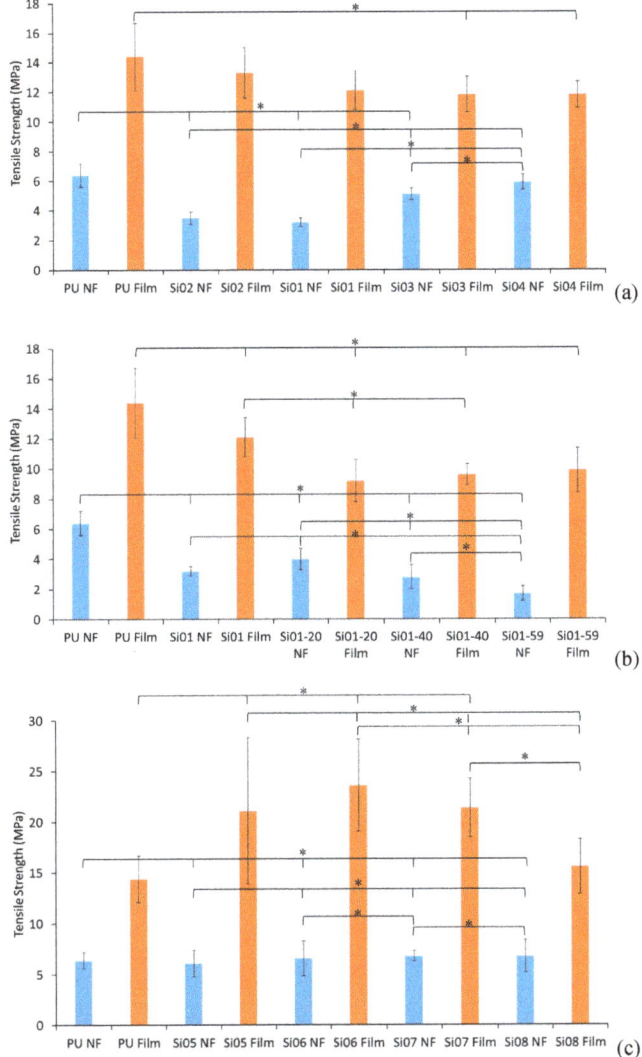

Figure 2. Comparison of tensile strength (MPa). "*" was statistically significant ($p < 0.05$) between each pair of samples. (**a**) Block-type PUSX nanofibers and films with various chain lengths, (**b**) Block-type PUSX nanofibers and films with various silicone concentrations, (**c**) Graft-type PUSX nanofibers and films with various chain lengths.

The tensile stress–strain curves of the electrospun PUSX nanofibers and the PUSX films are shown in Figure 3. Typical curves, each from a different structure of PUSX, were plotted for an obvious comparison. From Figure 3a, the differences before and after silicone modification could be easily observed. Si01-59 samples, with the highest silicone concentration in the block structure, show the lowest mechanical performances out of the fiber membranes, because of the low cohesive force of silicone. Instead, the tensile stress–strain curves of PUSX films showed very similar trends and much better mechanical performances, because of the random orientation of the molecular chains.

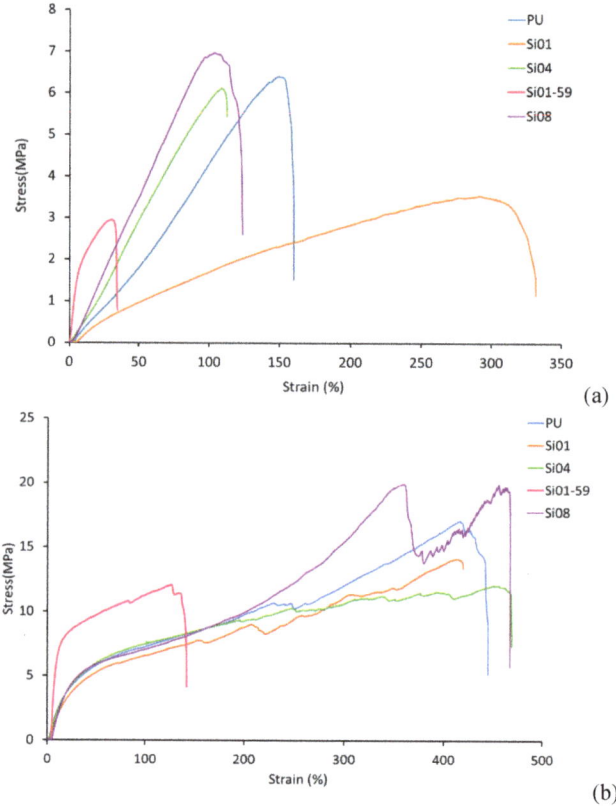

Figure 3. Comparison of stress–strain curves. (**a**) Stress–strain curves of the PUSX nanofibers, (**b**) Stress–strain curves of the PUSX films.

3.2.2. Elongation at Break

Referring to Figure 4, the elongation at break decreased with the increase of both silicone chain length and silicone concentration in block-type PUSX nanofibers. For the PUSX nanofibers with different silicone chain length, the increase of silicone chain length caused a decrease in fiber diameters under optimized conditions, which lead to the increase of entanglement and frictional resistance in the nanofibers. For PUSX nanofibers with different silicone concentrations, the increase of silicone concentration resulted in the decrease of polyurethane concentration in the molecular chains, and the high elongation property of polyurethane became difficult to observe. Meanwhile, the elongation at break of the graft-type PUSX nanofibers showed very similar results to each other for the same reason as with the tensile strength. As for the graft type, the elongation at break did not show an obvious trend, as the silicone groups on the side chains are not able to influence the nanofiber properties because the ratio of polyurethane and silicone in the molecular chain do not change with an increase of chain length. PUSX films have better tensile strengths than nanofibers, because of the porosity and fiber orientation of the nanofibers.

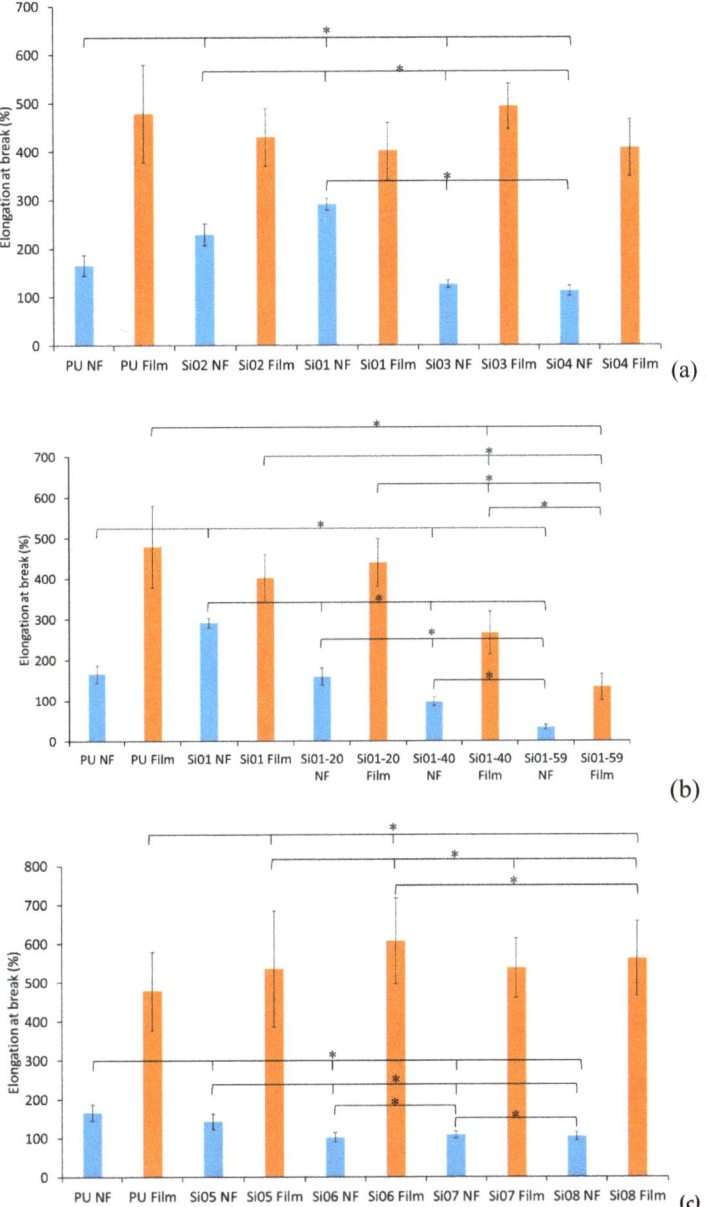

Figure 4. Comparison of elongation at break (%). "*" was statistically significant ($p < 0.05$) between each pair of samples. (**a**) Block-type PUSX nanofibers and films with various chain lengths, (**b**) Block-type PUSX nanofibers and films with various silicone concentrations, (**c**) Graft-type PUSX nanofibers and films with various chain lengths.

3.2.3. Young's Modulus

Referring to Figure 5, the Young's modulus increased with the increase of both the chain length and the silicone concentration in block-type PUSX nanofibers. As the silicone chain lengths and concentrations increased, the concentration of PU became lower and lower, and the characteristics of PU (with a high elongation at break) became difficult to observe, which meant that the samples were more elastic. Moreover, the existence of silicone also made it more difficult to change the shapes of the samples.

For both block- and graft-type PUSX nanofibers, all of the tensile test results showed that PUSX films have higher tensile strengths than nanofibers. This phenomenon can be explained by the orientation of the polymer chain, and the different structures of the nanofibers and films (porosity and density).

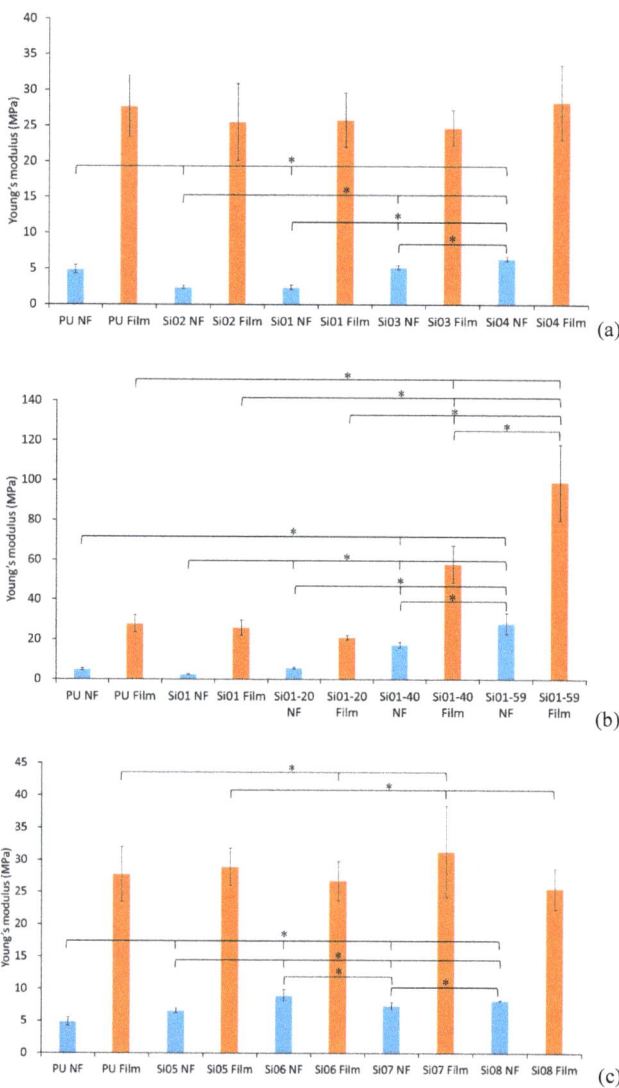

Figure 5. Comparison of Young's modulus (MPa). "*" was statistically significant ($p < 0.05$) between each pair of samples. (**a**) Block-type PUSX nanofibers and films with various chain lengths, (**b**) Block-type PUSX nanofibers and films with various silicone concentrations, (**c**) Graft-type PUSX nanofibers and films with various chain lengths.

3.3. Thermal Conductivity

Figure 6 shows the comparison of the thermal conductivity between block-type PUSX nanofibers and films and graft-type PUSX nanofibers and films. For the thermal conductivity analysis, there was no obvious trend of varying chemical structures such as varying silicone chain lengths, varying silicone concentrations and block or graft structures of PUSX. The thermal conductivity was influenced mostly by the shapes of the samples. Both block- and graft-type PUSX nanofibers have much lower conductivity than films, because of the pores keeping the air inside. The heat insulating property is proved here. This result can be explained by the heat-retaining property of nanofibers because of the high porosity.

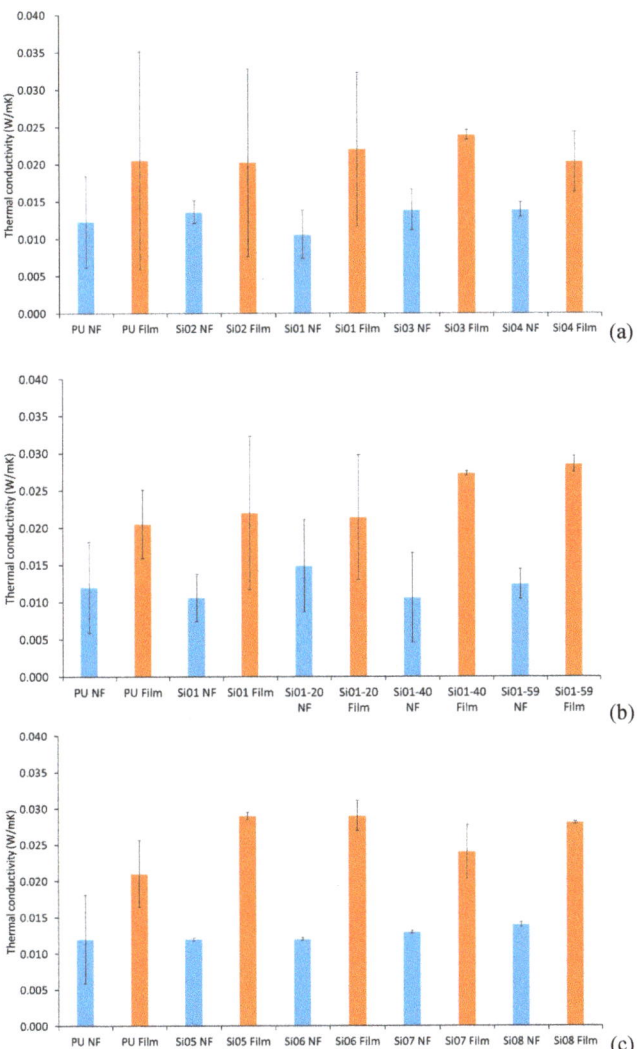

Figure 6. Comparison of the thermal conductivity (W/mK). (**a**) Block-type PUSX nanofibers and films with various chain lengths, (**b**) Block-type PUSX nanofibers and films with various silicone concentrations, (**c**) Graft-type PUSX nanofibers and films with various chain lengths.

It is worth mentioning that in solid objects, heat is transferred via conduction. For electrospun structures, much of the volume is made of empty spaces between fibers. Usually, the conduction along the fiber as part of the whole volume is negligible. But if the fibers becomes too compact, the packing density will be increased beyond an optimum standard, heat transfer via conduction through the solid fiber becomes significant, and the overall insulation property will be decreased. The insulation properties of the electrospun nanofibers are made more controllable due to their packing density compared to the films, to meet the demands of the market. As an ideal alternative to PUSX films, PUSX nanofibers provides good heat insulating properties and lightness in weight.

3.4. Water Retention and Water Contact Angle

3.4.1. Water Retention

In Table 3, there is a huge gap between the pure PU and PUSX nanofibers where the PU nanofibers showed the highest water retention of 280%, with statistical significance ($p < 0.05$).

Block-type PUSX nanofibers showed a decreasing trend in water retention with the increase of both silicone chain length and silicone concentration, because of the hydrophobicity of silicone. This also means that the water repellency of the PU nanofibers are able to be improved by introducing the silicone groups into the main chain. Meanwhile, compared with films, since nanofibers have a high porosity for holding water, the water retentions of the nanofibers are higher than films. Both the PU nanofibers and films have the highest water retentions. Higher hydrophobicity of silicone structure and moisture permeability were proven.

Meanwhile, for graft-type PUSX nanofibers, the silicone groups on the side-chain are not able to increase the hydrophobicity of the material, so that the water retention did not show any difference between the PU nanofibers and the graft-type PUSX nanofibers.

Table 3. Comparisons of water retention.

Block-Type PUSX with Different Silicone Chain Lengths	PU	Si02	Si01	Si03	Si04
Nanofiber Water Retention (%)	280.0 ± 40.0	27.0 ± 11.0	19.0 ± 8.1	11.0 ± 4.1	12.0 ± 2.5
Film Water Retention (%)	2.3 ± 1.5	2.7 ± 1.1	1.5 ± 0.2	5.8 ± 5.0	3.7 ± 1.1
Block-type PUSX with Different Silicone Concentrations	PU	Si01	Si01-20	Si01-40	Si01-59
Nanofiber Water Retention (%)	280.0 ± 40.0	19.0 ± 8.1	4.7 ± 3.1	2.7 ± 0.9	6.9 ± 6.0
Film Water Retention (%)	2.3 ± 1.5	1.5 ± 0.2	3.7 ± 0.7	2.7 ± 1.0	2.9 ± 2.3
Graft-Type PUSX	PU	Si05	Si06	Si07	Si08
Nanofiber Water Retention (%)	280.0 ± 40.0	169.0 ± 23.3	196.0 ± 28.8	165.0 ± 55.0	200.0 ± 13.5
Film Water Retention (%)	2.3 ± 1.5	2.3 ± 2.6	6.6 ± 7.5	6.8 ± 3.8	1.7 ± 0.9

3.4.2. Water Contact Angle

Figure 7 shows that the water contact angles of PU were much lower than PUSX for both the nanofibers and films. This phenomenon might be caused by the high hydrophobicity of the silicone structure. It also proved the results of water retention when comparing PU and PUSX materials. The PUSX nanofibers showed higher WCA than the films, because the surface of the nonwoven nanofiber membrane was much rougher than the films. The pores of the nanofibers make the surface microstructure similar to the lotus structure of Cassie's state (Figure 7h).

It is worth mentioning that for graft-type PUSX nanofibers, the water retention appeared to be higher compared with the block-type PUSX nanofibers, but the results of the water contact angle were similar. This might be caused by the different surfaces of the block-type and graft-type nanofibers. It is supposed that the silicone groups were distributed differently on the graft-type PUSX nanofibers than on the block-type PUSX nanofibers.

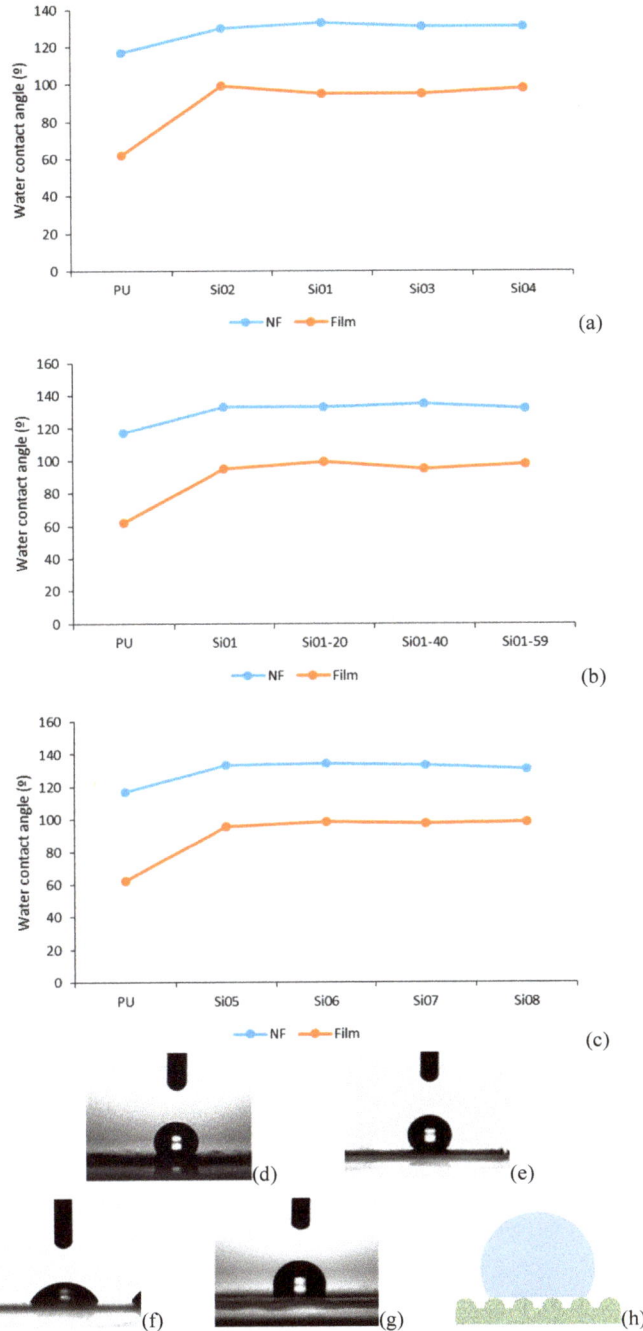

Figure 7. Comparison of water contact angle (WCA, degrees). (**a**) Block-type PUSX nanofibers and films with various chain lengths, (**b**) Block-type PUSX nanofibers and films with various silicone concentrations (**c**). WCA images of (**d**) PU nanofibers, (**e**) PUSX Si08 nanofibers, (**f**) PU films, (**g**) PUSX Si08 films, and (**h**) the Lotus microstructure (Cassie's state).

3.5. Cell Culture Studies

3.5.1. Cell Attachment and Cell Proliferation

To investigate the in vitro biocompatibility of the blended nanofibers, NIH3T3 cells were cultivated on to PUSX nanofibers and films of different structures. Cell attachment results were obtained and calculated after 3 h, and are showed in Figure 8. As the result, the number of adhered cells on block-type PUSX nanofibers became higher with the increase of silicone chain length. The reason for the increase of the fibroblast cells might be due to NIH3T3 cells being easily grown on the surface, which has a higher water repellency and hydrophobic surface characteristics. As discussed in Section 3.4, a higher water repellency is found with the increase of silicone chain length in block-type PUSX nanofibers. The cell attachment results of the PUSX films were not shown, due to the very low numbers after 3 h. Cells take a shorter time to adhere on nanofibers than on films, because of the porosity of the electrospun nanofibers. For cell attachment, PUSX nanofibers turned out to be more suitable than films.

Figure 9 represents the SEM images of the NIH3T3 fibroblast cells cultured for three days on different PUSX nanofibers and films with different structures. It can be seen that after three days of culture, there were more cells on the films than on the nanofibers, but the entanglement of the cells was totally different. The cells attached in the pores of the nanofibrous membranes with rough surfaces were much easier to manage as a scaffold for tissue engineering. Their stability was much higher than the cells attached onto the surfaces of the films. The reason might be that nanofibers have a fiber diameter of 400–700 nm, mimicking the extracellular matrix (ECM), as well as pores that help the cells to stay stable in the membranes. This work suggested that the PUSX nanofibers have an important advantage of being able to physically biomimic the natural ECM for tissue engineering applications, and cell ingrowth and cell encapsulation in the nanofibrous scaffolds are equally important. The architecture of a scaffold and the material used to play an important role in modulating tissue growth and response behavior of the cells that have been cultured onto the scaffold. In this regard, the scaffold should not only work as a substrate for cell attachment, growth, and proliferation, but also facilitate cell migration, ingrowth, and assembly into a stereo-structure. Referring to the SEM morphologies, the cells could attach onto PUSX nanofibers better than onto films, because the porosity makes the nanofibers more stereo than in films.

Figure 10 shows the cell proliferation results after one day, three days, five days, and seven days, respectively. The doubling time of NIH3T3 in the normal cell culture condition was between 20 to 26 h, which means that it requires 30 h to produce 10,000 cells. However, the conditions of the nanofibers can slow the processes down, because the structure allows less space for the cell to adhere as quickly as on the normal substrate. From the results, all 12 kinds of PUSX nanofibers were proven to be appropriate for cell proliferation, with a maximum cell number of around more than 10,000 on the fifth day. PUSX nanofibers can be applied in the biomedical field as a better alternative to PU nanofibers, and with controllable physical properties, as seen in the similar results of the cell proliferation test. As a result, this study, it was confirmed that biomedical materials of desired physical properties are able to be prepared by changing the structure without losing the same level of biocompatibility.

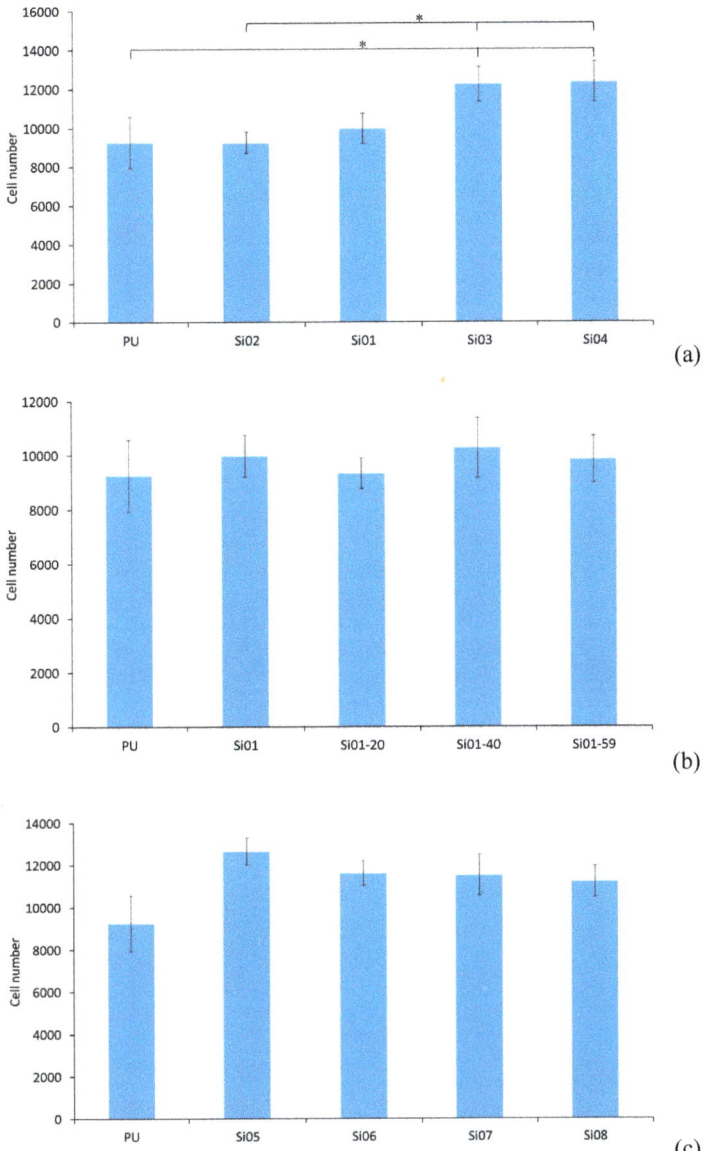

Figure 8. The attachment of NIH3T3 cells, (**a**) Block-type PUSX nanofibers with various chain lengths, (**b**) Block-type PUSX nanofibers with various silicone concentrations, (**c**) Graft-type PUSX nanofibers after cells have attached for three hours. "*" was statistically significant ($p < 0.05$) between each pair of samples.

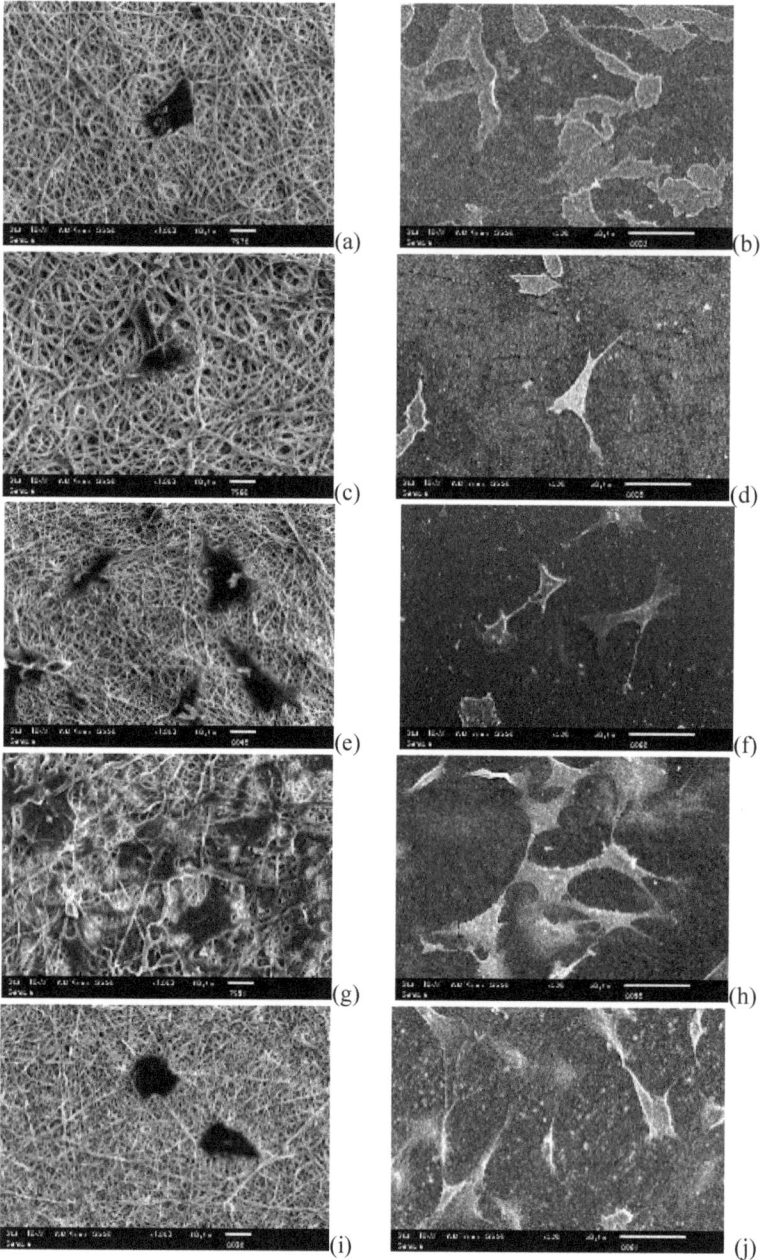

Figure 9. SEM images of NIH3T3 cells after culturing for three days on each sample. Cell attachment morphologies on PU nanofibers (**a**) and films (**b**), Si01 nanofibers (**c**) and films (**d**), Si01-59 nanofibers (**e**) and films (**f**), Si04 nanofibers (**g**) and films (**h**), Si08 nanofibers (**i**) and films (**j**). (Magnification of nanofibers 1000×, films: 500×).

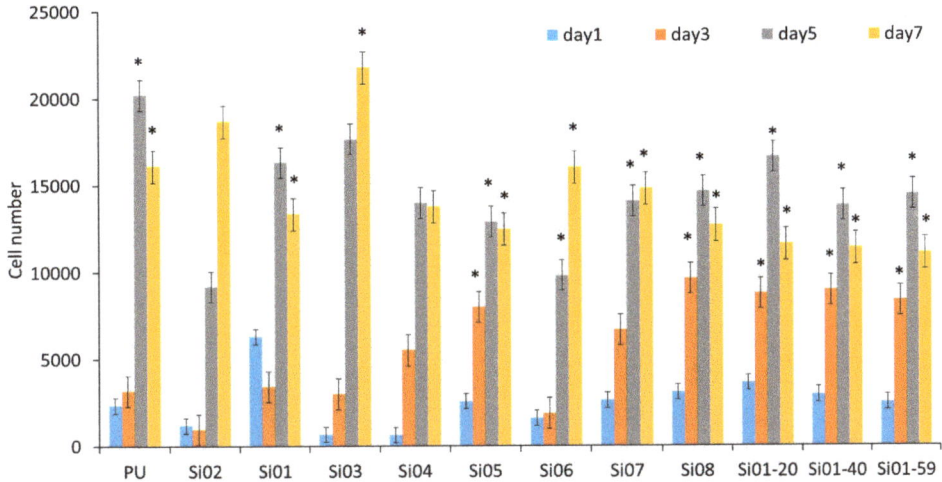

Figure 10. The proliferation of NIH3T3 cells on block-type PUSX nanofibers with various chain lengths, block-type PUSX nanofibers with various silicone concentrations, and graft-type PUSX nanofibers after cells were cultured for 1, 3, 5, and 7 days, respectively. "*" was statistically significant ($p < 0.05$) for each sample between 1 day and 3, 5, and 7 days.

3.5.2. Toxicity Evaluation

The results as displayed in Figure 11 showed that both the PUSX films and nanofibers caused less toxicity when they were in contact with cells. The cells were able to keep a spreading shape, and discrete intra-cytoplasmic granules with no cell lysis, which can be considered as a survival mechanism compared with the positive control (cells became round and layers were completely destroyed). The morphological grade of the cytotoxicity is supposed to be 0.

There was no obvious difference between the PUSX and PU materials, because of the biocompatibility of silicone. To the best of our knowledge, silicone has been extensively used in medical areas, in several products such as breast implants, contact lenses, lubricants, sealers, artificial cardiac tubes and valves, urethral and venous catheters, membranes for blood oxygenation, dialysis tubes, orthopedic applications, and facial reconstructions, because of its high biocompatibility [12,13]. The existence of the silicone group does not change the biocompatibility of the material. From these results, the PUSX materials showed a suitable biocompatibility for use as biomedical materials, such as waterproof bandages or scaffolds for tissue engineering.

Therefore, our novel nanofibrous membranes favored fibroblast cell attachment and growth by providing a stereo-structure environment that mimics the ECM, and which is considered to be biofriendly after toxicity evaluation. Indeed, there are a large amount of studies for developing biomedical materials for wound dressing and tissue engineering. For instance, Fenghua Xu et al. suggested tannic acid/chitosan/pullulan composite nanofibers, which showed synergistic antibacterial activity and the potential for deep and intricate wound healing. Compared with these materials, PUSX nanofibers show the controllable tensile strength of PU, which suggests potential applications for long-term tissue engineering, and hydrophobicity that is influenced by the silicone groups. As we know, wound infection is one of the main areas of concern in the management of the wound environment. Infection complicates treatment, and it impedes the healing process by damaging tissue, reducing wound tensile strength and inducing an undesirable inflammatory response [14–16]. PUSX nanofiber wound dressing may provide a hydrophobic effect, which helps to control bacteria by adsorbing bacteria onto the dressing surface. These results provide insight for the potential use of PUSX nanofibers for wound healing and tissue engineering into clinical practice in the future.

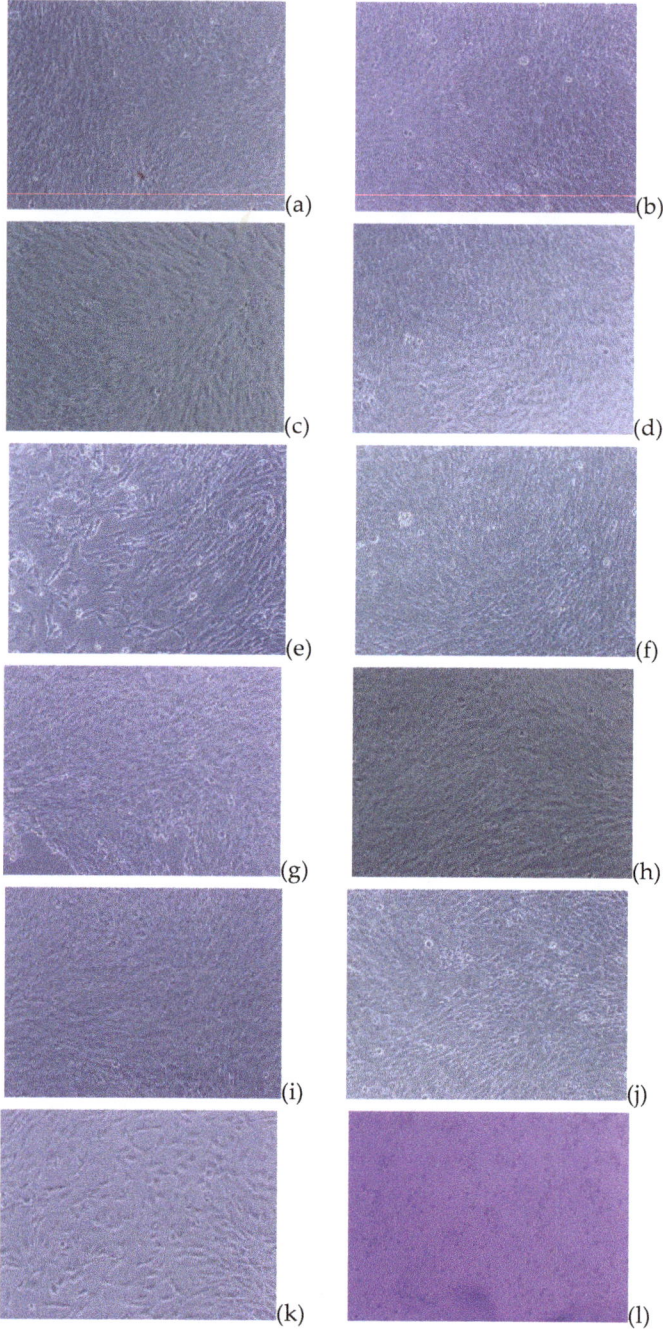

Figure 11. The toxicity evaluation of PU and PUSX samples. PU nanofibers (**a**) and films (**b**), Si01 nanofibers (**c**) and films (**d**), Si01-59 nanofibers (**e**) and films (**f**), Si04 nanofibers (**g**) and films (**h**), Si08 nanofibers (**i**) and films (**j**), compared to the negative control (**k**) and the positive control (**l**). (Magnification: 40×).

In the present study, the physical properties and biocompatible properties of PUSX nanofibers were investigated and compared with films. As a conclusion, for all of the analyses, physical properties, mechanical properties, water retention, and water contact angle (WCA) can be controlled and improved by adjusting the structure. Unfortunately, the graft-type PUSX did not show obvious changes in mechanical strength, because the side chains of silicone could not have as much influence as the main chain of PU does. Conversely, the graft-type PUSX nanofibers were the most similar alternatives to PU nanofibers in terms of mechanical properties, but with better water repellency according to the water contact angle results. Higher hydrophobicity and lower thermal conductivity were also found in the PUSX nanofibers, due to the unique advantages of nanofibers compared with films. This material can be expected to be applied in various fields. For instance, by controlling the silicone chain length and the concentration of block-type PUSX nanofibers, it can be applied in the medical field as bandages or scaffolds, the apparel field as outdoor goods and sportswear, and it can also be used for air or water filters.

In vitro biocompatible evaluation shows that cell proliferation can be performed on both the PUSX nanofibers and films. However, for cell attachment, the cells are not able to attach onto the PUSX films in a short time, nor entangle in the material, PUSX nanofibers were proven to be more appropriate for cell culture study. As we know, PU nanofibers have been developed as biomedical materials for several years. For instance, Lakshmi R. Lakshman et al. [17] and Afeesh R.Unnithan et al. [18] reported wound-dressing materials with antibacterial activity, which provided a basic understanding of the design for efficient PU nanofiber-based antibacterial wound dressing materials. Chang Hun Lee et al. [19] and Rui Chen et al. [20] demonstrated that electrospun PU nanofibers had the characteristics of a native extracellular matrix, and they may be used effectively as an alternative material for tissue engineering and functional biomaterials.

4. Conclusions

In this research, the physical properties and the biocompatibility of PUSX nanofibers were discussed and compared with PUSX films. Tensile strength increased with the increase of silicone chain length, and decreased with the increase of silicone concentration. The elongation at break decreased with the increase of both the silicone chain length and the silicone concentration in block-type PUSX nanofibers, while the Young's modulus showed the entirely opposite trend. Higher hydrophobicity and lower thermal conductivity were found in PUSX nanofibers, compared with PU nanofibers and films. An in vitro biocompatible evaluation shows that PUSX nanofibers are able to be applied in the cell culture field, and that it shows very little toxicity. Compared with films, PUSX nanofibers shows better cell attachment over a short period of time. Therefore, electrospun PUSX nanofibers with highly controllable physical properties can be expected as ideal alternatives to PU membranes applications of the biomedical field, such as for wound dressings and tissue engineering.

Author Contributions: Conceptualization, H.H., M.T., H.S., S.I., C.Y. and T.T.; Resources, H.H., M.T., H.S., and S.I.; Investigation, C.Y., S.R., R.O. and M.K.; Formal analysis, C.Y., S.R., R.O. and M.K.; Data curation, C.Y., H.S., H.H. and T.T.; Supervision, Y.T. and T.T.; Writing—original draft, C.Y.; Writing—review & editing, C.Y. and T.T.; Funding acquisition, C.Y. and T.T.

Funding: This work, including APC, was supported by a Grant-in-Aid for the Shinshu University Advanced Leading Graduate Program by the Ministry of Education, Culture, Sports, Science and Technology (MEXT), Japan.

Acknowledgments: The authors gratefully acknowledge financial support from Grant-in-Aid for the Shinshu University Advanced Leading Graduate Program by the Ministry of Education, Culture, Sports, Science and Technology (MEXT), Japan. In addition, the authors would like to expressly thank Shin-Etsu Chemical Co., Ltd (Tokyo, Japan) and Dainichiseika Color & Chemicals Mfg. Co., Ltd (Tokyo, Japan) for kindly providing the materials.

Conflicts of Interest: The authors declare no conflict of interest.

References

1. Subbiah, T.; Bhat, G.S.; Tock, R.W.; Parameswaran, S.; Ramkumar, S.S. Electrospinning of nanofibers. *J. Appl. Polym. Sci.* **2005**, *96*, 557–569. [CrossRef]
2. Viness, P.; Clare, D.; Yahya, E.C.; Charu, T.; Lomas, T.; Pradeep, K.; Lisa, C.T.; Valence, M.K.N. A Review of the Effect of Processing Variables on the Fabrication of Electrospun Nanofibers for Drug Delivery Applications. *J. Nanomater.* **2013**, *2013*, 789289.
3. Liu, X.; Ma, P.X. Polymeric scaffolds for bone tissue engineering. *Ann. Biomed. Eng.* **2004**, *32*, 477–486. [CrossRef] [PubMed]
4. Chen, Z.G.; Wang, P.W.; Wei, B.; Mo, X.M.; Cui, F.Z. Electrospun collagen–chitosan nanofiber: A biomimetic extracellular matrix for endothelial cell and smooth muscle cell. *Acta Biomater.* **2010**, *6*, 372–382. [CrossRef] [PubMed]
5. Wenying, L.; Stavros, T.; Younan, X. Electrospun Nanofibers for Regenerative Medicine. *Adv. Healthc. Mater.* **2012**, *1*, 10–25.
6. Kiliç, E.; Yakar, A.; Pekel, B.N. Preparation of electrospun polyurethane nanofiber mats for the release of doxorubicine. *J. Mater. Sci. Mater. Med.* **2018**, *29*, 8. [CrossRef] [PubMed]
7. Lionelli, G.T.; Lawrence, W.T. Wound dressing. *Surg. Clin. N. Am.* **2003**, *83*, 617–638. [CrossRef]
8. Thomas, S. Wound management and dressings. *Pharm. P* **1990**, *1*, 37–57.
9. Bhardwaj, N.; Kundu, S.C. Electrospinning: A fascinating fiber fabrication technique. *Biotechnol. Adv.* **2010**, *28*, 325–347. [CrossRef] [PubMed]
10. Chuan, Y.; Rino, O.; Mikihisa, K.; Toshihisa, T.; Hatsuhiko, H.; Masaki, T.; Hiromasa, S.; Shota, I.; Yoshitaka, K. Electrospinning of block and graft type silicone modified polyurethane nanofibers. *Nanomaterials* **2019**, *9*, 34. [CrossRef]
11. ISO10993-5, Biological Evaluation of Medical Devices, Part 5: Tests for In Vitro Cytotoxicity. Available online: https://www.iso.org/obp/ui/#iso:std:iso:10993:-5:ed-3:v1:en (accessed on 4 March 2019).
12. Quatela, V.C.; Chow, J. Synthetic facial implants. *Facial Plast. Surg. Clin. N. Am.* **2008**, *16*, 1–10. [CrossRef] [PubMed]
13. Chin, T.; Kobe, K.; Hyakusoku, H.; Uekusa, K.; Hirakawa, K.; Ohno, Y. Experimental analysis of silicone leakage. *J. Nippon Med. Sch.* **2009**, *76*, 109–112. [CrossRef] [PubMed]
14. Barry, W.; Hansen, D.L.; Burrell, R.E. The comparative efficacy of two antimicrobial barrier dressings: In vitro examination of two controlled release silver dressings. *Wounds* **1998**, *10*, 179–188.
15. Yin, H.Q.; Langford, R.; Burrell, R.E. Comparative evaluation of the antimicrobial activity of ACTICOAT antimicrobial barrier dressing. *J. Burn Care Rehabil.* **1999**, *20*, 195–200. [CrossRef] [PubMed]
16. Percival, S.L.; Bowler, P. Understanding the Effects of Bacterial Communities and Biofilms on Wound Healing. Available online: http://www.worldwidewounds.com/2004/july/Percival/Community-Interactions-Wounds.html (accessed on 4 March 2019).
17. Lakshmi, R.L.; Shalumon, K.T.; Sreeja, V.N.; Jayakumar, R.; Nair, S.V. Preparation of Silver Nanoparticles Incorporated Electrospun Polyurethane Nano-fibrous Mat for Wound Dressing. *J. Macromol. Sci. A Pure Appl. Chem.* **2010**, *47*, 1012–1018. [CrossRef]
18. Unnithan, A.R.; Barakat, N.A.; Pichiah, P.B.; Gnanasekaran, G.; Nirmala, R.; Cha, Y.S.; Jung, C.H.; El-Newehy, M.; Kim, H.Y. Wound-dressing materials with antibacterial activity from electrospun polyurethane–dextran nanofiber mats containing ciprofloxacin HCl. *Carbohydr. Polym.* **2012**, *90*, 1786–1793. [CrossRef] [PubMed]
19. Lee, C.H.; Shin, H.J.; Cho, I.H.; Kang, Y.M.; Kim, I.A.; Park, K.D.; Shin, J.W. Nanofiber alignment and direction of mechanical strain affect the ECM production of human ACL fibroblast. *J. Biomater.* **2005**, *26*, 1261–1270. [CrossRef] [PubMed]
20. Chen, R.; Huang, C.; Ke, Q.; He, C.; Wang, H.; Mo, X. Preparation and characterization of coaxial electrospun thermoplastic polyurethane/collagen compound nanofibers for tissue engineering applications. *Colloids Surf. B Biointerfaces* **2010**, *79*, 315–325. [CrossRef] [PubMed]

© 2019 by the authors. Licensee MDPI, Basel, Switzerland. This article is an open access article distributed under the terms and conditions of the Creative Commons Attribution (CC BY) license (http://creativecommons.org/licenses/by/4.0/).

Review

Electrospinning Nanofibers for Therapeutics Delivery

S. M. Shatil Shahriar [1,†], Jagannath Mondal [2,†], Mohammad Nazmul Hasan [2], Vishnu Revuri [2], Dong Yun Lee [3,*] and Yong-Kyu Lee [1,2,*]

1. Department of Chemical and Biological Engineering, Korea National University of Transportation, Chungju 27469, Korea; shatilshahriar26@gmail.com
2. Department of Green Bio Engineering, Korea National University of Transportation, Chungju 27469, Korea; jagannathrkm@gmail.com (J.M.); nazmulchemi@gmail.com (M.N.H.); vishnurevuri91@gmail.com (V.R.)
3. Department of Bioengineering, College of Engineering, and BK21 PLUS Future Biopharmaceutical Human Resources Training and Research Team, and Institute of Nano Science & Technology (INST), Hanyang University, Seoul 04763, Korea
* Correspondence: dongyunlee@hanyang.ac.kr (D.Y.L.); leeyk@ut.ac.kr (Y.-K.L.)
† These authors contributed equally to this work.

Received: 20 February 2019; Accepted: 22 March 2019; Published: 3 April 2019

Abstract: The limitations of conventional therapeutic drugs necessitate the importance of developing novel therapeutics to treat diverse diseases. Conventional drugs have poor blood circulation time and are not stable or compatible with the biological system. Nanomaterials, with their exceptional structural properties, have gained significance as promising materials for the development of novel therapeutics. Nanofibers with unique physiochemical and biological properties have gained significant attention in the field of health care and biomedical research. The choice of a wide variety of materials for nanofiber fabrication, along with the release of therapeutic payload in sustained and controlled release patterns, make nanofibers an ideal material for drug delivery research. Electrospinning is the conventional method for fabricating nanofibers with different morphologies and is often used for the mass production of nanofibers. This review highlights the recent advancements in the use of nanofibers for the delivery of therapeutic drugs, nucleic acids and growth factors. A detailed mechanism for fabricating different types of nanofiber produced from electrospinning, and factors influencing nanofiber generation, are discussed. The insights from this review can provide a thorough understanding of the precise selection of materials used for fabricating nanofibers for specific therapeutic applications and also the importance of nanofibers for drug delivery applications.

Keywords: electrospinning; nanofibers; fabrication; therapeutics; biomedical applications

1. Introduction

Nanotechnology—the use of nanomaterials for biomedical applications—is an emerging and promising paradigm in biomedical research. Nanomaterials with exceptional physiochemical properties, biocompatibility and minimal biological toxicity, can sense local biological environments and initiate cellular level reprogramming to achieve the desired therapeutic efficacy. Diverse zero dimensional (quantum dots, carbon dots, graphene quantum dots), one-dimensional (nanorods, nanowires, nanotubes, nanofibers) and two-dimensional (graphene oxide, transition metal dichalcogenides, transition metal oxide, MXens, etc.) nanomaterials along with nanosized particles are currently used for diagnosis, imaging and therapy [1–3]. Nanofibers are fiber shaped nanostructures, typically with two of their dimensions in nanoscale. Nanofibers have a high surface area-to-volume ratio with tunable porosity and can easily be functionalized with biological molecules. The choice of a wide variety of materials, such as natural and synthetic polymers, inorganic nanomaterials, composites and biomolecules as drugs for nanofiber fabrication makes them a robust and attractive candidate for many advanced biomedical applications [4–12]. These remarkable characteristics make nanofibers an ideal nanomaterial for energy generation and storage, water and environmental treatment, and healthcare and biomedical engineering applications.

Electrospinning is a unique and versatile technique that depends on the electrostatic repulsion between surface charges to constantly draw nanofibers from viscoelastic fluids. Polymers, ceramics, small molecules and their combinations are used as rich materials for the production of nanofibers. In addition to solid nanofibers, a secondary structure of nanofibers—including porous, hollow or core-sheath structures—has been manufactured and the surface of the structure can be functionalized with different molecular moieties, during or after the electrospinning process [13]. Electrospinning is the main method of choice for the large scale production of nanofibers due to its controllable diameter, easy handling, minimum consumption of solution and cost effectiveness. Electrospun nanofibers have various biomedical applications, such as wound dressings, drug and gene delivery tools, sensors and catalysts [14]. In this review, our focus is on addressing the application of electrospun nanofibers in therapeutics (drug/gene) delivery. Various therapeutics delivery systems have been investigated for efficient loading, releasing and accumulation of the therapeutics into the target site. However, electrospinning exhibits great flexibility for selecting diverse materials, drugs and genes (DNA, RNA etc.) for therapeutic applications [15].

We have reviewed the most efficient and recent therapeutic applications of electrospun based nanofibers and their future perspectives. Different electrospinning techniques are described here, along with their biomedical applications and suitability. Nanofibers are generally natural or synthetic polymers and they have vast therapeutic applications even though only a limited number reach clinical trials. The application of nanofibers as drug delivery carriers—especially for cancer drugs, antibacterial drugs, nonsteroidal anti-inflammatory drugs, cardiovascular agents, gastrointestinal drugs, antihistamine drugs, contraceptive drugs and palliative drugs delivery—is described here. Nanofibers have promising therapeutic applications for DNA, RNA and growth factors, for example, protein and steroid hormones delivery, which are included in this review. The current scenario for nanofiber based materials is mostly promising for wound dressing, sensor technology and for the catalysis of various reaction pathways in the laboratory [14]. The therapeutic application of nanofibers to drug or gene delivery is limited due to proper functionality, controlling capacity, toxicity and large-scale production limitations. However, the development of nanofiber based delivery systems is growing rapidly and many of them have shown excellent characteristics for future therapeutic delivery applications [15]. This review depicts the whole scenario around current therapeutic delivery systems based on nanofibers and their useful biomedical applications, as well as suggesting future perspectives for nanofiber based delivery systems, and provides relevant literature for researchers working in this area.

2. Electrospinning Techniques

Electrospinning is considered a promising, highly productive and simple method for fabricating the nanofibers of polymers, composites, and inorganic materials, including carbides, oxides, nitrides and hybrid composites. In the electrospinning technique, electrostatic forces are utilized to produce nanofibers from a polymer solution [16]. In general, the electrospinning setup consists of three main compartments, namely: (i) high voltage power supply; (ii) a spinneret; and (iii) a conductive collector, as shown in Figure 1. In the electrospinning process, a potential (kV) is applied between the spinneret and the collector. These parts conduct electricity and are separated at an optimum distance. When the applied electric field overcomes the surface tension of the droplet, a charged jet of polymer solution can be ejected from the tip of the needle. The jet grows longer and thinner, with an increasing high diameter loop, which results in the solidification of the polymer due to solvent evaporation. The solidified nanofibers are then collected on the target. In general, electrospun nanofibers are categorized in two ways; namely, random and aligned nanofibers [16]. Random nanofibers can be produced using a simple plate collector, while aligned nanofiber mats or uni-axial fiber bundles can be produced using a disc or cylinder, rotating at a high speed, as the collector, along the direction of rotation. The unique physical characteristics of electrospun nanofibers, such as high surface-to-volume ratio, controllable fiber diameters and surface morphologies (dense, hollow, and porous) and fibrous structures, can be altered by modulating parameters, for example: (i) the molecular weight of polymers and polymer solution properties (viscosity, conductivity, dielectric constant, and surface tension) [17,18]; (ii) The processing parameters (such as the electric potential, flow rate, feeding rate, distance between the capillary and collection, as well as using coaxial or triaxial needles for hollow, core–shell or multi-sheathed structures); and (iii) controlled post processing parameters (such as heating rates and heating temperatures, especially for inorganic materials) [19].

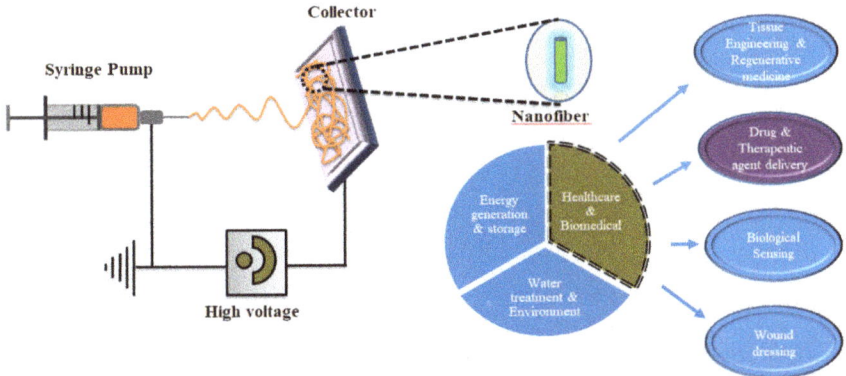

Figure 1. Schematic representation of a traditional electrospinning process and various healthcare and biomedical applications of nanofibers.

3. Types of Electrospinning

Although fibers produced by different electrospinning methods have attracted increasing attention in the field of biomedical applications, challenges persist for the selection of an appropriate method, as well as for optimizing multiple parameters to generate robust cargo loaded nanofibers [20]. Electrospinning techniques can be classified into five categories: (i) blend electrospinning; (ii) coaxial electrospinning; (iii) emulsion electrospinning; (iv) melt electrospinning; and (v) gas jet electrospinning. Blend electrospinning can be used to develop nanofibers with burst release, while co-axial and emulsion electrospinning can be used to generate core-shell nanofibers that can assist in sustained drug release [21]. Melt electrospinning is a cleaner fabrication method that has been used extensively to produce highly ordered electrospun nanofibers. However, melt electrospinning results in the generation of fibers with a larger diameter [22]. The basic mechanism of each electrospinning setup and the process parameters affecting the fiber morphology are discussed below.

3.1. Blend Electrospinning

Blend electrospinning is the conventional electrospinning technique, in which the drug/biomolecules are dispersed or dissolved in the polymer solution, resulting in the development of nanofibers with drugs/biomolecules dispersed throughout the fibers [23]. Although this method is simple in contrast with other electrospinning methods, the solvents used for the dispersion of bioactive molecules can lead to protein denaturation or loss of biological activity. In addition, the inherent charge of the biomolecules can often result in their migration on the jet surface and thereby result in their distribution on the surface of the nanofibers rather than the encapsulation of biomaterials within the fibers. The surface dispersion can be associated with the burst release of the biomolecules [24].

3.2. Co-Axial Electrospinning

Co-axial electrospinning is an improvement on conventional blend electrospinning, in which two nozzles are connected to the high voltage source rather than one nozzle. Two different solutions are loaded within each nozzle and are pumped out to generate nanofibers with core-shell morphologies [25]. Both synthetic and natural polymers can be used to develop core-shell nanofibers with improved physiochemical and biological properties. This method is advantageous over the blend electrospinning method, as they can protect as well as overcome the denaturation of drugs/biomolecules in the biological system [26]. In co-axial electrospinning, the biomolecules or drugs are situated in the inner jet and are co-spun with the polymers present in the outer jet. This results in protection of the cargo, as well as assisting their being sustained in the biological environments [27,28]. For example, Merkle et al. used polyvinyl alcohol (PVA) and gelatin for the fabrication of core-shell nanofibers [29]. The mechanical strength of the core PVA phase was improved by increasing the amount of gelatin in the shell. The authors have claimed that the gelatin shell augmented cell adhesion and fibroblast adhesion onto the surface of the PVA/gelatin fibers, compared to the PVA fibers. In addition, coaxial fibers can be surface-modified with biomolecules to improve their biofunctionality and enhance cell-surface interactions [30]. Apart from the design complexity, the viscoelasticity and interfacial tension of the core and shell polymers must be thoroughly monitored to generate core-shell nanofibers from co-axial electrospinning.

3.3. Emulsion Electrospinning

Emulsion electrospinning requires a setup similar to that of blend electrospinning, in which two immiscible solvents are simultaneously spun to generate core-shell nanofibers. Here, active bioactive molecules, along with surfactants, are initially allowed to form W/O emulsions and are later mixed with the polymer matrix solution [25]. During the fiber trajectory, the emulsion droplets are stretched into elliptical shape. Furthermore, the continuous phase solvent is swiftly evaporated, resulting in the viscosity gradient as well as droplet enrichment in the axial region [31]. This viscosity gradient between the polymer matrix and elliptical droplet guides the core material to settle within the fiber matrix rather than on the polymer surface. This method is relatively simple compared to co-axial electrospinning and also offers sustained release of the loaded cargo materials [32]. Applied voltage has a significant effect on the nanofiber diameters. Higher applied voltage ensues in the generation of nanofibers with a reduced diameter. In addition, other parameters like flowrate and spinning distance can also affect fiber morphology [33]. However, the interface tension between the organic and aqueous phases of the emulsion can damage the bioactive molecules loaded within the emulsion electrospun fibers [34].

3.4. Melt Electrospinning

Melt electrospinning has gained more attention in the electrospinning field, where toxicity and solvent accumulation is a concern. In this process, polymer melt is used instead of a solution, which is transformed from liquid to solid to achieve the desired product on cooling, rather than through solvent evaporation [22]. The polymer melt flow rate and homogeneous polymer melt conditions must be controlled to produce high quality fibers of uniform morphology, but with a broad range diameter. Polymer blends [35] and additives [36] have been used to reduce the average diameter of the fibers. The effect of the melt temperature can affect the structure and function of drugs, proteins and bioactive molecules loaded in the fibers [37]. The flow rate and melt viscosity of melt electrospinning can influence the characteristics of the resulting fibers. The surface wettability of melt electrospun fibers has improved through the formation of hydroxyl or peroxyl and N-containing functional groups, respectively, by using oxygen and ammonia plasma [35,38].

3.5. Gas Jet Electrospinning

Gas jet electrospinning is an improvement on the conventional melt electrospinning technique, in which the conventional electrospinning setup is additionally equipped with a gas jet device. The major limitation of melt electrospinning is that it requires definite control over the temperatures and therefore multiple heating zones must be placed to maintain the polymer melt. This results in generation of thicker nanofibers compared to the solutions. In this technique, the co-axial jet is surrounded by a tube feeding the heated gas, which can provide sufficient heat near the nozzle and delay the process of polymer solidification. For example, Zhmayev et al. spun polylactic acid (PLA) and showed a decrease in the diameter of the nanofibers obtained by gas jet electrospinning compared to normal electrospinning. Interestingly, the heated gas flow-rate had a significant effect on the diameter of the nanofiber. The increased gas flow rate can offer additional drag force to the jet surface which can result in the development of thinner nanofibers. Zhmayev et al.'s result showed a significant decrease in the diameter of the nanofiber, from 350 nm to 183 nm, when the gas flow rate was increased from 5.0 L/m to 15.0 L/m [39,40].

4. Therapeutics Delivery Systems

Most conventional drugs are hydrophobic and suffer from poor bio-distribution, solubility and stability in the biological system. Moreover, these drugs do not have the desired active targeting capabilities, which can result in non-specific systemic toxicity or faster elimination from the body, without achieving the desired therapeutic efficacy. Drug delivery systems (DDS) are approaches, formulations and technologies for transporting therapeutic agents to the targeted therapeutic site in the body [41]. The developed DDS technologies not only encapsulate the target drug/biomolecule but also attune their absorption, distribution, release and elimination with higher loading efficacy and safety. The drug release from DDS carriers depends on diffusion, degradation, swelling and affinity-based mechanisms [42]. As mentioned above, electrospun nanofibers are gaining significant attention as promising therapeutic nanocarriers. In addition, their impressive characteristics, including biocompatibility, biodegradability, and high therapeutic payload capacity, meet the prerequisites for a good therapeutic delivery candidate [13]. Employing electrospun nanofiber scaffolds as a therapeutic nanocarrier, different routes of administration (ROA) are being investigated. Drugs/therapeutics can be administered to any region/organ in the body by way of common routes, such as oral, parenteral (subcutaneous, intramuscular, intravenous, and intrathecal), sublingual/buccal, rectal, vaginal, ocular, nasal, inhalation, and transdermal, using electrospun nanofibers (Figure 2). Here, we present the common routes of drug administration by electrospun nanofiber [43].

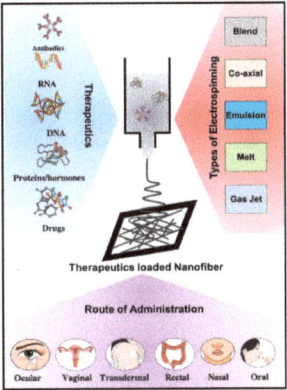

Figure 2. Types of electrospinning, different therapeutics-loaded nanofibers and their route of administrations.

4.1. Oral

Of all delivery routes, the oral route is considered the most preferable and convenient route of administration, which can overcome the problems associated with other routes of administration [44]. However, to achieve successful administration of therapeutics targeting the oral route of administration is a difficult task. Before designing a successful oral delivery system, scientists should consider the major key challenges, including the presence of acidic gastric juice in the stomach, along with proteases, mucosal barriers and intestinal retention, which can hamper the absorption of drug delivery systems into the body [45]. Electrospun nanofiber scaffolds offer a great opportunity to load and deliver both micro and macromolecules, targeting the oral route [46]. Some of the common features of electrospun nanofiber in oral drug delivery systems include targeted delivery of therapeutics, sustained release properties, high transfection efficiency, rapid onset of action and fascinating pharmacokinetic profiles. Another promising advantage of using electrospun nanofibers is that researchers can design any desirable release properties such as fast [47], controlled [48], biphasic [49] or delayed [50] releases of the drug. Various polymers are used for designing oral drug delivery systems utilizing electrospun nanofibers such as poly (lactic-co-glycolic acid) (PLGA), polyvinylpyrrolidone (PVP), poly(ethyleneoxide) (PEO), PVP/cyclodextrin, polyvinyl alcohol (PVA), polycaprolactone (PCL), PVP/ethyl cellulose, PVP/zein, Cellulose acetate, Eudragit L, hydroxypropyl methylcellulose (HPMC), Eudragit S, Eudragit S/Eudragit RS, and shellac [50].

4.2. Sublingual/Buccal

The oral mucosa has its own subdivisions according to its different regions, namely, buccal, sublingual and gingival. A few years ago, researchers were also interested in designing nanofiber based drug delivery systems for sublingual (under the tongue) or buccal (between the gums and teeth) routes of administration. The sublingual or buccal delivery systems usually allow the drug-loaded electrospun nanofibers to dissolve in the presence of mucus so that the drug can directly penetrate into the small blood vessels. Interestingly, these oromucosal routes of administration are the most studied sites for nanofiber-based therapeutics, offering versatile and multifunctional drugs, DNA, RNA, protein, peptide, growth factors or vaccine delivery platforms [51].

4.3. Rectal

Pediatric patients (aged under 6 months) usually cannot swallow any drug or food supplement. Under such circumstances, the rectal route could be the best alternative to oral drug delivery systems. In addition, the rectal route is also effective in cases of unconscious or vomiting patients. An electrospun nanofiber-based rectal drug delivery system is continually gaining popularity. In the treatment of post-operative peritoneal effusion following rectal/pelvic surgery, electrospun nanofibers could be considered a safe potential biocompatible, and biodegradable sealing fiber. The first clinical trial on the safety and sealing properties of electrospun nanofibers in lymphorrhea following pelvic surgery was published in 2014 [52]. The authors used a synthetic material (PuraMatrix) that consists of sixteen amino acid peptides to fabricate self-assembled nanofibers. A total of 20 colorectal cancer patients participated in the clinical trial. After a 2–3 month follow-up period, a significant reduction in post-operative drainage volumes was apparent in the experimental group compared with the control group. In another study, Modgill and co-workers investigated the permeability of penicillin from an extremely thin nanofiber scaffold through different biological membranes. The authors used PVA to fabricate ciprofloxacin-loaded ultra-thin nanofibers. In vitro permeability studies exhibited the potency of electrospun nanofibers compared with the plain drug. The PVA nanofibers revealed the highest ciprofloxacin permeability in the rectal mucosal membrane. The drug release study showed the controlled release behavior of ciprofloxacin from nanofibers in the rectal mucosal membrane, whereas the control group showed a high degree of fluctuations [53].

4.4. Vaginal

Electrospun nanofiber scaffolds targeting vaginal routes have recently been investigated. However, the acidic condition of vaginal mucosa (~pH 4) should be considered before designing vaginal drug delivery systems using nanofibers. For example, Brako et al. fabricated progesterone-encapsulated nanofibers targeting the vaginal route of administration. The authors used a mucoadhesive molecule (carboxymethylcellulose) for the fabrication of electrospun nanofibers. As-synthesized progesterone/carboxymethylcellulose electrospun nanofibers exhibited sustained release properties [54,55]. Another example of nanofiber-based vaginal drug delivery would be a study involving the anti-HIV drug (maraviroc), in which the authors spun maraviroc either with PVP or PEO and demonstrated fast drug dissolving properties upon contact with moisture [56].

4.5. Nasal

Currently, supramolecular peptides are used to fabricate electrospun nanofiber scaffolds. Electrospun supramolecular peptide nanofibers have proven their potential for intranasal vaccine delivery. For example, Si and coworkers developed intranasal administrated influenza vaccine using peptide nanofibers consisting of virus polymerase. This self-assembled nanofiber could elicit both humoral and cell mediated immune responses against the influenza virus. To be specific, nasal administrated nanofiber vaccine particles were initially ingested by antigen presenting cells in lung-draining mediastinal lymph nodes and activated both Th1 Th2 to produce the antibody against the given antigen. Furthermore, a high immune response was elicited without any use of vaccine adjuvant [57].

4.6. Ocular

Generally, various drops or ointments are used for ocular infection or inflammation. In these cases, those drugs are easily excreted from the eye very quickly. Therefore, to improve therapeutic efficacy and overcome the repetitive administration of an established ocular dosage, nanofiber-based ocular inserts have been considered and compared with eye drops or similar methods of dosage. Drug-loaded nanofiber scaffolds that allow the controlled release of incorporated drugs can be placed into the ocular mucosa. For example, the use of brimonidine tartrate (BT)-loaded electrospun nanofibers as ocular inserts were used to treat ocular infection and inflammation, which is described in detail in Section 5.1.9 [58].

4.7. Transdermal

Transdermal drug delivery systems (TDDS) deliver a drug locally, avoiding undesired drug distribution and possessing excellent skin permeability. Due to high solubility, great morphology and sustained drug release kinetics, electrospun nanofibers have been developed into TDDS. It has been reported that electrospun nanofibers could enhance solubility and form a transdermal patch for the biopharmaceutics classification systems II drugs with a high therapeutic efficiency and no cytotoxicity. Hydrophilic polymers with great permeation are the strength of TDDS, and several in vitro and in vivo evaluations suggest the possibility of using drug-loaded electrospun nanofibers in establishing TDDS [59].

5. Applications of Nanofibers in Therapeutics Delivery

Producing nanofibers with excellent physico-biochemical characteristics using an electrostatic spinning technique is considered to be an efficient multipurpose drug delivery strategy for delivering various therapeutic agents ranging from nanomedicines to macromolecules, including proteins and nucleic acids. Nanofibers can overcome several key challenges such as low solubility, loading efficiency, short circulation and the plasma half-life of drugs and are effective at improving the bioavailability of drug growth factors, DNA, RNA, and so forth. [60]. Using various biocompatible and biodegradable polymers (natural and synthetic or their hybrid combinations) in nanofiber production may overcome problems relating to the necessity of a second surgery to remove implants. In this section, we review the biomedical applications of electrospun nanofibers and delineate novel strategies employed to deliver drugs, biomolecules and nucleic acids by way of the electrospun nanofibers.

5.1. Nanofibers in Drug Delivery

5.1.1. Anticancer Agents

Discovering a new potential treatment strategy against cancer is very difficult due to several drawbacks of anticancer therapeutics such as imperfect solubility, impermanence in the circulatory bloodstream, poor accumulation in cancer cells, highly toxic to normal cells, low working efficiency in solid tumors and excess elimination profile [61]. To surpass such limitations, the nanofiber-based targeted drug delivery of anticancer therapeutics could be a promising strategy in the field of cancer nanomedicine. Electrospun nanofiber scaffolds have superior drug loading and transferring capabilities which not only amplify the therapeutic efficacy and potency of loaded drugs but also decrease undesirable side effects by ensuring high cellular accumulation of loaded drugs at the target site. These versatile drug carriers have proven to overcome the major limitations of traditional drug carriers that face anticancer drugs delivery, including poor therapeutic-loading capacity, non-targeted delivery, objectionable drug release and so on [62]. In addition, their sustained release properties downregulate the repetitive administration of drugs, thus improving patient compliance.

Recently, electrospun nanofibers have been widely used to ameliorate the solubility of anticancer drugs (Figure 3a). In this system, the drug-loaded nanofibers can circumscribe and crystallize the anticancer agents within the fiber, resulting in improvement of their dissolution rate in the biological system. For instance, water-insoluble anti-cancer drug Paclitaxel was loaded with surface modified mesoporous hollow stannic oxide nanofiber (SFNFP) in order to study whether electrospun nanofibers can improve the dissolution rate of Paclitaxel. The in vitro dissolution study results show that SFNFP exhibited an 8.34 times higher dissolution rate compared to naked Paclitaxel, over a period of 5 minutes. While the cumulative release rate of pure Paclitaxel was only $16.77 \pm 2.00\%$ after 1h, a high dissolution rate of $80.00 \pm 2.64\%$ was observed from SFNFP [63]. The release profiles of Paclitaxel from the SFNFP followed Noyes-Whitney's drug release profiles. Extreme-sleazy shells functionalized nanofibers as core-shell structure were synthesized using the coaxial electrospinning method to load and enhance the dissolution rate of water-insoluble drugs Quercetin and/or Tamoxifen Citrate, as another example. The core of the nanofibers was composed of PVP-K90 or PCL while Quercetin/Tamoxifen Citrate along with surfactant SDS and PVP-10 as hydrophilic moieties were used to form the shell. The in vitro results suggested a faster release of insoluble drugs from nanofibers in dissolution media within a 1 min period. To be specific, 16.14% and 15.15% of pristine Quercetin and TC were released in 0.5 h respectively, while drug loaded nanofibers released Quercetin/TC either immediately or in 1 min [64]. The reason for a fast release could be due to the uniform distribution of the model drug in the extremely thin (100 nm in diameter) outer layer of the core-shell nanofiber scaffolds, allowing a large contact surface area and a short diffusion distance. Therefore, multifunctional nanofibers are able to dramatically increase the release profile of poorly water-soluble anticancer drugs regardless of the drug's characteristics.

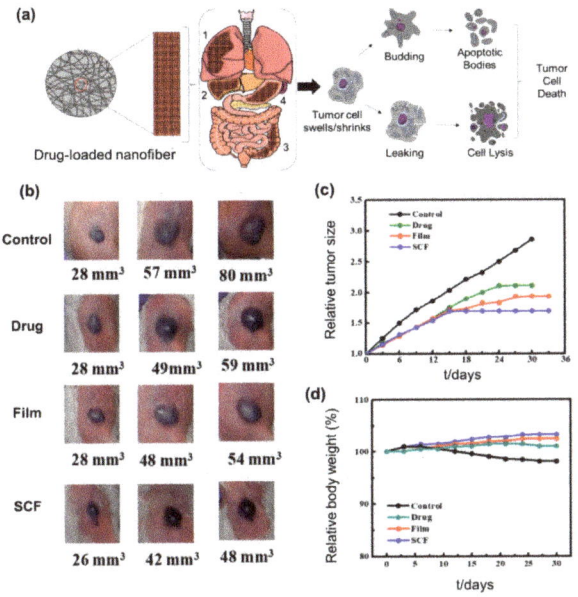

Figure 3. (a) The cytotoxicity effects of anticancer drug-loaded electrospun nanofibers targeting various deadly cancers. (Here, the numerical numbers 1, 2, 3, and 4 on the diagram represent the usages of nanofibers in lung, liver, gastric and colon cancer respectively.); (b) tumor dimension after treating with indicated patches (DXM-PLA); (c) quantitative tumor volume of mice treated with indicated scaffold/film/drug/control; (d) change of body weight of mice as a function of time upon treatment with indicated patch (Figure 3b–d reproduced from ref. [65] with permission from ACS Publication, 2019).

Even though polymeric nanoparticles (NPs) are used as a drug carrier to load anticancer drugs like Paclitaxel, the poor drug-loading efficiency of these systems minimizes the applications in the biological system [66]. On the other hand, nanofiber scaffolds using electrospinning facilitate higher drug loading because of large space and stereological honeycombed composition. Moreover, the choice of fabrication method for developing electrospun nanofibers, such as the coaxial process, emulsion methods, surface modification and blending, can tune/regulate the drug loading capacities of the nanofibers. For example, Xu and co-workers fabricated self assembleed Paclitaxel- Succinic Acid (PTX-SA) conjugate into supramolecular nanofibers. The loading efficiency of PTX in the as-synthesized nanoconstruct was more than 89%, which is considered the highest loading efficiency of PTX ever reported [67]. In addition, the controlled release of PTX from this nanoconjugate inhibited the proliferation of human lung adenocarcinoma cells in both in vitro and in vivo animal models. Compared to the burst release, the sustained release properties of cancer drug-loaded nanofiber scaffolds has great cytotoxic effects on tumor tissues. To understand these findings, Kumar et al. (2019) fabricated the anticancer drug dexamethasone (DXM)-loaded PLA nanofibers (DXM-PLA) in the form of a patch and applied it to a melanoma tumor in an experimental mice model. The sustained release properties of DXM from the as-synthesized nanopatch led to cytotoxic effects on cancer cells up to 85% within 3 days. Compared to the control groups, which resulted in 300% tumor growth, the DXM-PLA significantly prevented melanoma tumor cell proliferation and maintained the bodyweights of the animals (Figure 3b–d) [65].

Various reported scientific validations narrate that the circulation and plasma half-life of a drug nanocarrier highly depends on its size and shape. Geng et al. showed that uniquely shaped filomicelles like nanofibers can circulate in biological conditions for up to 7 days, which was 10 times greater than any type of polymeric nanoparticles, with which the drug can be eliminated from the bloodstream within 2 days. In addition, these filomicelles can selectively deliver the anticancer agent to the tumor tissues [68]. Furthermore, a versatile and scalable electrostatic spinning technique was employed to fabricate fiber rods with various sizes and shapes in order to confirm whether the shape and size of nanofibers can enhance the half-life, cancer tissue accumulation, cellular uptake, and tumor toxicity profiles of the anitcancer drug-loaded fiber rods. In this study, the electrospun fibers were treated under ultrasonication to fabricate fiber rods, where the lengths of the fiber rods were regulated by adding different volumes of sodium chloride (NaCl) void-precursors to the electrospun scaffolds. The experimental results revealed that when anticancer drug Doxorubicin was incorporated into fiber rods using PELA (RD$_{DOX}$, 500 nm) and these nanocomposites were injected into a tumor-bearing mice model, it showed 4 times more accumulation in the cancer site and was 3 times more stable in plasma level compared with microspheres. However, small length nanofibers of 2 micrometers in diameter exhibited the most powerful cancer cell apoptosis and necrosis activities with the highest resistance to metastasis [69].

To reduce the undesirable side effects and toxicity of anticancer drugs for normal tissues, electrospun nanofibers are considered for targeted and pH-mediated drug delivery to tumors. The surface modification of electrospun nanofibers with ligands that can target specific receptors overexpressed on the tumor, along with pH-dependent tunable drug release, can be optimal for targeted drug delivery. For example, a chitosan/PLA solution was used to encapsulate graphene oxide (GO), titanium dioxide (TiO$_2$) and a chemotherapy medication doxorubicin (Dox, $C_{27}H_{29}NO_{11}$) into nanocomplexes through electrospinning. As-prepared chitosan/PLA/GO/TiO$_2$/Dox fiber scaffolds showed an increased release of Dox in the acidic pH of a tumor microenvironment rather than in the physiological pH of 7.4 during a 200 h experiment. This characteristic would be imposed due to protonation of –NH$_2$ in Doxorubicin that can disintegrate the –H bond between doxorubicin and nanofibrous scaffolds resulting in higher drug release in cancerous tissues [70]. The in vitro cytotoxicity studies revealed that chitosan/PLA/GO/TiO$_2$/Dox fibers were biocompatible and did not expose any toxicity to normal cell lines. This research also evidenced that the cytotoxicity of nanofibers to tumor cells depends on the concentration of nanofibers—a higher concentration of nanofibers can increase both targetability and cytotoxicity to cancerous cells. A higher concentration of doxorubicin even appeared in the presence of an external magnetic field. As mentioned earlier, electrospun nanofibers can be developed with negligible toxicity towards normal cells by actively targeting the cancer cells. Heat shock protein (HSP 90) is highly overexpressed on lung cancer cell lines in humans. Researchers have developed novel stratgies by targeting HSP 90 to deliver anticancer drugs into tumor tissues of the lung [71]. Drugs like 17-DMAG (17-dimethylaminoethylamino-17-dimethoxy geldanamycin) can target both cancerous cells and the ATP-binding site of HSP 90, thus increasing the inhibition of both cancer cell proliferation in the lungs, and chaperoning the activities of HSP 90 and telomerase activity, respectively. Nevertheless, the unwanted side effects and extreme hepatotoxicity had minimized the use of 17-DMAG as a novel anticancer drug in lung cancer treatment [72]. To overcome such major problems, Mellatyar, and co-workers designed and developed 17-DMAG encapsulated PCL/PEG nanofibers via an electrospinning process. Their drug release profile revealed that about 96% 17-DMAG can release from PCL/PEG/17-DMAG fiber scaffolds after 6 h. In addition, the IC$_{50}$ values and MTT assay results proved the potency of PCL/PEG/17-DMAG nanofibers over 17-DMAG in A549 cells cytotoxicity. After 3 days, the free 17-DMAG can reduce the HSP 90 expression level and telomerase activity up to 48% and 71% respectively whereas the percentages of inhibition by PCL/PEG/17-DMAG nanofibers were 79% and 83%, respectively [73]. These promising characteristics of PCL/PEG/17-DMAG nanofibers are connected with its controlled release properties over an extended period of time. Nevertheless, nanofiber scaffolds are able to easily target and enter into cancer cells due to the irregular vascular composition of tumor tissues.

5.1.2. Antibiotics

Since Alexander Fleming discovered and developed a true antibiotic named penicillin, antibiotics have been the most commonly prescribed and used pharmaceutical agents to treat various bacterial infections. Despite all the favorable characteristics of antibiotics, their appropriate delivery routes, toxicological profile, poor water solubility and, most importantly, microbial resistance are major limitations to their therapeutic efficiency. Although several delivery approaches have been proposed over the past few decades, issues associated with poor antibiotic loading efficiency, systemic toxicity and drug release profile limited their translation into clinical settings. In this scenario, electrospun nanofibers are considered as an alternative for antibiotic delivery because the large surface area and tunable pore size offer maximum antibiotic loading capacity and encapsulation efficiency. In addition, the current generation of nano-based fibers can also regulate the sustained and controlled release activities of antibiotics, maximize the dissolution rate of poor water-soluble antimicrobial agents and minimize systemic toxicity Antibacterial drugs which are encapsulated into nanofibers usually exhibit antimicrobial actions by inhibiting cell wall synthesis, protein synthesis, DNA/RNA synthesis, mycolic-acid synthesis, and folic acid synthesis. Penicillin has been widely studied as a model antibiotic to load within nanofibers for testing antimicrobial actions. For instance, an aminopenicillin drug like amoxicillin (AMX) was initially loaded into nanomicelle as a hydrophobic antibacterial drug via a film dispersion hydration method and later coaxial electrospining was performed to load AMX-loaded nanomicelle into the core/shell nanofiber (AMX/NM/NF). Antibacterial assays showed that AMX/NM/NF can create an inhibition zone of 9.2 mm and 7.3 mm against *E. coli* and *S. aureus* respectively [74].

The first criteria for developing an effective delivery system is to ensure the excess release of the encapsulated drug into the physiological environment. Khorshidi et al. developed an electrospun scaffold for loading a second generation fluoroquinolone antibiotic (ciprofloxacin), which can result in ultrasound-assisted drug release. These alginate-emanated nanofibers revealed 3 times more drug release properties with ultrasonic stimuli at 15 W/cm^2 intensity and endowed higher percentages of bacterial DNA synthesis inhibition in both in *E. coli* and *S. aureus* [75]. The in vitro and in vivo studies further suggested that electrospun nanofiber scaffolds can enhance both the bactericidal and bacteriostatic activities of antibiotics. The use of metal ions and compounds, nanoparticles and salts are receiving burgeoning interest as antimicrobial agents. For example, iron oxide, silver, titanium dioxide, and zinc oxide either alone or in combination with other salts or ions are used as a core for generating antimicrobial nanofibers with bactericidal and/or bacteriostatic action. Among them, silver is considered the most potent antimicrobial agent due to its unique property of accumulating on the microbial cell wall and assisting in the arrest of the cell cycle and the denaturation of bacterial DNA [76]. Recently, Jatoi and co-workers proposed a new method to evaluate the antibacterial activities of silver nanoparticles and titanium dioxide by preparing cellulose acetate nanofibers, where TiO$_2$ was first bound with DOPA followed by the introduction of AgNPs to form TiO$_2$/AgNP. Finally, a TiO$_2$/AgNP-loaded cellulose acetate nanofiber scaffold (CA/TiO$_2$/AgNP) was fabricated with the electrospinning method. Their antimicrobial assays demonstrated that the as-synthesized CA/TiO$_2$/AgNP nanoparticles (10 wt.% of TiO$_2$/AgNP) showed almost 100% antibacterial activities against both *E. coli* and *S. aureus* for up to 3 days. In addition, CA/TiO$_2$/AgNP nanoparticles can overcome severe adverse effects (argyria or argyrosis) that are commonly associated with silver nanoparticles-based antimicrobial therapy [77]. Moreover, bacteria usually lose the integrity of their cell wall when they come into contact with highly charged nanofibers like cationic chitosan, thus resulting in cell lysis. This non-release antimicrobial system is a promising strategy because of their substantially prolonged activities outside of drug resistance [76].

The amount of antibiotic release and the effectiveness of its therapeutic window for effective treatment and preclusion of bacterial infections depend on the antibiotic loading efficiency in the carriers. In this regard, nanofiber-based drug delivery systems have gained more consideration for controlled drug release. This decreased and sustained dosing frequency of antibiotics can result in fewer side effects, high patient compliance, minimized fluctuation of antibiotics level in the bloodstream and overcome dose-dependent toxicity. In the traditional strategy of nanofiber-based drug delivery systems, the electrospinning process fabricates antibiotic-loaded nanofiber scaffolds either by making a solution of drug/nanofiber before electrospinning or by loading the antibiotic onto the large surface of the nanofibers through physical or chemical modifications as shown in Figure 4a. Both of these strategies can proficiently control the antibiotic release. One such example would be aminopenicillin (AMX) loaded PEGylated PLGA electrospun nanofibers, where AMX powder was blended with PEG-modified PLGA, followed by the fabrication of nanofiber scaffolds (AMX/PLGA-PEG) via electrospinning. The in vitro release profile exerted about 51.5% and 90.7% of AMX release within 2 and 48 h respectively from 1% AMX/PLGA-PEG nanofiber, and the drug releasing activities continued for more than 10 days. Due to the improvement of hydrophilicity in the AMX/PLGA-PEG nanofiber, faster and prolonged drug release properties were achieved. In addition, 1% AMX/PLGA-PEG nanofiber scaffolds can inhibit approximately 95.9% growth of penicillin resistance gram-positive *Streptococcus aureus* at an AMX concentration of 60 µg/mL. The hemolysis and anticoagulant in vitro experiments revealed that they have great hemocompatibility and cytocompatibility [78]. In another study, RuO_4 (ruthenium tetroxide) oxidized herringbone graphite carbon nanofibers (hGCNF) was prepared for surface labeling of antibiotics. The higher carboxylic acid groups of RuO_4-Oxidized hGCNF served as the binding site of antibiotics during the acyl substitution reaction. Finally, as-synthesized nanofiber scaffolds were covalently functionalized with tobramycin, amikacin, and ciprofloxacillin and this antibiotic-labeled RuO_4-oxidized hGCNF possessed magnificent antibacterial action versus *Pseudomonas aeruginosa* [79].

To control the antibacterial drug release pattern from the nanofiber-based carrier, another important method has recently been investigated in the fabrication of both multidrug loaded or core-shell electrospun nanofibers. Although both emulsion and coaxial electrospinning processes are generally used, coaxial electrospun core-shell nanofibers are more commonly used to regulate the sustained release profiles of the loaded drug (Figure 4b). Mainly, the drug molecules are encapsulated into the core region of this advanced core-shell structure of electrospun nanofibers whereas the outer shells act as a protector and also manage the release behaviors of the incorporated antibacterial drugs. Thus, the core portion and shell portion of a core-shell electrospun nanofiber scaffold can sustain the release profiles of the antibiotics. Due to the protective and biocompatible nature of the outer shell of antibacterial nanocarriers, the sensitive antimicrobial drug is more stable in blood plasma. This can further prolong its release characteristics and facilitate long term antibacterial activities without dose repetition. Apart from protecting the core region of the antibacterial drug, the shell membranes can function as an antibacterial agent that can generate bacterial resistance in biological environments. A shape memory polyurethane (core-shell) antibacterial nanofiber was synthesized with a coaxial electrospinning method in which a polycaprolactone-assisted shape memory polyurethane core was shelled with a pyridine presenting polyurethane. The results demonstrated that the developed core-shell nanofibers displayed enhanced antibacterial properties in both gram positive and gram negative bacteria [80]. On the other hand, the emulsion electrospinning process has also gained the attention of many researchers, in which the use of a single nozzle fabricated core-shell nanofiber scaffold is more simple and more advantageous over the coaxial electrospinning method. In addition, the production of biocompatible, prolonged release, and low toxic core-shell nanofibers with foaming free facilities allows researchers to employ emulsion electrospinning to develop nanofibers. For example, Chai and coworkers for the first time proposed antimicrobial core-shell nanofiber scaffolds that hold a colloidal particles emulsion by electrospinning. In their core-shell nanofibers (Van/OA-MION-PLA), the antibacterial drug vancomycin was in the core portion as a water phase and the polylactic acid solution served as the oil phase, while oleic acid coated magnetic iron oxide nanoparticles were used as an emulsifier. The in vitro vancomycin release study depicts that about 10% of burst redemption appeared within 5 h. Surprisingly, a prolonged release of vancomycin up to 25 days with approximately 57% of cumulative deliverance was observed from Van/OA-MION-PLA [81].

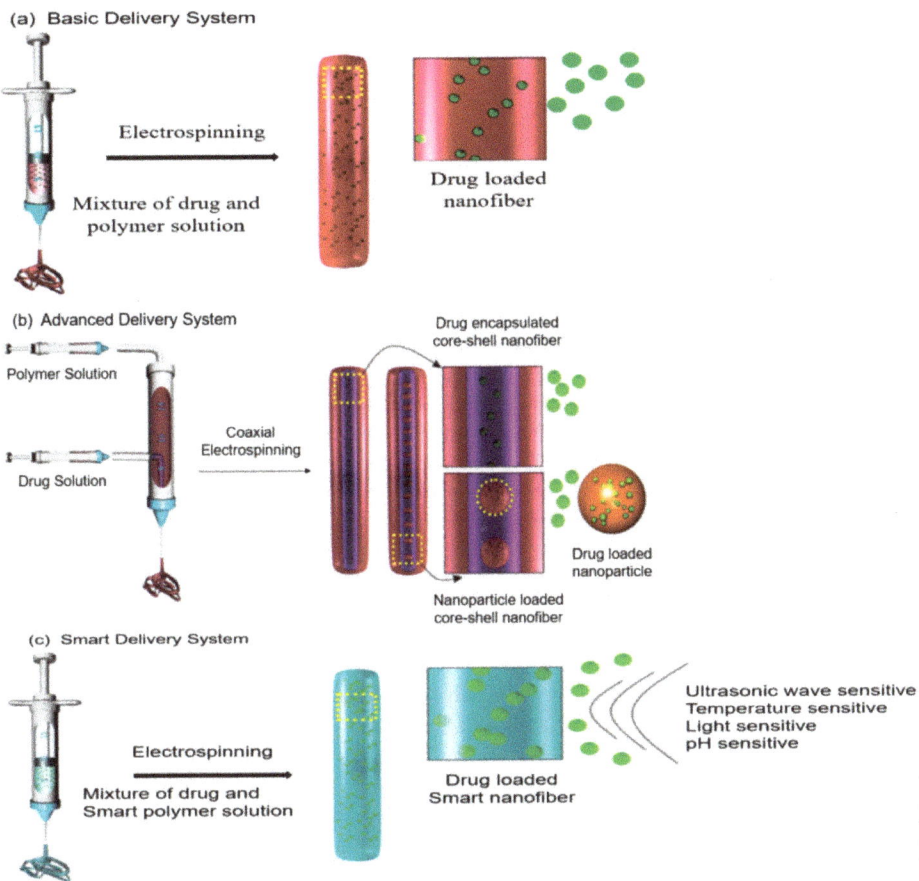

Figure 4. Drug loading strategies to nanofibers: (**a**) basic antibacterial delivery systems, (**b**) advanced antibacterial delivery systems (core–shell structure, nanoparticle decorated and multidrug loaded), and (**c**) smart delivery systems (stimuli responsive). (Reproduced from ref no [76] with permission from "Taylor and Francis Group", Ref: AF/IEDC/P19/0117).

Furthermore, in the smart antibiotic delivery system, electrospun nanofiber scaffolds are commonly used to control the release of drugs in response to various biological parameters including pH factor, temperature or UV-light sensitivity (Figure 4c). Recently Fakhri and coworkers investigated the UV-light and pH-responsive photocatalytic activity of Tungsten disulfide (TDS), TDS-chitosan and TDS-polycaprolactone nanofibers with respect to the degradation of the antimicrobial Neomycin. The antibacterial drug Neomycin conjugated with TDS-chitosan and TDS-polycaprolactone nanofibers revealed a comparatively good decomposition rate of Neomycin as well as an antimicrobial performance at pH 3 [82]. In other research, the pH-dependent antibiotic release Eudragit nanofiber mesh was fabricated via a coaxial electrospinning process. The drug release study indicated that the release rate of Tetracycline from a Tetracycline-loaded Eudragit nanofiber mesh depended on both the physiological pH value and the molar ratio of pure Eudragit and Eudragit L100 in an Eudragit nanofiber mesh. At pH 6 the Tetracycline release was much faster from the as-fabricated nanofiber mesh, but at pH 2 the release rate was very slow. In addition, at the high molar ratio of the Eudragit nanofiber mesh, Tetracycline release was optimum in both pH values [83]. In comparison with other drug nanocarriers, nanofiber scaffolds present more potential applications for antibacterial drug delivery with sustained release properties.

5.1.3. NSAIDs

Nonsteroidal anti-inflammatory drugs (NSAIDS) distinguish themselves from steroids and are popular worldwide due to their anti-inflammatory actions. In addition, they are a frequently prescribed medication as they are widely known to have pain relieving, anti-pyretic, and blood clotting activities. NSAIDs usually reduce the synthesis of prostaglandins by inhibiting the actions of biological enzymes cyclooxygenase-1 and cyclooxygenase-2. However, these drugs are not free from side and adverse effects. The severe side effects associated with NSAIDs include GI ulcers, heart attack and kidney disorders [84,85].

The amount of poorly water-soluble NSAIDs is increasing daily in the pharmaceutical industry. To achieve the desired goal and for improving treatment with NSAIDs, electrospinning techniques have already been introduced in this sector. Ibuprofen is a class of NSAIDs, generally suggested for pain, fever, inflammation, migraines, arthritis, and painful menstruation. In one study, Potrč and co-workers showed that ibuprofen-loaded PCL nanofibers can enhance the dissolution rate of this loaded drug in a biological environment where almost 100% ibuprofen was released from as-synthesized nanofibers within 4 h [41]. Naproxen sodium is another cyclooxygenase inhibitor of the NSAIDs class which is used to treat inflammation including fever and rheumatoid arthritis, pain and menstrual cramps. Naproxen is also used in combination with sumatriptan succinate (a member of the triptans family) to treat migraines or with proton pump inhibitors (i.e., esomeprazole) to avoid NSAIDs-assisted acidity. A few years ago, naproxen and its salt (naproxen sodium) had been electrospun with various hydrophilic (chitosan, PVA and polyacrylic acid) and hydrophobic (PCL) polymers with very good drug loading capacity for achieving rapid onset of action, avoiding hepatic fast pass metabolism and readily accessible through the sublingual route of administration. All developed nanofibers provided the burst release of naproxen in which almost 90% of the released drug from all nanofiber scaffolds was dissolved within 10 min in the acceptor phase. In addition, under the same conditions, no significant differences had been observed between the release profiles of naproxen and its salted form (naproxen sodium). Somehow, the hydrophobic PCL fiber mat exhibited a very quick release of naproxen [86].

Electrospinning techniques are not only applicable to fast releasing, dissolving and complete absorbtion of drugs but also these methods are used for easy swallowing and taste masking of various drugs in case of oral administration. Meloxicam is such a type of NSAID, which is used to treat inflammation in rheumatoid arthritis and osteoarthritis as well as pain. However, its bitter taste, difficulty to swallow, and low bioavailability due to incomplete absorption after peroral administration limited its usage, especially in children [87]. To improve these demerits, an oral dissolving formulation was developed employing the electrospinning process. The incorporation of a cyclic oligosaccharide (cyclodextrin) into PVP nanofibers improved the physical stability of mats whereas the addition of sweeteners avoided the bitter taste of meloxicam. The meloxicam-encapsulated PVP/cyclodextrin fiber mats were a nanometer in size with suitable tensile strength. Interestingly, the nanofiber mats released 100% of the meloxicam in the artificial saliva pH 6.8 within 120 min of administration whereas the tablet and powder dosage forms of meloxicam (marketed product) released only 30% in the same environment. These studies indicate that the developed oral formulation was capable of increasing the solubility, disintegration time, release, bioavailability, and palatability of the encapsulated drug. However, the nanofiber mats disintegrated within 1 min on contact with mouth saliva and, after that, the disintegrated fiber mats continuously released the drug in the gastrointestinal tract [88]. Thus, electrospun nanofiber scaffolds have their own potential to overcome the problem related to keeping the nanofiber mats in the oral cavity for a long time to complete release of the drug.

5.1.4. Cardiovascular Agents

Disorders related to heart and blood vessels are categorized as various cardiovascular diseases such as stroke, heart attack, hypertension, cardiomyopathy, heart arrhythmia, carditis, aortic aneurysms, peripheral artery disease, thromboembolic disease, venous thrombosis and rheumatic, valvular and congenital heart disease [89]. Among these disorders, coronary artery diseases like angina and myocardial infarction are the most common cardiovascular disorders. Nicorandil is widely used against angina or angina pectoris due to its agonistic properties to both ATP-sensitive K+ channel and a polyatomic ion, nitrate, channel [90]. The low bioavailability and slow onset of activities and the major side effects including excess turnover rate and mucosal ulceration have limited the usage of nicorandil as an anti-anginal agent. To overcome these limitations, nicorandil had been electrospun with polymeric nanofibers composed of riboflavin, hyaluronic acid, and PVA to prepare a sublingual dosage for treating angina pectoris. It was expected that the presence of riboflavin in the nanofiber scaffolds would cure mucosal ulceration whereas hyaluronic acid would ensure the quick recovery of inflammation in damaged tissue by reducing the amount of pro-inflammatory cytokines. However, this nano-sized drug-loaded fiber mat was able to sustain the controlled release of nicorandil over a prolonged period of time. The pharmacokinetic study revealed the maintenance of a therapeutic level over a longer period of time and about 4 times more biological half-life of the developed formulation in comparison with marketed nicorandil. Moreover, no mucosal ulceration had been evidenced by the histopathological study at the site of administration for the developed formulation [91].

Carvedilol is another cardiovascular drug that binds and blocks both alpha and beta-adrenergic receptors in an attempt to treat congestive heart failure. Potrč et al. (2015) researched electrospun PCL nanofiber scaffolds as a delivery carrier for the oral administration of poorly water-soluble carvedilol. It was observed that the average size of a drug-loaded PCL nanofiber is directly proportional to the amount of loaded drug and the crystallinity of the carvedilol decreased after encapsulation into the PCL nanofiber. The encapsulated drug was partly molecularly interspersed in the PCL nanofiber and in the formation of dispersed nanocrystals to a certain extent. It had been reported that up to 77% of carvedilol was released from the PCL electrospun nanofibers within only 4 h which indicated a significant improvement of the dissolution rate of this poorly water-soluble drug [41]. Hence, electrostatic spinning is a novel nanotechnology-based strategy especially for improving the dissolution rate of water-insoluble drugs.

5.1.5. Gastrointestinal Drugs

Drugs that are used against various gastrointestinal tract or gut or digestive system disorders or ailments to cure or prevent many severe symptoms of the esophagus, stomach, intestines (both small and large), rectum, and anus, are known as GIT drugs. GIT drugs include antidiarrheal, antiemetic, anti-ulcer agents, cathartics, cholagogues and choleretics, emetics, laxative, lipotropic agents, antibacterial and many other types which are frequently prescribed to control gastric juice, regulate gut motility, water flow and improve the digestion of patients [90]. Unfortunately, the inability of GIT drugs as a curative therapy may lead to surgery in the case of serious complications. Hence, finding a new method to improve the pharmacological action of GIT in human physiology is crucial, especially for serious diseases such as inflammatory bowel syndrome-assisted Crohn's disease, gastroesophageal reflux disease or acid reflux disease, irritable bowel syndrome, and peptic ulcer disease that can lead to stomach cancer.

Metoclopramide (4-amino-5-chloro-N-[2-(diethylamino) ethyl]-2-methoxybenzamide) is an antidopaminergic benzamide, pharmacologically used as a serotonin receptor agonist. Its inhibitory action on acetylcholinesterase exerts its prokinetic and anti-emetic effect due to its actions on contractility of colonic smooth muscle. Recently, Jaber and co-workers fabricated a core/shell nanofiber using PVA/PCL to load metoclopramide hydrochloride [92]. The release profile indicated an initial burst release of the loaded drug (about 55% of total release). The reason behind the initial burst release of metoclopramide may indicate the presence of micron or nano-sized pores in the PCL shell.

Designing the control release behavior of hydrophilic molecules such as protein, peptide, nucleic acid or even a drug is a very difficult task. Tiwari et al. proposed a new method by controlling the partition release of two-layer fiber matrix using a core-shell electrospun strategy where the polymer will serve as an outer layer and the encapsulated drug will be in the core. The authors used metoclopramide to represent a hydrophilic drug and loaded it into various monolithic fibers (PCL, PLLA, PLGA, and PVA) as well as core-shell nanofibers such as PVA/PCL, PVA/PLLA, and PVA/PLGA to investigate the control release behavior of metoclopramide loaded as-synthesized nanofibers. The drug release profile data suggested that the controlled release of hydrophilic entities is possible by using core/shell nanofibers and by verifying the physicochemical properties of core/shell solutions. The result also showed a clear difference according to the release characteristics between the monolithic fibers, which are made of hydrophilic and hydrophobic polymers, and core/shell fibers using PCL, PLLA, and PLGA 80/20 shell polymers. The monolithic fibers cannot control the initial burst release of the hydrophilic drug metoclopramide hydrochloride but core/shell electrospun nanofibers can easily regulate the controlled release of incorporated drugs. Thus, electrospun nanofibers would be a promising gastrointestinal drug carrier to achieve controlled release behavior and protect sensitive drugs in a biological pH [93].

5.1.6. Antihistamines

Antihistamine generally works by blocking the physiological activities of histamine, thus, antihistamines are used in the treatment of nasal congestion, sneezing, hives, seasonal hay fever and especially, to relieve the symptoms of various allergies such as dust allergy, cold allergy, allergic rhinitis, indoor and food allergies. For the time being, electrostatic nanofiber scaffolds are being used as a carrier of various antihistamine drugs. One experimental study was designed to incorporate the first-generation antihistamine (H1 receptor antagonist) chlorpheniramine maleate into glutinous rice starch combining polyvinyl alcohol (GRS/PVA) electrospun nanofibers to investigate a drug delivery carrier concept and control release properties of the nanofibers [94]. The hybrid nanofibers (GRS/PVA) offered a biphasic release of loaded-antihistamine in which 60% of initial release had taken place within first 10 min and reached the highest release at about 90% within 120 min of administration. The authors suggested this GRS/PVA nanofiber scaffold as a novel oral antihistamine drug carrier.

Development of fast dissolving delivery systems of therapeutics (FDDST) is an excellent and unique concept especially for those patients who have difficulties swallowing pharmaceutical solid dosage forms. FDDST offers very fast-dissolving solid oral dosage forms that take a few minutes for absorption in the patient's mouth. Thus, FDDST facilitates high drug bioavailability, therapeutic window and exemption of hepatic first-pass elimination. Previously, the electrospinning method was employed to design, develop and evaluate FDDST in which one study proved more than 80% of total ibuprofen release within 20 s [95] whereas another study showed the co-release of caffeine and vitamin B12 at about 100% and 40% respectively within 1 min [96]. However, Loratadine is another effective peripheral histamine H1 receptor inverse agonist and responsible for anti-allergic effects. Nevertheless, its low water solubility, bioavailability and rapid first-pass hepatic effect with mainly CYP3A4 and CYP2D6 (isoforms of cytochrome P450) metabolism systems reduce their usages. In this respect, electrospinning nanofiber-based Loratadine delivery would be the best solution. A few years ago, Akhgari et al. described the impact of few parameters such as the concentration of polymer and antihistamine, and the amount of feed ratio and the voltage on the first dissolving delivery systems for Loratadine. The authors prepared a Loratadine-encapsulated PVP nanofiber scaffold by way of the electrospinning technique and observed a lower feed ratio and low concentration of the polymeric solution with high voltage application produce nano-sized nanofibers with excellent uniformity. The results of this study showed that, to achieve quick solubility and release of antihistamine from as-prepared nanofibers, the fiber should be a nanometer in size and the antihistamine amount should be smaller [97].

Another example of an antihistamine is Diphenhydramine, a member of the ethanolamine class of histamine H1 receptor antagonists, which is mainly used to treat allergies, nausea, motion sickness, extrapyramidal symptoms and symptoms of a common cold by reversing the effects of histamine on capillaries. It is also used in parkinsonism due to its ability to act as a muscarinic acetylcholine receptor reverse agonist by crossing the blood-brain barrier. In addition, its usage as a sodium channel blocker introduced it as a local anesthetic [90]. However, to produce porous and fast-disintegrating nanofiber scaffolds targeting oral administration, diphenhydramine-incorporated nanofibers were directly electrospun onto a polymeric backing film of hydroxypropylmethylcellulose and glycerol [98]. The physicomechanical characterization data revealed the potency of nanofiber scaffolds in a nanofiber-based oral antihistamine delivery system with the following configurations: high encapsulation efficiency and very short disintegration times (12.8 s). This very short integration time may indicate the large surface area of PVA nanofibers, which were loaded with diphenhydramine.

5.1.7. Contraceptives

For the first time, a composite electrospun nanofiber was fabricated by free-surface electrospinning with various microscale geometries as a carrier of physicochemical diverse medicines including the contraceptive drug progestin levonorgestrel [99]. The as-fabricated PVA nanofibers were capable of encapsulating more than 80% of all matrix formulations except interwoven matrix where an artifact of the processing led to a calculated >150% encapsulation efficiencies for levonorgestrel. The authors checked the solid dispersion of levonorgestrel in the electrospun nanofibers by employing various differential scanning calorimetry analyses. After encapsulation of levonorgestrel, the thermograms of the PVA fabrics did not dramatically change even with high loading of levonorgestrel (17 wt.%). The in vitro release kinetics demonstrated slow and controlled release of levonorgestrel due to its highly hydrophobic nature, where levonorgestrel took 4 h to achieve 100% release from the as-synthesized composite microarchitectures. The idea is that these contraceptive-loaded nanofibers would be inserted into the vaginal mucosal environment to prevent unplanned pregnancy by releasing the appropriate dose of contraceptives into the local cells.

5.1.8. Palliative Drugs

Palliative treatment—also known as comfort or supportive care—aimed to relieve, reduce or control symptoms, side effects and adverse effects such as pain and sickness at any stage of serious diseases, thus enhancing the life expectancy and comfort of a patient even during the last stages of an illness.

Nagy et al. used lower molecular weight poly(vinyl alcohol) to fabricate the non-woven tissues of a nanofiber with a large surface area for developing an oral fast-dissolving dosage form by applying electrostatic spinning. After loading highly water-soluble Donepezil HCl (which is used to treat dementia-related Alzheimer's disease) into as-prepared nanofibers, the diameter of the electrospun nanofiber was 100–300 nm even when the polymer-drug ratio was 33 w/w%. Interestingly, in comparison with the release rate of the commercially available Aricept tablet (brand name of Donepezil HCl), the drug-loaded nanofiber took less than 30 s to release whereas the Aricept tablet took $\geq$30 min. This might be due to the availability of the nanofiber's large surface area which is directly proportional to the dissolution rate, corresponding to the Noyes and Whitney equation [100]. Hence, it can be said that organic solvent free electrospinning could be a promising option for manufacturing highly oral dissolving dosage forms for achieving instant drug release and quick onset action with high patient compliance, which is a basic requirement for palliative care policies.

5.1.9. Miscellaneous

Among various miscellaneous drugs, caffeine is the most used agent worldwide that binds to adenosine receptors by mimicking natural adenosine, affecting various body functions. At present, electrospinning techniques are being widely used to fabricate caffeine loaded nanofibers as a FDDST facilitating a high surface area and high porosity of nanofibers that can lead to fast-wetting features of the nanofiber surface to ensure quick release properties of incorporated drugs. In these respects, drugs are usually encapsulated into nanofibers either in amorphous or nanocrystal format, ensuring high solubility, dissolution, rapid onset of action and bioavailability of loaded drugs. A study was designed by electrospinning using PVA as the drug nanocarriers of caffeine (CA) and riboflavin (RFN) in which the dissolution time and wetting time were 1.5 s and 4.5 s respectively for both PVA-CA and PVA-RFN. In addition, the release percentages of CA and RFN from PVA filament-forming nanofibers were 100 and 40 respectively within 1 min [96]. In another study, paracetamol and caffeine-loaded PVP electrospun nanofiber scaffolds were fabricated as a FDDST model with more than 90% drug loading capacity in which both loaded drugs remained intact even after electrospinning. This study indicated a high disintegration rate of the drug-loaded nanofiber scaffolds within 0.5 s whereas it took less than 150 s to reach 100% dissolution rate. The drug-loaded nanofiber matrix revealed its potency over the pure standard drugs (caffeine and paracetamol individually) [101]. Moreover, it was evidenced that electrostatic spinning has the potential to fabricate such nanofibers that can ensure the ultrafast drug delivery and dissolution of water-insoluble miscellaneous drugs even through the oral route of administration [102].

Discovering an effective and potential ophthalmic drug delivery system (ODDS) is still a challenge due to the quick elimination of ocular drugs like eye drops. Thus, rapid peroneal losses of eye drops, even with the high volume of administration resulting in low bioavailability and very short half-life of applied drugs and a lower amount of ophthalmic drugs (1–3% of total volume), can reach into intraocular tissues via penetration through the cornea. To address these problems, Gagandeep et al. (2014), developed a nanopatch using PVA and PCL via electrostatic spinning to load and deliver a combination of timolol maleate and dorzolamide hydrochloride as a model drug against glaucoma. As-developed drug-loaded nanopatches were 200–400 d·nm with almost 100% drug entrapment efficiency—no interaction has been found between polymers and encapsulated drugs. Furthermore, the in vitro drug release experiment alluded an initial burst release of incorporated drug from nanopatches followed by controlled release behavior for up to 24 h [103]. The overall results reveal the efficiency of electrospun nanopatches in ocular drug delivery.

Drug enrichment on the surface of the polymer is only possible if the drug is blended with the polymer solution before electrostatic spinning. This system facilitates an initial burst release of the loaded-drug resulting in rapid onset of action, but it also decreases the potential lifetime of delivery carriers. Hence, to obtain desired results such as sustained therapy, local and controlled release, enhanced activity and retrenchment of side and adverse effects, core-shell nanofibers would be the best option. One study was designed to load brimonidine tartrate (BT) into a first dissolving dendrimer nanofibers (DNF) against glaucoma (Figure 5a). The in vivo intraocular pressure (IOP) was evaluated as an efficacy parameter for both BT and BT-DNF (Figure 5b). As shown in Figure 5c, the single dose responses of BT and BT-DNF were almost similar to each other regarding the therapeutic effects. Nevertheless, after daily treatment with BT and BT-DNF over a period of 3 weeks, BT-DNF was more effective in comparison with BT for reducing IOP (Figure 5d) [58]. In another study, a novel method was established, combining both coaxial electrospinning and emulsion electrospinning to fabricate a core/shell fiber matrix. As-proposed PLGA coaxial-emulsion electrospun fibers were loaded with an anticonvulsant class of drug—Levetiracetam—and the aim was to implant them in the brain to treat various seizures in adult patients and children with epilepsy. The constant and linear release kinetics of the incorporated lower molecular weight drug Levetiracetam from fiber scaffolds was noticed during 480 h of the study [104].

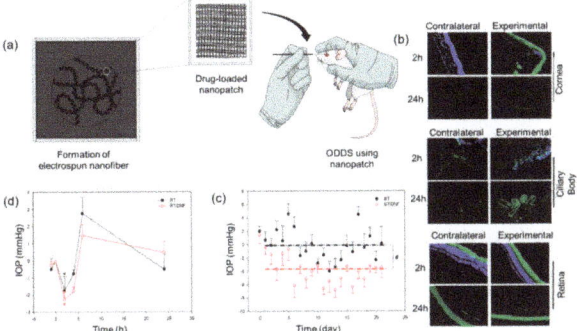

Figure 5. (**a**) An illustrative diagram of nanofiber-based ophthalmic drug delivery system; (**b**) Ocular deposition of dendrimer nanofibers (DNF). Brown Norway rats (BNR) (n = 3) received a DNF-FITC mat topically in the right eye (experimental eye), whereas the left eye received no treatment (contralateral eye); (**c**) In vivo single dose response. BNR (n = 4) received a single dose of brimonidine tartrate (BT) topically via saline eye drops or DNF; (**d**) In vivo 3-week daily dose response. Brown Norway rats (n = 4) received a daily dose of BT via saline eye drops or DNF for 3 weeks. (Here in, Figure 5b–d reproduced from ref. [58] with permission from ACS Publication, 2017).

5.2. DNA and RNA Delivery

The delivery of nucleic acid molecules such as DNA or interfering RNA, into a target cell to knockout/ knockdown mutated gene expressions, either by gene editing or impeding the mechanism of the mutated gene, has proven its novelty in the field of regenerative disease treatment. In this case, when a gene misbehaves in a protein, gene therapy is able to introduce the new copy of that specific gene for recuperating the function of that protein by adjusting the signal transduction pathway [105]. To date, the nucleic acid delivery system depends on the carriers and can be folded into two groups; a viral and a non-viral vector. As nucleic acid directly implanted into the cell does not work, researchers use vectors as their carrier. In viral vectors, useful viruses are genetically modified with a specific nucleic acid so that they can enter into the target cell and introduce new genetic material in the place of missing or faulty genes. However, viral vectors are very specific such as their delivery strategy depends on the type of tissues and gene, even they can carry a very small size of a gene that may sometime cause mutation. In comparison with a viral vector, non-viral vectors are more appreciable among scientists owing to their manageable toxicity, ability to carry various size of genes, large surface area, and large porosity [106]. Among all types of non-viral vectors, nanofiber scaffolds are widely used even though problems associated with electrospun nanofiber scaffolds, such as inappropriate nucleic acid encapsulation and transfection efficiency, are still unsolved. To overcome these limitations, several attempts have been introduced, for example, core/shell, surface modification, coating, encapsulation, incorporation, or interfacing electrostatic interaction, to protect the nucleic acid [107–132].

However, the encapsulation of a nucleic acid/polymer complex in the core of core/shell nanofiber scaffolds can protect the encapsulated DNA or interfering RNA from biological degradation and denaturation, as well as prolong their release up to several months by controlling it. For instance, non-knitted, membraneous nanofiber scaffolds were fabricated via the electrospinning procedure as a gene delivery carrier, containing plasmid DNA encapsulated with poly(lactide-co-glycolide) and poly(D,L-lactide)–poly(ethylene glycol) type biodegradable synthetic copolymers [107]. These promising nanofiber scaffolds can control the release behavior of plasmid DNA over 20 h of the study, whereas the burst release of plasmid DNA appeared within 2 h of the study. In addition, the cumulative release profile indicated up to 80% of plasmid DNA was released from nanofiber scaffolds. The released plasmid DNA had a high cellular transfection efficiency with specific protein encoding properties. In another study, various fiber mesh scaffolds were designed and prepared through the coaxial electrospinning method containing the plasmid DNA and non-viral gene delivery vector poly (ethyleneimine)-hyaluronic acid within the core and sheath polymer of poly(ethylene) glycol and poly(ε-caprolactone), respectively. The cumulative releases of plasmid DNA and the non-viral gene delivery vector from the core and sheath of fiber mesh scaffolds over a time period of almost 2 months showed a dramatic increase in transfection efficiency compared to the control group [108]. Though variously modified nanofiber scaffolds are used today to protect and encapsulate plasmid DNA-like genetic materials, the blending of plasmid DNA with an electrospun solution does not facilitate proper encapsulation and protection of biotherapeutics. In this case, the incorporated plasmid DNA is not uniformly distributed throughout the nanofiber scaffolds, which may hamper their release profile. Hence, surface modification with a cationic polymer could solve these issues. Kim et al. designed DNA-loaded surface modified nanofibers for epidermal gene delivery with a matrix metalloproteinases (MMPs) responsive control release behavior. Here, a MMPs-cleavable linker was used for surface modification of as-synthesized poly(ethylene) glycol/poly(ε-caprolactone) nanofiber meshes with linear polyethyleneimine (LPEI) so that the external MMPs can breakdown the MMPs-linker and facilitate the MMPs-responsive control release of the DNA-loaded polymer [119]. The release studies revealed that more than 80% DNA and almost 80% LPEI can release from the proposed nanofiber mashes over a 72 h time period whereas the transfection efficiency mainly depended on the charge ratio between DNA and LPEI rather than the amount of release.

To date, various natural and synthetic polymers have been used to fabricate electrospun nanofibers in order to allow the successful delivery of genetic materials to the target site. In addition, the hybrid blending of natural and synthetic polymers has also been introduced for the same purpose. In Table 1 we illustrate the usages, designs and advances of various nanofibers as delivery platforms for DNA and RNA, respectively.

Table 1. Electrospun nanofiber-based DNA/RNA delivery.

Design Strategy	Therapeutics	Polymer	Facilities	Ref
Core-Shell	pDNA (pCMVb encoding β-galactosidase)	PLA-PEG-PLA, PLGA	Prevent DNA degradation. Protect the bioactivity of DNA during electrospinning. Controlled and sustained release of DNA. High Transfection efficiency.	[126]
	pDNA encoding pCMV-EGFP	PEI-HA, PEG, PECL	Extended-release properties (>120 d). High transfection efficiency. Controllable release kinetics. High Expression of EGFP.	[108]
	AV encoding-gene for GFP	PECL, PEG	Controlled and porogen-assisted release behavior. High and localized transgene expression. Very low proliferation rate. Controlled virus exposure. The lower level of IL-1β, TNF-α, and IFN-α.	[127]
	pDNA (pEGFP-N$_2$ encoding-GFP)	PELA	Sustained the controlled release. High Transfection efficiency. Promising cell viability. Highly controlled spatiotemporal gene expression. Promote tissue regeneration.	[128]
	pDNA (pbFGF and pVEGF)	PELA, PEI	Sustain release properties (>28 d). Enhance cell attachment, viability and protein expression. Higher cellular transfection. Ensure extracellular secretion of collagen IV and laminin. Downregulate inflammation. Produce microvessels and mature blood vessels.	[129]
	Circular pDNAs; [pcDNA3.1/myc-His− (A), and pcDNA3.1/myc-His(−)/lacZ]	PEG-b-P4VP	Offers tetrasome like pathway for DNA delivery. Ensure highly controllable kinetics.	[130]

Table 1. Cont.

Design Strategy	Therapeutics	Polymer	Facilities	Ref
Coating, Encapsulation, Incorporation, or Interfacing	pDNA (PT7T3D-PacI containing BMP-2)	PLGA/Hap, Chitosan	Stable bioactivity of BMP-2 plasmid. Enhance Cell attachment ability. Low toxicity and immunological effects. Controlled release profile and bone regeneration capability. High DNA transfection efficiency.	[131,132]
	AAV r3.45 (cDNA encoding-GFP/CMVP)	ELP, PECL	Highly efficient cellular transduction. Influencing cell viability. Temperature-sensitive release properties.	[109]
	GAPDH siRNA	PCL, PEG	Control release of siRNA (>28 d). High cellular transfection. Up to 81% of GAPDH gene silencing efficiency. Local delivery of siRNA. Low toxicity. Tissue regeneration.	[110]
	Silencer® GAPDH siRNA	PCLEEP with transfection reagent	Sustained the controlled release of siRNA (>28 d). More than 97% release kinetics profile. Prominently induced the desired gene silencing.	[111]
	EGFP-specific siRNA duplexes	PLGA, Chitosan	Excellent release properties. Prolonged and efficient gene silencing	[112]
	GADPH siRNA	Zein	Offers proper encapsulation with intact bioactivity of siRNA. High loading efficiency. Ensure the sufficient release of siRNA. High cellular attachment and transfection efficiency. More significant gene knockdown.	[113]
Surface Modification	REST siRNA (siREST)	PCL	Controlled REST knockdown of specific neuronal cells. Prominently induced the desired gene silencing. Generate functional neurons as therapeutics.	[114]
	pDNA (pGL3 encoding luciferase)	PLA, PEI	High cultivation period (>5 d). Large surface-to-volume ratio and highly flexible. Manageable transgene expression.	[115]
	pDNA (Plasmid EGFP-N2)	HAp, PDLLA	Facilitate > 95% of the accumulated release (>14 d). High cell viability and density. High GFP expression.	[116]
	pDNA (pEGFP-N1)	PEG/PECL with LPEI	High transfection efficiency. MMp-responsive control release.	[119]
	siRNA	PEG-PECL with LPEI	MMP-2 responsive siRNA release. Excellence gene silencing effects. Facilitates neo-collagen accumulation at the wound sites. Wound recovery can restore to normal levels. Improve the prognosis of diabetic ulcers. Low toxicity.	[117]
	REST siRNA (siREST)	PECL	High loading efficiency. Prevent initial burst release. Enhance gene knockdown efficiency. Increase neuronal markers expressions. Reduce glial cell commitment.	[118]
	hTERT siRNA	$ZnGa_2O_4$: Cr	Increase the siRNA concentration in-situ. High cellular transfection efficiency. Enhance gene silencing effects.	[120]
	REST siRNA (siREST, s11932)	PCL	Low toxic. Prominently induced the desired gene silencing.	[121]
	Silencer® COLA1 siRNA (siCOLA1)	PCLEEP	Prolonged the availability of siRNA ($\geq$30 d). High cellular transfection efficiency. Ensure genetic intervention.	[122]
Electrostatic Interaction	pDNA (pCMVβ encoding β-galactosidase)	PLGA, PLA-PEG	Up to 80% of intact released of pDNA (>20 d). Capable of cell transfection and bioactivity. Able of tissue regeneration. Capable to deliver a combination of genes in a controllable sequence.	[107]
	siRNA	Palmitoyl-GGGAAAKRK peptide (P-G3A3KRK)	Provide localized targeted gene delivery. Ensure intracellular uptake. Enhance siRNA residence time in the brain region. Successfully downregulate the specific gene expression and increase the apoptosis rate. Ensure genetic intervention.	[123]

Table 1. Cont.

Design Strategy	Therapeutics	Polymer	Facilities	Ref
DCEF	pDNA (pEGFP-C3)	Alginate/PCL	Improve gene immobilization. Enhance cell adhesion, viability, and proliferation. High Transfection efficiency. Biocompatible. Tissue Regeneration.	[124]
Controlling fibrous capsule formation	Silencer® COL1A1 siRNA (siCOL1A1)	PCLEEP	Localized and sustained delivery (>28 d). High loading efficiency. Downregulates COL1A1 both in vitro and in vivo. Enhance cellular uptake. Prominently induced the desired gene silencing.	[125]

pDNA: plasmid DNA, PEG: polyethylene glycol, PLA: poly(lactide), PDLLA: poly(D,L-lactide), PLGA: poly(lactide-co-glycolide), PEI: poly (ethyleneimine), HA: hyaluronic acid, PCL: poly(caprolactone), PECL: poly(ε-caprolactone), PELA: poly(DL-lactide)–poly(ethylene glycol), Hap: hydroxyapetite, P4VP: poly(4-vinyl pyridine), LPEI: linear polyethyleneimine, ELP: elastin-like polypeptides, siRNA: short interfering RNA, PEG: polyethylene glycol, PLGA: poly(lactide-co-glycolide), PCL: poly(caprolactone), PECL: poly(ε-caprolactone), LPEI: linear polyethyleneimine, PCLEEP: poly(ε-caprolactone-co-ethyl ethylene phosphate), ZnGa$_2$O$_4$:Cr: chromium-doped zinc gallate.

Nanofiber-based siRNA delivery is also in the spotlight due to its promising mechanism for silencing specific gene expression, targeting various diseases. Those genes have the ability to develop genetic mutations and block the secretion of inhibitory factors resulting in excess cell proliferation that may lead to cancer. To date, various electrospun nanofiber scaffolds have been used as a siRNA nanocarrier to deliver siRNA into the physiological system. Among them, PCL is the most used nanofiber scaffold that facilitates high siRNA loading efficiency, local delivery of siRNA, control release behavior, higher cellular transfection, manageable toxicity and maximum gene silencing properties [114].

However, the hydrophobic nature and slow degradation properties of PCL are responsible for the slow release properties of encapsulated siRNA [110,121]. To overcome this issue, surface modification, the formation of the block polymer and the polydopamine coating of PCL types strategies are taken into consideration which are briefly described in Table 1. Peptide-based nanofiber scaffolds have also been introduced as an siRNA nanocarrier targeting neurodegenerative disease. Their unique characteristics facilitate targetwise siRNA release and accumulation, a high residence period of siRNA in the brain region, a promising gene silencing profile, and ensure genetic intervention [123]. In addition, for the first time, scientists proposed a zein nanofiber-inspired siRNA delivery system that ensures proper siRNA encapsulation and the sufficient release of siRNA, high loading efficiency, cellular attachment, and transfection of siRNA. This zein-based electrospun nanofiber successfully preserves the efficiency of siRNA [113]. Various nanoparticles, except for PCL, such as PEG (polyethylene glycol), PCLEEP (poly(ε-caprolactone-co-ethyl ethylene phosphate)), ZnGa2O4:Cr (chromium-doped zinc gallate), P-G3A3KRK (Palmitoyl-GGGAAAKRK peptide), Zein, PECL (poly(ε-caprolactone)), PEG-b-P4VP ((Poly(ethylene glycol)-b-poly(4-vinylpyridine)), PLGA (poly(lactide-co-glycolide)), LPEI (linear polyethyleneimine), ELP (elastin-like polypeptides), PDLLA (poly(D,L-lactide)), and many more are used as a nucleic acid carrier [111–113,120,122,123,125] and are summarized in Table 1.

5.3. Growth Factor

Growth factors are naturally occurring endogenous signaling molecules (i.e., proteins or steroid hormones) which are efficient for inciting cellular development, differentiation, regeneration and proliferation by binding to cellular receptors. Recently, marked progress has been made in preparing several types of ultrathin electrospun nanofiber scaffolds to successfully load and deliver several growth factors (Figure 6a) as well as regenerative medicines, owing to the unique characteristics of nanofibers such as large surface area, porous formation, high loading capacity, easy access, and cost-effectiveness [133]. Notably, the large surface area to volume ratio of electrospun nanofibers facilitates cell adhesion, loading, storage and release of growth factors [134]. Those parameters are required for guiding cellular behaviors and transmitting signals that regulate proliferation, differentiation, metabolism, and apoptosis of cells, including extracellular matrix deposition in tissue engineering.

The quick inactivation and very short biological half-life of growth factors are the most important factors hindering their effectual delivery. In addition, it has been noted that the growth factors' carrier should have slow and sequential releasing properties. The establishment of growth factors into nanofiber scaffolds play very crucial roles in repairing and regenerating damaged tissues by mimicking signal transduction from cell to cell, or from cell to its extracellular matrix [135]. Either electrospinning procedures alone (for example, bending, emulsion, and coaxial electrospinning) or in combination with other traditional methods such as hydrogel (electrospinning) are ranked at the top among all nanofiber fabrication techniques for their controlled and sustained release properties. On the other hand, surface modified nanofibers promote high loading efficiency, protection of incorporated growth factors and the binding of growth factors to cell receptors. Those surface modified nanofibers exhibit biochemical and morphological uniformity to natural tissues.

Recently, the polysaccharide matrix is being used as a superior nanofiber material to fabricate a growth factor carrier owing to their low toxicity, biocompatibility, and extracellular matrix mimicking capability. However, introducing the sulfation state to the alginate scaffolds ensures the appointed binding and loading of growth factor into the polymer backbone, as well as maintenance of its control release properties. For instance, Mohammadi S, et al. fabricated electrospun scaffolds by blending alginate sulfate and polyvinyl alcohol to load and deliver transforming growth factor-beta1 (TGF-β1) for tissue engineering applications. For the first time, this research revealed that more successful binding, encapsulation and sustain release of heparin-like proteins from electrospun scaffolds could be achieved by the addition of an affinity site to the alginate [134]. A part of mammalian polysaccharides—glycosaminoglycans (heparin-like substance)—would be another example of an excellent nanofiber matrix because of its efficiency in conjugating with GFs by successfully interacting with its (GFs) anionic sulfated backbones and mimicking a pathway similar to ECM.

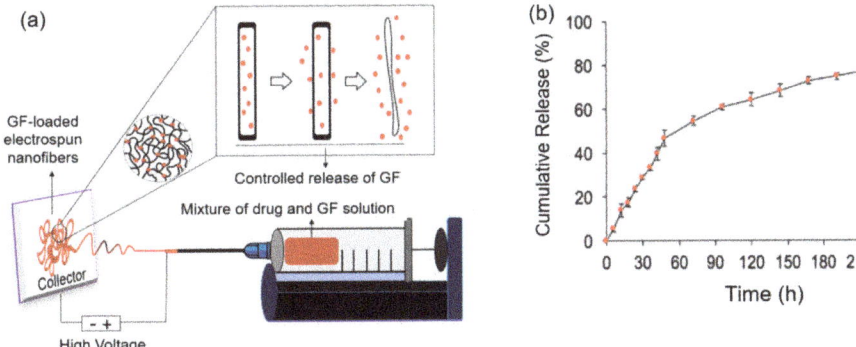

Figure 6. Schematic of the synthesis of growth factor loaded electrospun nanofiber with the controlled release properties (**a**), and VEGF (vascular endothelial growth factor) release profile with respect to time from PCL nanofibers (Reproduced from ref no [123] with permission from ACS Publications, 2018) (**b**).

Usually, the traditional nanofiber scaffolds-based GFs delivery methods such as dip coating and covalent binding are associated with low entrapment efficiency and primary burst release properties. Currently, researchers are looking for new, innovative and stable strategies by focusing on controlled and sustained release properties. To control the release properties of encapsulated GFs, the usage of gelatin particles in nanofiber fabrication is becoming popular. Generally, the in vitro degradation time of gelatin particles is about 240 h, so researchers tend to use gelatin particle-mediated nanofiber scaffolds as a promising GF nanocarrier. The basic advantages of using gelatin particles are their slow degradation and controlled release properties in a biological environment because of introducing cross-linking treatment during fabrication. An example would be the work of Jiang and co-workers who prepared PCL fibrous scaffolds containing vascular endothelial growth factor (VEGF)-incorporated gelatin particles which were able to exhibit both diffusion and degradation mediated release for a long period of time (Figure 6b), differentiate mesenchymal stem cells from endothelial cells, sustain the durability of the tubular composition, and form new blood vessels among endothelial cells [136]. The integration of various sustained release characteristics within nanofiber scaffolds protects the encapsulated GFs from biological degradation, generates cellular signal transduction as well as repairs and regenerates damage.

Biocompatible and biodegradable polymeric nanoparticles and biomimetic hydrogel have gained more attention for encapsulating various types of growth factors and confirming their controlled and prolonged releasing [133]. VEGF and bone morphogenetic protein 2 (BMP2) were incorporated and injected to improve the angiogenesis and regeneration of bone in rabbits [133]. Another preclinical study suggested that VEGF-loaded and hMSC (human bone marrow stroma cells)-seeded PDLA nanofiber scaffolds are able to regenerate blood vessels and enhance bone growth [137]. Interestingly, to achieve dual actions from two separate GFs at a time, core/shell nanofibers or emulsion electrospinning methods are widely used in tissue engineering applications. Liu et al. prepared bicomponent scaffolds with the emulsion electrospinning method for delivering two GFs at the same time in nerve tissue engineering. In this study, the nerve growth factor was encapsulated in poly(D,L-lactic acid) (PDLLA) nanofiber scaffolds whereas the glial cell line-derived neurotrophic factor was incorporated in poly(lactic-co-glycolic acid) (PLGA) electrospun scaffolds [138]. This dual factor core/shell electrospun nanofiber scaffold not only revealed its sustained release and degradation-dependent weight loss properties over 42 days but also ensured the high entrapment efficiency and the intact bioactivity of GFs. Apart from degradation-based controlled release behavior of electrospun scaffolds, providing the affinity sites and/or covalent binding of heparin to the surface of nanofibers is also established. The covalent binding of heparin-like proteoglycans on the surface of the nanofibers preserves the biological integrity of GFs and promotes the sustained release properties of the scaffolds. The aminated PLGA was immobilized with VEGF in the presence of heparin via EDC/NHS coupling to facilitate the sustained release properties of the encapsulated growth factor [139]. The presence of hydroxyl (–OH) and carboxylic acid (–COOH) groups on the surface of the heparin simplifies its modification with both GFs and nanofiber scaffolds. Heparin also increased the binding capacity of VEGF to VEGF receptors on the cell surface.

The time period of sustained release of GFs from its carrier should be proportional to the time period of GFs capability to preserve the nanofiber scaffolds; therefore, this important factor is considered during the design of nanofiber-mediated GFs delivery. Employing several electrospinning techniques, the preservation of GFs' bioactivity is achieved for the proliferation and migration of specific cells. The heterogeneous electrospun matrix made of biocompatible poly(L-lactide-co-caprolactone) and poly(2-ethyl-2-oxazoline) was able to protect the bioactivity of dual angiogenic basic fibroblast growth factor and vascular endothelial growth factor during and after their physical blending. It is also capable of mimicking the topography and chemical cues of native cardiac tissue structures [140]. The structural integrity and bioactivity of encapsulated GFs remained the same over the total period of sustained release in comparison with the pure GFs. In another study, for the first time, Naskar et al. reported the anteriority of using natural polymer silk protein fibroin reinforced synthetic carbon nanofiber scaffolds for preserving bioactivity and controlling the sustained release kinetics of BMP-2 and TGF-β1 GFs over a prolonged period of 50 days [141]. The primary attachment of GFs and the regeneration, excess proliferation, and differentiation of both osteoblastic cells and MSC proved the structural integrity of encapsulated GFs in the nanofiber scaffolds. Moreover, a few linear polysaccharides (i.e., chitosan) are very popular for encapsulating GFs (transforming growth factor beta-1). The major challenge associated with the polysaccharide-based GFs delivery system is their short biological half-life. Hence, they are further composited with nanofiber scaffolds in order to improve the long term presence of GFs in stem cells through maintaining their sustained release from electrospun scaffolds [142]. Previously reported data proved the efficiency of electrospinning methods to load and achieve the sustained release properties of two different GFs at the same time in tissue engineering applications, but now it is also possible to accomplish the distinct release kinetics and spatiotemporal delivery of two different GFs from the same scaffolds with various layered configurations by applying several electrospinning techniques. The combination of emulsion, sequential, high speed and double-sourced double power electrospinning were employed to fabricate tri and tetra layered nanofiber scaffolds using poly(lactic-co-glycolic acid) and poly(D,L-lactic acid) [135]. This glial cell line-derived growth factor and nerve growth factor incorporated multilayer electrospun nanofibers was able to maintain various release actions of these GFs due to aligned fiber coating in each layer with a controlled thickness. The cumulative release studies of encapsulated-GFs indicated distinct release properties from different sites of tri-layered and tetra-layered nanofiber scaffolds over a period of 42 days.

6. Conclusions and Future Outlook

Electrospinning is a promising technology for manufacturing nanofibers from the laboratory to industrial level. Nanomaterials possess a large surface area and enhanced porosity, which are advantages for drug delivery and many other biomedical applications. Nanofibers can be formed using three different techniques: electrospinning, self-assembly, and phase separation. Electrospinning is the most widely used technique because of its promising results. Electrospun nanofibers have attracted much attention due to their biocompatibility, adhesiveness, sterile nature and their efficiency in diverse applications. Currently, nanofibers are considered suitable candidates for scaffold materials, wound dressing materials, drug delivery systems, filtration membranes, catalysts for reduction and so on. Nanofibers are used for medical applications, batteries and fuel cells as a new material.

Though the application of nanofibers is versatile, still more development and research is required for the proper use of nanofibers, especially for biomedical therapeutic delivery. The application of nanofibers in drug delivery, medical implants or tissue engineering is still very limited and most of the materials cannot reach the stage of clinical trials. A thorough investigation on the scaling up of nanofiber technology is required, to make it widely available.

Author Contributions: S.M.S.S. & J.M. contributed equally in this manuscript. S.M.S.S., J.M., M.N.H., V.R., D.Y.L. & Y.-K.L. designed this review. M.N.H. & V.R. gathered research articles, and helped in drawing schemes and tables. D.Y.L. & Y.-K.L. supervised this project.

Funding: This research was supported by The Leading Human Resource Training Program of Regional Neo industry through the National Research Foundation of Korea (NRF) funded by the Ministry of Science, ICT and future Planning (grant number) (NRF-2016H1D5A1910188) and (NRF-2018R1D1A1A09083269). Also, this research was partially supported by Creative Materials Discovery Program through the National Research Foundation of Korea (NRF) funded by the Ministry of Science, ICT and Future Planning (NRF-2017M3D1A1039289). In addition, this study was partially supported by a grant from the Korea Health Technology R&D Project through the Korea Health Industry Development Institute (KHIDI), funded by the Ministry of Health & Welfare, Republic of Korea (grant number: HI18C0453).

Conflicts of Interest: The authors declare no conflicts of interest.

References

1. Tiwari, J.N.; Tiwari, R.N.; Kim, K.S. Zero-dimensional, one-dimensional, two-dimensional and three-dimensional nanostructured materials for advanced electrochemical energy devices. *Prog. Mater. Sci.* **2012**, *57*, 724–803. [CrossRef]
2. Yong, K.T.; Yu, S.F. AlN nanowires: Synthesis, physical properties, and nanoelectronics applications. *J. Mater. Sci.* **2012**, *47*, 5341–5360.
3. Lim, C.T. Synthesis, optical properties, and chemical-biological sensing applications of one-dimensional inorganic semiconductor nanowires. *Prog. Mater. Sci.* **2013**, *58*, 705–748.

4. Hassanzadeh, P.; Kharaziha, M.; Nikkhah, M.; Shin, S.R.; Jin, J.; He, S.; Sun, W.; Zhong, C.; Dokmeci, M.R.; Khademhosseini, A.; et al. Chitin nanofiber micropatterned flexible substrates for tissue engineering. *J. Mater. Chem. B* **2013**, *1*, 4217–4224. [CrossRef]
5. Behrens, A.M.; Casey, B.J.; Sikorski, M.J.; Wu, K.L.; Tutak, W.; Sandler, A.D.; Kofinas, P. In situ deposition of PLGA nanofibers via solution blow spinning. *ACS Macro Lett.* **2014**, *3*, 249–254. [CrossRef]
6. Shah, S.; Yin, P.T.; Uehara, T.M.; Chueng, S.T.D.; Yang, L.; Lee, K.B. Guiding stem cell differentiation into oligodendrocytes using graphene-nanofiber hybrid scaffolds. *Adv. Mater.* **2014**, *26*, 3673–3680. [CrossRef]
7. Yang, X.; Zou, W.; Su, Y.; Zhu, Y.; Jiang, H.; Shen, J.; Li, C. Activated nitrogen-doped carbon nanofibers with hierarchical pore as efficient oxygen reduction reaction catalyst for microbial fuel cells. *J. Power Sources* **2014**, *266*, 36–42. [CrossRef]
8. Shang, M.; Wang, W.; Sun, S.; Gao, E.; Zhang, Z.; Zhang, L.; O'Hayre, R. The design and realization of a large-area flexible nanofiber-based mat for pollutant degradation: An application in photocatalysis. *Nanoscale* **2013**, *5*, 5036–5042. [CrossRef] [PubMed]
9. Cheng, L.; Ma, S.Y.; Wang, T.T.; Li, X.B.; Luo, J.; Li, W.Q.; Mao, Y.Z.; Gz, D.J. Synthesis and characterization of SnO_2 hollow nanofibers by electrospinning for ethanol sensing properties. *Mater. Lett.* **2014**, *131*, 23–26. [CrossRef]
10. Wu, Q.; Tran, T.; Lu, W.; Wu, J. Electrospun silicon/carbon/titanium oxide composite nanofibers for lithium ion batteries. *J. Power Sources* **2014**, *258*, 39–45. [CrossRef]
11. Liu, Y.; Zhao, L.; Li, M.; Guo, L. TiO_2/CdSe core-shell nanofiber film for photoelectrochemical hydrogen generation. *Nanoscale* **2014**, *6*, 7397–7404. [CrossRef] [PubMed]
12. Shi, H.; Zhou, M.; Song, D.; Pan, X.; Fu, J.; Zhou, J.; Ma, S.; Wang, T. Highly porous SnO_2/TiO_2 electrospun nanofibers with high photocatalytic activities. *Ceram. Int.* **2014**, *40*, 10383–10393. [CrossRef]
13. Xue, J.; Xie, J.; Liu, W.; Xia, Y. Electrospun Nanofibers: New Concepts, Materials, and Applications. *Acc. Chem. Res.* **2017**, *50*, 1976–1987. [CrossRef]
14. Thenmozhi, S.; Dharmaraj, N.; Kadirvelu, K.; Kim, H.Y. Electrospun nanofibers: New generation materials for advanced applications. *Mater. Sci. Eng. B Solid-State Mater. Adv. Technol.* **2017**, *217*, 36–48. [CrossRef]
15. Hu, X.; Liu, S.; Zhou, G.; Huang, Y.; Xie, Z.; Jing, X. Electrospinning of polymeric nanofibers for drug delivery applications. *J. Control. Release* **2014**, *185*, 12–21. [CrossRef]
16. Sridhar, R.; Lakshminarayanan, R.; Madhaiyan, K.; Barathi, V.A.; Limh, K.H.C.; Ramakrishna, S. Electrosprayed nanoparticles and electrospun nanofibers based on natural materials: Applications in tissue regeneration, drug delivery and pharmaceuticals. *Chem. Soc. Rev.* **2015**, *44*, 790–814. [CrossRef]
17. Niu, C.; Meng, J.; Wang, X.; Han, C.; Yan, M.; Zhao, K.; Xu, X.; Ren, W.; Zhao, Y.; Xu, L.; et al. General synthesis of complex nanotubes by gradient electrospinning and controlled pyrolysis. *Nat. Commun.* **2015**, *6*, 1–9. [CrossRef]
18. Ren, X.; Ying, P.; Yang, Z.; Shang, M.; Hou, H.; Gao, F. Foaming-assisted electrospinning of large-pore mesoporous ZnO nanofibers with tailored structures and enhanced photocatalytic activity. *RSC Adv.* **2015**, *5*, 16361–16367. [CrossRef]
19. Peng, S.; Li, L.; Hu, Y.; Srinivasan, M.; Cheng, F.; Chen, J.; Ramakrishna, S. Fabrication of Spinel One-Dimensional Architectures by Single-Spinneret Electrospinning for Energy Storage Applications. *ACS Nano* **2015**, *9*, 1945–1954. [CrossRef] [PubMed]
20. Ma, F.; Zhang, N.; Wei, X.; Yang, J.; Wang, Y.; Zhou, Z. Blend-electrospun poly(vinylidene fluoride)/polydopamine membranes: Self-polymerization of dopamine and the excellent adsorption/separation abilities. *J. Mater. Chem. A* **2017**, *5*, 14430–14443. [CrossRef]
21. Bhattarai, R.S.; Bachu, R.D.; Boddu, S.H.S.; Bhaduri, S. Biomedical Applications of Electrospun Nanofibers: Drug and Nanoparticle Delivery. *Pharmaceutics* **2018**, *11*, 5. [CrossRef]
22. Muerza-Cascante, M.L.; Haylock, D.; Hutmacher, D.W.; Dalton, P.D. Melt Electrospinning and Its Technologization in Tissue Engineering. *Tissue Eng. Part B Rev.* **2014**, *21*, 187–202. [CrossRef]
23. Steyaert, I.; Van Der Schueren, L.; Rahier, H.; De Clerck, K. An alternative solvent system for blend electrospinning of polycaprolactone/chitosan nanofibres. *Macromol. Symp.* **2012**, *321–322*, 71–75. [CrossRef]
24. Schoolaert, E.; Steyaert, I.; Vancoillie, G.; Geltmeyer, J.; Lava, K.; Hoogenboom, R.; De Clerck, K. Blend electrospinning of dye-functionalized chitosan and poly(ε-caprolactone): Towards biocompatible pH-sensors. *J. Mater. Chem. B* **2016**, *4*, 4507–4516. [CrossRef]

25. Nikmaram, N.; Roohinejad, S.; Hashemi, S.; Koubaa, M.; Barba, F.J.; Abbaspourrad, A.; Greiner, R. Emulsion-based systems for fabrication of electrospun nanofibers: Food, pharmaceutical and biomedical applications. *RSC Adv.* **2017**, *7*, 28951–28964. [CrossRef]
26. Kai, D.; Liow, S.S.; Loh, X.J. Biodegradable polymers for electrospinning: Towards biomedical applications. *Mater. Sci. Eng. C* **2014**, *45*, 659–670. [CrossRef]
27. Liao, I.C.; Chew, S.Y.; Leong, K.W. Aligned core–shell nanofibers delivering bioactive proteins. *Nanomedicine* **2006**, *1*, 465–471. [CrossRef] [PubMed]
28. Yu, D.G.; Chian, W.; Wang, X.; Li, X.Y.; Li, Y.; Liao, Y.Z. Linear drug release membrane prepared by a modified coaxial electrospinning process. *J. Membr. Sci.* **2013**, *428*, 150–156. [CrossRef]
29. Merkle, V.M.; Zeng, L.; Slepian, M.J.; Wu, X. Core-shell nanofibers: Integrating the bioactivity of gelatin and the mechanical property of polyvinyl alcohol. *Biopolymers* **2014**, *101*, 336–346. [CrossRef]
30. Vaidya, P.; Grove, T.; Edgar, K.J.; Goldstein, A.S. Surface grafting of chitosan shell, polycaprolactone core fiber meshes to confer bioactivity. *J. Bioact. Compat. Polym.* **2015**, *30*, 258–274. [CrossRef]
31. McClellan, P.; Landis, W.J. Recent Applications of Coaxial and Emulsion Electrospinning Methods in the Field of Tissue Engineering. *BioRes. Open Access* **2016**, *5*, 212–227. [CrossRef]
32. Luo, X.; Xie, C.; Wang, H.; Liu, C.; Yan, S.; Li, X. Antitumor activities of emulsion electrospun fibers with core loading of hydroxycamptothecin via intratumoral implantation. *Int. J. Pharm.* **2012**, *425*, 19–28. [CrossRef]
33. Zhang, X.; Wang, M. Effects of Emulsion Electrospinning Parameters on the Morphology and Structure of Core-Shell Structured PLLA Fibers. *Adv. Mater. Res.* **2011**, *410*, 386–389. [CrossRef]
34. Wang, C.; Tong, S.N.; Tse, Y.H.; Wang, M. Conventional Electrospinning vs. Emulsion Electrospinning: A Comparative Study on the Development of Nanofibrous Drug/Biomolecule Delivery Vehicles. *Adv. Mater. Res.* **2011**, *410*, 118–121. [CrossRef]
35. Yoon, Y.I.; Park, K.E.; Lee, S.J.; Park, W.H. Fabrication of microfibrous and nano-/microfibrous scaffolds: Melt and hybrid electrospinning and surface modification of poly(L-lactic acid) with plasticizer. *BioMed Res. Int.* **2013**, *2013*, 309048. [CrossRef]
36. Dalton, P.D.; Klinkhammer, K.; Salber, J.; Klee, D.; Möller, M. Direct in Vitro Electrospinning with Polymer Melts. *Biomacromolecules* **2006**, *7*, 686–690. [CrossRef] [PubMed]
37. Hutmacher, D.W.; Dalton, P.D. Melt Electrospinning. *Chem. Asian J.* **2011**, *6*, 44–56. [CrossRef]
38. Kim, S.J.; Jeong, L.; Lee, S.J.; Cho, D.; Park, W.H. Fabrication and surface modification of melt-electrospun poly(D,L-lactic-co-glycolic acid) microfibers. *Fibers Polym.* **2013**, *14*, 1491–1496. [CrossRef]
39. Zhmayev, E.; Cho, D.; Joo, Y.L. Nanofibers from gas-assisted polymer melt electrospinning. *Polymer* **2010**, *51*, 4140–4144. [CrossRef]
40. Esfahani, H.; Jose, R.; Ramakrishna, S. Electrospun ceramic nanofiber mats today: Synthesis, properties, and applications. *Materials* **2017**, *10*, 1238. [CrossRef] [PubMed]
41. Potrč, T.; Baumgartner, S.; Roškar, R.; Planinšek, O.; Lavrič, Z.; Kristl, J.; Kocbek, P. Electrospun polycaprolactone nanofibers as a potential oromucosal delivery system for poorly water-soluble drugs. *Eur. J. Pharm. Sci.* **2015**, *75*, 101–113. [CrossRef]
42. Fu, A.S.; von Recum, H.A. Affinity-Based Drug Delivery. In *Engineering Polymer Systems for Improved Drug Delivery*, 1st ed.; Wiley: New Jersey, NJ, USA, 2014; pp. 429–452.
43. Karim Haidar, M.; Eroglu, H. Nanofibers: New Insights for Drug Delivery and Tissue Engineering. *Curr. Top. Med. Chem.* **2017**, *17*, 1564–1579. [CrossRef] [PubMed]
44. Jager, J.; Obst, K.; Lohan, S.B.; Viktorov, J.; Staufenbiel, S.; Renz, H.; Unbehauen, M.; Haag, R.; Hedtrich, S.; Teutloff, C.; et al. Characterization of hyperbranched core-multishell nanocarriers as an innovative drug delivery system for the application at the oral mucosa. *J. Period. Res.* **2018**, *53*, 57–65. [CrossRef]
45. Caffarel-Salvador, E.; Abramson, A.; Langer, R.; Traverso, G. Oral delivery of biologics using drug-device combinations. *Curr. Opin. Pharmacol.* **2017**, *36*, 8–13. [CrossRef] [PubMed]
46. Moroz, E.; Matoori, S.; Leroux, J.-C. Oral delivery of macromolecular drugs: Where we are after almost 100 years of attempts. *Adv. Drug Deliv. Rev.* **2016**, *101*, 108–121. [CrossRef] [PubMed]
47. Reda, R.I.; Wen, M.M.; El-Kamel, A.H. Ketoprofen-loaded Eudragit electrospun nanofibers for the treatment of oral mucositis. *Int. J. Nanomed.* **2017**, *12*, 2335–2351. [CrossRef] [PubMed]
48. Laha, A.; Sharma, C.S.; Majumdar, S. Sustained drug release from multi-layered sequentially crosslinked electrospun gelatin nanofiber mesh. *Mater. Sci. Eng. C* **2017**, *76*, 782–786. [CrossRef]

49. Lee, H.; Xu, G.; Kharaghani, D.; Nishino, M.; Song, K.H.; Lee, J.S.; Kim, I.S. Electrospun tri-layered zein/PVP-GO/zein nanofiber mats for providing biphasic drug release profiles. *Int. J. Pharm.* **2017**, *531*, 101–107. [CrossRef]
50. Hamori, M.; Nagano, K.; Kakimoto, S.; Naruhashi, K.; Kiriyama, A.; Nishimura, A.; Shibata, N. Preparation and pharmaceutical evaluation of acetaminophen nano-fiber tablets: Application of a solvent-based electrospinning method for tableting. *Biomed. Pharmacother.* **2016**, *78*, 14–22. [CrossRef]
51. Akhgari, A.; Shakib, Z.; Sanati, S. A review on electrospun nanofibers for oral drug delivery. *Mashhad Univ. Med. Sci.* **2017**, *4*, 197–207.
52. Kondo, Y.; Nagasaka, T.; Kobayashi, S.; Kobayashi, N.; Fujiwara, T. Management of peritoneal effusion by sealing with a self-assembling nanofiber polypeptide following pelvic surgery. Hepato-gastroenterology. *Hepatogastroenterology* **2014**, *61*, 349–353. [PubMed]
53. Modgill, V.; Garg, T.; Goyal, A.K.; Rath, G. Permeability study of ciprofloxacin from ultra-thin nanofibrous film through various mucosal membranes. *Artif. Cells Nanomed. Biotechnol.* **2016**, *44*, 122–127. [CrossRef] [PubMed]
54. Brako, F.; Thorogate, R.; Mahalingam, S.; Raimi-Abraham, B.; Craig, D.Q.M.; Edirisinghe, M. Mucoadhesion of Progesterone-Loaded Drug Delivery Nanofiber Constructs. *ACS Appl. Mater. Interfaces* **2018**, *10*, 13381–13389. [CrossRef] [PubMed]
55. Brako, F.; Raimi-Abraham, B.T.; Mahalingam, S.; Craig, D.Q.M.; Edirisinghe, M. The development of progesterone-loaded nanofibers using pressurized gyration: A novel approach to vaginal delivery for the prevention of pre-term birth. *Int. J. Pharm.* **2018**, *540*, 31–39. [CrossRef]
56. Ball, C.; Woodrow, K.A. Electrospun solid dispersions of Maraviroc for rapid intravaginal preexposure prophylaxis of HIV. *Antimicrob. Agents Chemother.* **2014**, *58*, 4855–4865. [CrossRef]
57. Si, Y.; Wen, Y.; Kelly, S.H.; Chong, A.S.; Collier, J.H. Intranasal delivery of adjuvant-free peptide nanofibers elicits resident CD8+ T cell responses. *J. Control. Release* **2018**, *282*, 120–130. [CrossRef] [PubMed]
58. Lancina, M.G.; Singh, S.; Kompella, U.B.; Husain, S.; Yang, H. Fast Dissolving Dendrimer Nanofiber Mats as Alternative to Eye Drops for More Efficient Antiglaucoma Drug Delivery. *ACS Biomater. Sci. Eng.* **2017**, *3*, 1861–1868. [CrossRef]
59. Kamble, R.N.; Gaikwad, S.; Maske, A.; Patil, S.S. Fabrication of electrospun nanofibres of BCS II drug for enhanced dissolution and permeation across skin. *J. Adv. Res.* **2016**, *7*, 483–489. [CrossRef] [PubMed]
60. Imani, R.; Yousefzadeh, M.; Nour, S. Functional Nanofiber for Drug Delivery Applications BT. In *Handbook of Nanofibers*; Barhoum, A., Bechelany, M., Makhlouf, A., Eds.; Springer International Publishing: Cham, Switzerland, 2018; pp. 1–55. ISBN 978-3-319-42789-8.
61. Torres-Martinez, E.; Cornejo Bravo, M.; Serrano Medina, A.; Gonzz, P.; Lizeth, G.; Villarreal, G. A Summary of Electrospun Nanofibers as Drug Delivery System: Drugs Loaded and Biopolymers Used as Matrices. *Curr. Drug Deliv.* **2018**, *15*, 1360–1374. [CrossRef]
62. Fan, Y.; Moon, J.J. Nanoparticle Drug Delivery Systems Designed to Improve Cancer Vaccines and Immunotherapy. *Vaccines* **2015**, *3*, 662–685. [CrossRef]
63. Lv, H.; Wu, C.; Liu, X.; Bai, A.; Cao, Y.; Shang, W.; Hu, L.; Liu, Y. As a Targeting Drug Carrier to Improve the Antitumor Effect of Paclitaxel for Liver Cancer Therapy. *Biomed Res. Int.* **2018**, *2018*, 11. [CrossRef]
64. Li, J.-J.; Yang, Y.-Y.; Yu, D.-G.; Du, Q.; Yang, X.-L. Fast dissolving drug delivery membrane based on the ultra-thin shell of electrospun core-shell nanofibers. *Eur. J. Pharm. Sci.* **2018**, *122*, 195–204. [CrossRef]
65. Kumar, S.; Singh, A.P.; Senapati, S.; Maiti, P. Controlling Drug Delivery Using Nanosheet-Embedded Electrospun Fibers for Efficient Tumor Treatment. *ACS Appl. Bio Mater.* **2019**, *2*, 884–894. [CrossRef]
66. Li, X.; Lu, X.; Xu, H.; Zhu, Z.; Yin, H.; Qian, X.; Li, R.; Jiang, X.; Liu, B. Paclitaxel/Tetrandrine Coloaded Nanoparticles Effectively Promote the Apoptosis of Gastric Cancer Cells Based on "Oxidation Therapy". *Mol. Pharm.* **2012**, *9*, 222–229. [CrossRef]
67. Xu, H.; Lu, X.; Li, J.; Ding, D.; Wang, H.; Li, X.; Xie, W. Superior antitumor effect of extremely high drug loading self-assembled paclitaxel nanofibers. *Int. J. Pharm.* **2017**, *526*, 217–224. [CrossRef]
68. Geng, Y.; Dalhaimer, P.; Cai, S.; Tsai, R.; Tewari, M.; Minko, T.; Discher, D.E. Shape effects of filaments versus spherical particles in flow and drug delivery. *Nat. Nanotechnol.* **2007**, *2*, 249–255. [CrossRef]
69. Zhang, H.; Liu, Y.; Chen, M.; Luo, X.; Li, X. Shape effects of electrospun fiber rods on the tissue distribution and antitumor efficacy. *J. Control. Release* **2016**, *244*, 52–62. [CrossRef] [PubMed]

70. Samadi, S.; Moradkhani, M.; Beheshti, H.; Irani, M.; Aliabadi, M. Fabrication of chitosan/poly(lactic acid)/graphene oxide/TiO$_2$ composite nanofibrous scaffolds for sustained delivery of doxorubicin and treatment of lung cancer. *Int. J. Biol. Macromol.* **2018**, *110*, 416–424. [CrossRef]
71. Wang, C.; Zhang, Y.; Guo, K.; Wang, N.; Jin, H.; Liu, Y.; Qin, W. Heat shock proteins in hepatocellular carcinoma: Molecular mechanism and therapeutic potential. *Int. J. Cancer* **2016**, *138*, 1824–1834. [CrossRef] [PubMed]
72. Mellatyar, H.; Talaei, S.; Pilehvar-Soltanahmadi, Y.; Barzegar, A.; Akbarzadeh, A.; Shahabi, A.; Barekati-Mowahed, M.; Zarghami, N. Targeted cancer therapy through 17-DMAG as an Hsp90 inhibitor: Overview and current state of the art. *Biomed. Pharmacother.* **2018**, *102*, 608–617. [CrossRef] [PubMed]
73. Mellatyar, H.; Talaei, S.; Pilehvar-Soltanahmadi, Y.; Dadashpour, M.; Barzegar, A.; Akbarzadeh, A.; Zarghami, N. 17-DMAG-loaded nanofibrous scaffold for effective growth inhibition of lung cancer cells through targeting HSP90 gene expression. *Biomed. Pharmacother.* **2018**, *105*, 1026–1032. [CrossRef]
74. Chen, X.; Cai, J.; Ye, D.; Wu, Y.; Liu, P. Dual controlled release nanomicelle-in-nanofiber system for long-term antibacterial medical dressings. *J. Biomater. Sci. Polym. Ed.* **2019**, *30*, 64–76.
75. Khorshidi, S.; Karkhaneh, A. On-demand release of ciprofloxacin from a smart nanofiber depot with acoustic stimulus. *J. Biosci.* **2018**, *43*, 959–967. [CrossRef] [PubMed]
76. Calamak, S.; Shahbazi, R.; Eroglu, I.; Gultekinoglu, M. An overview of nanofiber-based antibacterial drug design. *Expert Opin. Drug Discov.* **2017**, *12*, 391–406. [CrossRef] [PubMed]
77. Jatoi, A.W.; Kim, I.S.; Ni, Q.-Q. Cellulose acetate nanofibers embedded with AgNPs anchored TiO$_2$ nanoparticles for long term excellent antibacterial applications. *Carbohydr. Polym.* **2019**, *207*, 640–649. [CrossRef]
78. Zhang, L.; Wang, Z.; Xiao, Y.; Liu, P.; Wang, S.; Zhao, Y.; Shen, M.; Shi, X. Electrospun PEGylated PLGA nanofibers for drug encapsulation and release. *Mater. Sci. Eng. C* **2018**, *91*, 255–262. [CrossRef]
79. Rotella, M.; Briegel, A.; Hull, J.; Lagalante, A.; Giuliano, R. Synthesis and Antibacterial Activity of Antibiotic-Functionalized Graphite Nanofibers. *J. Nanomater.* **2015**, *2015*, 10. [CrossRef]
80. Zhuo, H.T.; Hu, J.L.; Chen, S.J. Coaxial electrospun polyurethane core-shell nanofibers for shape memory and antibacterial nanomaterials. *Express Polym. Lett.* **2011**, *5*, 182–187. [CrossRef]
81. Cai, N.; Han, C.; Luo, X.; Chen, G.; Dai, Q.; Yu, F. Fabrication of Core/Shell Nanofibers with Desirable Mechanical and Antibacterial Properties by Pickering Emulsion Electrospinning. *Macromol. Mater. Eng.* **2017**, *302*, 1600364. [CrossRef]
82. Fakhri, A.; Gupta, V.K.; Rabizadeh, H.; Agarwal, S.; Sadeghi, N.; Tahami, S. Preparation and characterization of WS2 decorated and immobilized on chitosan and polycaprolactone as biodegradable polymers nanofibers: Photocatalysis study and antibiotic-conjugated for antibacterial evaluation. *Int. J. Biol. Macromol.* **2018**, *120*, 1789–1793. [CrossRef] [PubMed]
83. Son, Y.J.; Kim, Y.; Kim, W.J.; Jeong, S.Y.; Yoo, H.S. Antibacterial Nanofibrous Mats Composed of Eudragit for pH-Dependent Dissolution. *J. Pharm. Sci.* **2015**, *104*, 2611–2618. [CrossRef]
84. Bally, M.; Dendukuri, N.; Rich, B.; Nadeau, L.; Helin-Salmivaara, A.; Garbe, E.; Brophy, J.M. Risk of acute myocardial infarction with NSAIDs in real world use: Bayesian meta-analysis of individual patient data. *BMJ* **2017**, *357*, j1909. [CrossRef] [PubMed]
85. Lanas, A.; Chan, F.K.L. Peptic ulcer disease. *Lancet* **2017**, *390*, 613–624. [CrossRef]
86. Vrbata, P.; Berka, P.; Stránská, D.; Doležal, P.; Musilová, M.; Čižinská, L. Electrospun drug loaded membranes for sublingual administration of sumatriptan and naproxen. *Int. J. Pharm.* **2013**, *457*, 168–176. [CrossRef]
87. Mennella, J.A.; Spector, A.C.; Reed, D.R.; Coldwell, S.E. The bad taste of medicines: Overview of basic research on bitter taste. *Clin. Ther.* **2013**, *35*, 1225–1246. [CrossRef]
88. Samprasit, W.; Akkaramongkolporn, P.; Ngawhirunpat, T.; Rojanarata, T.; Kaomongkolgit, R.; Opanasopit, P. Fast releasing oral electrospun PVP/CD nanofiber mats of taste-masked meloxicam. *Int. J. Pharm.* **2015**, *487*, 213–222. [CrossRef]
89. Mendis, S.; Puska, P.; Norrving, B. *Global Atlas on cardiovascular Disease Prevention and Control*; World Health Organization: Geneva, Switzerland, 2011.
90. Brunton, L.L.; Lazo, J.S.; Parker, K.L. *Goodman & Gilman's Pharmacological Basis of Therapeutics*, 11th ed.; La Jolla, C., Ed.; McGRAW-Hill Medical Publishing Division: New York, NY, USA, 2006; ISBN 0071608915.

91. Garg, T.; Goyal, A.K.; Rath, G. Development, optimization, and characterization of polymeric electrospun nanofiber: A new attempt in sublingual delivery of nicorandil for the management of angina pectoris. *Artif. Cells Nanomed. Biotechnol.* **2016**, *44*, 1498–1507.
92. Jaber, B.M.; Petroianu, G.A.; Rizvi, S.A.; Borai, A.; Saleh, N.A.; Hala, S.M.; Saleh, A.M. Protective effect of metoclopramide against organophosphate-induced apoptosis in the murine skin fibroblast L929. *J. Appl. Toxicol.* **2018**, *38*, 329–340. [CrossRef] [PubMed]
93. Tiwari, S.K.; Tzezana, R.; Zussman, E.; Venkatraman, S.S. Optimizing partition-controlled drug release from electrospun core–shell fibers. *Int. J. Pharm.* **2010**, *392*, 209–217. [CrossRef]
94. Jaiturong, P.; Sirithunyalug, B.; Eitsayeam, S.; Asawahame, C.; Tipduangta, P.; Sirithunyalug, J. Preparation of glutinous rice starch/polyvinyl alcohol copolymer electrospun fibers for using as a drug delivery carrier. *Asian J. Pharm. Sci.* **2018**, *13*, 239–247. [CrossRef]
95. Yu, D.-G.; Shen, X.-X.; Branford-White, C.; White, K.; Zhu, L.-M.; Annie Bligh, S.W. Oral fast-dissolving drug delivery membranes prepared from electrospun polyvinylpyrrolidone ultrafine fibers. *Nanotechnology* **2009**, *20*, 55104. [CrossRef]
96. Li, X.; Kanjwal, M.A.; Lin, L.; Chronakis, I.S. Electrospun polyvinyl-alcohol nanofibers as oral fast-dissolving delivery system of caffeine and riboflavin. *Colloids Surf. B Biointerfaces* **2013**, *103*, 182–188. [CrossRef] [PubMed]
97. Akhgari, A.; Ghalambor Dezfuli, A.; Rezaei, M.; Kiarsi, M.; Abbaspour, M. The Design and Evaluation of a Fast-Dissolving Drug Delivery System for Loratadine Using the Electrospinning Method. *Jundishapur J. Nat. Pharm. Prod.* **2016**, *11*, e33613. [CrossRef]
98. Dott, C.; Tyagi, C.; Tomar, L.K.; Choonara, Y.E.; Kumar, P.; Du Toit, L.C.; Pillay, V. A mucoadhesive electrospun nanofibrous matrix for rapid oramucosal drug delivery. *J. Nanomater.* **2013**, *2013*, 3. [CrossRef]
99. Blakney, A.K.; Krogstad, E.A.; Jiang, Y.H.; Woodrow, K.A. Delivery of multipurpose prevention drug combinations from electrospun nanofibers using composite microarchitectures. *Int. J. Nanomed.* **2014**, *9*, 2967–2978. [CrossRef] [PubMed]
100. Nagy, Z.K.; Nyúl, K.; Wagner, I.; Molnár, K.; Marosi, G. Electrospun water soluble polymer mat for ultrafast release of Donepezil HCl. *Express Polym. Lett.* **2010**, *4*, 763–772. [CrossRef]
101. Illangakoon, U.E.; Gill, H.; Shearman, G.C.; Parhizkar, M.; Mahalingam, S.; Chatterton, N.P.; Williams, G.R. Fast dissolving paracetamol/caffeine nanofibers prepared by electrospinning. *Int. J. Pharm.* **2014**, *477*, 369–379. [CrossRef] [PubMed]
102. Borbás, E.; Balogh, A.; Bocz, K.; Müller, J.; Kiserdei, É.; Vigh, T.; Sinkó, B.; Marosi, A.; Halász, A.; Dohányos, Z.; et al. In vitro dissolution–permeation evaluation of an electrospun cyclodextrin-based formulation of aripiprazole using µFlux™. *Int. J. Pharm.* **2015**, *491*, 180–189. [CrossRef]
103. Garg, T.; Malik, B.; Rath, G.; Goyal, A.K. Development and characterization of nano-fiber patch for the treatment of glaucoma. *Eur. J. Pharm. Sci.* **2014**, *53*, 10–16.
104. Viry, L.; Moulton, S.E.; Romeo, T.; Suhr, C.; Mawad, D.; Cook, M.; Wallace, G.G. Emulsion-coaxial electrospinning: Designing novel architectures for sustained release of highly soluble low molecular weight drugs. *J. Mater. Chem.* **2012**, *22*, 11347–11353. [CrossRef]
105. Hanna, E.; Rémuzat, C.; Auquier, P.; Toumi, M. Gene therapies development: Slow progress and promising prospect. *J. Mark. Access Health Policy* **2017**, *5*, 1265293. [CrossRef]
106. Nayerossadat, N.; Ali, P.; Maedeh, T. Viral and nonviral delivery systems for gene delivery. *Adv. Biomed. Res.* **2012**, *1*, 27. [CrossRef] [PubMed]
107. Luu, Y.K.; Kim, K.; Hsiao, B.S.; Chu, B.; Hadjiargyrou, M. Development of a nanostructured DNA delivery scaffold via electrospinning of PLGA and PLA–PEG block copolymers. *J. Control. Release* **2003**, *89*, 341–353. [CrossRef]
108. Saraf, A.; Baggett, L.S.; Raphael, R.M.; Kasper, F.K.; Mikos, A.G. Regulated non-viral gene delivery from coaxial electrospun fiber mesh scaffolds. *J. Control. Release* **2010**, *143*, 95–103. [CrossRef] [PubMed]
109. Lee, S.; Kim, J.-S.; Chu, H.S.; Kim, G.-W.; Won, J.-I.; Jang, J.-H. Electrospun nanofibrous scaffolds for controlled release of adeno-associated viral vectors. *Acta Biomater.* **2011**, *7*, 3868–3876. [CrossRef]
110. Cao, H.; Jiang, X.; Chai, C.; Chew, S.Y. RNA interference by nanofiber-based siRNA delivery system. *J. Control. Release* **2010**, *144*, 203–212. [CrossRef]
111. Rujitanaroj, P.; Wang, Y.-C.; Wang, J.; Chew, S.Y. Nanofiber-mediated controlled release of siRNA complexes for long term gene-silencing applications. *Biomaterials* **2011**, *32*, 5915–5923. [CrossRef]

112. Chen, M.; Gao, S.; Dong, M.; Song, J.; Yang, C.; Howard, K.A.; Kjems, J.; Besenbacher, F. Chitosan/siRNA Nanoparticles Encapsulated in PLGA Nanofibers for siRNA Delivery. *ACS Nano* **2012**, *6*, 4835–4844. [CrossRef] [PubMed]
113. Karthikeyan, K.; Krishnaswamy, V.R.; Lakra, R.; Kiran, M.S.; Korrapati, P.S. Fabrication of electrospun zein nanofibers for the sustained delivery of siRNA Fabrication of electrospun zein nanofibers for the sustained delivery of siRNA. *J. Mater. Sci. Mater. Med.* **2015**, *26*, 101. [CrossRef] [PubMed]
114. Low, W.C.; Rujitanaroj, P.-O.; Lee, D.-K.; Messersmith, P.B.; Stanton, L.W.; Goh, E.; Chew, S.Y. Nanofibrous scaffold-mediated REST knockdown to enhance neuronal differentiation of stem cells. *Biomaterials* **2013**, *34*, 3581–3590. [CrossRef]
115. Sakai, S.; Yamada, Y.; Yamaguchi, T.; Ciach, T.; Kawakami, K. Surface immobilization of poly (ethyleneimine) and plasmid DNA on electrospun poly (L-lactic acid) fibrous mats using a layer-by-layer approach for gene delivery. *J. Biomed. Mater. Res. Part A* **2009**, *88*, 281–287. [CrossRef]
116. Zou, B.; Liu, Y.; Luo, X.; Chen, F.; Guo, X.; Li, X. Electrospun fibrous scaffolds with continuous gradations in mineral contents and biological cues for manipulating cellular behaviors. *Acta Biomater.* **2012**, *8*, 1576–1585. [CrossRef]
117. Kim, H.S.; Yoo, H.S. Matrix metalloproteinase-inspired suicidal treatments of diabetic ulcers with siRNA-decorated nanofibrous meshes. *Gene Ther.* **2012**, *20*, 378. [CrossRef]
118. Low, W.C.; Rujitanaroj, P.-O.; Lee, D.-K.; Kuang, J.; Messersmith, P.B.; Chan, J.K.Y.; Chew, S.Y. Mussel-Inspired Modification of Nanofibers for REST siRNA Delivery: Understanding the Effects of Gene-Silencing and Substrate Topography on Human Mesenchymal Stem Cell Neuronal Commitment. *Macromol. Biosci.* **2015**, *15*, 1457–1468. [CrossRef] [PubMed]
119. Kim, H.S.; Yoo, H.S. MMPs-responsive release of DNA from electrospun nanofibrous matrix for local gene therapy: In vitro and in vivo evaluation. *J. Control. Release* **2010**, *145*, 264–271. [CrossRef] [PubMed]
120. Qin, L.; Yan, P.; Xie, C.; Huang, J.; Ren, Z.; Li, X.; Best, S.; Cai, X.; Han, G. Gold nanorod-assembled ZnGa$_2$O$_4$:Cr nanofibers for LED-amplified gene silencing in cancer cells. *Nanoscale* **2018**, *10*, 13432–13442. [CrossRef]
121. Chooi, W.H.; Ong, W.; Murray, A.; Lin, J.; Nizetic, D.; Chew, S.Y. Scaffold mediated gene knockdown for neuronal differentiation of human neural progenitor cells. *Biomater. Sci.* **2018**, *6*, 3019–3029. [CrossRef]
122. Pinese, C.; Lin, J.; Milbreta, U.; Li, M.; Wang, Y.; Leong, K.W.; Chew, S.Y. Sustained delivery of siRNA/mesoporous silica nanoparticle complexes from nanofiber scaffolds for long-term gene silencing. *Acta Biomater.* **2018**, *76*, 164–177. [CrossRef] [PubMed]
123. Mazza, M.; Hadjidemetriou, M.; de Lázaro, I.; Bussy, C.; Kostarelos, K. Peptide Nanofiber Complexes with siRNA for Deep Brain Gene Silencing by Stereotactic Neurosurgery. *ACS Nano* **2015**, *9*, 1137–1149. [CrossRef]
124. Hu, W.-W.; Ting, J.-C. Gene immobilization on alginate/polycaprolactone fibers through electrophoretic deposition to promote in situ transfection efficiency and biocompatibility. *Int. J. Biol. Macromol.* **2019**, *121*, 1337–1345. [CrossRef]
125. Rujitanaroj, P.; Jao, B.; Yang, J.; Wang, F.; Anderson, J.M.; Wang, J.; Chew, S.Y. Controlling fibrous capsule formation through long-term down-regulation of collagen type I (COL1A1) expression by nanofiber-mediated siRNA gene silencing. *Acta Biomater.* **2013**, *9*, 4513–4524. [CrossRef]
126. Liang, D.; Luu, Y.K.; Kim, K.; Hsiao, B.S.; Hadjiargyrou, M.; Chu, B. In vitro non-viral gene delivery with nanofibrous scaffolds. *Nucleic Acids Res.* **2005**, *33*, e170. [CrossRef] [PubMed]
127. Liao, I.-C.; Chen, S.; Liu, J.B.; Leong, K.W. Sustained viral gene delivery through core-shell fibers. *J. Control. Release* **2009**, *139*, 48–55. [CrossRef] [PubMed]
128. Yang, Y.; Li, X.; Cheng, L.; He, S.; Zou, J.; Chen, F.; Zhang, Z. Core–sheath structured fibers with pDNA polyplex loadings for the optimal release profile and transfection efficiency as potential tissue engineering scaffolds. *Acta Biomater.* **2011**, *7*, 2533–2543. [CrossRef]
129. He, S.; Xia, T.; Wang, H.; Wei, L.; Luo, X.; Li, X. Multiple release of polyplexes of plasmids VEGF and bFGF from electrospun fibrous scaffolds towards regeneration of mature blood vessels. *Acta Biomater.* **2012**, *8*, 2659–2669. [CrossRef]
130. Wang, W.; Zhang, K.; Chen, D. From Tunable DNA/Polymer Self-Assembly to Tailorable and Morphologically Pure Core–Shell Nanofibers. *Langmuir* **2018**, *34*, 15350–15359. [CrossRef]
131. Nie, H.; Ho, M.-L.; Wang, C.-K.; Wang, C.-H.; Fu, Y.-C. BMP-2 plasmid loaded PLGA/HAp composite scaffolds for treatment of bone defects in nude mice. *Biomaterials* **2009**, *30*, 892–901. [CrossRef]

132. Nie, H.; Wang, C.-H. Fabrication and characterization of PLGA/HAp composite scaffolds for delivery of BMP-2 plasmid DNA. *J. Control. Release* **2007**, *120*, 111–121. [CrossRef]
133. Zhang, W.; Wang, X.; Wang, S.; Zhao, J.; Xu, L.; Zhu, C.; Zeng, D.; Chen, J.; Zhang, Z.; Kaplan, D.L.; et al. The use of injectable sonication-induced silk hydrogel for VEGF165 and BMP-2 delivery for elevation of the maxillary sinus floor. *Biomaterials* **2011**, *32*, 9415–9424. [CrossRef] [PubMed]
134. Mohammadi, S.; Ramakrishna, S.; Laurent, S.; Shokrgozar, M.A.; Semnani, D.; Sadeghi, D.; Bonakdar, S.; Akbari, M. Fabrication of Nanofibrous PVA/Alginate-Sulfate Substrates for Growth Factor Delivery. *J. Biomed. Mater. Res. Part A* **2019**, *107*, 403–413. [CrossRef]
135. Liu, C.; Li, X.; Xu, F.; Cong, H.; Li, Z.; Song, Y.; Wang, M. Spatio-temporal release of NGF and GDNF from multi-layered nanofibrous bicomponent electrospun scaffolds. *J. Mater. Sci. Mater. Med.* **2018**, *29*, 102. [CrossRef] [PubMed]
136. Jiang, Y.-C.; Wang, X.-F.; Xu, Y.-Y.; Qiao, Y.-H.; Guo, X.; Wang, D.-F.; Li, Q.; Turng, L.-S. Polycaprolactone Nanofibers Containing Vascular Endothelial Growth Factor-Encapsulated Gelatin Particles Enhance Mesenchymal Stem Cell Differentiation and Angiogenesis of Endothelial Cells. *Biomacromolecules* **2018**, *19*, 3747–3753. [CrossRef] [PubMed]
137. Geiger, F.; Lorenz, H.; Xu, W.; Szalay, K.; Kasten, P.; Claes, L.; Augat, P.; Richter, W. VEGF producing bone marrow stromal cells (BMSC) enhance vascularization and resorption of a natural coral bone substitute. *Bone* **2007**, *41*, 516–522. [CrossRef] [PubMed]
138. Liu, C.; Wang, C.; Zhao, Q.; Li, X.; Xu, F.; Yao, X.; Wang, M. Incorporation and release of dual growth factors for nerve tissue engineering using nanofibrous bicomponent scaffolds Incorporation and release of dual growth factors for nerve tissue engineering using nano fi brous bicomponent scaffolds. *Biomed. Mater.* **2018**, *13*, 044107. [CrossRef]
139. Lü, L.; Deegan, A.; Musa, F.; Xu, T.; Yang, Y. The effects of biomimetically conjugated VEGF on osteogenesis and angiogenesis of MSCs (human and rat) and HUVECs co-culture models. *Colloids Surf. B Biointerfaces* **2018**, *167*, 550–559. [CrossRef] [PubMed]
140. Lakshmanan, R.; Kumaraswamy, P.; Krishnan, U.M.; Sethuraman, S. Engineering a growth factor embedded nanofiber matrix niche to promote vascularization for functional cardiac regeneration. *Biomaterials* **2016**, *97*, 176–195. [CrossRef] [PubMed]
141. Naskar, D.; Ghosh, A.K.; Mandal, M.; Das, P.; Nandi, S.K.; Kundu, S.C. Dual growth factor loaded nonmulberry silk fibroin/carbon nanofiber composite 3D scaffolds for in vitro and in vivo bone regeneration. *Biomaterials* **2017**, *136*, 67–85. [CrossRef] [PubMed]
142. Ardeshirylajimi, A.; Ghaderian, S.M.-H.; Omrani, M.D.; Moradi, S.L. Biomimetic scaffold containing PVDF nanofibers with sustained TGF-β release in combination with AT-MSCs for bladder tissue engineering. *Gene* **2018**, *676*, 195–201. [CrossRef]

© 2019 by the authors. Licensee MDPI, Basel, Switzerland. This article is an open access article distributed under the terms and conditions of the Creative Commons Attribution (CC BY) license (http://creativecommons.org/licenses/by/4.0/).

Review

Electrospun Nanofibers: Recent Applications in Drug Delivery and Cancer Therapy

Rafael Contreras-Cáceres [1,2], Laura Cabeza [3,4,5], Gloria Perazzoli [3,4,5], Amelia Díaz [1], Juan Manuel López-Romero [1], Consolación Melguizo [3,4,5],* and Jose Prados [3,4,5]

1. Department of Organic Chemistry, Faculty of Science, University of Málaga, 29071 Málaga, Spain; rafcontr@ucm.es (R.C.-C.); amelia@uma.es (A.D.); jmromero@uma.es (J.M.L.-R.)
2. Department of Chemistry of Pharmaceutical Science, Faculty of Pharmacy, Complutense University of Madrid, 28040 Madrid, Spain
3. Institute of Biopathology and Regenerative Medicine (IBIMER), Biomedical Research Center (CIBM), University of Granada, 18100 Granada, Spain; lautea@ugr.es (L.C.); gperazzoli@ugr.es (G.P.); jprados@ugr.es (J.P.)
4. Instituto de Investigación Biosanitaria ibs.GRANADA, 18012 Granada, Spain
5. Department of Anatomy and Embryology, Faculty of Medicine, University of Granada, 18016 Granada, Spain
* Correspondence: melguizo@ugr.es; Tel.: +34-958-228819

Received: 11 March 2019; Accepted: 19 April 2019; Published: 24 April 2019

Abstract: Polymeric nanofibers (NFs) have been extensively reported as a biocompatible scaffold to be specifically applied in several researching fields, including biomedical applications. The principal researching lines cover the encapsulation of antitumor drugs for controlled drug delivery applications, scaffolds structures for tissue engineering and regenerative medicine, as well as magnetic or plasmonic hyperthermia to be applied in the reduction of cancer tumors. This makes NFs useful as therapeutic implantable patches or mats to be implemented in numerous biomedical researching fields. In this context, several biocompatible polymers with excellent biocompatibility and biodegradability including poly lactic-co-glycolic acid (PLGA), poly butylcyanoacrylate (PBCA), poly ethylenglycol (PEG), poly (ε-caprolactone) (PCL) or poly lactic acid (PLA) have been widely used for the synthesis of NFs using the electrospun technique. Indeed, other types of polymers with stimuli-responsive capabilities has have recently reported for the fabrication of polymeric NFs scaffolds with relevant biomedical applications. Importantly, colloidal nanoparticles used as nanocarriers and non-biodegradable structures have been also incorporated by electrospinning into polymeric NFs for drug delivery applications and cancer treatments. In this review, we focus on the incorporation of drugs into polymeric NFs for drug delivery and cancer treatment applications. However, the principal novelty compared with previously reported publications is that we also focus on recent investigations concerning new strategies that increase drug delivery and cancer treatments efficiencies, such as the incorporation of colloidal nanoparticles into polymeric NFs, the possibility to fabricate NFs with the capability to respond to external environments, and finally, the synthesis of hybrid polymeric NFs containing carbon nanotubes, magnetic and gold nanoparticles, with magnetic and plasmonic hyperthermia applicability.

Keywords: electrospun nanofibers; cancer treatment; drug release; nanomedicine; biocompatible polymers; hyperthermia

1. Polymeric Nanofibers in Biomedicine: General Overview

The improvement of chemotherapeutic treatments in cancer patients is seriously limited by the difficulty of increasing the ability of drugs to specifically target tumor cells, thus reducing their toxicity in healthy cells. However, these antitumor molecules cannot increase their therapeutic response, which leads to poor prognosis and results in serious health problems, due to the prevalence of these several pathologies. Consequently, further research is needed to find new therapeutic approaches that improve the prognosis of patients affected by several types of cancers, thus decreasing the possible mortality rate. In this context, new strategies are appearing concerning the development of systems with the capability to encapsulate chemotherapeutic drugs and act as vehicles to be delivered in a specific area in a higher extension, thus avoiding the previously mentioned limitations and improving their efficiency, specifically in tumor cells [1,2]. According to the data from the World Health Organization, cancer caused 9.6 million deaths in 2018, and is one of the leading causes of death worldwide. The most common tumors are lung, breast and colorectal cancer [3]. For this reason, it is important to develop new nanoformulations that allow the improvement of cancer treatments and therefore the survival of these patients.

In general, drug delivery systems are nanostructures that can be loaded with small molecules or macromolecules, thus acting as vehicles of specific compounds to be used in a pharmaceutical administration process. Nowadays they represent one of the most promising challenges in the improvements of biomedical investigations [4]. Such materials are able to transport a chemotherapeutic molecule to a desired area, thus increasing the drug concentration, to be subsequently released in a controlled manner. Among a great number of nanoformulations, such as liposomes [5], micelles [6], Pickering emulsions [7], dendrimers [8] or nanoemulsions [9], polymeric nanoparticles (NPs) have been extensively reported as drug delivery systems to be applied in, for example, the chemotherapeutic treatment of solid tumors [9]. These NPs can be composed by synthetic polymers such as polylactic acid (PLA), poly lactic-co-glycolic acid (PLGA) or polyethylene glycol (PEG) [10,11], which in principle can be administered as colloidal systems and dispersed in a solution by means of intravenous administration. However, during the last years, apart from colloidal structures, polymeric nanofibers (NFs) have been reported as a scaffold with the ability to encapsulate antitumor drugs for biomedical investigations, including drug delivery and cancer treatments [12,13]. Among various techniques available for NFs fabrication, electrospinning is simple and produces NFs with high interconnected pores in the nanoscale range [14], having also a large surface area-to-volume ratio, high interfiber porosity, low hindrance for mass transfer, flexible handling, adjustable morphology, and high mechanical strength, which make NFs useful as therapeutic patches or mats for biomedical applications [15,16], indeed they are used in the fabrication of non-woven fibers with diameters ranging from a few nanometers to microns.

In the electrospun technique, when a strong electrostatic field is applied to a polymer solution held in a syringe, the pendent droplet of the polymer solution is deformed into a Taylor cone [17]. When the electric force overcomes the surface tension of the droplet, one or multiple charged jets are ejected from the tip of the droplet. As the jet moves towards a collecting metal screen, the solvent evaporates, and a nonwoven fabric mat is formed on the screen. This technique is able to fabricate fibers with diameter in the order of nanometers. During the last years, by using this technique, a great number of polymers have been used for the generation of biocompatible NFs scaffolds [18]. Biodegradable synthetic polymers such as PCL or PLGA, as well PEG and PLA can be used as substrates for the fabrication of NFs by using electrospun approaches [19]. Importantly, during the last few years, other types of polymers with stimuli-responsive behavior have been investigated as NFs scaffolds [20]. Two examples are poly(N-Isopropylacrylamide) (pNIPAM) and poly(4-vinylpyridine), which are the most investigated thermo- and pH-responsive polymers, respectively. By using this technique, the fiber diameter can be modulated by several polymer solution properties as viscosity, elasticity, polymer concentration and conductivity, the electric field strength, the distance between the injector and the metal collector, or other external parameters as temperature and humidity. Importantly, these NFs can incorporate and accumulate chemotherapeutic molecules by means of two main approaches [21]; (i) Blend electrospinning, which is based on mixing a drug with a polymeric solution prior to electrospinning process or (ii) Coaxial electrospinning, which is basically a simultaneous co-spinning of two polymeric liquids. The general system for this "core/shell" electrospinning is based on two needles, structured in a coaxial manner Indeed, drugs can be incorporated into the NFs through physical adsorption (involving electrostatic interactions) [22] or covalent bond [23], as well as the aforementioned co-axial electrospinning or mixing with the polymer solution. By applying some of the previously mentioned approaches, electrospinning also offers the possibility to fabricate hybrid composites NFs. These composites structures are produced by the incorporation of other systems with specific properties into the polymeric NFs as CNTs, magnetic nanoparticles or metal nanoparticles. Among several reported reviews concerning the fabrication of polymeric NFs by electrospinning for drug delivery purposes [24–26], the principal novelty of this review is that we include recent advances for the fabrication and application of polymeric NFs by electrospinning not only focused in polymeric NFs. We also include recent investigation in 3 major items: (i) the incorporation of particles as nanocarriers into the NFs (vesicles, micelles or silica particles), which are able to increase drug accumulations, (ii) the fabrication of polymeric NFs with the capability to respond to external environments, and (iii) the generation of hybrid systems structured as polymeric NFs containing CNTs, magnetic and gold nanoparticles for hyperthermia applicability.

2. Electrospun Nanofibers for Drug Delivery Application

2.1. Free drug-Loaded Nanofibers

As was previously mentioned, coaxial electrospinning is based on co-spinning two liquids in a core/shell structure. Figure 1 shows a typical coaxial electrospinning setup used for the fabrication of core/shell NFs. As can be observed, it is composed by 2 syringe pumps. In this particular case, the injector is formed by a coaxial needle where the inner part contains a paclitaxel (PTX) solution and the outer part contains the polymer [27]. This technique has been applied for the incorporation of PTX into poly(L-lactic acid-co-ε-caprolactone), P(LLA-CL) (75:25)NFs, thus resulting in PTX loaded P(LLA-CL) NFs. Paclitaxel is a chemotherapeutic drug which is extensively used in breast, ovarian, lung, bladder and prostate cancer. Huang et al. [27] prepared a PTX solution by dissolving this chemotherapeutic drug in 2,2,2-trifluoroethanol. Then, co-axial electrospinning was used to directly introduce the PTX solution into a polymeric P(LLA-CL) shell. They prepared various core/shell NFs with tunable diameter, which was controlled by the polymer concentration and flow rate between the 2 solutions. Drug delivery investigations exhibited a short burst of PTX during 24 h followed by a very slow release for the following 60 days. Indeed, these PTX-IN-P(LLA-CL) NFs also inhibited the activity of HeLa cells. Paclitaxel was also introduced into PLGA nNFs mats fabricated by blend electrospinning for controlled drug delivery [28]. In this case, the drug release investigation was performed for the in vitro treatment of C6 glioma. The polymer fiber diameter was controlled by using different polymer concentrations, as well as different amounts of an organic salt named tetrabutylammonium tetraphenylborate (TATPB). They obtained NFs with dimension from several tens nanometers to 10 mm, and after the addition of organic salts, the NFs diameter were decreased up to 30 nm. For PTX-IN-PLGA NFs the encapsulation efficiency was higher that 90%. In vitro release profiles confirmed a sustained PTX release for more than 60 days. Their results also indicated that the density of C6 glioma cells was much lower after administration of different concentrations of PTX-IN-PLGA NFs, as compared to the control and blank PLGA NFs after 72 h.

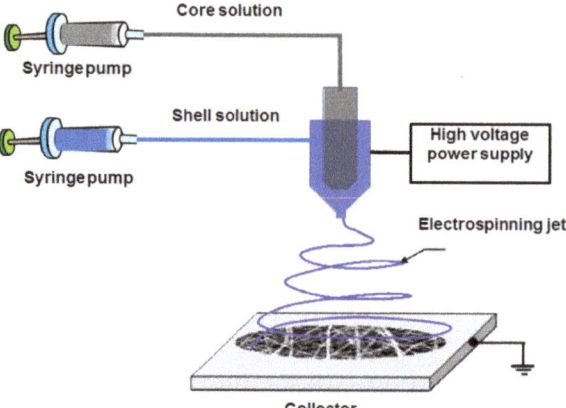

Figure 1. Schematic representation of the coaxial electrospinning setup. In this example the core solution is composed by the PTX dissolved in 2,2,2-trifluoroethanol and the shell solution is the poly(L-lactic acid-co-ε-caprolactone) polymer. Reprinted with permission from reference [27]. Copyright Wiley Online Library, 2009.

Low-water soluble molecules as PTX are not the only option that has been introduced into NFs. A hydrophilic antibiotic drug, as MefoxinR, has also been incorporated into polymeric NFs by using PLGA and a mixture of PLGA/PEG-b-PLA/PLA (80:15:5) through blend electrospinning [29]. These authors demonstrated that the morphology and density of the prepared NFs depended on the drug concentration, which is basically produced by the different conductivity provided by the ionic salt during the electrospinning process. They obtained interesting antimicrobial effects on Staphylococcus aureus cultures when reaching a maximum dosage after 1 h. Indeed, when the amphiphilic block copolymer (PEG-b-PLA) with a ratio of 85:15 was used during electrospinning, the fabricated NFs were able to reduce the cumulative amount of the released drug at shorter times, and prolonged the drug release rate at longer times (up to a 1-week period). Blend electrospinning was also used for the encapsulation of three chemotherapeutic drugs into poly(L-lactide)(PLLA) NFs: PTX, doxorubicin (DOX) and DOX hydrochloride [30]. In this case, the influence of solubility and compatibility of drugs with the polymer was investigated, and it was associated with the drug delivery behavior. In these scaffolds, after drug incorporation, the degradation of PLLA fibers was monitored in the presence of the enzyme proteinase K, following a drug release with a nearly zero-order kinetics. Cisplatin is another compound that is used in chemotherapy for the treatment of liver cancer, although it has little effectiveness. Zhang et al. [16] used an electrospun system that consisted of five layers composed of PLA and the drug interleaved in layers. The drug was located in the even, second and fourth layers, while the PLA was located in the odd, first, third and fifth layers. In this study, fruitful results were obtained for in vivo investigations since the layered structure allows a continuous release of the drug. Consequently, a reduction in the toxicity during the treatment and a longer half-life of the mice was obtained. NFs can be also used for attenuating the side effects of the used drug. This is the case of the study of Singh et al. [31] These authors introduced Docetaxel (DOC) into polyvinyl alcohol (PVA) fiber using the electrospinning method. The aim of the study was to prevent inflammation, extravasation and other side effects of chemotherapy in the treatment of oral cancer. Because of that, they designed a mucoadhesive nano-carrier DOC-PVA and made in vitro studies with positive results, showing that anticancer drugs can be successfully used for local administration with polymeric NF.

Apart from chemotherapeutic drugs and antibiotics, Lovastain, a commonly used drug that reduces the cholesterol level and the risk of heart attack, has been incorporated into biocompatible NFs for drug delivery purposes. Zhu et al. [32] used blend electrospinning to introduce these biomedical properties of lovastatin into biodegradable and biocompatible PLLA NFs with drug delivery capability.

Lovastatin was fully dissolved with PLLA by using hexafluoro-isopropanol as a solvent and at weight percentages of lovastatin in PLLA of 0%, 5%, and 10%. Interestingly, authors found that lovastatin values of 5 wt% or 10 wt% improved the NFs properties for alignment and surface smoothness, while also enhancing the NFs diameter. These Lovastain-IN-PLLA NFs reached high drug entrapment efficiency, ranging from 72% to 82%. The in vitro drug delivery investigations confirmed a release behavior in two stages. Initially, fast release was produced during the first day, and a slower release was measured that reached a plateau after 7 days. By using a cylinder collector during the electrospinning, they also fabricated PLLA films. These authors compared the drug delivery capabilities of PLLA NFs with PLLA films, and they found a higher release rate for fibers compared with films.

2.2. Nanocarriers-IN-Nanofibers

Nanocarriers have gained attention in drug delivery due to their ability to act as vehicles in the transport and delivery of different drugs. It is important to remark that nanocarriers are not only interesting in terms of vehiculization, they are also relevant because after introducing the drug into the nanocarrier, the amount of chemotherapeutic molecule to be delivered is considerably increased. It is accepted that a colloidal system is formed by a complex fluid where a certain substance (disperse phase) remains immersed in another substance (solvent). Accordingly, a colloidal system is formed by a disperse phase containing solid particles that are dispersed in a liquid. In this context, vesicles, micelles, microgels, or emulsions are colloidal structures where the disperse phase is a solid particle which is dispersed into a certain solvent. The possibility to introduce chemotherapeutic drugs into these colloidal particles has been extensively exploited in drug delivery applications and cancer treatments. Importantly, as these particles are in the range of nanometers, they can be incorporated into the human body by intravenous administration. However, it is important to mention that the simple introduction of drugs into some nanocarriers normally leads to inevitable burst drug release. To overcome this limitation, some improvements have been found, for example, by the incorporation of these nanocarriers into polymeric NFs, which improves drug delivery and the applicability of cancer treatments.

2.2.1. Vesicles and Micelles as Nanocarriers

Vesicles and micelles have been used as drug nanocarriers to be incorporated into polymeric NFs. An important advantage of this type of colloidal particles is the fact that they possess in their structure two different environments (hydrophobic or hydrophilic) that can be exploited in, for example, the incorporation of two different drugs, thus performing dual drug delivery. Li et al. [33] investigated a dual drug delivery approach by using two types of drugs, 5-FU and paenolum. 5-Fluorouracil is a hydrophilic chemotherapeutic drug principally used in colon cancer treatment, and paenolum is a hydrophobic molecule used to prevent blood platelet clotting, thus having anti-inflammatory properties. Initially, the vesicles were fabricated by mixing two surfactants, cetyltrimethylammonium bromide (CTAB) and sodium dodecylbenenesulfonate (SDBS), which were able to trap the aforementioned drugs. As is represented in Figure 2A, hydrophilic 5-FU were encapsulated within the aqueous inner part of the vesicle, and the hydrophobic paenolum was situated into the external bilayer of the vesicle. Then, after nanocarriers fabrication and drug encapsulation, they were mixed with a PEO solution, and blend electrospinning was performed to obtain a core/shell scaffold, with the drug-loaded vesicles as core introduced into PEO NFs, Figure 2B. The release investigations concluded that the hydrophilic drug was released in an increased manner when the molar ratio of CTAB/SDBS was higher. In contrast, the hydrophobic drug showed a decrease in the release capability as the molar ratio of surfactants was increased. Figure 2A,B shows a schematic representation for the incorporation of dual drug-loaded vesicles into PEO NFs. The vesicle is initially introduced into the mixture of drugs, and then electrospinning is used to form the vesicle-IN-PEO NFs scaffolds. More recently, the same authors fabricated three different vesicles that were used as nanocapsules systems introduced into polymeric NFs for drug delivery purposes. These vesicles were composed of didodecyldimethylammonium bromide, cetyl trimethyl ammonium bromide (CTAB)/sodium dodecyl benzene sulfonate (SDBS) (7/3) and CTAB/SDBS (3/7). After that, these vesicles were introduced into nanocapsules fabricated by a mixture of sodium alginate and chitosan. PEO NFs were obtained by blend electrospinning, and they were composed by PEO, containing a mixture of chitosan/sodium alginate with the vesicles as a template. In this work 5-FU was chosen as a model chemotherapeutic molecule to be incorporated during vesicles fabrication. The drug release behavior was followed by UV-visible spectroscopy. As was expected, the different drug-delivery systems showed different release rates and pH-responsive behaviors.

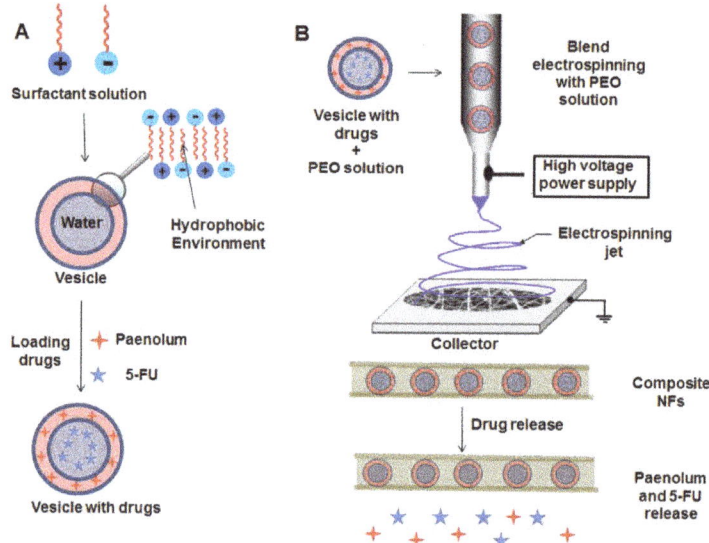

Figure 2. Schematic representation of the preparation of hydrophilic/hydrophobic electrospun composite fibers. (**A**) Accumulation of the drug mixture into the vesicle and (**B**) incorporation into the NF by electrospinning and drug release. Reprinted with permission from reference [33]. Copyright American Chemical Society, 2015.

Other types of linear polymer, as pluronic F127, were also used for release investigations, along with with tissue regeneration. Electrospinning was chosen for the fabrication of a scaffold containing high molecular weight PCL NFs containing pluronic (F127) vesicles, which were delivered by exploiting the slow dissolution of PCL into glacial acetic acid. The vesicles were fabricated by pluronic F127 self-assembled with low-molecular weight PCL in a tetrahydrofuran-water mixture [34]. The authors were able to tune the vesicle size from 1 to 10 μm in diameter. Time-dependent stability of the vesicles in glacial acetic acid was determined before the electrospinning process. The electrospun membrane was found to be composed of pluronic F127/PCL vesicles within a PCL mat with a fiber diameter between 50–300 nm. Authors proposed that the most probable condition for the vesicles generation is the non-solubility driven self-assembly and stabilization of PCL and F127 into bilayers in the tetrahydrofuran (THF)-water mixture. By using this method, the amphiphilic polymer is dissolved in a water miscible organic solvent and mixed at a high speed with water, leading to the rapid precipitation of polymers into nanoscale particles. Drug delivery of a model molecule as rhodamine-B (introduced into the polymer network) for this composite pluronic F127-IN-PCL showed an important reduction in the release rate of this molecule, when it was compared to the free vesicles. Indeed, the systems containing vesicles in the membrane presented an enhanced hydrophilicity compared to the control PCL membrane. Apart from drug delivery behavior, this increased surface hydrophilicity was exploited for increasing the cell viability of L929 cells on the membrane [34]. Another study with PCL was carried out by Yohe et al. [35] using the N-38 anticancer drug for the colorectal cancer cell line HT-29. They electrospun a mesh using 10% of a hydrophobic poly(glycerolmonostearate-co-ε-caprolactone) (PGC-C18) and 90% of PCL loaded with the N-38, showing promising results in the cytotoxicity of the cell line.

Micelles introduced into polymeric NFs have been also used for dual drug delivery. Hu et al. [36] fabricated colloidal structures formed by a block co-polymer composed by methoxypoly(ethylene glycol)-block-poly(L-lactide). Into these colloidal particles, a water soluble chemotherapeutic drug (5-FU) and a lipophilic drug as cefradine (an antibiotic active against Gram positive bacteria) were introduced. The external shell of the NFs was fabricated by a mixture of chitosan and PEO, and the micelles loading drugs were introduced into NFs by blend electrospinning. The NFs without micelles were uniform and smooth, with diameters in the range of 100–500 nm. However, after drug-loaded micelles incorporation within the NFs, the surface of the NFs was relatively rough, with diameters in the range of 200–800 nm, with some black spheres with diameters of approximately 150 nm in the NFs. By using this drug loaded nanocarrier-IN-NFs, they performed release investigations that revealed a low burst release tendency of 5-FU and Cefradin. Indeed, this system was able to reduce the activity of HepG-2 cells with good cell viability after 3 days of incubation. Zhang et al. [37] fabricated micelles composed by triblock copolymers as poly(L-glutamic acid)-b-poly(propyleneoxide)-b-poly(L-glutamic acid) which were able to deliver low water-soluble drugs, such as PTX, at clinically relevant doses. These micelles containing PTX were used as drug loaded nanocarrier which were grafted onto NFs scaffolds of (poly(L-lactide-co-ε-caprolactone) (PLCL):fibrinogen; 2:1 (w/w)) by blend electrospinning. Yang et al. [38] fabricated an implantable device structured as drug-loaded micelles-IN-NFs for controlled drug delivery. In this case authors used DOX introduced into micelles composed by PCL-PEG copolymer. The implantable devices were fabricated by coaxial electrospinning, with the core composed by the DOX-loaded PCL-PEG micelles introduced into a PVA solution dissolved in distillated water. The outer shell layer of the NFs was composed by genipin cross-linked gelatin. Importantly, these authors also functionalized the micelles with folic acid to specifically target tumor cells. The drug release investigations demonstrated that the implantable device reduced the drug dose, the frequency of administration and side effect of chemotherapeutic drugs while maintaining highly therapeutic efficacy against solid tumors.

2.2.2. Silica Particles as Nanocarriers

Dual drug delivery has been also performed by using two different types of colloidal nanoparticles as mesoporous silica nanoparticles (MSNs) and hydroxyapatite nanoparticles. These colloidal systems were used as nanocarriers for the incorporation of DOX HCl and the topoisomerase inhibitor hydroxycamptothecin [39]. Initially, the mixture of mesoporous silica particles and hydroxyapatite nanoparticles individually containing both chemotherapeutic drugs was a mixture within a polymer solution of PLGA to be introduced into NFs as a core/shell structure by electrospinning. This dual anticancer biocompatible system improved the mechanical capacity as well as the thermal stability of the NFs. In addition, when in vitro investigation was performed, the dual micelle-IN-NFs system provided a sustained and controlled drug release and an improved capacity for inhibiting HeLa cells growth. Using a similar approach, Qiu et al. [40] fabricated a drug-loaded implantable device for the treatment of a tissue defect after tumor resection. They used MSNs as nanocarriers for the incorporation of the anticancer drug DOX hydrochloride. These MSNs colloidal systems with incorporated chemotherapeutic drugs (DOX@MSNs) were introduced into PLLA NFs generated by electrospinning, thus obtaining a drug-loaded NFs scaffold DOX@MSNs-IN-NFs. Initially they confirmed the successful introduction of DOX-loaded MSNs into the PLLA NFs by UV-vis spectroscopy, and then several nanocomposite systems with different MSNs and DOX contents were fabricated. Figure 3 shows the schematic representation for the fabrication of DOX-loaded MSNs-IN-PLLA NFs. Optimal results concerning particles distribution that also improved thermal stability were found by using PLLA/1.0% DOX and 10% MSNs NFs. These authors investigated the in vitro antitumor efficacy against HeLa cells, and they found high DOX-loading capacities. Due to this fact, the drug was released in a sustained and prolonged manner, with a higher in vitro antitumor efficacy compared with free MSNs particles. Thus, these fabricated composite NFs mats are highly promising as a local implantable device for potential postsurgical cancer treatment. Yuan et al. [41] propose a drug delivery system which can release anti-tumor drugs in two phases. They designed a NFs scaffold for breast-conserving therapy after breast cancer by using DOX-loaded MSNs into an electrospun PLLA nanofibrous scaffold. In vivo results (mice) showed a significantly inhibition in the tumor growth, making its use promising in a coadjuvant therapy against this tumor type.

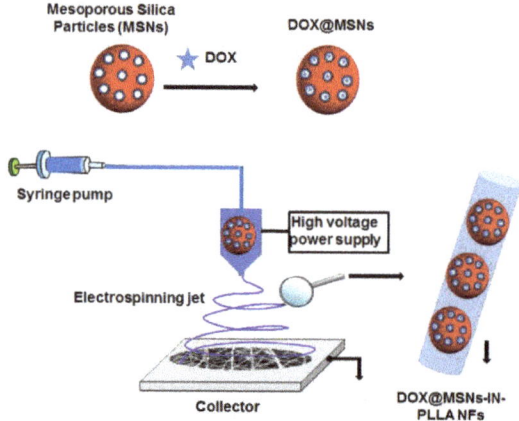

Figure 3. Schematic illustration for the process of fabrication of DOX@MSNs-IN-NFs electrospun composite NFs and the location of DOX in the fiber [40].

2.2.3. Gelatin Nanoparticles as Nanocarriers

Song et al. [42] used gelatin nanospheres (GNs) as colloidal systems to improve the antibacterial effects of silk NFs membranes. They prepared two types of GNs by adding different amounts of glutaraldehyde into the suspension during the synthesis of GNs. The size and distribution of the GNs into the NFs, fabricated by electrospinning and using a mixture of PEO and silk fibroin, was monitored by fluorescent labeling. Positively charged drugs as vancomycin and colistin were used as a model for controlled drug delivery investigations. Vancomycin is an antibacterial compound that inhibits the synthesis of the bacterial cell wall, and colistin is an apolypeptide that is effective against most of the Gram-negative bacilli. These authors used both blend and coaxial electrospinning to fabricate GNs-IN-PEO/silk NFs, and they examined the antibacterial effect by introducing the mentioned antibacterial drugs. By using the NGs synthesized with the highest amount of glutaraldehyde, the fabricated NFs supplied a more sustained release of vancomycin compared with pure GNs. Indeed, apart from being totally cytocompatible, the NFs showed excellent and sustained antibacterial effects against Staphylococcus aureus. Lai et al. [43] fabricated collagen (Col) and hyaluronic acid (HA) NFs with the aim to release a series of growth factors directly embedded in the NFs or encapsulated in the gelatin nanoparticles (GNs) by using electrospinning technology. The fabricated GNs-IN-Col/HA NFs showed mechanical properties that mimicked human natural skin. The designed GNs-IN-Col/HA NFs were able to release growth factors in a slow controlled manner for up to 1 month. From the above, the electrospun Col-HA-GN composite nanofibrous skin substitute with a stage-wise release pattern of multiple angiogenic factors could be a promising bioengineered construct for chronic wound healing in skin tissue regeneration. They also used several release patterns for GNs-IN-Col/HA

NFs with 4 different growth factors and demonstrated their potential capability to deliver multiple bioactive molecules.

Electrospun PLGA and PLGA/gelatin NFs embedded with MSNs were synthesized, obtaining MSNs distributed in the core of the fiber. PLGA and MSNs contributed to increase the hydrophobicity of electrospun NFs and the gelatin contributed to increase the mechanical properties of this scaffold. With a final size of approximately 267 nm, these nanoformulations are synthesized to provide a very suitable microenvironment for the adhesion, growth and migration of stem cells involved in nervous tissue regeneration [44,45]. Aytac et al. [46] synthesized electrospun gelatin NFs to vehiculate ciprofloxacin (CIP) and hydroxypropyl-beta-cyclodextrin (HPβCD)-inclusion complex (IC), which is normally used to improve the physico-chemical properties of some drugs as well as their bioavailability. This IC achieved an increase of the solubility and wettability of the nanoformulations, leading to their fast dissolution and therefore to a fast release of CIP transported in gelatin NFs, which can be an important property in certain situations where a rapid release of drugs is required. With this type of gelatin-based scaffolds, in addition to drug transport and use for tissue engineering, the migration of certain cells can also be induced or favored. This is the case in the study of Piran et al. [47] that synthesized electrospun three-layered scaffold with plasma enriched with growth factor to promote the migration and growth of fibroblast, which is a very important aspect in the regeneration of wounds.

2.2.4. Stimuli-Responsive Nanoparticles as Nanocarriers

Some stimuli-responsive colloidal particles have been used as nanocarriers to be introduced into polymeric NFs for drug delivery applications. For example, Gong el al. [48] introduced redox-responsive nanoparticles into a polymeric scaffold to be incorporated into the body with the aim to deliver a growth factor (morphogenic protein BMP-2) used of bone regeneration. The strategy to fabricate redox-sensitive NFs with a core/shell structure consisted of a blend of PCL and redox responsive c-6A PEG-PCL nanogel with –S–S– bond on the outer shell. This redox-sensitive shell was able to respond to the change of the glutathione concentration and thus regulate the BMP-2 release for in vitro and in vivo investigations.

Light-responsive nanoparticles, as TiO_2 and incorporated into a polyacrylonitrile (PAN)/multiwalled carbon nanotube composite NFs have been reported for the photocatalytic degradation of pharmaceutical molecules as Ibuprofen, Cetirizine, and Naproxen [49]. Visible light (0.1 W/cm^2) irradiation was employed to investigate the drug degradation. The photocatalytic degradation of molecules using TiO_2-IN-PAC/MWCNT NFs was higher compared with TiO_2-IN-PAC NFs under visible light irradiation. A total degradation of drugs molecules was performed at 200, 50, and 90 min, respectively under visible light.

Another interesting strategy was developed by Elashnikov et al. [50]. They were able to release antimicrobial molecules from thermos-responsive microgels introduced into PLLA NFs. Crystal violet (CV) was incorporated within the polymer network of temperature-responsive PNIPAM microgels used as nanocarriers. Then, blend electrospinning was performed using PLLA as a biocompatible polymer, thus resulting in composite PLLA NFs with incorporated pNIPAM particles containing antimicrobial CV. They investigated the controlled drug delivery behavior of these core/shell NFs by UV-vis spectroscopy, which was produced after modification of the external temperature below and above the lower critical solution temperature (LCST). The antibacterial activity was investigated against gram-negative Escherichia coli (E. coli) and gram-positive Staphylococcus epidermidis (S. epidermidis). Authors demonstrated that the temperature-responsive release of antibacterial CV possessed remarkable antibacterial activity. This activity showed higher inhibition zones at temperatures above the LCST, with its size dependent on the polymers ratio and temperature.

2.3. Stimuli-Responsive Nanofibers

As was mentioned, synthetic polymers have been extensively used as scaffolds in the synthesis of NFs for drug delivery investigations. However, during the last years, important researching efforts have been focused on the use of stimuli-responsive systems [20]. These structures can undergo changes in response to external stimulus, as temperature, pH, ion strength, or solvent nature [51–53]. Most of the recent studies in the area of switchable drug release have been dedicated to the creation of systems for drug encapsulation based on two types of polymer [54]: pH-responsive and thermo-responsive polymers.

2.3.1. pH-Responsive Nanofibers

The acidic environment found in tumor tissues can be employed as a way to specifically target the release of antitumor drugs at the tumor in response to a change in pH by the use of nanoformulations sensitive to pH changes [55]. Illangakoon et al. [56] used ES100, which is an anionic co-polymer constituted by metacrylic acid and methylmethacrylate, for the fabrication of pH-responsive nanofiber for the delivery investigation of 5-FU. Co-axial electrospinning was carried out with a core composed of poly(vinylpirrolidone), ethyl cellulose (EC) and the 5-FU drug, and the shell was formed from pH-responsive ES100. The drug release investigation demonstrated a controlled drug release developed at pH 1, reaching a maximum of 80% drug release after 2 h, produced by the diffusion of 5-FU through the pores of the ES100 polymer. At this pH, the polymeric fibers were fragmented, supplying an increased 5-FU delivery. Indeed, these authors fabricated NFs with a core/shell structure made of ES100 for the shell and Eudragit L100 (EL100) for the core to allow the controlled release at certain pH conditions, and controlled this release using the cover thickness [57]. Tran et al. [58] introduced ibuprofen into pH- and thermo-responsive polymers for controlled drug delivery investigations. Ibuprofen was initially mixed with a polymeric solution of PCL as the control experiment and also into poly(Nisopropylacrylamide-co-methacrylic acid) (pNIPAM-co-MAA) to fabricate stimuli-responsive NFs by blend electrospinning. As comparative results, when PCL NFs were investigated as a drug delivery system for ibuprofen, the fabricated NFs did not show significant drug release behavior at temperatures between 22–40 °C or pH from 1.7–7.4. However, NFs generated from pNIPAM-co-MAA were able to diffuse ibuprofen in a linear and controllable manner when the temperature was above the lower critical solution temperature (LCST) of pNIPAM-co-MAA (33 °C), as well as at pH lower that the pKa of carboxylic acids (pH 2). However, when the drug delivery experiments were performed at room temperature, the release rate was radically increased by closely ten times, compared to the release behavior at higher temperature and lower pH. NFs of cationic chitosan and poly(acrylic acid) (PAA) were synthetized with different levels of Cs deacetylation, showing that the mechanical properties of these NFs are determined by both the pH and the level of deacetylation, which could be useful in biomedical applications such as the transport and release of drugs [59]. Gelatin and poly(lactide-co-ε-caprolactone) (PLCL) were used to synthetize NFs that were loaded with ciprofloxacin and sodium bicarbonate with a response to low pH of gelatin/sodium bicarbonate fibers, whereas the hydrophobic PLCL had no sensitivity to pH [60]. These NFs not only showed good biocompatibility in fibroblasts (L929), they were also able to stimulate cell growth compared to untreated cells. Functionalized electrospun PCL scaffolds sensitive to pH changes, were loaded with DOX and tested at different pH levels (from 7.4 to 2.5) finding the highest drug release (90%–95%) at the lowest pH levels [61]. This was observed in the human embryonic kidney cells (HEK) treated with these scaffolds with a cell viability at pH 6 lower than those obtained at pH 7.2.

Thixotropic silk NFs hydrogels were loaded with DOX and it was designed to release the drug in the tumor site due to their thixotropy capability [62]. This allows the hydrogel inoculation that then solidifies at the specific site and releases the drug in response to certain pH conditions. In the in vitro studies carried out on the human breast cancer cell line MDA-MB-231, it could be observed that the DOX-loaded hydrogels were more suitable for a long treatment, since even after 10 days they continued to inhibit cell growth, in contrast with free DOX. In the in vivo studies in breast tumors bearing BALB/c nude mice, it was observed that after the inoculation of the liquid hydrogel it solidified around the tumor, finding remains of the hydrogel even 5 weeks after the inoculation. In a similar way, in the in vivo studies after the inoculation, the decrease in tumor volume was similar with the treatment of free DOX and DOX transported by the hydrogels in the first weeks. However, at the fifth week, significant differences were observed in the volume and weight of the tumor treated with DOX-loaded hydrogels compared to free DOX, reaching reductions of approximately 1.5 times for both parameters, see Figure 4.

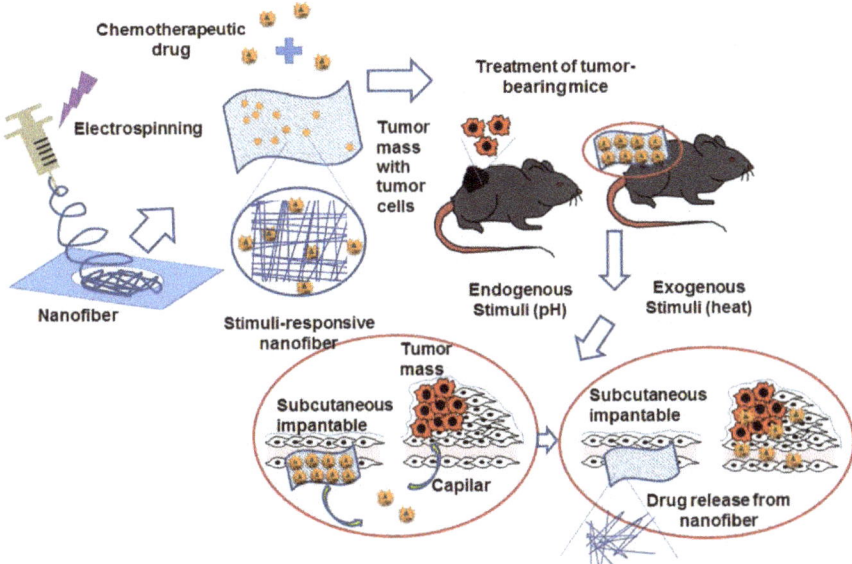

Figure 4. Stimuli-responsive NFs. Once NFs are synthesized by the electrospinning process and loaded with the antitumor drug, treatment may be applied in an experimental mouse model that carries a specific type of tumor. Once the treatment has been inoculated, an internal stimulus, such as the low pH present in the tumor tissues, or an external stimulus such as a temperature rise, stimulate the release of the drug at the specific site of the tumor, thus applying the treatment on tumor cells. Reprinted with permission from reference [62]. Copyright American Chemical Society, 2016.

2.3.2. Thermo-Responsive Nanofibers

In general, there are various protocols for synthesizing nanoformulations sensitive to temperature that are suitable for their possible use in biomedicine, because they can be administered by injection and may be degraded. These properties are useful in the transport of drugs for cancer targeting and controlled release such as degradable NFs fabricated by an electrospinning technique [63]. Slemming-Adamsen et al. [64] presented a novel approach to introduce DOX into thermoresponsive pNIPAM-NHS/gelatin NFs by cross-linking with 1-ethyl-3-(3-dimethyl-aminopropyl)-1-carbodiimide hydrochloride (EDC) and N-hydroxysuccinimide (NHS). This strategy consisted of a mixture with a solution of pNIPAM-NHS/gelatin acting as a shell with another mixture of EDC and NHS in the presence of DOX. By using this approach, EDC initiates the conjugation by bonding with a carboxyl group of the polymer. Then, the EDC-polymer conjugate is able to react with a primary amine, or, NHS, replacing EDC with the amine ester linkage. Finally, NHS can be replaced by a primary amine, linking the carboxyl-polymer with the amine-polymer. This mixture was electrospun to obtain cross-linked pNIPAM/gelatin NFs containing an anticancer drug that can be released in a controlled manner. The DOX-IN-pNIPAM NFs showed thermo-responsive swelling/deswelling properties. Indeed, the fabricated cross-linked NFs were able to release DOX when the temperature was raised above the LCST and were able to reduce the viability of human cervical cancer cells [64].

Zhang et al. [65] fabricated a core/shell structure formed by polylactic acid PLA as a core using electrospinning, and then a thermoresponsive pNIPAM shell was incorporated by UV photo-polymerization.

Initially, biodegradable PLA NFs were fabricated by electrospinning in the presence of Combretastatin A4 (CA4), a tubulin polymerization inhibitor which was used as the model drug was produced. These fabricated PLA NFs were introduced into a pNIPAM solution in presence of the crosslinker (N,N′-methylenbisacrylamide). After exposed to UV radiation, the drug-loaded PLA core was coated and cross-linked with a pNIPAM shell. The composite NFs exhibited different wettability confirmed by water contact angle measurements at temperatures below or above the lower critical solution temperature (LCST) of pNIPAM. Most importantly, in vitro drug release investigations demonstrated a difference drug release when the temperature was at 25 or 40 °C. For example, the pNIPAM shell could limit the release rate of CA4 below the LCST, however, above the LCST, the rate of drug release increased significantly. Cicotte et al. [66] used thermos-responsive pNIPAM films fabricated by electrospinning to exploit a rapid reversible adhesion of mammalian cells, thus performing cell attachment and detachment using pNIPAM scaffolds. These authors modified various parameters during the electrospinning process such as the needle gauge, collection time, and molecular weight of the polymer. Two types of cells were investigated for reversible attachment of pNIPAM mats that provided potential results by seeding mammalian cells from standard cell lines (MC3T3-E1) as well as cancerous tumor (EMT6) cells. Once attached, the temperature of the cells and mats was changed to ~25 °C, resulting in the extremely rapid swelling of the pNIPAM NFs. The authors found that pNIPAM mats fabricated using small and dense fibers fabricated from high molecular weight pNIPAM polymers are extremely appropriate as a rapid release method for cell sheet harvesting. Recently, new nanoformulations that allow the release of drugs in a dual way in response to both temperature and pH stimuli have been designed. This is the case of poly(N-isopropylacrylamide-co-acrylic acid) NFs in a passive thermoplastic polyurethane (TPU) which are sensitive to pH and temperature. Consequently, by varying these two parameters, the movement in terms of direction and size can be modulated, which could be interesting in several biomedical applications, such as drug release [67]. In another study, a thermo-sensitive polymer, PNIPAAm and a pH-sensitive polymer, Eudragit® L100-55 (EL100-55), were synthetized and made NFs by electrospinning [68]. These NFs showed sensibility to pH and temperature and a release of ketaprofen that are dependent on these parameters without toxicity against fibroblast, even at high concentrations. Another example is a fiber mixture of poly(N-vinylcaprolactam) and ethyl cellulose (EC) in the case of temperature-dependent release and EL100 fibers for pH-dependent release, synthesized by twin-jet electrospinning [69]. This mixture of fibers showed a sustained release of the non-steroidal anti-inflammatory drug ketoprofen in response to pH and changes in temperature, also showing a very good biocompatibility in fibroblasts. The study of biocompatibility of these NFs is noteworthy because it is an essential property that nanomaterials which are intended for therapeutic use must comply with. For this purpose, fibroblasts (L929) were seeded on cover slips that were previously sterilized and where fibers were directly slectrospun. After 1, 3 and 5 days of exposure, cytotoxicity is determined by the MTT assay. NFs are inclined to show good biocompatibility, but some types may be more appropriate than others, such as thermosensitive fibers made of poly(di(ethylene glycol) methyl ether methacrylate) (PDEGMA) synthesized and electrospun into fibers using EC, which showed great biocompatibility even after 5 days of exposure and better in vitro biocompatibility than other nanoformulations such as EC/NIPAM [70,71]. Another type of stimuli-responsive NFs that could be used for the transport of antitumor drugs is electrospun self-immolative polymer (SIP)/polyacrylonitrile (PAN) fibers [72], which depolymerize surprisingly rapidly in response to an external stimulus, producing an instantaneous release of the transported molecules at the right time.

2.4. Hybrid Nanofibers

2.4.1. CNTs/Nanofibers

Carbon nanotubes have been used as platforms to improve mechanical, structural and drug delivery properties in NFs synthesized by electropun. In a recent investigation carried out by Qi et al. [73],

DOX was chemically incorporated onto the surface of MWCNTs. After being optimized, the drug encapsulated up to 83.7% and the dispersion of MWCNTS@DOX particles was mixed with a PLGA polymer solution at 3 different amounts of DOX, relative to PLGA, to fabricate a composite NF by electrospinning. Importantly, the incorporation of MWCNTs into PLGA NFs did not alter the structure of the PLGA NFs, instead it improved their mechanical properties. In vitro viability assay demonstrated that the developed DOX-loaded MWCNTs-IN-PLGA composite NFs were totally cytocompatible with L929 cells. The drug delivery investigation confirmed that this composite system was able to reduce burst DOX release, and it also allowed a continued DOX release over 42 days. This composite structure structured as CNTs-IN-NFs was also used for Yu el al. for the incorporation of DOX HCl thus creating a hybrid NFs mat by blend electrospinning. In this case CNTs were also used as carriers of DOX and they were introduced within PLGA NFs by electrospinning, thus creating a composite nanofibrous mat. In this investigation, authors modified the amount of CNTs in the final mixture polymer-CNTs during electrospinning. The in vitro antitumor efficacy against HeLa cells was investigated, resulting in a DOX release by a sustained and prolonged manner, which effectively inhibited growth of HeLa cells. Figure 5 shows a schematic representation for the fabrication of DOX-loaded CNTs-IN-PLGA NFs by blend electrospinning.

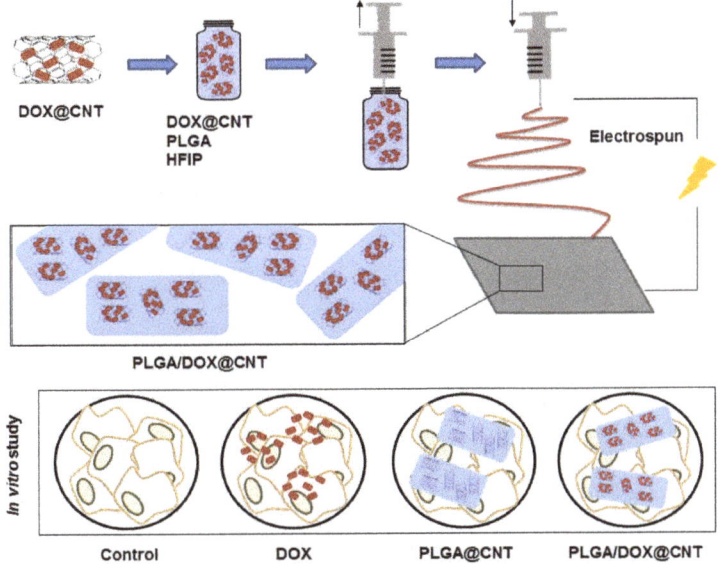

Figure 5. Schematic illustration for the fabrication process of PLGA/DOX-IN-CNTs electrospun composite NFs. The morphology and diameter distributions of PLGA and PLGA/DOX@CNTs composite NFs [74].

Zhang et al. [75] fabricated a new class of highly porous NFs by using PLA as the principal polymer scaffold and PEO as a porogen. Different concentrations of CNTs were incorporated onto the fibers to allow self-sealing behavior, which was carried out by photothermal conversion after light irradiation. Basically, they developed a strategy to fabricate porous PLA/CNT fibers with controlled pore sizes. Indeed, CNTs were used in the current study to perform the pore self-closure through their photothermal conversion ability. Zhang et al. investigated several ratio polymer/CNTs during the electrospinning process and they found that the fibers containing 0.4 mg/mL of CNTs showed the optimum encapsulation efficiency of model biomacromolecules such as dextran, bovine serum albumin, and nucleic acid. The pores of the surface were reversibly reopened by PLA degradation, reaching a stable release of biomacromolecules after encapsulation. They showed morphological changes of PLA/CNT fibers with 0, 0.2, 0.4, and 0.8 mg/mL CNT concentrations. At a CNT concentration of 0.2 mg/mL, only a small amount of CNTs could be seen on the fibers surface. As the concentration was increased to 0.4 mg/mL, much more CNTs were observed, which seemed to be dispersed homogeneously around the nanopores on the fibers. Nevertheless, with further increase of the CNT concentration to 0.8 mg/mL, significant agglomeration was observed. After trapping the drug molecules, a controlled and sustained release from the fibers over extended periods of time was investigated. An extended release of the molecules for over the period of 15 days was achieved, with a relatively low initial burst release within the first 24 h. Indeed, in these structures the amount of CNT introduced into the nanofiber was relevant to control parameters as morphology, structure, thermal/mechanical capability, degradation, and cell viability.

It is well-known that CNTs have the capacity to acts as thermal generators by absorbing near-infrared radiation (NIR). Zhang et al. [76] combined MWCNTs and DOX in a methanol/chloroform solution with PLLA dispersed in CHCl$_3$ to fabricate NFs by the electrospinning technique. Without NIR irradiation, DOX release was extremely restricted. However, when NIR irradiation (2 W/cm^2) was applied on DOX-loaded MWCNTs-IN-PLLA NFs burst release of the loaded DOX was observed at the time of 2 h with an amount of ~20% during the 30 min irradiation period. Indeed, these authors demonstrated that after NIR radiation, the temperature of the tumor area in contact with the NFs was significantly increased. Apart from that, these multifunctional NFs showed increased cytotoxicity both in vitro and in vivo for Hela cancer cells through the combination of photothermal induced hyperthermia and drug delivery. Barzegar et al. [77] introduced graphene into PVA NFs by electrospinning, thus creating grapheme/PVA hybrid NFs. Figure 6 shows scanning electron microscope (SEM) images where uniform hollow PVA NFs containing graphene dispersion within the NFs can be appreciated. The synthesized polymer reinforced NFs have potential biomedical materials for drug delivery.

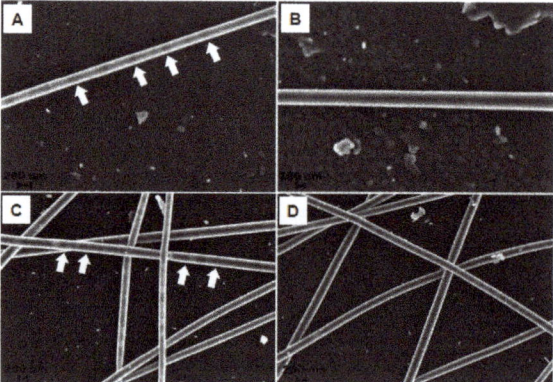

Figure 6. SEM micrographs of PVA/(graphene foam and expanded graphite) NFs at 2 kV operating voltage for (**A**) solution with 0.02 g GF concentration, (**B**) solution with 0.08 g GF concentration, (**C**) solution with 0.02 g EG concentration and (**D**) solution with 0.08 g EG. Reprinted with permission from reference [77]. Copyright Elsevier, 2015.

In recent years, there has been an increase in the number of investigations related to the use of electrospun NFs combined with conductive nanomaterials, as biosensors, due to their promising applications [78]. Electrospun NFs in combination with CNTs have also been investigated as biosensors for the detection of early stages of pancreatic cancer by the use of the biomarker CA19-9. Electrospun NFs made of poly(allylamine hydrochloride) and polyamide 6 were covered with MWCNTs, and CA19-9 detection were determined by electrochemical impedance spectroscopy [79]. This biosensor showed a good sensitivity for the antigen without the rest of the blood components could suppose interference in the detection. This high sensitivity was achieved through antibody-antigen irreversible adsorption. In addition, this system was able to distinguish between patients with higher levels of the biomarker in blood and those who had it in lower concentration.

However, concerning toxicology studies, it has been reported that CNTs produce some adverse effects, due principally to the incorporation of catalytic metal impurities during their fabrication [80]. For this reason, a general use of CNTs in the biomedical field has generated some doubts about their security and possible toxicity. For example, long-term accumulations of CNTs were produced in lungs and liver of mice after exposure during 90 days [81]. Some reported investigations detailed the negative effects of these CNTs in biological systems. An alternative to reduce toxicity is the use of surface functionalized CNTs. Specifically, a recent investigation showed that pure CNTs had negative effect on the systemic immunity, producing more inflammation and immunosuppression as compared to some surface modified CNTs, such as PEG-modified CNTs [82].

2.4.2. Magnetic Nanoparticles/Nanofibers

Magnetic nanoparticles are another type of colloidal particles widely used in biomedical applications. The systems, in the range of nanometers, are able to respond to external magnetic fields. This property has been extensively exploited in the biomedical field for localized drug delivery, because they can be transported to a specific area by the application of an external magnetic field. Apart from conventional drug delivery systems used in nanomedicine for the reduction of solid tumors, magnetic hyperthermia treatments [83–85], in which magnetic or magnetic-derived NPs are administered to tumors, and then a local heat is supplied under alternating magnetic field (AMF) application, have attracted an ever increasing interest due to their improved precision for cancer therapy [86–88]. The basis of using hyperthermia as a treatment modality for cancer is the high sensitivity of cancer cells to temperatures in the range from 41 to 45 °C, in contrast to normal cells. However, the use of free magnetic nanoparticles has certain limitations such as poor tumor targetability, high variability in the amount of magnetic nanoparticles administered to the tumor, as well as the transfer of magnetic nanoparticles into the healthy tissues close to the tumor. To overcome this disadvantage, magnetic NPs have been also used in electrospinning to fabricate hybrid NFs for drug delivery and cancer treatment purposes.

It is important to mention that a significant advantage of magnetic nanoparticles is that they present a reduced toxicity, while also being accepted by the human body. Indeed, once they are placed into the cells they degraded reasonably quickly [89]. The degradation of magnetic nanoparticles into iron and oxygen is performed inside the lysosomes of macrophages, and it is influenced by several parameters as the presence of hydrolytic enzymes, a low pH, as well as proteins related with the iron metabolism. In particular, iron oxides nanoparticles are able to degrade in vivo by iron mobilization and some other published routes [90]. Most importantly, magnetite is an iron oxides derivative approved by FDA for in vivo investigation [91].

Feng et al. [92] prepared a mixture of Fe_3O_4 NP and graphene oxide sheets containing functional groups to be introduced into PAN NFs for guiding cellular application. They used blend electrospinning to fabricate a hybrid GO/Fe_3O_4-IN-PAN as short-fibers (SFs). These synthesized NFs films were cut into small pieces, and after being dispersed in tert-butanol solution, they demonstrated a strong magnetic capability. As guiding cellular behavior, breast cancer cells were cultured on the surface of these magnetic SFs, and due to the external GO on the surface of SFs they promoted adhesion of cell membrane proteins and good biocompatibility. Indeed, guided cellular behavior by magnetic actuation with the help of magnetic SFs was performed. Huang et al. [93] introduced 50 nm iron oxide nanoparticles into polystyrene (PS) NFs scaffolds. Fe_3O_4 NPs were initially dispersed and ultrasonicated in THF, this dispersion was a mixture with a PS solution to proceed with blend electrospinning. Upon applying an AMF on these hybrid Fe_3O_4-IN-PS NFs, an important heating process occurred due to the high loading capacity of the fibers. Using this hybrid structure, the fabricated magnetic fibers can be heated several times without loss of heating capacity or releasing magnetic nanoparticles. Indeed, these authors functionalized the surface of the hybrid NFs structure with collagen in order to increase cell attachment. In vivo investigations were performed by using Human SKOV-3 ovarian cancer cells, which were incorporated onto the fibers. After the application of an AMF during 10 min to the mats, the cancer cells deposited on the Fe_3O_4-in-fibers were eliminated. This methodology possesses two important advantages to be implemented for in vivo investigations, as the fibers can be loaded with magnetic nanoparticles in a controlled manner and the composite scaffolds can be localized in the body by magnetic resonance imaging (MRI). Sasikala et al. [94] fabricated an implantable hybrid magnetic NFs device to be applied in both magnetic hyperthermia, upon an AMF, and cancer cell-specific drug release, to perform a synergistically cancer therapy. A borate-containing anticancer drug was investigated known as bortezomib (BTZ), a protease inhibitor frequently used in chemotherapeutic administration. This device was fabricated by blend electrospinning, by a mixture that initially was a Fe_3O_4 NPs dispersion and a PLGA solution. Then, a shell of polydopamine was grown through a simple immersion method to be used as a shell-mimicking mussel adhesive. The mussel-inspired magnetic NFs with numerous catechol moieties were able to bind and release borate-containing anti-cancer drugs. Figure 7 includes transmission electron microscopy (TEM) image of free Fe_3O_4 NPs as well as field emission scanning electron microscopy (FESEM) and TEM images of electrospun PLGA NFs. The FESEM images confirmed that hybrid NFs shows good fiber morphology and some aggregation of the Fe_3O_4 NPs inside of the NFs. They investigated the effect of repeated hyperthermia application in murine breast cancer (4T1) cell lines by using the BTZ-loaded Fe_3O_4-IN-PLGA NFs. After three hyperthermia cycles (15 min/24 h) were applied to the BTZ-loaded Fe_3O_4-IN-PLGA NFs, an improved antitumor efficacy was obtained.

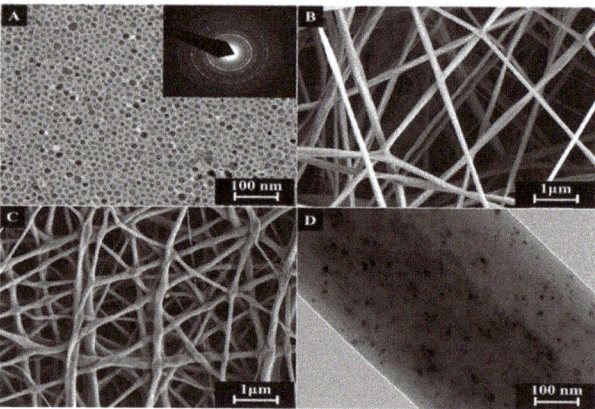

Figure 7. (A) TEM image of the iron oxide NPs (the inset shows the corresponding selected-area electron diffraction pattern), (B,C) FESEM images of electrospun PLGA (poly lactic-co-glycolic acid) NFs and magnetic NF matrix, respectively, (D) TEM image of MNF. Reprinted with permission from reference [94]. Copyright Elsevier, 2016.

In other reported cases, polyurethane NFs combined with superparamagnetic iron nanoparticles (γ-Fe_2O_3) have been synthesized for cancer treatment by hyperthermia, which were able to reach a temperature increase of up to 43 °C in 70s by the application of an AMF. This promising result was achieved by making the electrospinning process much more precise with the addition of a conical aluminum auxiliary electrode [95]. Radmansouri et al. [96] designed DOX hydrochloride-loaded electrospun chitosan/cobalt ferrite/titanium oxide NFs to treat melanoma cells (B16F10) by means of chemotherapy and hyperthermia at the same time. NFs were made of chitosan by the electrospinning process and were combined with cobalt ferrite nanoparticles and titanium oxide nanoparticles (used to modulate the temperature increase). Higher cell death was observed when hyperthermia and chemotherapy were combined, achieving a synergistic effect, which enhances the cytotoxic effect and allows a reduction of side effects.

2.4.3. Gold Nanoparticles/Nanofibers

It is well-known that some non-spherical gold nanoparticles (Au NPs) display large near infrared (NIR) resonances that can be used to induce both hyperthermia and drug delivery when they are irradiated with the appropriate wavelength [97]. As non-invasive therapy, NIR radiation is crucial for biomedical uses because it penetrates tissue more deeply, but it is absorbed less than other types of radiation. NIR hyperthermia is a minimally-invasive oncological treatment strategy in which photon energy is selectively administered and converted into sufficient heat to induce cellular injury [98–100]. An important property concerning Au NPs is that by varying their size and shape, the surface plasmon absorption can be tuned from ultraviolet (UV) to infrared (IR) wavelengths. Recently, the potential uses of gold nanoparticles in NIR-hyperthermia have been reported using a variety of noble metal nanostructures, including gold nanoshells [99,101], gold nanorods [102–104], and recently, gold nanocages [105]. The potential toxic impact of AuNPs has been discussed. The particles size, surface chemistry and the presence of functional groups may play a relevant role in cell toxic effects [106]. Some studies have reported that cationic Au NPs are toxic while anionic AuNPs are non-toxic for cells [107]. The toxicity is caused by the electrostatic interaction of Au NPs with the negatively charged bilayer of the cellular membrane. Several investigations have also confirmed that some modified Au NPs (as PEG-modified Au NPs) are non-toxic at the dose that is effective for in vivo drug delivery [108]. Recently, the synthetic toxicity of AuNPs capped with polyethylenimine (PEI) and PEGylated anisamide has been tested for in vivo investigations, obtaining changes in blood cells by hemocytometer. These results demonstrated nonsignificant differences between hematological toxicity of these modified NPs and controls (saline serum) In addition, an extensive analysis of the tissue injury was carried out using gold nanoparticles prepared with PEG and DOX. In this case, Au3 treatment did not induce histopathologically observable differences in mice (including among others, heart, lung, stomach, intestine, liver, pancreas, kidney, spleen, skeletal muscle, brain, spinal cord) from those treated with saline serum, thus indicating no systemic toxicity [109].

In this context, Zhang et al. [97] incorporated gold nanorods (AuNRs) into pNIPAM NFs in order to create a hybrid composite with fast thermal/optical response and structural integrity by electrospinning. They prepared a mixture of the pNIPAM polymer at 12 wt% and AuNRs in THF, then electrospinning was performed to achieve the hybrid AuNRs-IN-pNIPAM NFs. The photothermal property of these metal nanoparticles and the thermo-responsive property of pNIPAM were demonstrated. They obtained a NFs heating from room temperature to 34.5 °C after 1 s of laser application, and a further increase to 60 °C in 5 s of irradiation. Figure 8 shows TEM images of both the AuNRs dispersion and the hybrid NFs where the metal nanoparticles are incorporated onto the pNIPAM surface, a photograph of the hybrid composite is also included.

Poly (ε-caprolactone diol) based polyurethane solutions were used to synthesize NFs by electrospinning and they were combined with gold nanoparticles [110]. This nanoformulation was loaded with temolozolamide and was designed as a potential implant that allows a continued release of the antitumor drug for the treatment of glioblastoma multiforme. In fact, in U-87 MG human glioblastoma cell line this nanoformulation achieves a greater cell death overtime in contrast with free temozolamide (25% more) that practically does not modify the percentage of cell proliferation. In the same way as with electrospun NFs made of poly(allylamine hydrochloride) and polyamide 6 covered with MWCNTs, these NFs were also covered with gold nanoparticles for the detection of the pancreatic cancer biomarker CA19-9. In this case, lower biomarker detection thresholds were obtained with the use of gold nanoparticles (1.57 U mL^{-1}) compared with the use of MWCNTs (1.84 U mL^{-1}) [79].

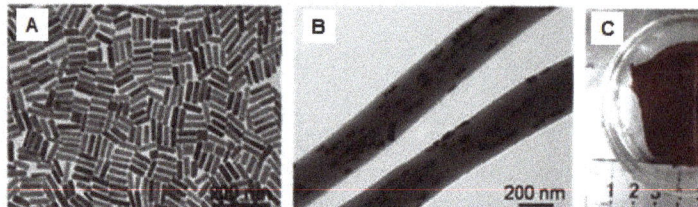

Figure 8. TEM images of (**A**) Au nanorods (AuNRs), (**B**) AuNRs/PNIPAM electrospun fibers and (**C**) photograph of AuNRs/PNIPAM composite film immersed in water. Reprinted with permission from reference [97]. Copyright ACS Publications, 2017.

3. Conclusions

Electrospinning is a technique used worldwide that allows the fabrication of polymeric NFs in the range of micro- and nanometers. Biocompatible and biodegradable synthetic polymers are essential structures that have improved chemotherapeutic treatments in the biomedical field. During the last years they have been extensively and specifically reported as systems with important benefits in important fields, including drug delivery and cancer treatments. In this review we have summarized recent advantage for the fabrication of NFs by electrospinning focused on drug delivery applications and cancer treatments (magnetic and plasmonic hyperthermia). Two methods are principally used for the fabrication of NFs by electrospinning: blend and coaxial electrospinning. Based on these methodologies, we reported recent approaches for the fabrication of drugs-IN-NFs with drug delivery and cancer treatment applications. We not only included a direct introduction of the drug dissolved within the polymer, nowadays, a very useful strategy is the incorporation of colloidal particles used as vehicles into the NFs during electrospinning. These particles are able to increase the amount of drug into the NFs to be released in a constant and controlled manner. In this sense, in this review we have also included polymers with stimuli-responsive behavior, obtaining NFs with the ability to increase the drug delivery capability in the function of an external stimulus (temperature or pH). Indeed, electrospinning also offers the possibility to fabricate hybrid NFs, structured as polymeric NFs with systems such as CNTs, graphene oxide, or even magnetic or metallic nanoparticles. Consequently, we also pointed out in the relevance of hybrid NFs in drug delivery improvements as well as in the reduction of solid tumors through magnetic and plasmonic hyperthermia.

Author Contributions: Conceptualization, R.C.-C. and J.P.; revision and supervision, R.C.-C., C.M., J.M.L.-R., J.P.; contribution to the oncological application, figures and discussion, L.C. and G.P.; contribution to nanoplatform types, figures and discussion, A.D.; funding acquisition, J.M.L.-R. and C.M.

Funding: This research was funded by the Comunidad de Madrid, Spain fellowship "Atracción de Talento Investigador" (2018-T1/IND-10736), Consejería de Salud de la Junta de Andalucía (project PI-0476-2016 and PI-0102-2017) and CICYT, Spain (project CTQ16-76311).

Acknowledgments: We thank D. Antonio Ramírez (Department of Anatomy and Embryology) for his administrative assistance.

Conflicts of Interest: The authors declare no conflict of interest.

References

1. Courtney, C.M.; Goodman, S.M.; McDaniel, J.A.; Madinger, N.E.; Chatterjee, A.; Nagpal, P. Photoexcited quantum dots for killing multidrug-resistant bacteria. *Nat. Mater.* **2016**, *15*, 529–534. [CrossRef] [PubMed]
2. Lyutakov, O.; Goncharova, I.; Rimpelova, S.; Kolarova, K.; Svanda, J.; Svorcik, V. Silver release and antimicrobial properties of PMMA films doped with silver ions, nano-particles and complexes. *Mater. Sci. Eng. C Mater. Biol. Appl.* **2015**, *49*, 534–540. [CrossRef]

3. Bray, F.; Ferlay, J.; Soerjomataram, I.; Siegel, R.L.; Torre, L.A.; Jemal, A. Global cancer statistics 2018: GLOBOCAN estimates of incidence and mortality worldwide for 36 cancers in 185 countries. *CA. Cancer J. Clin.* **2018**, *68*, 394–424. [CrossRef] [PubMed]
4. Dubey, N.; Letourneau, P.C.; Tranquillo, R.T. Neuronal contact guidance in magnetically aligned fibrin gels: Effect of variation in gel mechano-structural properties. *Biomaterials* **2001**, *22*, 1065–1075. [CrossRef]
5. Mickova, A.; Buzgo, M.; Benada, O.; Rampichova, M.; Fisar, Z.; Filova, E.; Tesarova, M.; Lukas, D.; Amler, E. Core/shell nanofibers with embedded liposomes as a drug delivery system. *Biomacromolecules* **2012**, *13*, 952–962. [CrossRef] [PubMed]
6. Kataoka, K.; Harada, A.; Nagasaki, Y. Block copolymer micelles for drug delivery: Design, characterization and biological significance. *Adv. Drug Deliv. Rev.* **2001**, *47*, 113–131. [CrossRef]
7. Chevalier, Y.; Bolzinger, M.-A. Emulsions stabilized with solid nanoparticles: Pickering emulsions. *Colloids Surf. Physicochem. Eng. Asp.* **2013**, *439*, 23–34. [CrossRef]
8. Zheng, Y.; Li, S.; Weng, Z.; Gao, C. Hyperbranched polymers: Advances from synthesis to applications. *Chem. Soc. Rev.* **2015**, *44*, 4091–4130. [CrossRef] [PubMed]
9. Soppimath, K.S.; Aminabhavi, T.M.; Kulkarni, A.R.; Rudzinski, W.E. Biodegradable polymeric nanoparticles as drug delivery devices. *J. Control. Release* **2001**, *70*, 1–20. [CrossRef]
10. Ji, W.; Sun, Y.; Yang, F.; van den Beucken, J.J.J.P.; Fan, M.; Chen, Z.; Jansen, J.A. Bioactive electrospun scaffolds delivering growth factors and genes for tissue engineering applications. *Pharm. Res.* **2011**, *28*, 1259–1272. [CrossRef]
11. Danhier, F.; Ansorena, E.; Silva, J.M.; Coco, R.; Le Breton, A.; Préat, V. PLGA-based nanoparticles: An overview of biomedical applications. *J. Control. Release* **2012**, *161*, 505–522. [CrossRef] [PubMed]
12. Sill, T.J.; von Recum, H.A. Electrospinning: Applications in drug delivery and tissue engineering. *Biomaterials* **2008**, *29*, 1989–2006. [CrossRef]
13. Bhardwaj, N.; Kundu, S.C. Electrospinning: A fascinating fiber fabrication technique. *Biotechnol. Adv.* **2010**, *28*, 325–347. [CrossRef]
14. Li, W.-J.; Laurencin, C.T.; Caterson, E.J.; Tuan, R.S.; Ko, F.K. Electrospun nanofibrous structure: A novel scaffold for tissue engineering. *J. Biomed. Mater. Res.* **2002**, *60*, 613–621. [CrossRef] [PubMed]
15. Venugopal, J.; Prabhakaran, M.P.; Low, S.; Choon, A.T.; Zhang, Y.Z.; Deepika, G.; Ramakrishna, S. Nanotechnology for nanomedicine and delivery of drugs. *Curr. Pharm. Des.* **2008**, *14*, 2184–2200. [CrossRef]
16. Zhang, Y.; Liu, S.; Wang, X.; Zhang, Z.; Jing, X.; Zhang, P.; Xie, Z. Prevention of local liver cancer recurrence after surgery using multilayered cisplatin-loaded polylactide electrospun nanofibers. *Chin. J. Polym. Sci.* **2014**, *32*, 1111–1118. [CrossRef]
17. Gañán-Calvo, A.M.; Dávila, J.; Barrero, A. Current and droplet size in the electrospraying of liquids. Scaling laws. *J. Aerosol Sci.* **1997**, *28*, 249–275. [CrossRef]
18. Agarwal, S.; Wendorff, J.H.; Greiner, A. Progress in the field of electrospinning for tissue engineering applications. *Adv. Mater. Deerfield Beach Fla* **2009**, *21*, 3343–3351. [CrossRef]
19. Frenot, A.; Chronakis, I.S. Polymer nanofibers assembled by electrospinning. *Curr. Opin. Colloid Interface Sci.* **2003**, *8*, 64–75. [CrossRef]
20. Oh, J.K.; Drumright, R.; Siegwart, D.J.; Matyjaszewski, K. The development of microgels/nanogels for drug delivery applications. *Prog. Polym. Sci.* **2008**, *33*, 448–477. [CrossRef]
21. Chew, S.Y.; Wen, Y.; Dzenis, Y.; Leong, K.W. The role of electrospinning in the emerging field of nanomedicine. *Curr. Pharm. Des.* **2006**, *12*, 4751–4770. [CrossRef]
22. Lee, J.; Yoo, J.J.; Atala, A.; Lee, S.J. The effect of controlled release of PDGF-BB from heparin-conjugated electrospun PCL/gelatin scaffolds on cellular bioactivity and infiltration. *Biomaterials* **2012**, *33*, 6709–6720. [CrossRef] [PubMed]
23. Pertici, V.; Martrou, G.; Gigmes, D.; Trimaille, T. Synthetic Polymer-based Electrospun Fibers: Biofunctionalization Strategies and Recent Advances in Tissue Engineering, Drug Delivery and Diagnostics. *Curr. Med. Chem.* **2018**, *25*, 2385–2400. [CrossRef]
24. Torres-Martinez, E.J.; Cornejo Bravo, J.M.; Serrano Medina, A.; Pérez González, G.L.; Villarreal Gómez, L.J. A Summary of Electrospun Nanofibers as Drug Delivery System: Drugs Loaded and Biopolymers Used as Matrices. *Curr. Drug Deliv.* **2018**, *15*, 1360–1374. [CrossRef] [PubMed]
25. Son, Y.J.; Kim, W.J.; Yoo, H.S. Therapeutic applications of electrospun nanofibers for drug delivery systems. *Arch. Pharm. Res.* **2014**, *37*, 69–78. [CrossRef] [PubMed]

26. Yu, D.-G.; Zhu, L.-M.; White, K.; Branford-White, C. Electrospun nanofiber-based drug delivery systems. *Health* **2009**, *01*, 67. [CrossRef]
27. Huang, H.-H.; He, C.-L.; Wang, H.-S.; Mo, X.-M. Preparation of core-shell biodegradable microfibers for long-term drug delivery. *J. Biomed. Mater. Res. A* **2009**, *90*, 1243–1251. [CrossRef] [PubMed]
28. Kim, K.; Luu, Y.K.; Chang, C.; Fang, D.; Hsiao, B.S.; Chu, B.; Hadjiargyrou, M. Incorporation and controlled release of a hydrophilic antibiotic using poly(lactide-co-glycolide)-based electrospun nanofibrous scaffolds. *J. Control. Release* **2004**, *98*, 47–56. [CrossRef] [PubMed]
29. Xie, J.; Wang, C.-H. Electrospun micro- and nanofibers for sustained delivery of paclitaxel to treat C6 glioma in vitro. *Pharm. Res.* **2006**, *23*, 1817–1826. [CrossRef] [PubMed]
30. Zeng, J.; Yang, L.; Liang, Q.; Zhang, X.; Guan, H.; Xu, X.; Chen, X.; Jing, X. Influence of the drug compatibility with polymer solution on the release kinetics of electrospun fiber formulation. *J. Control. Release* **2005**, *105*, 43–51. [CrossRef]
31. Singh, H.; Sharma, R.; Joshi, M.; Garg, T.; Goyal, A.K.; Rath, G. Transmucosal delivery of Docetaxel by mucoadhesive polymeric nanofibers. *Artif. Cells Nanomed. Biotechnol.* **2015**, *43*, 263–269. [CrossRef] [PubMed]
32. Zhu, Y.; Pyda, M.; Cebe, P. Electrospun fibers of poly(L-lactic acid) containing lovastatin with potential applications in drug delivery. *J. Appl. Polym. Sci.* **2017**, *134*, 45287. [CrossRef]
33. Li, W.; Luo, T.; Yang, Y.; Tan, X.; Liu, L. Formation of controllable hydrophilic/hydrophobic drug delivery systems by electrospinning of vesicles. *Langmuir* **2015**, *31*, 5141–5146. [CrossRef]
34. Nair, B.P.; Vaikkath, D.; Mohan, D.S.; Nair, P.D. Fabrication of a microvesicles-incorporated fibrous membrane for controlled delivery applications in tissue engineering. *Biofabrication* **2014**, *6*, 045008. [CrossRef]
35. Yohe, S.T.; Herrera, V.L.M.; Colson, Y.L.; Grinstaff, M.W. 3D superhydrophobic electrospun meshes as reinforcement materials for sustained local drug delivery against colorectal cancer cells. *J. Control. Release* **2012**, *162*, 92–101. [CrossRef]
36. Hu, J.; Zeng, F.; Wei, J.; Chen, Y.; Chen, Y. Novel controlled drug delivery system for multiple drugs based on electrospun nanofibers containing nanomicelles. *J. Biomater. Sci. Polym. Ed.* **2014**, *25*, 257–268. [CrossRef]
37. Zhang, X.; Chen, B.; Fu, W.; Fang, Z.; Liu, Z.; Lu, W.; Shi, Z.; Chen, L.; Chen, T. The research and preparation of a novel nano biodegradable polymer external reinforcement. *Appl. Surf. Sci.* **2011**, *258*, 196–200. [CrossRef]
38. Yang, G.; Wang, J.; Wang, Y.; Li, L.; Guo, X.; Zhou, S. An implantable active-targeting micelle-in-nanofiber device for efficient and safe cancer therapy. *ACS Nano* **2015**, *9*, 1161–1174. [CrossRef]
39. Chen, M.; Feng, W.; Lin, S.; He, C.; Gao, Y.; Wang, H. Antitumor efficacy of a PLGA composite nanofiber embedded with doxorubicin@MSNs and hydroxycamptothecin@HANPs. *RSC Adv.* **2014**, *4*, 53344–53351. [CrossRef]
40. Qiu, K.; He, C.; Feng, W.; Wang, W.; Zhou, X.; Yin, Z.; Chen, L.; Wang, H.; Mo, X. Doxorubicin-loaded electrospun poly(L-lactic acid)/mesoporous silica nanoparticles composite nanofibers for potential postsurgical cancer treatment. *J. Mater. Chem. B* **2013**, *1*, 4601–4611. [CrossRef]
41. Yuan, Z.; Pan, Y.; Cheng, R.; Sheng, L.; Wu, W.; Pan, G.; Feng, Q.; Cui, W. Doxorubicin-loaded mesoporous silica nanoparticle composite nanofibers for long-term adjustments of tumor apoptosis. *Nanotechnology* **2016**, *27*, 245101. [CrossRef] [PubMed]
42. Song, J.; Klymov, A.; Shao, J.; Zhang, Y.; Ji, W.; Kolwijck, E.; Jansen, J.A.; Leeuwenburgh, S.C.G.; Yang, F. Electrospun Nanofibrous Silk Fibroin Membranes Containing Gelatin Nanospheres for Controlled Delivery of Biomolecules. *Adv. Healthc. Mater.* **2017**, *6*, 1700014. [CrossRef] [PubMed]
43. Lai, H.-J.; Kuan, C.-H.; Wu, H.-C.; Tsai, J.-C.; Chen, T.-M.; Hsieh, D.-J.; Wang, T.-W. Tailored design of electrospun composite nanofibers with staged release of multiple angiogenic growth factors for chronic wound healing. *Acta Biomater.* **2014**, *10*, 4156–4166. [CrossRef] [PubMed]
44. Mehrasa, M.; Asadollahi, M.A.; Ghaedi, K.; Salehi, H.; Arpanaei, A. Electrospun aligned PLGA and PLGA/gelatin nanofibers embedded with silica nanoparticles for tissue engineering. *Int. J. Biol. Macromol.* **2015**, *79*, 687–695. [CrossRef] [PubMed]
45. Mehrasa, M.; Asadollahi, M.A.; Nasri-Nasrabadi, B.; Ghaedi, K.; Salehi, H.; Dolatshahi-Pirouz, A.; Arpanaei, A. Incorporation of mesoporous silica nanoparticles into random electrospun PLGA and PLGA/gelatin nanofibrous scaffolds enhances mechanical and cell proliferation properties. *Mater. Sci. Eng. C Mater. Biol. Appl.* **2016**, *66*, 25–32. [CrossRef] [PubMed]
46. Aytac, Z.; Ipek, S.; Erol, I.; Durgun, E.; Uyar, T. Fast-dissolving electrospun gelatin nanofibers encapsulating ciprofloxacin/cyclodextrin inclusion complex. *Colloids Surf. B Biointerfaces* **2019**, *178*, 129–136. [CrossRef]

47. Piran, M.; Shiri, M.; Soufi Zomorrod, M.; Esmaeili, E.; Soufi Zomorrod, M.; Vazifeh Shiran, N.; Mahboudi, H.; Daneshpazhouh, H.; Dehghani, N.; Hosseinzadeh, S. Electrospun triple-layered PLLA/gelatin. PRGF/PLLA scaffold induces fibroblast migration. *J. Cell. Biochem.* **2019**. [CrossRef] [PubMed]
48. Gong, T.; Liu, T.; Zhang, L.; Ye, W.; Guo, X.; Wang, L.; Quan, L.; Pan, C. Design Redox-Sensitive Drug-Loaded Nanofibers for Bone Reconstruction. *ACS Biomater. Sci. Eng.* **2018**, *4*, 240–247. [CrossRef]
49. Mohamed, A.; Salama, A.; Nasser, W.S.; Uheida, A. Photodegradation of Ibuprofen, Cetirizine, and Naproxen by PAN-MWCNT/TiO2-NH2 nanofiber membrane under UV light irradiation. *Environ. Sci. Eur.* **2018**, *30*, 47. [CrossRef] [PubMed]
50. Elashnikov, R.; Slepička, P.; Rimpelova, S.; Ulbrich, P.; Švorčík, V.; Lyutakov, O. Temperature-responsive PLLA/PNIPAM nanofibers for switchable release. *Mater. Sci. Eng. C Mater. Biol. Appl.* **2017**, *72*, 293–300. [CrossRef] [PubMed]
51. Pelton, R.H.; Chibante, P. Preparation of aqueous latices with N-isopropylacrylamide. *Colloids Surf.* **1986**, *20*, 247–256. [CrossRef]
52. Zelzer, M.; Todd, S.J.; Hirst, A.R.; McDonald, T.O.; Ulijn, R.V. Enzyme responsive materials: Design strategies and future developments. *Biomater. Sci.* **2012**, *1*, 11–39. [CrossRef]
53. Fernández-Nieves, A.; Fernández-Barbero, A.; Vincent, B.; de las Nieves, F.J. Charge Controlled Swelling of Microgel Particles. *Macromolecules* **2000**, *33*, 2114–2118. [CrossRef]
54. Plamper, F.A.; Richtering, W. Functional Microgels and Microgel Systems. *Acc. Chem. Res.* **2017**, *50*, 131–140. [CrossRef]
55. Fan, Y.; Chen, C.; Huang, Y.; Zhang, F.; Lin, G. Study of the pH-sensitive mechanism of tumor-targeting liposomes. *Colloids Surf. B Biointerfaces* **2017**, *151*, 19–25. [CrossRef]
56. Illangakoon, U.E.; Yu, D.-G.; Ahmad, B.S.; Chatterton, N.P.; Williams, G.R. 5-Fluorouracil loaded Eudragit fibers prepared by electrospinning. *Int. J. Pharm.* **2015**, *495*, 895–902. [CrossRef]
57. Han, D.; Steckl, A.J. Selective pH-Responsive Core-Sheath Nanofiber Membranes for Chem/Bio/Med Applications: Targeted Delivery of Functional Molecules. *ACS Appl. Mater. Interfaces* **2017**, *9*, 42653–42660. [CrossRef] [PubMed]
58. Tran, T.; Hernandez, M.; Patel, D.; Wu, J. Temperature and pH Responsive Microfibers for Controllable and Variable Ibuprofen Delivery. Available online: https://www.hindawi.com/journals/amse/2015/180187/ (accessed on 6 March 2019).
59. Zhang, R.-Y.; Zaslavski, E.; Vasilyev, G.; Boas, M.; Zussman, E. Tunable pH-Responsive Chitosan-Poly(acrylic acid) Electrospun Fibers. *Biomacromolecules* **2018**, *19*, 588–595. [CrossRef] [PubMed]
60. Sang, Q.; Williams, G.R.; Wu, H.; Liu, K.; Li, H.; Zhu, L.-M. Electrospun gelatin/sodium bicarbonate and poly(lactide-co-ε-caprolactone)/sodium bicarbonate nanofibers as drug delivery systems. *Mater. Sci. Eng. C Mater. Biol. Appl.* **2017**, *81*, 359–365. [CrossRef]
61. Jassal, M.; Boominathan, V.P.; Ferreira, T.; Sengupta, S.; Bhowmick, S. pH-responsive drug release from functionalized electrospun poly(caprolactone) scaffolds under simulated in vivo environment. *J. Biomater. Sci. Polym. Ed.* **2016**, *27*, 1380–1395. [CrossRef]
62. Wu, H.; Liu, S.; Xiao, L.; Dong, X.; Lu, Q.; Kaplan, D.L. Injectable and pH-Responsive Silk Nanofiber Hydrogels for Sustained Anticancer Drug Delivery. *ACS Appl. Mater. Interfaces* **2016**, *8*, 17118–17126. [CrossRef] [PubMed]
63. Sivakumaran, D.; Bakaic, E.; Campbell, S.B.; Xu, F.; Mueller, E.; Hoare, T. Fabricating Degradable Thermoresponsive Hydrogels on Multiple Length Scales via Reactive Extrusion, Microfluidics, Self-assembly, and Electrospinning. *J. Vis. Exp.* **2018**, *16*, e54502. [CrossRef]
64. Slemming-Adamsen, P.; Song, J.; Dong, M.; Besenbacher, F.; Chen, M. In Situ Cross-Linked PNIPAM/Gelatin Nanofibers for Thermo-Responsive Drug Release. *Macromol. Mater. Eng.* **2015**, *300*, 1226–1231. [CrossRef]
65. Zhang, H.; Niu, Q.; Wang, N.; Nie, J.; Ma, G. Thermo-sensitive drug controlled release PLA core/PNIPAM shell fibers fabricated using a combination of electrospinning and UV photo-polymerization. *Eur. Polym. J.* **2015**, *71*, 440–450. [CrossRef]
66. Cicotte, K.N.; Reed, J.A.; Nguyen, P.A.H.; De Lora, J.A.; Hedberg-Dirk, E.L.; Canavan, H.E. Optimization of electrospun poly(N-isopropyl acrylamide) mats for the rapid reversible adhesion of mammalian cells. *Biointerphases* **2017**, *12*, 02C417. [CrossRef] [PubMed]

67. Liu, L.; Bakhshi, H.; Jiang, S.; Schmalz, H.; Agarwal, S. Composite Polymeric Membranes with Directionally Embedded Fibers for Controlled Dual Actuation. *Macromol. Rapid Commun.* **2018**, *39*, e1800082. [CrossRef] [PubMed]
68. Li, H.; Sang, Q.; Wu, J.; Williams, G.R.; Wang, H.; Niu, S.; Wu, J.; Zhu, L.-M. Dual-responsive drug delivery systems prepared by blend electrospinning. *Int. J. Pharm.* **2018**, *543*, 1–7. [CrossRef] [PubMed]
69. Li, H.; Liu, K.; Williams, G.R.; Wu, J.; Wu, J.; Wang, H.; Niu, S.; Zhu, L.-M. Dual temperature and pH responsive nanofiber formulations prepared by electrospinning. *Colloids Surf. B Biointerfaces* **2018**, *171*, 142–149. [CrossRef]
70. Hu, J.; Li, H.-Y.; Williams, G.R.; Yang, H.-H.; Tao, L.; Zhu, L.-M. Electrospun Poly(N-isopropylacrylamide)/Ethyl Cellulose Nanofibers as Thermoresponsive Drug Delivery Systems. *J. Pharm. Sci.* **2016**, *105*, 1104–1112. [CrossRef]
71. Li, H.; Liu, K.; Sang, Q.; Williams, G.R.; Wu, J.; Wang, H.; Wu, J.; Zhu, L.-M. A thermosensitive drug delivery system prepared by blend electrospinning. *Colloids Surf. B Biointerfaces* **2017**, *159*, 277–283. [CrossRef]
72. Han, D.; Yu, X.; Chai, Q.; Ayres, N.; Steckl, A.J. Stimuli-Responsive Self-Immolative Polymer Nanofiber Membranes Formed by Coaxial Electrospinning. *ACS Appl. Mater. Interfaces* **2017**, *9*, 11858–11865. [CrossRef]
73. Qi, R.; Tian, X.; Guo, R.; Luo, Y.; Shen, M.; Yu, J.; Shi, X. Controlled release of doxorubicin from electrospun MWCNTs/PLGA hybrid nanofibers. *Chin. J. Polym. Sci.* **2016**, *34*, 1047–1059. [CrossRef]
74. Yu, Y.; Kong, L.; Li, L.; Li, N.; Yan, P. Antitumor Activity of Doxorubicin-Loaded Carbon Nanotubes Incorporated Poly(Lactic-Co-Glycolic Acid) Electrospun Composite Nanofibers. *Nanoscale Res. Lett.* **2015**, *10*, 1044. [CrossRef]
75. Zhang, J.; Zheng, T.; Alarçin, E.; Byambaa, B.; Guan, X.; Ding, J.; Zhang, Y.S.; Li, Z. Porous Electrospun Fibers with Self-Sealing Functionality: An Enabling Strategy for Trapping Biomacromolecules. *Small Weinh. Bergstr. Ger.* **2017**, *13*, 1701949. [CrossRef]
76. Zhang, Z.; Liu, S.; Xiong, H.; Jing, X.; Xie, Z.; Chen, X.; Huang, Y. Electrospun PLA/MWCNTs composite nanofibers for combined chemo- and photothermal therapy. *Acta Biomater.* **2015**, *26*, 115–123. [CrossRef]
77. Barzegar, F.; Bello, A.; Fabiane, M.; Khamlich, S.; Momodu, D.; Taghizadeh, F.; Dangbegnon, J.; Manyala, N. Preparation and characterization of poly(vinyl alcohol)/graphene nanofibers synthesized by electrospinning. *J. Phys. Chem. Solids* **2015**, *77*, 139–145. [CrossRef]
78. Sapountzi, E.; Braiek, M.; Chateaux, J.-F.; Jaffrezic-Renault, N.; Lagarde, F. Recent Advances in Electrospun Nanofiber Interfaces for Biosensing Devices. *Sensors* **2017**, *17*, 1887. [CrossRef]
79. Soares, J.C.; Iwaki, L.E.O.; Soares, A.C.; Rodrigues, V.C.; Melendez, M.E.; Fregnani, J.H.T.G.; Reis, R.M.; Carvalho, A.L.; Corrêa, D.S.; Oliveira, O.N. Immunosensor for Pancreatic Cancer Based on Electrospun Nanofibers Coated with Carbon Nanotubes or Gold Nanoparticles. *ACS Omega* **2017**, *2*, 6975–6983. [CrossRef] [PubMed]
80. Firme, C.P.; Bandaru, P.R. Toxicity issues in the application of carbon nanotubes to biological systems. *Nanomed. Nanotechnol. Biol. Med.* **2010**, *6*, 245–256. [CrossRef]
81. Catalán, J.; Siivola, K.M.; Nymark, P.; Lindberg, H.; Suhonen, S.; Järventaus, H.; Koivisto, A.J.; Moreno, C.; Vanhala, E.; Wolff, H.; et al. In vitro and in vivo genotoxic effects of straight versus tangled multi-walled carbon nanotubes. *Nanotoxicology* **2016**, *10*, 794–806. [CrossRef]
82. Zhang, T.; Tang, M.; Zhang, S.; Hu, Y.; Li, H.; Zhang, T.; Xue, Y.; Pu, Y. Systemic and immunotoxicity of pristine and PEGylated multi-walled carbon nanotubes in an intravenous 28 days repeated dose toxicity study. *Int. J. Nanomed.* **2017**, *12*, 1539–1554. [CrossRef]
83. Jordan, A.; Scholz, R.; Maier-Hauff, K.; van Landeghem, F.K.H.; Waldoefner, N.; Teichgraeber, U.; Pinkernelle, J.; Bruhn, H.; Neumann, F.; Thiesen, B.; et al. The effect of thermotherapy using magnetic nanoparticles on rat malignant glioma. *J. Neurooncol.* **2006**, *78*, 7–14. [CrossRef]
84. Rabias, I.; Tsitrouli, D.; Karakosta, E.; Kehagias, T.; Diamantopoulos, G.; Fardis, M.; Stamopoulos, D.; Maris, T.G.; Falaras, P.; Zouridakis, N.; et al. Rapid magnetic heating treatment by highly charged maghemite nanoparticles on Wistar rats exocranial glioma tumors at microliter volume. *Biomicrofluidics* **2010**, *4*, 024111. [CrossRef]
85. Shinkai, M.; Yanase, M.; Suzuki, M.; Honda, H.; Wakabayashi, T.; Yoshida, J.; Kobayashi, T. Intracellular hyperthermia for cancer using magnetite cationic liposomes. *J. Magn. Magn. Mater.* **1999**, *194*, 176–184. [CrossRef]

86. Hervault, A.; Thanh, N.T.K. Magnetic nanoparticle-based therapeutic agents for thermo-chemotherapy treatment of cancer. *Nanoscale* **2014**, *6*, 11553–11573. [CrossRef]
87. Kobayashi, T. Cancer hyperthermia using magnetic nanoparticles. *Biotechnol. J.* **2011**, *6*, 1342–1347. [CrossRef] [PubMed]
88. Santhosh, P.B.; Ulrih, N.P. Multifunctional superparamagnetic iron oxide nanoparticles: Promising tools in cancer theranostics. *Cancer Lett.* **2013**, *336*, 8–17. [CrossRef]
89. Müller, R.H.; Maaβen, S.; Weyhers, H.; Specht, F.; Lucks, J.S. Cytotoxicity of magnetite-loaded polylactide, polylactide/glycolide particles and solid lipid nanoparticles. *Int. J. Pharm.* **1996**, *138*, 85–94. [CrossRef]
90. Weissleder, R.; Bogdanov, A.; Neuwelt, E.A.; Papisov, M. Long-circulating iron oxides for MR imaging. *Adv. Drug Deliv. Rev.* **1995**, *16*, 321–334. [CrossRef]
91. Kaminski, M.D.; Rosengart, A.J. Detoxification of blood using injectable magnetic nanospheres: A conceptual technology description. *J. Magn. Magn. Mater.* **2005**, *293*, 398–403. [CrossRef]
92. Feng, Z.-Q.; Shi, C.; Zhao, B.; Wang, T. Magnetic electrospun short nanofibers wrapped graphene oxide as a promising biomaterials for guiding cellular behavior. *Mater. Sci. Eng. C Mater. Biol. Appl.* **2017**, *81*, 314–320. [CrossRef]
93. Huang, C.; Soenen, S.J.; Rejman, J.; Trekker, J.; Chengxun, L.; Lagae, L.; Ceelen, W.; Wilhelm, C.; Demeester, J.; Smedt, S.C.D. Magnetic Electrospun Fibers for Cancer Therapy. *Adv. Funct. Mater.* **2012**, *22*, 2479–2486. [CrossRef]
94. Sasikala, A.R.K.; Unnithan, A.R.; Yun, Y.-H.; Park, C.H.; Kim, C.S. An implantable smart magnetic nanofiber device for endoscopic hyperthermia treatment and tumor-triggered controlled drug release. *Acta Biomater.* **2016**, *31*, 122–133. [CrossRef] [PubMed]
95. Song, C.; Wang, X.-X.; Zhang, J.; Nie, G.-D.; Luo, W.-L.; Fu, J.; Ramakrishna, S.; Long, Y.-Z. Electric Field-Assisted in Situ Precise Deposition of Electrospun γ-Fe_2O_3/Polyurethane Nanofibers for Magnetic Hyperthermia. *Nanoscale Res. Lett.* **2018**, *13*, 273. [CrossRef] [PubMed]
96. Radmansouri, M.; Bahmani, E.; Sarikhani, E.; Rahmani, K.; Sharifianjazi, F.; Irani, M. Doxorubicin hydrochloride—Loaded electrospun chitosan/cobalt ferrite/titanium oxide nanofibers for hyperthermic tumor cell treatment and controlled drug release. *Int. J. Biol. Macromol.* **2018**, *116*, 378–384. [CrossRef]
97. Zhang, C.-L.; Cao, F.-H.; Wang, J.-L.; Yu, Z.-L.; Ge, J.; Lu, Y.; Wang, Z.-H.; Yu, S.-H. Highly Stimuli-Responsive Au Nanorods/Poly(N-isopropylacrylamide) (PNIPAM) Composite Hydrogel for Smart Switch. *ACS Appl. Mater. Interfaces* **2017**, *9*, 24857–24863. [CrossRef]
98. Nolsøe, C.P.; Torp-Pedersen, S.; Burcharth, F.; Horn, T.; Pedersen, S.; Christensen, N.E.; Olldag, E.S.; Andersen, P.H.; Karstrup, S.; Lorentzen, T. Interstitial hyperthermia of colorectal liver metastases with a US-guided Nd-YAG laser with a diffuser tip: A pilot clinical study. *Radiology* **1993**, *187*, 333–337. [CrossRef]
99. O'Neal, D.P.; Hirsch, L.R.; Halas, N.J.; Payne, J.D.; West, J.L. Photo-thermal tumor ablation in mice using near infrared-absorbing nanoparticles. *Cancer Lett.* **2004**, *209*, 171–176. [CrossRef]
100. Sultan, R.A. Tumour ablation by laser in general surgery. *Lasers Med. Sci.* **1990**, *5*, 185–193. [CrossRef]
101. Hirsch, L.R.; Stafford, R.J.; Bankson, J.A.; Sershen, S.R.; Rivera, B.; Price, R.E.; Hazle, J.D.; Halas, N.J.; West, J.L. Nanoshell-mediated near-infrared thermal therapy of tumors under magnetic resonance guidance. *Proc. Natl. Acad. Sci. USA* **2003**, *100*, 13549–13554. [CrossRef]
102. Huang, X.; El-Sayed, I.H.; Qian, W.; El-Sayed, M.A. Cancer cell imaging and photothermal therapy in the near-infrared region by using gold nanorods. *J. Am. Chem. Soc.* **2006**, *128*, 2115–2120. [CrossRef]
103. Huang, X.; Jain, P.K.; El-Sayed, I.H.; El-Sayed, M.A. Gold nanoparticles: Interesting optical properties and recent applications in cancer diagnostics and therapy. *Nanomedicine* **2007**, *2*, 681–693. [CrossRef]
104. Huff, T.B.; Tong, L.; Zhao, Y.; Hansen, M.N.; Cheng, J.-X.; Wei, A. Hyperthermic effects of gold nanorods on tumor cells. *Nanomedicine* **2007**, *2*, 125–132. [CrossRef]
105. Hu, M.; Chen, J.; Li, Z.-Y.; Au, L.; Hartland, G.V.; Li, X.; Marquez, M.; Xia, Y. Gold nanostructures: Engineering their plasmonic properties for biomedical applications. *Chem. Soc. Rev.* **2006**, *35*, 1084–1094. [CrossRef]
106. Gerber, A.; Bundschuh, M.; Klingelhofer, D.; Groneberg, D.A. Gold nanoparticles: Recent aspects for human toxicology. *J. Occup. Med. Toxicol. Lond. Engl.* **2013**, *8*, 32. [CrossRef]
107. Goodman, C.M.; McCusker, C.D.; Yilmaz, T.; Rotello, V.M. Toxicity of gold nanoparticles functionalized with cationic and anionic side chains. *Bioconjug. Chem.* **2004**, *15*, 897–900. [CrossRef] [PubMed]

108. Cho, W.-S.; Cho, M.; Jeong, J.; Choi, M.; Cho, H.-Y.; Han, B.S.; Kim, S.H.; Kim, H.O.; Lim, Y.T.; Chung, B.H.; et al. Acute toxicity and pharmacokinetics of 13 nm-sized PEG-coated gold nanoparticles. *Toxicol. Appl. Pharmacol.* **2009**, *236*, 16–24. [CrossRef] [PubMed]
109. Du, Y.; Xia, L.; Jo, A.; Davis, R.M.; Bissel, P.; Ehrich, M.F.; Kingston, D.G.I. Synthesis and Evaluation of Doxorubicin-Loaded Gold Nanoparticles for Tumor-Targeted Drug Delivery. *Bioconjug. Chem.* **2018**, *29*, 420–430. [CrossRef] [PubMed]
110. Irani, M.; Sadeghi, G.M.M.; Haririan, I. The sustained delivery of temozolomide from electrospun PCL-Diol-b-PU/gold nanocompsite nanofibers to treat glioblastoma tumors. *Mater. Sci. Eng. C Mater. Biol. Appl.* **2017**, *75*, 165–174. [CrossRef]

© 2019 by the authors. Licensee MDPI, Basel, Switzerland. This article is an open access article distributed under the terms and conditions of the Creative Commons Attribution (CC BY) license (http://creativecommons.org/licenses/by/4.0/).

Review

Sputtering of Electrospun Polymer-Based Nanofibers for Biomedical Applications: A Perspective

Hana Kadavil, Moustafa Zagho, Ahmed Elzatahry and Talal Altahtamouni *

Materials Science and Technology Program, College of Arts and Sciences, Qatar University, P.O. Box 2713, Doha, Qatar; hk1002981@student.qu.edu.qa (H.K.); mmsalah@qu.edu.qa (M.Z.); aelzatahry@qu.edu.qa (A.E.)
* Correspondence: taltahtamouni@qu.edu.qa; Tel.: +974-4403-6809

Received: 24 October 2018; Accepted: 13 November 2018; Published: 8 January 2019

Abstract: Electrospinning has gained wide attention recently in biomedical applications. Electrospun biocompatible scaffolds are well-known for biomedical applications such as drug delivery, wound dressing, and tissue engineering applications. In this review, the synthesis of polymer-based fiber composites using an electrospinning technique is discussed. Formerly, metal particles were then deposited on the surface of electrospun fibers using sputtering technology. Key nanometals for biomedical applications including silver and copper nanoparticles are discussed throughout this review. The formulated scaffolds were found to be suitable candidates for biomedical uses such as antibacterial coatings, surface modification for improving biocompatibility, and tissue engineering. This review briefly mentions the characteristics of the nanostructures while focusing on how nanostructures hold potential for a wide range of biomedical applications.

Keywords: electrospinning; sputtering; drug delivery; wound dressing; biocompatibility; tissue engineering

1. Introduction

In the last 20 years, the emergence of nanotechnology has drawn much attention to the electrospinning process. This process is used for the preparation of polymer Nano-micro fibers, and it has great importance in the biomedical industry due to its cost-effectiveness, scalability, versatility, and simplicity. The process was developed in 1901 by JF Cooley and WJ Morton, but had slower development over the subsequent 100 years. In more recent years, Reneker [1] developed fiber preparation from an organic polymer, which created a new field of science for the formulation of fiber diameter ranging between 1×10^{-9} and 1×10^{-6} m.

Electrospinning devices include four main components (Figure 1): high power supply, syringe pump, syringe needle with solutions, and a collector for fiber deposition. The electric field is applied between the needle and the collector, where the positive electrode is connected to the needle and the negative electrode to the collector. Hence, when voltage is applied, the repulsive charge accumulates at the tip of the needle, which is shaped in the form of a hemisphere [2]. When the repulsive charge overcomes the surface tension, it leads to the formation of a Taylor cone. This directs the polymer solution to the negative electrode, as the collector, and produces fibers. The solvent from the polymer solution is evaporated, and the polymer solution is deposited on the collector as dry fibers ranging in size from nanometers to micrometers [3].

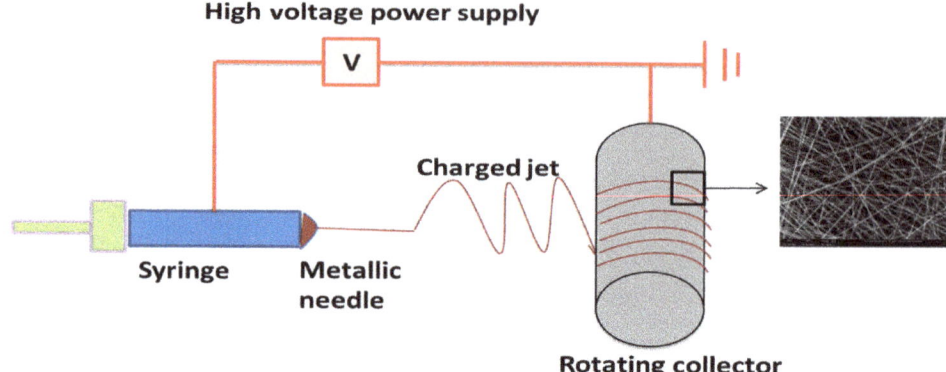

Figure 1. Schematic representation of electrospinning apparatus with a rotating collector. Reproduced with permission from [Coatings]. Copyright MDPI, 2014.

1.1. Electrospinning Parameters

There are many parameters controlling the electrospinning process. These include solution parameters and processing parameters. The solution parameters include the concentration of the polymer solution, molecular weight, viscosity, surface tension, solvents, and conductivity/surface charge density. Additionally, processing parameters such as voltage, collector/needle distance, flow rate and the diameter of the syringe, along with ambient parameters like humidity and temperature, also play a major role in the fabrication of nanofibers for electrospinning process [4]. Customized electrospun fibers are produced by changing the parameters above [5]. Understanding these parameters is necessary in order to optimize the fiber structure.

1.1.1. Solution Parameters

The solution parameters are important criteria for the production of uniform fibers. Of these, concentration is the most powerful tool for governing the morphology of the electrospun nanofibers. Electrospraying occurs when the concentration of the solution is too low, resulting in the development of micro- or nanoparticles [6]. However, smooth fibers are established when the concentration is increased [7]. However, increase in the concentration beyond a certain point will lead to the formation of a ribbon-like morphology. Simultaneously, higher molecular weight also leads to the ribbon-like structure for the nanofibers, even at a lower concentration of fibers [8]. Surface tension is also an important parameter, and depends on solvent composition. It has been noted that a reduction in the surface tension of the solution can lead to smooth fibers at a fixed concentration due to a smaller electric field being required to overcome the surface tension for the formation of uniform Taylor cones [9,10]. Concurrently, viscosity also performs a critical role in the electrospinning process. A lower-viscosity polymeric solution cannot be electrospun, but higher-viscosity solutions also cause problematic ejection jets in the solution. Therefore, viscosity adjustment is also an important factor for continuous and uniform fiber formation with continuous Taylor cones [11]. Additionally, conductivity also plays an important role in fiber formation. Conductivity depends on the type of polymers, solvents, and addition of salts. It has been found in the literature that, the increment in the conductivity leads to thinner and more uniform fiber formation [12]. All solution parameters are linked with one another [13]. It follows that solution optimization is an important step for the design of uniform and continuous fiber morphologies in the electrospinning process.

1.1.2. Processing Parameters

Processing parameters are also an important factor in the electrospinning process. These include voltage, tip–collector distance, flow rate, type of collector, and electric field. Voltage influences the anatomical morphology of the nanofibers due to the dependency on fluid flow dynamics [14]. Many authors have found a non-linear relationship between fiber diameters and voltage [15]. The level of impact of voltage on the fiber diameter is known; however, polymer concentration and tip-to-collector distance play a more significant role than voltage [16]. Tip-to-collector distance has shown direct proportionality with fiber diameter due to solvent evaporation at larger distances exhibiting thinner fibers, and vice versa [17]. Furthermore, flow rate also plays a key role in fiber morphology. As the flow rate increases, the fiber diameter increases directly with thicker beads [13]. The type of collector also influences the fiber morphology. There are different types of collectors, including pins [13], girded bars, and rotating rods/wheels [18]. It has been noted that a smaller collection area shows a negative effect on fiber morphology with respect to the formation of beads [19]. The electric field plays a very important role in the electrospinning process. The electric field can be approximated as the voltage divided by the distance between the needle and the collector [20–22].

1.1.3. Ambient Parameters

Environmental conditions also affect the morphology of the electrospun fibers. For instance, temperature, pressure, and humidity play a unique role. For example, electrospinning under high vacuum may lead to higher electric fields, and thus the formation of larger fiber diameters [23]. On the other hand, Supaphol et al. observed thinner fiber of polyamide-6 at 60 °C than at 30 °C [24], suggesting that an increase in temperature leads to a reduction in fiber diameter [24]. In addition, surface pores became apparent when electrospinning in an atmosphere with relative humidity higher than 30% [25].

The purpose of this review is to highlight the benefits of employing electrospun and sputtered electrospun polymer nanofibers in current biomedical applications connected to electrospinning and sputtering technologies. Furthermore, a novel approach is discussed for the processing of fibrous materials for different biomedical applications by combining electrospinning and sputtering technologies.

2. Biomedical Applications of Polymer-Based Electrospun Nanofibers

It is worth noting that polymer-based composites are widely used in a variety of different applications [26–37]. The fibers prepared by the electrospinning process have a high surface-to-volume ratio, adjustable porosity, tailored composition, and other favorable properties. To take advantage of this, a wide variety of polymers can be electrospun, including natural polymers, synthetic polymers, and biodegradable polymers. These micro/nanofibrous polymers have several advantages, including the fact that the fiber scaffold mimics the extracellular matrix, thereby enhancing cell adhesion, proliferation, migration, and differentiation. On the other hand, the pore size of electrospun membranes is too small to house cells inside the pores, and the cells spread out on the surface of the material [38]. This makes possible the release of biofactors such as drugs, proteins, and genes, as well as promoting nutrient and oxygen diffusion and waste removal. Also, the morphology of electrospun nanofibers—including core/shell, hollow, nanowire-micro tubers, and three-dimensional fiber scaffold morphologies—can be modified by changing the parameters of the electrospinning process. Thus, these beneficial factors make electrospun nanofibers suitable for biomedical applications such as drug delivery, tissue engineering, and wound healing.

2.1. Drug Delivery

The idea of drug delivery emerged in the 1970s for the controlled release of drug for treatment [39]. The high surface area and porosity of polymer fibers has attracted great attention in recent years for use as a drug carrier. The use of the electrospinning technique can modify polymer fiber morphology and bulk properties. In this process, polymer nanofibers loaded with drugs are synthesized for drug delivery. Drugs ranging from antibiotic and anticancer agents to proteins, aptamer, DNA [40] and RNA [41] have been incorporated into nanofibers. The release mechanism of drugs in polymer fibers can be altered by changing the type of drug loadings employed, which include co-axial electrospinning, emulsion electrospinning, multiple layers, blended electrospinning, co-electrospinning, etc. However, co-axial electrospinning and multi-layered electrospun fibers have shown great application in drug delivery due to a sustained release of the drug, rather than an initial burst release of drug from the fiber scaffolds. Therefore, recent advancement in the field of electrospinning for drug delivery will be discussed in the proceeding paragraphs. The discussion of electrospinning in drug delivery will be broken down into categories of drug loading types, drug loading materials, types of drugs, and mathematical modeling of drug delivery systems, as explained in detail below.

2.1.1. Drug Loading Types

Electrospinning has different drug loading types, which determine the diverse structure and drug release kinetics. The drug loading procedure in electrospinning can be executed in different ways, including co-electrospinning, multi electrospinning, side-by-side electrospinning, co-axial, surface immobilization and emulsion electrospinning (Figure 2). In co-electrospinning, the drug molecules are mixed with the polymer solution before electrospinning. These electrospun fibers provide a uniform distribution of drugs/biomolecules and high drug/biomolecule loading. However, the biomolecule properties can be negatively affected when they are directly exposed to high voltage. On the other hand, blend electrospinning and side-by-side electrospinning help to solve the issue of drug and molecule solubility in common solvents. Moreover, multi-jets with more than two spinnerets represent a way of protecting the bioactivity of the drug. In addition, surface immobilization is another method, in which drug molecules are covalently bonded with the scaffolds via chemical or physical immobilization methods. In these chemical methods, the surfaces of the nanofibers are changed by introducing amines, carboxyl, hydroxyl or thiol; the physical methods involve the incorporation of Van der Waals, electrostatic and hydrophobic interactions. These methods of immobilization retain biomolecular activity, but the drug molecules in all electrospinning processes exhibit burst release kinetics. To overcome this, co-axial and emulsion electrospinning processes were introduced. Co-axial and emulsion electrospinning processes have been gaining increasing interest recently due to their promising ability to shield the biomolecules with the core and to minimize the drawback of the initial burst. They provide sustained release of the drug by minimizing the initial burst release by controlling the thickness and composition of the shells. The core/shell structure is generated utilizing a single-nozzle electrospinning unit employing emulsion input, commonly named emulsion electrospinning [42]. Another means of drug loading for sustained release of a drug is layer by layer, via the addition of drug in between the electrospun scaffolds; controlled release is promoted by the shield provided by the layer of polymer scaffolds. Therefore, co-axial electrospinning and multilayer electrospinning drug loading types provide a more sustained and controlled release of the drug.

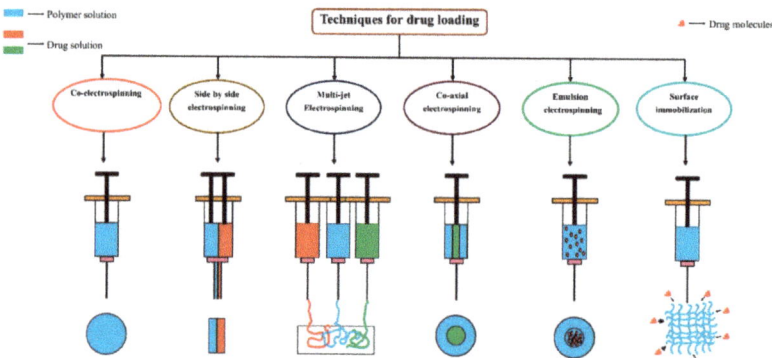

Figure 2. Schematic representation of different types of electrospinning processes. Reproduced with permission from [42]. Royal Society of Chemistry, 2015.

2.1.2. Multiple Layered Fiber Mats

Multilayered fiber mats provide controlled release of a drug through the layer-by-layer stacking of a nanofiber sandwich, with drug loaded in between. This type of design is straightforward, easily controllable, and with a simple fabrication process as compared to the core/shell design. The drug release mechanism of multilayer fibrous mats can be controlled by adjusting the thickness of the outer layer, the amount of drug loaded, the porosity of the scaffolds, etc. The design of a core/shell structure is a complicated process, in one sense, due to the diverse electrical and rheological properties, such as conductivity and surface tension, of the core and shell polymer materials [43]. Hence, due to the difficulty of fabricating a core/shell design, electrospun fiber mats are not able to achieve sustained repeatability; additionally, controlled release of drug from the structure is difficult to be investigated efficiently. GeunHyung Kim [44] prepared polycaprolactone (PCL)-PEO-PCL layered fiber mats, and drug delivery was examined with various thicknesses of PCL outer layer. It was shown that burst release can be avoided by increasing the thickness of the PCL layer, as well as by incorporating antimicrobial peptide HPA3NT3, which does not lose its biological activities (Figure 3).

On the other hand, sustained release of the drug haloperidol was investigated by changing the hydrophobicity of the scaffolds. Therein, polyvinyl alcohol (PVA)-methylated b-cyclodextrin was incorporated with PLA and PLGA. The addition of b-cyclodextrin reduced the fiber degradation rate of PVA [45]. It was noted that when the hydrophobicity of the scaffold was increased, the release of the hydrophilic drug was sustained in a controlled manner, while polyester polymers released the drug by means of hydrolysis. The blending of the hydrophobic and hydrophilic drug will minimize the toxicity caused by the burst release of the drug. This type of combination can be applied for hydrophobic and heat-sensitive drugs, due to the simplicity of the process.

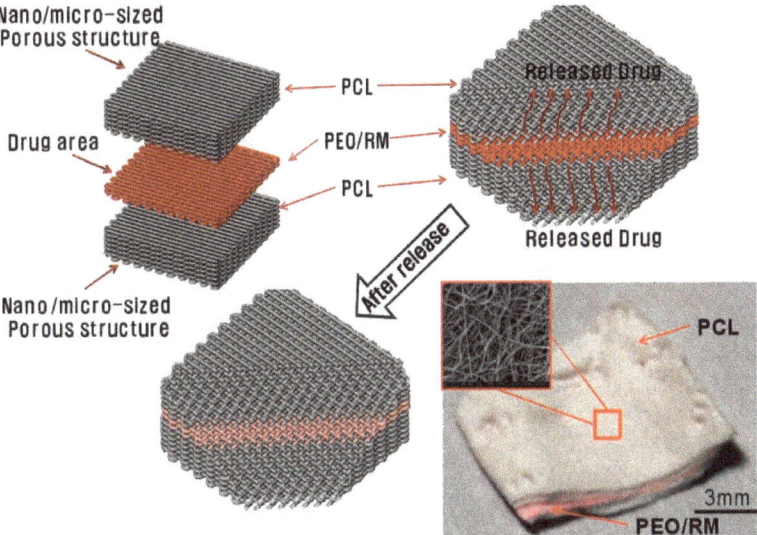

Figure 3. Schematic of the new drug release system consisting of two different electrospun mats. The inner and outer parts of the mat were PEO/rhodamine and PCL fibers, respectively. The red color shows that the rhodamine was well embedded in the PEO/rhodamine mat. Reproduced with permission from [44]. Springer Nature, 2010.

The drug delivery of ibuprofen from sandwich-layered fiber mats was studied, and its mathematical modeling was elaborated by using applying the power law, the Higuchi equation, and Fick's second law [46]. The mathematical modeling suggests that the thickness of the fiber mats have a greater impact on drug delivery than the concentration of the loaded drug. Here, PLA was successfully electrospun by incorporating ibuprofen drug in between the two layers of PLA. Finally, according to the type of treatment, the drug loading can be changed by altering the thickness of the layers for controlled release of the drug. Dave Wei-Chih Chen et al. [47] studied drug delivery of vancomycin, gentamicin, and lidocaine for wound-healing applications. In their research, they successfully mixed PLGA/collagen on the outer layer and PLGA loaded with a drug in the middle layer. The drugs vancomycin and gentamicin were released in high concentrations from the biodegradable polymer scaffold. However, lidocaine showed a release time of up to 3 weeks. The bioactivities of the drug were shown to exhibit 40–100% efficiency, and it was concluded that this scaffold was suitable for boosting the wound-healing process in the initial stage of the wound.

2.1.3. Drug Loading Materials

Varieties of polymers can be electrospun into diverse designs for drug delivery applications, taking account of polymer–drug compatibility and their ability to be molded to fit a range of delivery routes. When designing an optimized drug delivery system, there are many polymer factors to be considered. For instance, biocompatibility, biodegradability, mechanical properties and hydrophilicity [48]. There are many polymer varieties, such as natural and synthetic polymers, that are used for designing drug delivery systems [49,50]. A diverse range of drugs have been loaded into delivery systems, including growth factors, DNA, proteins, inhibitors, and antibiotics [33–35].

Electrospinning processes can be applied to synthetic polymers easily and with great flexibility. However, synthetic polymers affect cell affinity due to their hydrophobic nature and the smooth surfaces of their cell recognition sites. On the other hand, natural polymers show enhanced biocompatibility, and some exhibit antibacterial properties and better clinical functionality.

The group of natural polymers includes cellulose, chitosan, chitin, dextrose, collagen, silk, gelatin, etc. [51]. Lee et al. investigated the features of different polysaccharides upon electrospinning, as well as their biomedical applications, such as drug delivery, wound dressings and enzyme immobilization [52]. The studied polysaccharidses included cellulose, chitosan, alginate, chitin, starch, hyaluronic acid, dextran, and heparin. Chitosan polymer had anticancer properties due to its polycationic nature.

The quartininized form of chitosan is well known for its improved in vitro anticancer ability against Hep3B, HeLa and SW480 cells [53]. However, natural polymers lack mechanical strength, and have a relatively sudden degradation rate due to their hydrophilic nature, inhibiting their use in long-term drug delivery process. In addition, the disadvantages of immunogenicity, batch-to-batch differences, limited availability, expensive production and vulnerability to cross-contamination all limit their clinical application [54].

On the other hand, the limitations of natural polymers could be overcome in application through the use of synthetic polymers, which mainly include biodegradable polymers such as PCL, PVA, polylactic acid (PLA) and Polyglycolic acid (PLGA). These synthetic polymers can be degraded via enzymolysis or hydrolysis. These materials are therefore of great importance in drug delivery, as drug delivery for the tissue regeneration process can take time; also, tissue regeneration can occur [55]. The rate of degradation depends on the sustained release of the drug, such that the degradation rate can be controlled by changing parameters such as the ratio of amorphous to crystalline segments of polymers and polymer blend compositions [41,42]. Synthetic polymers have many advantages in comparison to natural polymers, as they are inexpensive, have excellent mechanical properties and tunable degradation, as well as exhibiting great durability. However, they also have disadvantages, such as lack of cell-specific recognition sites due to their smooth and hydrophobic surfaces.

The production of novel composite fibers through the combination of synthetic and natural polymers could reduce the disadvantages [56,57]. The combination of natural and synthetic polymers would help in the formation of a fiber that was the same as the extracellular matrix, with outstanding mechanical properties and adjustable biodegradability. For example, PLGA-gelatin was fabricated by blending electrospinning for the drug delivery of fenbufen (FBF) [58]. These blended scaffolds have optimized mechanical properties, degradation rates and bioactivites. However, the drug release profile could be controlled by increasing the volume of PLGA in the blend. This would make the scaffolds more hydrophobic, resulting in a slower degradation rate. In another paper, composite scaffolds were prepared through a combination of PCL-gelatin, resulting, because PCL is a hydrophobic polymer, in tunable hydrophobicity, degradation rate, and mechanical properties.

Simultaneously, gelatin provided cellular attachment and adhesion of bone marrow derived from human mesenchymal stem cells (hMSCs). Thus, these types of tunable properties could result in promising scaffolds for drug delivery applications and tissue engineering systems [59]. While designing a system for the sustained release of a drug, many factors contribute to the efficient release of drug from the polymer scaffolds. These elements include the degradation and wettability of the polymer scaffolds, the type of drug and the drug loading type.

For the sustained release of the drug, the most important factor is the drug loading type. There are many types of loading, including co-axial electrospinning and multilayer electrospinning, which shows a controlled release of the drug over a longer term. The sustained release of the drug depends on the following factors in coaxial electrospinning: the thickness of the shell layer, porosity, degradation rate of the shell fiber, the hydrophobicity of the scaffolds, etc. On the other hand, in multilayered electrospinning, the drug release kinetics depends on the scaffold porosity, the thickness of the outer layer, the hydrophilicity of the scaffold, etc. The following sections describe co-axial electrospinning and multilayered electrospun scaffolds prepared by PVA hydrophilic and PCL hydrophobic polymers incoporating various drugs.

Polycaprolactone (PCL)

Polycaprolactone (PCL) is a hydrophobic polyester polymer widely studied in electrospinning. PCL has wide biomedical applications due to its biocompatibility, biodegradability, mechanical properties, non-toxicity, low cost, and low melting point. Commercially available PCL has a molecular weight ranging between 3000 and 85,000 g/mol. PCL is a hydrophobic molecule. Hence, it dissolves in solvents like chloroform, acetone, acetic acid, dichloromethane, toluene, methanol, benzene and tetrachloride [10]. The properties of PCL, including biodegradability, cytotoxicity and degradation rate, have been studied elaborately with respect to short- and long-duration implants [60,61]. Degradation of PCL is non-enzymatic, and occurs by means of hydrolysis. PCL fibers have been widely studied as a drug carrier in drug delivery.

Polyvinyl Alcohol (PVA)

Polyvinyl alcohol (PVA) is a semicrystalline hydrophilic polymer that is easily soluble in water. The solubility in water gives PVA wide applicability in drug delivery [62,63]. PVA is a biocompatible, biodegradable and easily electrospinnable polymer. PVA has been used as a sacrificing template for the preparation of non-electrospinnable polymers. However, PVA alone cannot be used for drug delivery due to its water solubility. PVA was fused with chitosan to improve its biocompatibility and cell attachment [47,48]. Gelatin electrospun with PVA was used as a template for improved gelatin fibers [64]. However, PVA has poor mechanical properties. Therefore, many scientists have tried to study composite materials that might enhance the mechanical properties of PVA [64]. To avoid the burst release of drugs, Zupančič et al. synthesized core/shell nanofibers with a poly(methylmethacrylate) (PMMA) shell and a monolithic PVA core, or novel core/shell nanofibers with a blended PMMA/PVA core loaded with ciprofloxacin hydrochloride (CIP) [65]. The combination of PVA with PCL polymer has gained much attention recently, because the addition of PCL might enhance its mechanical characteristics. Therefore, the study of PCL/PVA as a multilayer scaffolds for the sustained and controlled release of drug is described below.

Combined PCL/PVA

Multilayered structures have gained much attention due to their versatility and controlled release of drugs. The drug was studied as a middle layer, and the outer layer requires the controlled release of antibiotics. For instance, Liu et al. [66] prepared a novel scaffold by integrating a 3D bioprinting platform and electrospinning in order to study multiple drug delivery. Here, PVA blended with gentamicin sulfate and co-axial PVA-DFO/PCL was fused layer by layer to form a 3D scaffold for osteointegration and sustained drug release. Burst release was noted for gentamicin sulfate, but the sustained and controlled release of DFO due to the presence of a vertical gradient of sodium alginate/gelatin in the scaffold give the DFO a gradient mode of release. Therefore, a combination of 3D bioprinting and electrospinning can be used to prepare functional gradient scaffolds. In another study, the release of the model drugs tetracycline hydrochloride (TC-HCL) and phenytoin sodium from PVA-PCL-PVA multilayered electrospun nanofibers was reported [67]. Hydrophilic and hydrophobic polymers were prepared layer by layer by incorporating multiple drugs such as PHT-Na with OVA and TC-HCL with PCL, respectively. 87% of the TC-HCL was released from a single fiber, and only 47% was released from the multilayer scaffolds. The release kinetics mechanism was Fickian diffusion, and the release profile corresponded to the Korsmeyer-Peppas equation. These materials had great application in wound dressing mats. Multilayered electrospun fiber scaffolds have great importance in drug delivery.

2.1.4. Types of Applied Drugs

Antibiotics and Antibacterial Agents

Antibiotics and antibacterial agents have been incorporated for the enhancement of scaffold properties. Ignatova et al. studied the use of several antibiotics in electrospun scaffolds and their application for wound dressings [68]. The antibiotics included tetracycline hydrochloride, ciprofloxacin, moxifloxacin, levofloxacin and antibacterial agents (for example, 8-hydroquioline derivatives, benzalkonium chloride, itraconazole, fusidic acid, and silver nanoparticles (Ag NPs)). Gentamicin sulfate-loaded PLGA and gelatin were also studied for the continuous release of drugs [69]. The results showed that 70/30 PLGA/gelatin nanofiber scaffolds exhibited a gradual release of the drug over the first 15 h, rather than a burst release effect, indicating that this is a promising scaffold for wound healing applications. On the other hand, the drug release profile was studied for a polyethylene covinyl acetate and PLA blend scaffold in which tetracycline hydrochloride was the model drug [70]. The drug delivery release profile depends on the type of fiber and percentage of drug content. The 50/50 blend provided about 5% release of tetracycline hydrochloride within 5 h, with a regulated and smooth release thereafter. Additionally, 25 wt% exhibited a more rapid release than 5 wt% due to the surface segregation of tetracycline, which dissolves quickly.

Zhang et al. [71] electrospun nylon 6 nanofibers and electrosprayed TiO_2 NPs onto them to fabricate highly porous photocatalytic TiO_2 NP-decorated nanofibers with excellent antibacterial behaviors. Moreover, they also prepared solution-blown soy protein nanofibers decorated with Ag NPs. Another type of antibacterial electrospun nanofiber prepared from sodium alginate (SA)/PVA was discussed by Shalumon et al. [72]. Incorporating ZnO NPs increased the diameter of the prepared fibers. Antibacterial examinations confirmed that the processed mats displayed inhibition of both bacterial strains for all contents of ZnO NPs, and that the inhibition increased with an increase in the ZnO NP content [72].

Unnithan et al. prepared uniform nanofibers of polyurethane–dextran loaded with ciprofloxacin drug. The cell attachment and viability were improved after adding dextran to the polyurethane. The nanofibers displayed a good antibacterial activity for both Gram-positive and Gram-negative bacteria [73]. In addition, a biocompatible composite based on chitosan/collagen exhibited high liquid absorption and good antibacterial activity [74].

Anticancer Agents

Not only antibiotics, but also many other types of drugs, such as anticancer drugs, have been applied to the scaffolds of electrospun mats for chemotherapy. Diverse anticancer drugs, such as doxorubicin (Dox), paclitaxel (PTX), dichloroacetate and platinum complexes have been incorporated into the electrospun fibers for localized postoperative chemotherapy sessions. For instance, Xu et al. fabricated PEG-PLLA-loaded electrospun fibers via an water-in-oil emulsion method in which the aqueous phase was the hydrophilic drug, and the oily phase was the chloroform solution of PEG-PLLA [75]. The drug was well and uniformly dispersed in the PEG-PLLA fibers by using the electrospinning technique. In the same way, they successfully incorporated hydrophobic Paclitaxel (PTX) and DOX, which were simultaneously added to the nanofiber scaffolds via the emulsion electrospinning method; subsequently, multiple drug delivery was studied [76]. In contrast, Xe et al. prepared an electrospun scaffold of (30/70) PLA/PLGA blended fiber with the addition of cisplatin, and the results showed a 90% encapsulation efficiency; the sustained release of drug was noted for 75 days in the in vitro treatment of glioma [77].

Protein, DNA, RNA and Other Growth Factors

Over time, electrospinning has improved, thereby propagating many new and innovative ideas for biomedical applications. Blend electrospinning and co-axial electrospinning have been developed with the combination of protein, DNA, RNA and growth factors with the electrospun fiber mats for biomedical applications. The main challenge faced in this type of design is the loss in the bioactivity of the drug incorporated. Therefore, it is mandatory to optimize the material and electrospinning parameters for efficient results. Hence, the processes of blend electrospinning and co-axial electrospinning have drawn more interest towards this specific type of drug addition. Co-axial electrospinning is more efficient for protecting the bioactivity of the drug than blend electrospinning. Chew et al. encapsulated the human nerve growth factor, with BSA as a carrier, in polymers such as PCL and poly(ethyl ethylene phosphate) [78]. The results showed that there was a partial bioactive retention of the hNGF when the PC12 cell line was introduced to the scaffolds. There was a consistent release of hNGF for around three months, without burst release. The same group studied the release of small interfering RNA (snRNA) and transfection reagent (TKO) on electrospun fibers of copolymer caprolactone and ethyl ethylene phosphate (PCLEEP) [79]. The results showed a sustained release of siRNA for around 28 days. The copolymerization of ethyl ethylene phosphate with PCL led to improvements in the delivery rate of siRNA, as well as in gene knockdown efficiency, when compared to PCL alone. In co-axial electrospinning, the bioactive components are incorporated inside the core and are protected by the shell polymer. Hence, bioactivity can be protected from the electrospinning environment and the biological environment. Saraf and co-workers studied the incorporation of plasmid DNA (pDNA) into the core and shell polymers with non-viral gene carrier poly(ethleamine)-hyalouric acid (PEI-HA) [80]. The gene release was observed to last around 60 days by altering parameters such as the concentration of pDNA and the molecular weight of the core in order to control the transfer efficiency of the pDNA. The bioactivity of the drug could be controlled by the new design suggested by Mickova et al. [81]. They proposed the addition of liposomes to the core, which are able to hold the bioactive ingredients and protect their activity for effective action by shielding the lipid sphere from the electrospinning process.

2.1.5. Mechanisms of Drug Release

The release of drug from the scaffolds takes place via three mechanisms: desorption from the surface, diffusion through the fibers, and fiber degradation [82]. These three processes can occur simultaneously, which impacts the release kinetics throughout the entire process. Figure 4 provides a schematic representation of the drug release behavior of different types of drug loading. When the fiber is immersed in the aqueous media, the desorption mechanism occurs for drug on the surface of, as well as drug present inside of, the nanopores of the nanofibers [83]. Of these three mechanisms, desorption is undergone by drug on the surface of the polymer; therefore, burst release is observed. This burst release is due to the direct interaction of the medium with the polymer surface. Because burst release of a drug is not useful, surface modification is carried out, which is the main physical modification implemented for the controlled and sustained release of the drug to the environment.

For example, Srikar et al. [84] embedded Rhodamine 610 chloride fluorescent dye in PCL/PMMA nanofibers to investigate the release of water-soluble compounds from electrospun polymer nanofibers. Furthermore, Gandhi et al. examined the release of serum albumin (BSA) and an anti-integrin antibody (AI) from electrospun PCL nanofibers [85]. The mechanism of release was observed to be dominated by desorption from the PCL surface. The two-stage desorption-controlled release of fluorescent dye Rhodamine B and vitamin B2 (riboflavin) from solution-blown and electrospun poly(ethylene terephthalate) (PET) nanofibers containing porogens was reported by Khansari et al. [86].

The second type of kinetics is the diffusion mechanism, whereby the concentration gradient causes the release of the drug into the medium. Herein, the diffusion process reduces the initial burst release and promotes a controlled and sustained release of the drug. Co-axial and emulsion electrospinning methods can exhibit this type of release kinetics. In emulsion electrospinning, drug droplets are well dispersed in the polymer solution before electrospinning [87]. A core/shell fibrous morphology comprises a core consisting of macromolecule aggregates in the aqueous phase, and a shell consisting of the polymers [87].

Finally, the third type of release mechanism is the degradation of the outer surface. For instance, using a low-degradability polymer as the shell will result in the sustained release of the drug due to the low degradation rate. In this sense, the mechanism of drug release kinetics can be optimized depending on the polymer incorporated and the type of electrospinning process. PCL is a low-biodegradability polymer; however, PVA is a highly biodegradable polymer. Therefore, a combination of these two polymers could provide a better drug release profile.

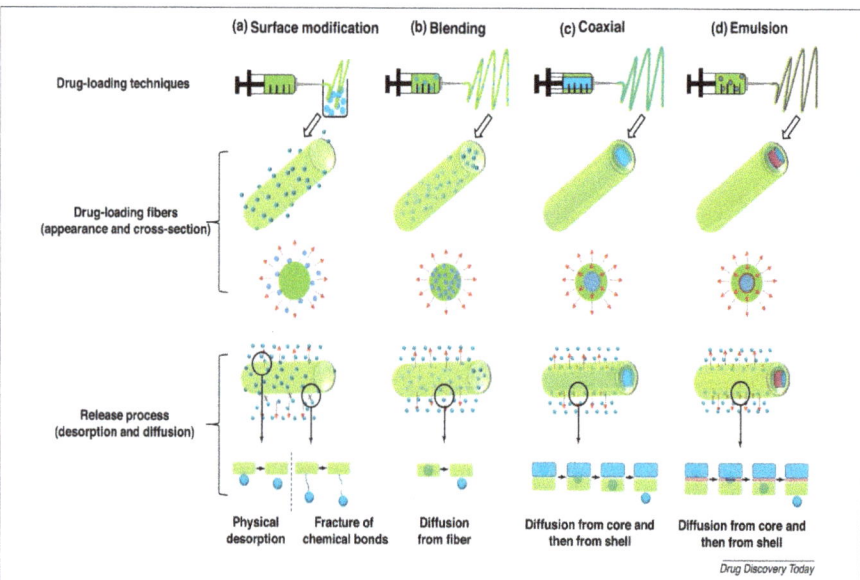

Figure 4. Drug loading and release (desorption and diffusion) from polymeric micro/nanofibers fabricated by (**a**) surface modification; (**b**) blending; (**c**) coaxial; and (**d**) emulsion electrospinning. The green color stands for the polymer, blue for drugs, and maroon for the surfactant. The red arrows represent the direction of the drug release. Reproduced with permission from [87]. Elsevier, 2017.

2.2. Tissue Engineering

Electrospinning techniques have received much consideration in tissue engineering. The electrospun fiber structure possesses many characters suitable for tissue engineering, including favorable mechanical properties, high surface-to-volume ratio, and adjustable porosity. Tissue engineering is an advanced field of science which merges applied engineering and bioscience in order to construct biomaterials that recover, sustain or improve the biological activities of injured tissues [88]. For efficient tissue engineering processes, three parameters need to be considered: seeding and attachments of cells, biomaterial scaffolds, and the addition of cell signaling factors. Of these, the biomaterial scaffold is a major parameter, and should mimic natural extracellular matrix with having sufficient mechanical properties, biocompatibility, biodegradability, high surface area, and high interpore connectivity. These criteria contribute to cell proliferation, differentiation, and migration. To this end, electrospun fibers can be prepared efficiently and cost-effectively to produce suitable candidate scaffolds for tissue engineering. Nowadays, several electrospun fiber mats are being prepared and studied for tissue engineering—with and without the addition of biological agents or growth factors—for wound healing, bond construction, and nerve tissue regeneration.

2.2.1. Skin Tissue Engineering/Wound Healing

In recent years, wound healing and skin tissue engineering have frequently been researched. Acute wounds normally heal in a very orderly and efficient process characterized by four distinct, but overlapping phases: hemostasis, inflammation, proliferation, and remodeling [89]. In the first phase, wound healing is stabilized by different cells, growth factors, and cytokines. The host cells and bacteria are removed by macrophages in the inflammation step. Then, fibroblasts migrate in to begin the proliferative phase and deposit new extracellular matrix. The new collagen matrix then becomes cross-linked and organized during the final remodeling phase. Therefore, effective wound dressing material is mandatory for proper treatment of wounds. Wound healing scaffolds should have good biocompatibility, mechanical properties and the capacity to prevent fluid evaporation from the injured site. Furthermore, it should provide a site for cell epithelization and inhibit infection [90].

Hence, the ability of cell attachment to electrospun fiber scaffolds plays an important role in the efficiency of engineered wound dressing scaffolds. The material and manufacturing process is important for the preparation of ideal wound dressings mats. Electrospinning is an ideal manufacturing process for wound dressing mats due to the above-mentioned advantages, such as biocompatibility, biodegradability, hydrophilic surface, porosity, and so on. Moreover, nanofiber scaffolds offer better clearing of exudates from the injured site, and manage both the loss of water from and the diffusion of oxygen in and out of the wound site [91]. There are natural (like collagen, gelatin, chitosan) and synthetic biodegradable polymers (PCL, PLGA, PGA, PLA, PVA, etc.) that can be molded together to form scaffolds. An electrospun scaffold is prepared in combinations of different natural and synthetic biodegradable polymers loaded with antibacterial and wound healing factors. The polymers can employ co-electrospinning, as well as blended, co-axial and multilayer electrospinning. Syed Mahdi Saeed [92] prepared a multilayered fiber mat loaded with curcumin as an active antibacterial component with novel PCL-PVA-PCL multilayered electrospun fibers. The results showed that multilayered PCL-PVA-curcumin-PCL illustrated better exudate absorbance than a pristine dressing at the incision. In the same vein, it indicates that 16% loaded curcumin displays antibacterial activity without killing the cell viability. Antibacterial properties can be built up in the scaffold with the addition of antibacterial agents or antibiotics. Silver is a well-known antibacterial agent, because it can mitigate the DNA replication of bacteria [93]. Khodkar and Ebrahimi [94] successfully prepared PVA/PCL core/shell fibers loaded with Ag NPs in the core for wound dressing applications. Fibers loaded with silver showed lower porosity, as well as an improved water vapor transmission rate (WVTR) and greater angle of contact. These scaffolds are suitable for long-term antibacterial activity (*Escherichia coli* and *Staphylococcus aureus*) because of the sustained and controlled release of the Ag NPs in the core/shell structure. On the other hand, PCL/PVA has been co-electrospun by loading silver sulfadiazine (SSD) as a drug for wound dressing mats [93]. PCL and PVA loaded with SSD were prepared successfully. The effect of different weight % of SSD on cell toxicity and mechanical and antibacterial properties was studied. Higher SSD concentrations were correlated with improved antibacterial ability, and cellular attachment, as well as proliferation, were observed. Fibronectin coatings can improve the biocompatibility of scaffolds loaded with SSD. Therefore, 5 wt% SSD-loaded co-electrospun PVA/PCL showed better antibacterial and reasonable cell proliferation and differentiation. Recently, Online et al. fabricated PVA merged with monodispersed Ag NPs and PCL loaded with Ascorbyl palmitate (AP) by dual-spinneret electrospinning [95]. The NIH-3T3 fibroblast cells were seeded on the scaffold mats, and it was shown that AP inhibits the toxic effects of Ag NPs on cell proliferation. It should also be noted that antibacterial tests confirmed the inhibition of gram-negative and gram-positive *Escherichia coli* (*E. coli*) and *Staphylococcus aureus* (*S. aureus*), respectively. Wound healing tests and histological observation concluded that this material provided a promising candidate for future biomedical applications [96]. Porosity and surface wettability are important parameters which determine the healing process. Xin Liu has electrospun PVA, PCL, PAN, and PVDF-HFP incorporating wool protein and Ag for wound dressing mats. Hydrophilic membranes have been shown to be an efficient remedy for wounds in comparison to hydrophobic membranes. Porosity, for the purpose of oxygen diffusion, also leads to an improved wound healing process. However, the wound healing process for diabetic ulcers is time-consuming due to the lack of efficient blood supply resulting from the higher amount of sugar in the blood. These processes lead to a long inflammatory stage, defective angiogenesis and blocked fibroblast proliferation.

Recently, Wang et al. [97] fabricated silk fibroin (SF)/GO nanofibers for wound healing applications. It was emphasized that graphene oxide enhanced the biocompatibility and antibacterial properties of SF composite nanofibers [97].

2.2.2. Bone Tissue Engineering Applications

Bone is a strong rigid organ that plays an essential role in our body. It protects our vital interior organs, movements, manufactures white blood cells and red blood cells, and also stores minerals [98]. Bone extracellular matrix mainly consists of organic and inorganic components, such as collagen and hydroxyapatite (HAp). Incorporation of these components results in suitable scaffolds for bone tissue engineering applications. The architecture of the electrospun scaffolds, including microstructure, porosity and surface properties, plays an essential role in successful bone regeneration [99]. Electrospun fibers should offer better mechanical properties to support the structure, and provide space for osteochondral adhesion, proliferation, and differentiation. Hence, the development of an ideal scaffold for tissue regeneration could be achieved by using a porous ceramic material, lamellar material and a fiber matrix material for imrpoved biological and physical properties. Subramanian Uma Maheshwari developed a scaffold comprised of a polymer–ceramic combination in a PCL/PVA bilayer scaffold blended with HAp NPs [100]. (PVA-PCL)-HAp has an improved porosity of around 64%, as well as hydrophilicity of around 141%. Also, MTT assay studies with MG-63 osteoblast cells had better cell adhesion and proliferation, which indicates promise for application in tissue regeneration. However, the incorporation of growth factors (GFs) or drug to the scaffold is also crucial for enhancing the regrowth of broken bones. Many GFs, including bone morphogenetic protein-2 (BMP-2) and VEGF, have been added to electrospun scaffolds in order to achieve long-lasting sustained release of GFs to mimic the natural healing process. For instance, co-axial electrospun of collagen-PCL incorporating BMP-2 and dexamethasone (DEX) have shown a more controlled release of GFs, thereby encouraging the osteogenic expression of human mesenchymal stromal cells (hMSCs) [101]. In this design, the shell layer was loaded with DEX, and the core incorporated BMP-2. Dual drug release was exhibited, in which DEX showed a fast release. However, BMP-2 demonstrated a sustained release over 22 days. This scaffold provides an efficient healing process, as well as osteogeneration.

On the other hand, incorporation of stem cells into the biomaterials is also a novel approach for tissue regeneration of the cells. For instance, Abbas Shafie has studied, in vitro and in vivo, cartilage tissue regeneration from rabbit bone marrow mesenchymal stem cells (BM-MSC) seeded on the electrospun scaffold of PVA/PCL nanofibers [102]. In vitro, the MTT assay showed that the scaffolds supported the chondrogenic differentiation of MSC. In vivo, the scaffold with and without MSC loading was implanted in rabbit full-thickness cartilage defects. To study cartilage regeneration, histological and semi-quantitative grading was executed. The results showed that scaffold seeded with MSC enhanced the healing process in comparison to non-seeded scaffolds. These results indicate that PVA/PCL scaffold seeded with MSC is suitable for grafts for articular cartilage repair.

Recently, the effect of polyacrylonitrile/MoS_2 nanofibers on the growth behavior of bone marrow mesenchymal stem cells (BMSCs) was discussed by Wu et al. [103]. The nanofibers were realized to enhance the contact of BMSCs with each other, to enhance cellular behavior, and also to provide positive promotion of regulation of cellular proliferation [103]. In addition, for guided spinal fusion, an injectable and thermosensitive hydrogel made of collagen/n-HA/BMP-2@PCEC/PECE enclosed in poly(D,L-lactide) (PDLLA) nanofibrous membranes was made by Qu et al. [104]. This system restricted the escape factor in order to maintain osteogenesis in the desired position [104].

2.2.3. Skeletal Muscle Regeneration

Skeletal muscle makes up around 40% of the human body. Skeletal muscle is made of various fibers, with diameters ranging from 10 to 80 μm [105]. These fibers are unidirectional and produce an enormous amount of force during contraction [106]. If a muscle cell gets injured or wounded, it will not be possible to contract, and satellite cells are switched on in order to perform their muscle cell regeneration activities. However, this healing process can create scar tissue and block muscle function [106]. Many efforts have been made to study the initial steps of muscle regeneration, such as autologous muscle transplant, satellite cells, exogenousmyogenic cells, and myoblasts, but these methods have met with limited success [107]. Therefore, long-term denervation and severe injuries can lead to the loss of skeletal muscle function.

Muscle tissue engineering materials require better contraction ability and mechanical properties [108]. Muscle cell adhesion and proliferation have been studied using both mechanical properties and electric stimulus in cell culture. Mckeon-Fischer K D prepared co-axial electrospun fibers with a PCL core and a modified outser shell layer comprising multiwalled carbon nanotubes (MWCNT) and a blend of (83/17, 60/40, 50/50, and 40/60) poly(acrylic acid/poly(vinyl alcohol) (PAA/PVA) [109]. All four components were electrically conductive, although the scaffold was not actuated when an electric field was applied. The best results occurred at 20 V. MTA assay in soleus and vastus lateralis (VL) muscles extracted from rats showed that 0%, 0.14% and 0.7% concentrations of MWCNT in the scaffold were non-toxic for cells over a four-week period. Based on the different percentages of blend solutions, 40/60 PAA/PVA in the outer layer illustrated a higher number cells than other scaffolds. The scaffold has tensile properties that are higher than those of skeletal muscle. Further modification of these scaffolds for contraction, rather than bending, could lead to promising scaffolds for artificial muscle applications.

2.2.4. Nerve Tissue Engineering

Electro-conducting polymers such as polypyrrole (PPy), polyaniline (PANI), polythiophene (PT), poly(3,4-ethylene dioxythiophene) (PEDOT)) show attractive electrical and optical phenomena. Thus, they have been researched in the past few decades for various applications such as microelectronics, actuators and polymer batteries [110]. Electro-conducting polymers that possess the advantages of biocompatibility and good conductivity can be applied as biosensors and tissue engineering scaffolds [111]. Electrospun electro-conductive polymers are an excellent tool for electrically stimulating neurons and for nerve tissue engineering, as well as for application in neural prostheses for therapeutical function [88,89,112]. Schmidt et al. first studied PC12 cells using the polypyrrole (PPy) electroconductive polymer, recognizing the growth of PC12 cells on the PPy thin film, they enhanced the neurite outgrowth from the cells; these results suggest significant application of these type of scaffolds for nerve tissue regeneration [113]. Many studies have proposed the improvement of the electro-conducting polymer for nerve tissue regeneration applications by adding cell adhesive [114], neurotrophins [115], and topographical features [116]. Jae Y Lee prepared electrospun nanofibers coated with the conductive polymer PPy for nerve tissue engineering applications [117]. PPy-PLGA showed improved growth of rat pheochromocytoma 12 PC12 cells and hippocampal neurons compared to non-coated PLGA as a control. This suggests that PPy-PLGA could be used for nerve tissue engineering applications. Simultaneously, electrical stimulus studies on the scaffold indicated that a stimulus of 10 mV/cm improved the neurites such that they were 40–50% longer, as well as exhibiting 40–90% greater neuron formation, compared to the same scaffolds with no stimulus. Moreover, aligned scaffolds show greater neurite elongation and formation than randomly oriented PPy-PLGA fiber scaffolds. The good results for electric stimulus suggest that biocompatible polymers prepared by electrospinning have significant advantages in biomedical applications such as nerve tissue engineering.

3. Combination of Electrospinning and Sputtering Technologies for Biomedical Applications

3.1. Setup—Operating Principles of Sputtering Technology

Sputtering is the ejection of atoms by the bombardment of a liquid or solid target by energetic particles, typically ions [118]. These ejected atoms are then deposited on the substrate [118]. Figure 5 presents a schematic representation of sputtering technology. Argon gas is commonly used as a sputtering gas. The ejected fragments are accumulated on the substrate by adjusting the distance between the target and substrate. The number of atoms excited from the surface per incident ion is known as sputtering yield, 'S'. The value of S depends on many parameters, including target material composition, experimental geometry, binding energy, and the properties of incident ions. In addition, there are also experimental parameters, such as voltage and current. In a conventional sputtering machine, the cathode is connected to the target and the anode is connected to the substrate, with the plasma in between them.

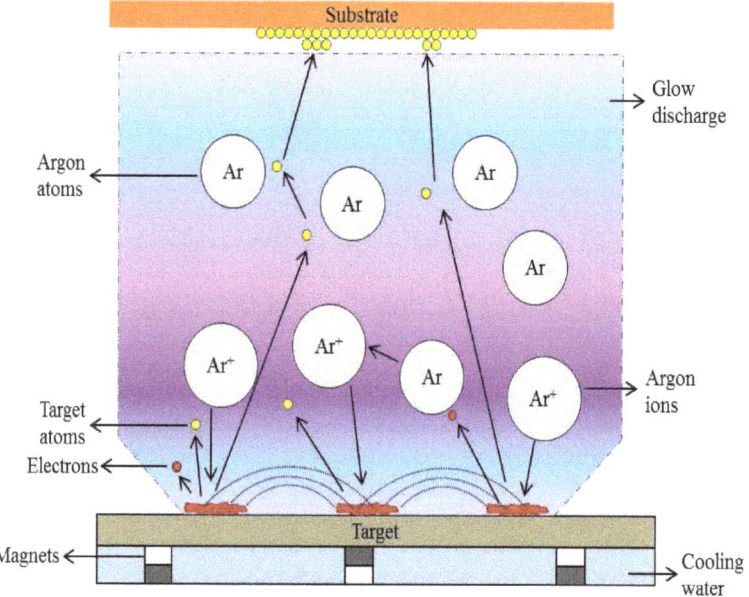

Figure 5. Schematic representation of sputtering technology [119].

The direct current (DC) and radio frequency (RF) sputtering processes differ mainly with respect to the power supply installed. In the DC sputtering process, metals are only used as a target for coating. However, in the case of RF, insulators are used as a target for coating purposes. This is mainly achieved by providing the RF potential to the target. In practice, when insulators are used as the target, their charges are accumulated on the surface of the target after striking positive ions. This would make the target surface inaccessible for the further bombardment of ions. Therefore, to inhibit this process, both positive ions and electrons are bombarded directly onto the insulator target [120]. This is achieved by applying a RF power supply, which gives enough energy for the oscillating electron in the presence of an alternating field to originate ionizing collision, and a self-preserved discharge is maintained. However, the incorporation of magnetron sputtering into DC/RF sputtering could increase the sputtering yield by applying a strong magnetic field along the sputtering target to confine the plasma to the nearby target surface, on which the electric field $\vec{E}$ and the magnetic field $\vec{B}$ are used for electronic motion. Electrons experience the well-known Lorentz force in the magnetic field, and the electric field force in the electric field, leading to the circular motion of electrons close to the target material, enhancing the collision with the target, and improving the sputtering yield. The sputtering process is accomplished under vacuum to avoid oxygen or other gas contamination that might cause impurities to form on the substrate surface. The Factors affecting the plasma sputtering process are depicted in Figure 6.

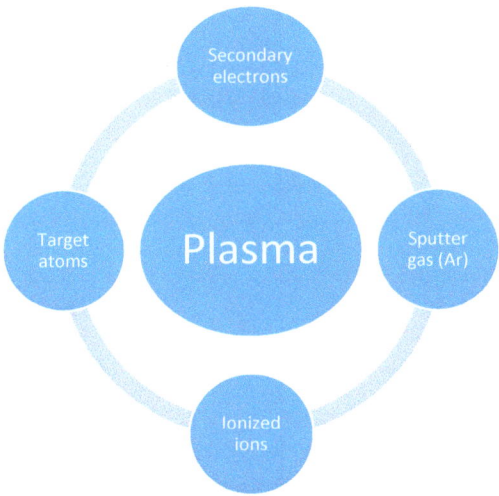

Figure 6. Factors affecting the plasma sputtering technique.

3.2. Biomedical Applications of Sputtered Electrospun Polymer-Based Nanofibers

Plasma technology can be used to improve the surface properties of polymers without changing their bulk characters. Plasma-treated polymers have found wide application in diverse fields, such as the automobile, microelectronics, chemical and biomedical industries [121]. Polymer surface properties such as hydrophobicity, roughness, chemical structure, conductivity, etc. can be modified for various applications. Plasma treatment can affect the polymer surfaces through micro-etching, organic contamination, cross-linking, surface chemistry modification, and surface coating with a specific target material [122]. The biomaterials should possess good mechanical and surface characteristics that are appropriate for the biological environment. For instance, for cell adhesion, the polymer surface should have low surface free energy, surface roughness, and hydrophilicity. Plasma treatment via magnetron sputtering technology has been implemented to coat the surfaces of polymers to form biomaterials suitable for biomedical applications such as antibacterial, biocompatibility, and tissue engineering.

Plasma sputtering technology includes both thermal and non-thermal deposition processes. However, non-thermal deposition processes are highly recommended for polymers, because they do not damage the bulk properties of the polymer. Magnetron sputtering is the technique used for coating the polymer surface. Magnetron sputtering is a technology that was developed during the 1970s, and it is a high-speed and low-temperature technique for preparing a strong and uniform adhesion film on the surface of polymers, ceramics and composite materials [123]. However, argon gas is commonly used, because it does not damage the target due to its nobility. The full process is quick and requires only a low temperature, while offering a high film forming rate and strong film adhesion [124]. For instance, composite microfibers of Poly(methyl methacrylate)/organically modified montmorillonite (O-MMT) were manufactured by electrospinning with the incorporation of emulsion polymerization [125]. Here, the prepared composite microfibers of PMMA-O-MMT were magnetron sputter-coated with Titanium dioxide (TiO_2). The results showed that the deposited anatase-TiO_2 and rutile-TiO_2 exhibited better surface wettability without damaging the PMMA-O-MMT compound. These composite fibers have a UV absorption of 254 nm. Therefore, it induces the photocatalytic degradation of the model compound methylene blue. Thus, these materials provide a promising application in dye wastewater treatment.

Polymer microspheres [126], thin films [127], and fibers [128] have been coated with Ag [107,108,129], Cu [130], Ti [131], TiO_2 [132], gold (Au) [133], hydroxyapatite (HAP), tricalcium phosphate (TCP) [134], amorphous calcium pyrophosphate (CPP) [134], and dicalcium phosphate dihydrate (DCPD) [134] for different biomedical applications.

3.2.1. Antibacterial Coatings

The attachment of bacteria to the surface of a polymer can lead to the formation of biofilm. Therefore, biofilm-resistant polymers are an essential factor for the medical field. Biofilm resistance could be imbued in the polymer through the addition of antibacterial agents on the surface of the polymer to prevent bacterial adhesion. Materials such as medical textiles, wound dressings, prostheses and implant materials should display antibacterial activity for efficient biological activity. Antibacterial properties are an essential parameter to take into account for wound dressing. Antibacterial activities are promoted through the addition of some antibacterial components to the fabrics. There are many components, including both inorganic and organic (drugs), as well as metals. Inorganic agents include TiO_2, carbon nanotubes, and Ag, Zn, ZnO_2, Cu, Ga, and Au NPs [135]. Organic agents such as Triclosan inhibit the development of micro-organisms using electrochemical activity to disrupt their cell walls [136].

Among the inorganic antibacterial components, Ag NPs have been well studied [137]. Ag NPs were added to electrospun fibers via Ag ions through the wetting process [138,139], silver sulfaazide [93], etc. The wetting process for the addition of Ag to the matrix has many disadvantages, such as uneven distribution of NPs, use of reducing agents that are toxic, and the difficulty of controlling the size of NPs—depending on the strong and weak reducing agents used [140]. However, the most efficient way of introducing NPs to the surface of polymers or fabrics is by using plasma technology. Plasma technology provides more uniform deposition, less use of resources, and a simpler process for the coating of antibacterial material such as Ag, Si, Cu, etc., onto the surface of the polymer than the wetting process. Sputtering, known as physical vapor deposition, has been used effectively in the coating of a number of thin films for electronics applications. Therefore, sputter-coating of metals to enhance antibacterial properties can be performed with the addition of many target materials, including Ag, Ag/Si, Cu, Ti, etc. Therefore, more studies are required to compare the antibacterial properties of various materials incorporated in bioresorbable polymers.

Silver (Ag)

Silver is a transition metal in the periodic table. Silver-related compounds or NPs have a biocidal effect on around 16 species of bacteria, because of its toxic effect on microorganisms [119,120,141,142]. Thus, silver is coated on medical devices for antibacterial applications [143]. At low concentrations,

Ag NPs show good antibacterial efficiency [144]. Moreover, the lack of toxic effect of Ag NPs on human monocytes cell lines indicates the possible application of Ag in the fabrication of medical devices.

The mechanism of antibacterial activity of Ag on microorganisms has not yet been well studied. It has been shown that, in *E. coli*, AgNP-treated bacteria exhibit some pits on the cell wall and an accumulation of Ag in the cellular membrane. This type of membrane exhibits an increase in permeability. Bacterial DNA loses its replication ability and cellular proteins and becomes denatured by binding Ag ions or NPs to the functional group of the protein [145]. On the other hand, some authors have reported that Ag NPs would denature the cellular proteins required for cellular nutrient transport and damage the cell membrane or cell wall, enhancing cell permeability and ultimately leading to cell death [142]. It has also been noted that the antibacterial efficiency of Ag depends on its shape. Ag NPs with a {111} lattice basal plane (representing the cubic structure lattice pattern) display more robust antibacterial action than spherical and rod-shaped NPs and silver ions [146].

Silver has excellent antibacterial properties. Silver treatment is well known for its application in wound dressing materials. The silver is incorporated as Ag NPs by introduction through $AgNO_3$ using a reducing agent. However, this type of silver incorporation leads to burst release of silver from the material, resulting in a very high concentration of Ag in the wound. This is followed by the sudden reduction in silver because of both bacterial consumption and reaction with other compounds present in the wound beds, such as phosphates, chlorine, and proteins. Therefore, silver release in the wound should take place in a controlled manner. Also, the silver nitrate present is a hypotonic; hence, it can cause a strong electrolyte imbalance, which could damage the wound site and produce gross systemic inequality, which could kill patients with extensive burns who require large doses of silver. However, silver sulfaazide was developed in order to minimize the side effects of using silver nitrate. However, the removal of silver sulfaazide cream from the wound surface is performed by scraping, and this could result in a highly painful dressing procedure for the patients. Moreover, sulfaazide does not show any hypotonic effect. Therefore, it is necessary to develop a better process for delivering silver that is efficient, involves introduction over a prolonged period, acts against many ranges of bacteria, requires only a few changes of the wound dressing, and never interferes with the wound healing process. With this in mind, sputtering is a new field of surface coating of wound dressing materials for extended release of silver with potent antibacterial properties. The optimization of the sputtering process is an essential criterion for better antibacterial properties.

It is also noted that silver is the best candidate for wound curing applications, because it reduces inflammation [147], impedes contraction, and improves cell epithelialization [148]. Ag NPs exhibit cellular toxicity, and this leads to a decrease the biocompatibility of the scaffolds [149]. However, the amount of Ag in the scaffolds can be used to optimize the antibacterial effect and cellular toxicity of the Ag.

Antibacterial coatings on medical textiles are an important tool for avoiding infections during the surgical process [150]. Silver is coated onto textiles via different techniques. Silver-coated textiles are limited in application because of their reduced durability. Therefore, strong adhesion of silver on fabrics can be obtained by using the sputter-coating technique [150]. Cotton fabric with antibacterial properties could have a variety of applications. Silver is the most commonly used material for enhancing the antibacterial properties of cellulosic fibers [150]. Wet and dry methods can be used to incorporate silver particles. The wet method changes the bulk properties of textiles and also has a negative impact on the environment. However, dry processes such as sputter-coating are eco-friendly processes than only change the surface of the matrix. Therefore, Ag was incorporated into cotton matrixes of various thicknesses in order to study the antibacterial properties, the release of the Ag in water, etc. The results suggested that Ag shows antibacterial activity against *Staphylococcus aureus*, *Escherichia coli* and *Candida albicans* [150]. In addition, sputter-coating also improved the water contact angle of the cotton fabrics. Thus, antibacterial properties could be added to the nanofibers using the sputtering technique. Simultaneously, Chen et al. sputtered PET fabrics using high-power impulse magnetron sputtering, which provides a highly concentrated plasma, so that these fabrics will support adhered films [151]. Examination of antimicrobial activity revealed that a silver film that is deposited for more than 1 min displays excellent bactericidal (>0) and bacteriostatic (>2.0) effects, based on JIS standards. Furthermore, the coated fabrics showed the capacity to retain antibacterial properties over 20 cycles of washing, indicating the long-term durability for the materials.

The wound dressing mats of polymer sheet and electrospun fiber scaffolds were sputter-coated with Ag and their antibacterial properties studied. Liu et al. examined the influence of magnetron sputter-coating of nanosilver on polyetheretherketone (PEEK) and investigated the resulting cytotoxicity and antibacterial properties [152]. PEEKs were sputter-coated with Ag 3 nm, 6 nm, 9 nm and 12 nm NPs (Figure 7). The antibacterial properties and bacterial adhesion to the surface were studied. Homogeneous nanosilver was coated on the surface; an increase in the water contact angle was observed, and there was no cytotoxicity for the CCK-8. In addition, the coating also provided excellent adhesion of bacteria to the PEEK and improved antibacterial activity towards *Streptococcus mutans* and *Staphylococcus aureus*.

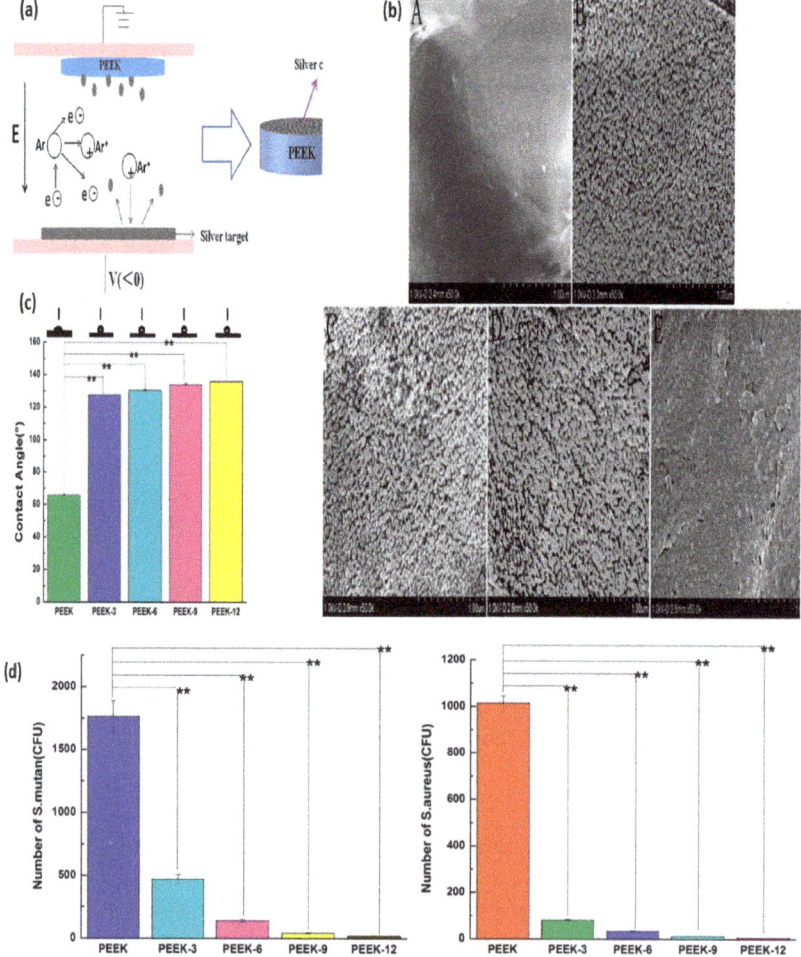

Figure 7. (**a**) Schematic representation of PEEK film coated with nanosilver via sputter-coating; (**b**) SEM images of PEEK at different thicknesses of Ag (3, 6, 9, 12 nm); (**c**) Water contact angle of the coated thin film; (**d**) Antibacterial activity of the PEEK/Ag composite material. Reproduced with permission from [152]. Elsevier, 2017.

The combination of electrospinning and sputtering technology can result in many novel composite fibers with diverse applications in the biomedical field, such as for wound dressing mats with excellent biocompatibility and antibacterial properties. The electrospun microfibers were coated with Ag by DC magnetron sputtering [129]. The electrospun scaffolds of poly(glycerol sebacate)/poly(3-caprolactone) (PGS/PCL) were coated with Ag, and their antibacterial properties and silver release behavior were studied. PGS/PCL showed good mechanical and thermal behavior due to the increase in fiber diameter and the decrease in fiber pore size when sputter-coated with Ag. The fiber scaffolds demonstrated a gradual release of Ag, contributing to antibacterial activity. Therefore, this material could find appropriate application in wound dressing and bandages. Moreover, prosthetic implants also require antibacterial properties in order to avoid infection after surgery. With the objective of avoiding abdominal infections after implanting prostheses for hernia repair, Muzio et al. [153] prepared polypropylene prostheses coated with a silver-silica composite (Ag/SiO_2) layer. The prepared mesh hernia prostheses (CMC) consisted of two layers of microporous light mesh and a thin transparent film of polypropylene. The Ag/SiO_2 composite was sputter-coated onto the CMC meshes and the microporous mesh layer alone. The sputtering process was optimized via addition in order to test biocompatibility and antibacterial properties. In addition, it is noted that sputter-coating with CMC improved the antibacterial properties, but reduced biocompatibility. However, the sputter-coated meshes alone showed good antibacterial properties and biocompatibility. In addition, fiber meshes coated with Ag/SiO_2 enhanced the growth of seeded fibroblast without causing apoptosis or necrosis of the fibroblast; in addition, the meshes also exhibited good antibacterial properties.

Copper (Cu)

In addition to Ag, electrospun scaffolds have been sputter-coated with Cu to improve their antibacterial properties [132,133]. Cu is cheaper than Ag; therefore, Cu coating can provide economical wound dressing mats [154,155]. A PLA scaffold was DC magnetron sputter-coated with copper (Cu) [130]. The PLLA scaffold had increased hydrophobicity, proportional to plasma treatment time. Antibacterial testing concluded that the modified composite scaffold had a bacteriostatic effect in which bacteria were reduced by 30% and 50% [130]. In addition, it was also found that copper has a stronger antibacterial impact than copper oxide. Therefore, this type of composite material could be used for economical wound dressing mats with antibacterial effect.

Eichornia crassipes, commonly known as water hyacinth, is a natural fiber that has found significant applications in recent years [156]. It was sputter-coated with copper (Cu) to study the antibacterial properties. The results revealed that the Cu-coated fibers showed better bacterial inhibition towards *Escherichia coli* (*E. coli*) and *Staphylococcus aureus* (*S. aureus*) compared to pure water hyacinth fibers. In addition, the incorporation of the Cu coating improved the hydrophobicity of the fiber, thereby enhancing the antibacterial activity.

Titanium Dioxide (TiO_2)

Electrospun chitosan (Cs) nanofibers were sputtered with TiO_2 particles by plasma-enhanced chemical vapor deposition using a planar magnetron device [132]. The physiochemical interaction of the Cs and Cs-TiO_2 was studied computationally using the Gaussian software package, followed by experimentally examining the antibacterial properties of the materials. The results revealed that Cs-TiO_2 behaves as a single composite unit by forming a C-O-C glycosidic bond in the glucose ring. The Cs-TiO_2 composite had an improved structure and reactivity because of the reduced HUMO-LUMO energy gap, larger dipole moment, and lower ionization potential when compared to pure Cs. Finally, the Cs-TiO_2 composites exhibited antibacterial activity, with inhibition zones approximately 11.5 mm in diameter.

3.2.2. Surface Modification for Enhancing Biocompatibility

Surface modification via coating can improve the cell adhesion, proliferation, and differentiation; this reduces the risk of thrombosis and imparts bactericidal properties to the stent. In the literature, it has been shown that plasma surface treatment of L-PLA and PCL polymers improves the surface roughness, as well as reducing the surface free energy for better cell attachment of diverse cells to the surface of the polymer mesh [157]. Magnetron-sputtered polymer sheets offer reduced cytotoxicity and better cell viability for biomedical applications. The surface characteristics have been studied for polymer thin films, as well as for electrospun fiber scaffolds. For example, Staszek et al. [158] reported the cytotoxicity of glycerin sputtered with different noble metals such as Au, Ag, palladium (Pd) and platinum (Pt). The results suggested that they had prepared Au, Ag, Pd, and Pt NPs with sizes of 6.1 ± 1.0 nm, 4.2 ± 0.9 nm, 2.5 ± 0.6 nm and 1.9 ± 0.4 nm, respectively. In addition, Pt and Pd demonstrated great cytotoxicity for the 6 cells lines tested (human cells from hepatocarcinoma (HepG2), human keratinocytes (HaCaT), mouse macrophages (RAW264.7), mouse embryonic fibroblasts (L929 and NIH3T3), and cells from Chinese hamster ovary (CHO-K1)), and lower cytotoxicity was noted for Ag and Au after 24, 48 and 72 h.

Consequently, surface modification of hydrophobic polymers via plasma treatment can enhance wettability. E.N. Bolbasov studied the surface modification of PLA and PCL bioresorbable polymers via radio frequency thermal glow discharge plasma using hydroxyapatite as a target in the presence of Ar+ as plasma [127]. The results indicated that the PLA and PCL surfaces showed enhanced biocompatibility for cell line EA-hy926 attachment to the surface. Surface free energy and surface roughness were improved by long exposure to plasma treatment. In addition, plasma sputtering technology can enhance surface roughness, improving cell attachment onto the thin polymers fibers [159]. The PLLA polymer thin film was RF magnetron-sputtered with hydroxyapatite target. This coating led to an increase in biocompatibility with the cells of bone marrow multipotent mesenchymal stromal cells. This was mainly due to the increase in the surface roughness of the PLLA film resulting from the plasma coating, in addition to the enhancement of calcium and phosphorous caused by the hydroxyapatite target. Surface modification of polymer films enhances biocompatibility and reduces cell toxicity.

Furthermore, biodegradable PLA polymer was prepared via electrospinning, and surface modification was implemented by RF magnetron sputtering. The electrospun PLA scaffold was sputter-coated with hydroxyapatite (HAP), tricalcium phosphate (TCP), amorphous calcium pyrophosphate (CPP) and dicalcium phosphate dihydrate (DCPD) [134]. It was found that all prepared fibers showed cytotoxicity because of the production of a toxic compound on the fiber surface, as well as the fact that the fiber surface had been devastated due to the extended plasma treatment.

On the other hand, the same team of scientists worked on PCL scaffold fibers that were magnetron sputter-coated with titanium targets (Figure 8) [131]. They found that hydrophilicity improved with an increase in plasma treatment time. In addition, increasing the number of pores on the fiber structure did not affect the mean fiber diameter. As plasma treatment time was increased, the adhesion of cells improved. Consequently, cell viability decreased when plasma treatment time reached 9 min.

In another study, poly (L-lactic) acid (PLLA) scaffold was sputter-coated with titanium target under a nitrogen atmosphere [160]. The pure PLLA did not show any changes in its physiomechanical properties. Biocompatibility testing in in vivo rat models indicated that there was no severe tissue reaction after around three months for the implemented subcutaneous tissue. Finally, the replacement of scaffolds from the recipient tissue depends on plasma treatment time.

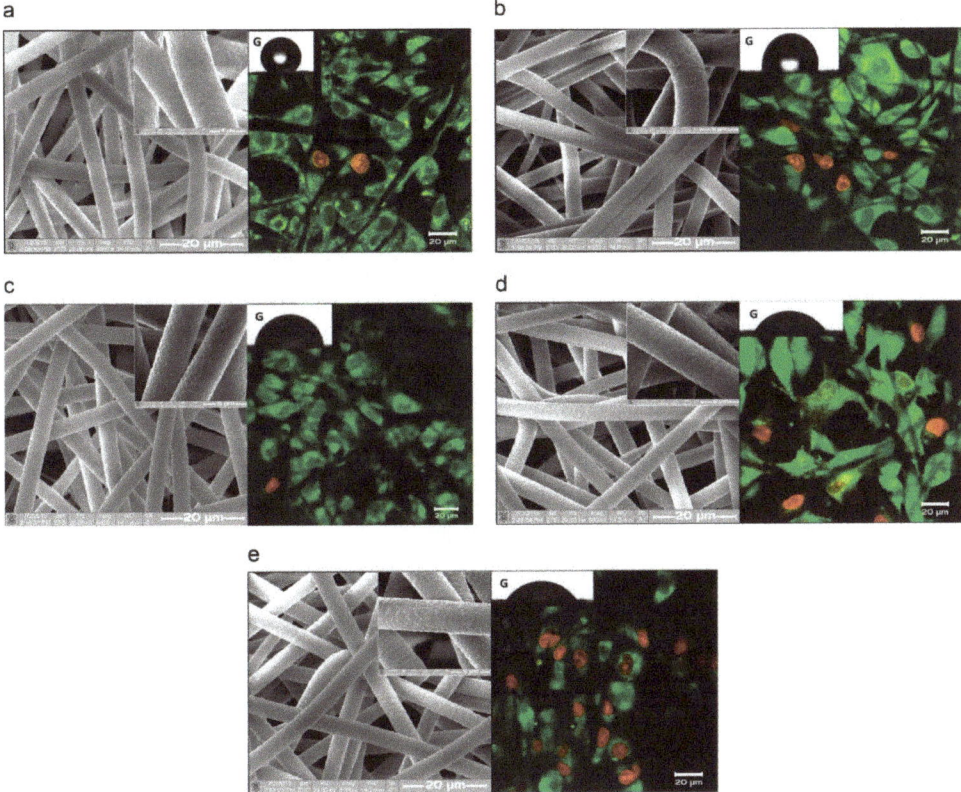

Figure 8. SEM images, fluorescent map of labeled cells (live green cells (acridine orange), orange nucleus of dead cells (ethidium bromide) and wettability for PCL samples that were (**a**) untreated; and treated in plasma for (**b**) 1 min; (**c**) 3 min; (**d**) 6 min; and (**e**) 9 min. Reproduced with permission from [131]. Elsevier, 2016.

3.2.3. Tissue Engineering

Plasma technology has also emerged recently for use in tissue engineering applications such as vascular grafting, stem cell therapy, and artificial muscle sputter-coated with conductive Au. For this purpose, highly biocompatible polymers have provided a platform for cell adhesion, proliferation, and differentiation. To this end, stem cells are added to the polymer scaffold to provide a better tissue regeneration environment. Stem cell therapy is a new platform that may act as an alternative to many complicated surgical procedures. Stems cell-loaded materials have gained much attention recently [126]. Lee et al. [126] reported the use of PCL microspheres sputter-coated with Au as a platform for differentiating cardiomyogenic cells from human embryonic stem cells. They sputtered the PCL microspheres for 5 min, and then incorporated the human embryonic stem cells (hESCs). It was noted that these composites showed a higher cardiac differentiation, because Au acted as the mediator for gene expression on day 4 and day 14 [126].

Moreover, PLLA and PEG fibers were electrospun and sputter-coated with calcium phosphate for bone tissue engineering applications. Here, simple combination of the electrospinning and sputtering techniques is feasible for the fabrication of biopolymer scaffolds for biomedical applications [161].

Many novel composite materials have emerged due to the fusion of two valuable techniques, such as the electrospinning and sputtering techniques [162]. Innovative materials have been studied for vascular tissue engineering. These composite materials were prepared using the electrospinning and sputtering technique. To this end, PCL and PHBV were incorporated at a ratio of 1:2 (v/v) and sputter-coated with Ti. Firstly, the sputtering process was optimized so that it would not damage the macrostructure of the scaffolds. The biocompatibility of the prepared composite mats was studied with hybridoma of the endothelial cells of the human umbilical vein and human lung carcinoma (EA.hy.926 cell line). The results showed that cell adhesion was improved for Ti-coated scaffolds, and that they exhibited better proangiogenic activity.

Furthermore, a novel approach was applied to process fibrous scaffolds for artificial muscles or human body smart devices by combining electrospinning and sputtering technologies [133]. A core/shell structure was made by first electrospinning the PMMA in optimized form to obtain a uniform fiber; later, the PMMA was coated with Au to induce conductivity and obtain suitable mechanical properties in the scaffolds. Subsequently, polyaniline (PANI) was coated onto the scaffolds via in situ electrochemical polymerization, starting with aniline and using sulfuric acid as an oxidizing agent (Figure 9). PANI-coated metalized fiber scaffolds in a structure similar to the core/shell structure showed fascinating electrochromic properties, in which color changes occurred when the applied voltage was switched from 0 to 1 V, and vice versa. In vitro biocompatibility testing revealed good cell adhesion, with a better result shown when tested on human amniotic fluid stem cells than on eukaryotic cells. Therefore, this type of web could be used to prepare smart artificial muscle devices via a versatile and straightforward preparation technique using electrospinning and sputtering.

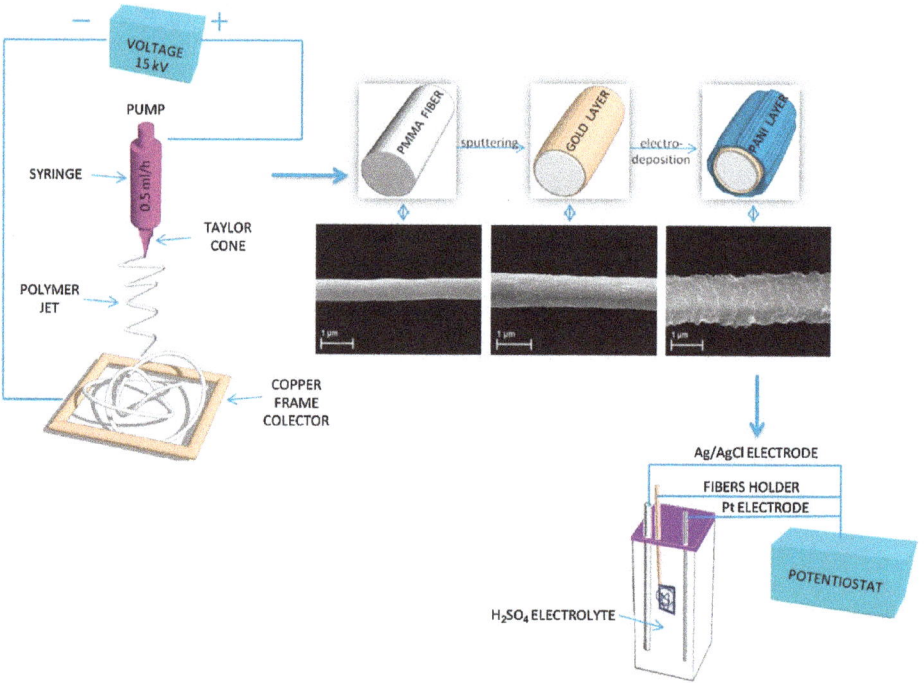

Figure 9. Schematic representation of electrospinning and sputtering for muscle tissue engineering applications. Reproduced with permission from [133]. Elsevier, 2016.

4. Conclusions

Throughout this review, new insights for biomedical applications have been addressed, focusing predominantly on the promising benefits of employing sputtered electrospun polymer-based nanofibers. It is evident from the number of ineffective conventional treatments that there is a desperate necessity for distinct and unique therapies. As addressed extensively in this paper, combining sputtering and electrospinning technologies has the potential to play a critical function in different biomedical applications such as antibacterial coatings, surface modification for enhancing biocompatibility, and tissue engineering. Investigated by means of various nanostructure studies, the above-mentioned concepts can be used in an attempt to strengthen overall therapeutic behavior. As represented through the numerous reports addressed, therapies including silver and copper nanoparticles have the potential to be applied directly in different biomedical applications. Overall, it is clear that the field of combined sputtering and electrospinning technologies is progressing at an incredibly fast rate, providing promising behaviors for various biomedical applications.

Author Contributions: H.K. and M.Z. wrote the manuscript; A.E. and T.A. supervised the writing and reviewed the manuscript.

Funding: This work was funded by Qatar University, grant number GCC-2017-007 and the publication of this article was funded by the Qatar National Library.

Conflicts of Interest: All authors of this review article declare no conflicts of interest.

References

1. Doshi, J.; Reneker, D.H. Electrospinning process and applications of electrospun fibers. *J. Electrostat.* **1995**, *35*, 151–160. [CrossRef]
2. Geoffrey, T. Electrically driven jets. *Proc. R. Soc. Lond. A Math. Phys. Sci.* **1969**, *313*, 453–475.
3. Greiner, A.; Wendorff, J.H. Electrospinning: A fascinating method for the preparation of ultrathin fibers. *Angew. Chem. Int. Ed. Engl.* **2007**, *46*, 5670–5703. [CrossRef] [PubMed]
4. Landau, O.; Rothschild, A.; Zussman, E. Processing-Microstructure-Properties Correlation of Ultrasensitive Gas Sensors Produced by Electrospinning. *Chem. Mater.* **2009**, *21*, 9–11. [CrossRef]
5. Teo, W.E.; Inai, R.; Ramakrishna, S. Technological advances in electrospinning of nanofibers. *Sci. Technol. Adv. Mater.* **2011**, *12*, 013002. [CrossRef] [PubMed]
6. Deitzel, J.; Kleinmeyer, J.; Harris, D.; Beck Tan, N. The effect of processing variables on the morphology of electrospun nanofibers and textiles. *Polymer* **2001**, *42*, 261–272. [CrossRef]
7. Guo, C.; Zhou, L.; Lv, J. Effects of expandable graphite and modified ammonium polyphosphate on the flame-retardant and mechanical properties of wood flour-polypropylene composites. *Polym. Polym. Compos.* **2013**, *21*, 449–456. [CrossRef]
8. Zhao, Y.Y.; Yang, Q.B.; Lu, X.F.; Wang, C.; Wei, Y. Study on correlation of morphology of electrospun products of polyacrylamide with ultrahigh molecular weight. *J. Polym. Sci. Part B Polym. Phys.* **2005**, *43*, 2190–2195. [CrossRef]
9. Yang, Q.; Zhenyu, L.I.; Hong, Y.; Zhao, Y.; Qiu, S.; Wang, C.E.; Wei, Y. Influence of solvents on the formation of ultrathin uniform poly(vinyl pyrrolidone) nanofibers with electrospinning. *J. Polym. Sci. Part B Polym. Phys.* **2004**, *42*, 3721–3726. [CrossRef]
10. Qin, X.; Wu, D. Effect of different solvents on poly(caprolactone)(PCL) electrospun nonwoven membranes. *J. Therm. Anal. Calorim.* **2012**, *107*, 1007–1013. [CrossRef]
11. Sukigara, S.; Gandhi, M.; Ayutsede, J.; Micklus, M.; Ko, F. Regeneration of Bombyx mori silk by electrospinning—Part 1: Processing parameters and geometric properties. *Polymer* **2003**, *44*, 5721–5727. [CrossRef]
12. Huang, C.; Chen, S.; Lai, C.; Reneker, D.H.; Qiu, H.; Ye, Y.; Hou, H. Electrospun polymer nanofibres with small diameters. *Nanotechnology* **2006**, *17*, 1558–1563. [CrossRef] [PubMed]
13. Li, Z.; Wang, C. *One-Dimensional Nanostructures*, 1st ed.; Springer: Berlin/Heidelberg, Germany, 2013; ISBN 978-3-642-36427-3.
14. Jacobs, V.; Anandjiwala, R.D.; Maaza, M. The influence of electrospinning parameters on the structural morphology and diameter of electrospun nanofibers. *J. Appl. Polym. Sci.* **2009**, *115*, 3130–3136. [CrossRef]
15. Areias, A.C.; Gómez-Tejedor, J.A.; Sencadas, V.; Alió, J.; Ribelles, J.L.G.; Lanceros-Mendez, S. Assessment of parameters influencing fiber characteristics of chitosan nanofiber membrane to optimize fiber mat production. *Polym. Eng. Sci.* **2012**, *52*, 1293–1300. [CrossRef]

16. Yördem, O.S.; Papila, M.; Menceloğlu, Y.Z. Effects of electrospinning parameters on polyacrylonitrile nanofiber diameter: An investigation by response surface methodology. *Mater. Des.* **2008**, *29*, 34–44. [CrossRef]
17. Homayoni, H.; Ravandi, S.A.H.; Valizadeh, M. Electrospinning of chitosan nanofibers: Processing optimization. *Carbohydr. Polym.* **2009**, *77*, 656–661. [CrossRef]
18. Xu, C.Y.; Inai, R.; Kotaki, M.; Ramakrishna, S. Aligned biodegradable nanofibrous structure: A potential scaffold for blood vessel engineering. *Biomaterials* **2004**, *25*, 877–886. [CrossRef]
19. Um, I.C.; Fang, D.; Hsiao, B.S.; Okamoto, A.; Chu, B. Electro-Spinning and Electro-Blowing of Hyaluronic Acid. *Biomacromolecules* **2004**, *5*, 1428–1436. [CrossRef]
20. Kong, C.S.; Lee, T.H.; Lee, S.H.; Kim, H.S. Nano-web formation by the electrospinning at various electric fields. *J. Mater. Sci.* **2007**, *42*, 8106–8112. [CrossRef]
21. Yang, Y.; Jia, Z.; Liu, J.; Li, Q.; Hou, L.; Wang, L.; Guan, Z. Effect of electric field distribution uniformity on electrospinning. *J. Appl. Phys.* **2008**, *103*, 104307. [CrossRef]
22. Samadian, H.; Mobasheri, H.; Hasanpour, S.; Faridi Majidi, R. Electrospinning of Polyacrylonitrile Nanofibers and Simulation of Electric Field via Finite Element method. *Nanomed. Res. J.* **2017**, *2*, 87–92. [CrossRef]
23. Reneker, D.H.; Chun, I. Nanometre diameter fibres of polymer, produced by electrospinning. *Nanotechnology* **1996**, *7*, 216–223. [CrossRef]
24. Supaphol, P.; Mit-Uppatham, C.; Nithitanakul, M. Ultrafine electrospun polyamide-6 fibers: Effect of emitting electrode polarity on morphology and average fiber diameter. *J. Polym. Sci. Part B Polym. Phys.* **2005**, *43*, 3699–3712. [CrossRef]
25. Casper, C.L.; Stephens, J.S.; Tassi, N.G.; Chase, D.B.; Rabolt, J.F. Controlling Surface Morphology of Electrospun Polystyrene Fibers: Effect of Humidity and Molecular Weight in the Electrospinning Process. *Macromolecules* **2004**, *37*, 573–578. [CrossRef]
26. Al-Marri, M.J.; Masoud, M.S.; Nassar, A.M.G.; Zagho, M.M.; Khader, M.M. Synthesis and Characterization of Poly(vinyl alcohol): Cloisite® 20A Nanocomposites. *J. Vinyl Addit. Technol.* **2017**, *23*, 181–187. [CrossRef]
27. Kabalan, L.; Zagho, M.M.; Al-Marri, M.J.; Khader, M.M. Experimental and theoretical studies on the mechanical and structural changes imposed by the variation of clay loading on poly(vinyl alcohol)/cloisite® 93A nanocomposites. *J. Vinyl Addit. Technol.* **2018**. [CrossRef]
28. Hussein, E.A.; Zagho, M.M.; Nasrallah, G.K.; Elzatahry, A.A. Recent advances in functional nanostructures as cancer photothermal therapy. *Int. J. Nanomed.* **2018**, *13*, 2897–2906. [CrossRef]
29. Zagho, M.M.; Khader, M.M. The Impact of Clay Loading on the Relative Intercalation of Poly (Vinyl Alcohol)—Clay Composites. *J. Mater. Sci. Chem. Eng.* **2016**, *4*, 20–31. [CrossRef]
30. Zagho, M.M.; Al, M.; Almaadeed, A.; Majeed, K. Role of TiO2 and Carbon Nanotubes on Polyethylene, and Effect of Accelerated Weathering on Photo Oxidation and Mechanical Properties. *J. Vinyl Addit. Technol.* **2018**. [CrossRef]
31. Zagho, M.M.; Elzatahry, A. Recent Trends in Electrospinning of Polymer Nanofibers and their Applications as Templates for Metal Oxide Nanofibers Preparation. In *Electrospinning—Material, Techniques, and Biomedical Applications*; Haider, S., Ed.; InTech: Rijeka, Croatia, 2016; pp. 3–24.
32. Majeed, K.; AlMaadeed, M.A.A.; Zagho, M.M. Comparison of the effect of carbon, halloysite and titania nanotubes on the mechanical and thermal properties of LDPE based nanocomposite films. *Chin. J. Chem. Eng.* **2018**, *26*, 428–435. [CrossRef]
33. Zagho, M.M.; Hussein, E.A.; Elzatahry, A.A. Recent Overviews in Functional Polymer Composites for Biomedical Applications. *Polymers* **2018**, *10*, 739. [CrossRef]
34. Khan, M.I.; Zagho, M.M.; Shakoor, R.A. A Brief Overview of Shape Memory Effect in Thermoplastic Polymers, Smart Polymer Nanocomposites. In *Smart Polymer Nanocomposites*; Ponnamma, D., Sadasivuni, K.K., Cabibihan, J.-J., Al-Maadeed, M.A.A., Eds.; Springer Series on Polymer and Composite Materials; Springer International Publishing: Cham, Switzerland, 2017; pp. 281–301, ISBN 978-3-319-50423-0.
35. Dawoud, H.D.; Altahtamouni, T.M.; Zagho, M.M.; Bensalah, N. A brief overview of flexible CNT/PANI super capacitors. *J. Mater. Sci. Nanotechnol.* **2017**, *1*, 23–36.
36. Zagho, M.M.; AlMaadeed, M.A.A.; Majeed, K. Thermal Properties of TiO2 NP/CNT/LDPE Hybrid Nanocomposite Films. *Polymers* **2018**, *10*, 1270. [CrossRef]
37. Al-Enizi, A.M.; Zagho, M.M.; Elzatahry, A.A. Polymer-Based Electrospun Nanofibers for Biomedical Applications. *Nanomaterials* **2018**, *8*, 259. [CrossRef] [PubMed]

38. Gómez-Tejedor, J.A.; Van Overberghe, N.; Rico, P.; Ribelles, J.L.G. Assessment of the parameters influencing the fiber characteristics of electrospun poly(ethyl methacrylate) membranes. *Eur. Polym. J.* **2011**, *47*, 119–129. [CrossRef]
39. Liu, M.; Duan, X.; Li, Y.; Yang, D.; Long, Y. Electrospun nano fi bers for wound healing. *Mater. Sci. Eng. C-Mater.* **2017**, *76*, 1413–1423. [CrossRef] [PubMed]
40. Zhang, J.; Duan, Y.; Wei, D.; Wang, L.; Wang, H.; Gu, Z.; Kong, D. Co-electrospun fibrous scaffold–adsorbed DNA for substrate-mediated gene delivery. *J. Biomed. Mater. Res. Part A* **2011**, *96A*, 212–220. [CrossRef] [PubMed]
41. Cao, H.; Jiang, X.; Chai, C.; Chew, S.Y. RNA interference by nanofiber-based siRNA delivery system. *J. Control. Release* **2010**, *144*, 203–212. [CrossRef] [PubMed]
42. Balaji, A.; Vellayappan, M.V.; John, A.A.; Subramanian, A.P.; Jaganathan, S.K.; Supriyanto, E.; Razak, S.I.A. An insight on electrospun-nanofibers-inspired modern drug delivery system in the treatment of deadly cancers. *RSC Adv.* **2015**, *5*, 57984–58004. [CrossRef]
43. Kim, G.H.; Min, T.; Park, S.A.; Kim, W.D. Coaxially electrospun micro/nanofibrous poly(ε-caprolactone)/eggshell-protein scaffold. *Bioinspir. Biomim.* **2008**, *3*, 016006. [CrossRef]
44. Kim, G.; Yoon, H.; Park, Y. Drug release from various thicknesses of layered mats consisting of electrospun polycaprolactone and polyethylene oxide micro/nanofibers. *Appl. Phys. A Mater. Sci. Process.* **2010**, *100*, 1197–1204. [CrossRef]
45. Fathi-Azarbayjani, A.; Chan, S.Y. Single and multi-layered nanofibers for rapid and controlled drug delivery. *Chem. Pharm. Bull.* **2010**, *58*, 143–146. [CrossRef]
46. Immich, A.P.S.; Arias, M.L.; Carreras, N.; Boemo, R.L.; Tornero, J.A. Drug delivery systems using sandwich configurations of electrospun poly(lactic acid) nanofiber membranes and ibuprofen. *Mater. Sci. Eng. C* **2013**, *33*, 4002–4008. [CrossRef] [PubMed]
47. Chen, D.W.C.; Liao, J.Y.; Liu, S.J.; Chan, E.C. Novel biodegradable sandwich-structured nanofibrous drug-eluting membranes for repair of infected wounds: An in vitro and in vivo study. *Int. J. Nanomed.* **2012**, *7*, 763–771. [CrossRef]
48. Ahlin Grabnar, P.; Kristl, J. The manufacturing techniques of drug-loaded polymeric nanoparticles from preformed polymers. *J. Microencapsul.* **2011**, *28*, 323–335. [CrossRef] [PubMed]
49. Zamani, M.; Prabhakaran, M.P.; Ramakrishna, S. Advances in drug delivery via electrospun and electrosprayed nanomaterials. *Int. J. Nanomed.* **2013**, *8*, 2997–3017. [CrossRef]
50. Chen, S.; Wang, G.; Wu, T.; Zhao, X.; Liu, S.; Li, G.; Cui, W.; Fan, C. Silver nanoparticles/ibuprofen-loaded poly(L-lactide) fibrous membrane: Anti-infection and anti-adhesion effects. *Int. J. Mol. Sci.* **2014**, *15*, 14014–14025. [CrossRef] [PubMed]
51. Mele, E. Electrospinning of natural polymers for advanced wound care: Towards responsive and adaptive dressings. *J. Mater. Chem. B* **2016**, *4*, 4801–4812. [CrossRef]
52. Lee, K.Y.; Jeong, L.; Kang, Y.O.; Lee, S.J.; Park, W.H. Electrospinning of polysaccharides for regenerative medicine. *Adv. Drug Deliv. Rev.* **2009**, *61*, 1020–1032. [CrossRef]
53. Toshkova, R.; Manolova, N.; Gardeva, E.; Ignatova, M.; Yossifova, L.; Rashkov, I.; Alexandrov, M. Antitumor activity of quaternized chitosan-based electrospun implants against Graffi myeloid tumor. *Int. J. Pharm.* **2010**, *400*, 221–233. [CrossRef]
54. Garg, T.; Rath, G.; Goyal, A.K. Biomaterials-based nanofiber scaffold: Targeted and controlled carrier for cell and drug delivery. *J. Drug Target.* **2015**, *23*, 202–221. [CrossRef] [PubMed]
55. Gunn, J.; Zhang, M. Polyblend nanofibers for biomedical applications: Perspectives and challenges. *Trends Biotechnol.* **2010**, *28*, 189–197. [CrossRef] [PubMed]
56. Kim, K.; Yu, M.; Zong, X.; Chiu, J.; Fang, D.; Seo, Y.-S.; Hsiao, B.S.; Chu, B.; Hadjiargyrou, M. Control of degradation rate and hydrophilicity in electrospun non-woven poly(d,l-lactide) nanofiber scaffolds for biomedical applications. *Biomaterials* **2003**, *24*, 4977–4985. [CrossRef]
57. Behrens, A.M.; Kim, J.; Hotaling, N.; Seppala, J.E.; Kofinas, P.; Tutak, W. Rapid fabrication of poly(DL-lactide) nanofiber scaffolds with tunable degradation for tissue engineering applications by air-brushing. *Biomed. Mater.* **2016**, *11*, 035001. [CrossRef] [PubMed]
58. Meng, Z.X.; Xu, X.X.; Zheng, W.; Zhou, H.M.; Li, L.; Zheng, Y.F.; Lou, X. Preparation and characterization of electrospun PLGA/gelatin nanofibers as a potential drug delivery system. *Colloids Surf. B* **2011**, *84*, 97–102. [CrossRef] [PubMed]

59. Binulal, N.S.; Natarajan, A.; Menon, D.; Bhaskaran, V.K.; Mony, U.; Nair, S.V. PCL-gelatin composite nanofibers electrospun using diluted acetic acid-ethyl acetate solvent system for stem cell-based bone tissue engineering. *J. Biomater. Sci. Polym. Ed.* **2014**, *25*, 325–340. [CrossRef] [PubMed]
60. Zhu, X.; Ni, S.; Xia, T.; Yao, Q.; Li, H.; Wang, B.; Wang, J.; Li, X.; Su, W. Anti-neoplastic cytotoxicity of SN-38-loaded PCL/Gelatin electrospun composite nanofiber scaffolds against human glioblastoma cells in vitro. *J. Pharm. Sci.* **2015**, *104*, 4345–4354. [CrossRef] [PubMed]
61. Gomes, S.R.; Rodrigues, G.; Martins, G.G.; Roberto, M.A.; Mafra, M.; Henriques, C.M.R.; Silva, J.C. In vitro and in vivo evaluation of electrospun nanofibers of PCL, chitosan and gelatin: A comparative study. *Mater. Sci. Eng. C* **2015**, *46*, 348–358. [CrossRef]
62. Alhosseini, S.N.; Moztarzadeh, F.; Mozafari, M.; Asgari, S.; Dodel, M.; Samadikuchaksaraei, A.; Kargozar, S.; Jalali, N. Synthesis and characterization of electrospun polyvinyl alcohol nanofibrous scaffolds modified by blending with chitosan for neural tissue engineering. *Int. J. Nanomed.* **2012**, *7*, 25–34. [CrossRef]
63. Mojtaba, K.; Hamid, M. Electrospinning, mechanical properties, and cell behavior study of chitosan/PVA nanofibers. *J. Biomed. Mater. Res. Part A* **2015**, *103*, 3081–3093. [CrossRef]
64. Ba, L.N.T.; Ki, M.Y.; Ho-Yeon, S.; Byong-Taek, L. Fabrication of polyvinyl alcohol/gelatin nanofiber composites and evaluation of their material properties. *J. Biomed. Mater. Res. Part B Appl. Biomater.* **2010**, *95B*, 184–191. [CrossRef]
65. Zupančič, Š.; Sinha-Ray, S.; Sinha-Ray, S.; Kristl, J.; Yarin, A.L. Controlled Release of Ciprofloxacin from Core–Shell Nanofibers with Monolithic or Blended Core. *Mol. Pharm.* **2016**, *13*, 1393–1404. [CrossRef] [PubMed]
66. Liu, Y.; Yu, H.; Liu, Y.; Liang, G.; Zhang, T.; Hu, Q. Dual Drug Spatiotemporal Release from Functional Gradient Scaffolds Prepared Using 3D Bioprinting and Electrospinning. *Polym. Eng. Sci.* **2016**, *56*, 170–177. [CrossRef]
67. Askari, P.; Zahedi, P.; Rezaeian, I. Three-layered electrospun PVA/PCL/PVA nanofibrous mats containing tetracycline hydrochloride and phenytoin sodium: A case study on sustained control release, antibacterial, and cell culture properties. *J. Appl. Polym. Sci.* **2016**, *133*, 43309. [CrossRef]
68. Ignatova, M.; Rashkov, I.; Manolova, N. Drug-loaded electrospun materials in wound-dressing applications and in local cancer treatment. *Exp. Opin. Drug Deliv.* **2013**, *10*, 469–483. [CrossRef] [PubMed]
69. Dwivedi, C.; Pandey, H.; Pandey, A.C.; Ramteke, P.W. Fabrication and assessment of gentamicin loaded electrospun nanofibrous scaffolds as a quick wound healing dressing material. *Curr. Nanosci.* **2015**, *11*, 222–228. [CrossRef]
70. Kenawy, E.R.; Bowlin, G.L.; Mansfield, K.; Layman, J.; Simpson, D.G.; Sanders, E.H.; Wnek, G.E. Release of tetracycline hydrochloride from electrospun poly(ethylene-co-vinylacetate), poly(lactic acid), and a blend. *J. Control. Release* **2002**, *81*, 57–64. [CrossRef]
71. Zhang, Y.; Lee, M.W.; An, S.; Sinha-Ray, S.; Khansari, S.; Joshi, B.; Hong, S.; Hong, J.-H.; Kim, J.-J.; Pourdeyhimi, B.; et al. Antibacterial activity of photocatalytic electrospun titania nanofiber mats and solution-blown soy protein nanofiber mats decorated with silver nanoparticles. *Catal. Commun.* **2013**, *34*, 35–40. [CrossRef]
72. Shalumon, K.T.; Anulekha, K.H.; Nair, S.V.; Nair, S.V.; Chennazhi, K.P.; Jayakumar, R. Sodium alginate/poly(vinyl alcohol)/nano ZnO composite nanofibers for antibacterial wound dressings. *Int. J. Biol. Macromol.* **2011**, *49*, 247–254. [CrossRef]
73. Unnithan, A.R.; Barakat, N.A.M.; Pichiah, P.B.T.; Gnanasekaran, G.; Nirmala, R.; Cha, Y.-S.; Jung, C.-H.; El-Newehy, M.; Kim, H.Y. Wound-dressing materials with antibacterial activity from electrospun polyurethane–dextran nanofiber mats containing ciprofloxacin HCl. *Carbohydr. Polym.* **2012**, *90*, 1786–1793. [CrossRef]
74. Wang, C.-C.; Su, C.-H.; Chen, C.-C. Water absorbing and antibacterial properties of N-isopropyl acrylamide grafted and collagen/chitosan immobilized polypropylene nonwoven fabric and its application on wound healing enhancement. *J. Biomed. Mater. Res. Part A* **2008**, *84A*, 1006–1017. [CrossRef] [PubMed]
75. Xu, X.; Chen, X.; Xu, X.; Lu, T.; Wang, X.; Yang, L.; Jing, X. BCNU-loaded PEG–PLLA ultrafine fibers and their in vitro antitumor activity against Glioma C6 cells. *J. Control. Release* **2006**, *114*, 307–316. [CrossRef] [PubMed]
76. Xu, X.; Chen, X.; Wang, Z.; Jing, X. Ultrafine PEG-PLA fibers loaded with both paclitaxel and doxorubicin hydrochloride and their in vitro cytotoxicity. *Eur. J. Pharm. Biopharm.* **2009**, *72*, 18–25. [CrossRef] [PubMed]
77. Xie, J.; Ruo, S.T.; Wang, C.H. Biodegradable microparticles and fiber fabrics for sustained delivery of cisplatin to treat C6 glioma in vitro. *J. Biomed. Mater. Res. A* **2008**, *85*, 897–908. [CrossRef] [PubMed]

78. Chew, S.Y.; Wen, J.; Yim, E.K.F.; Leong, K.W. Sustained release of proteins from electrospun biodegradable fibers. *Biomacromolecules* **2005**, *6*, 2017–2024. [CrossRef] [PubMed]
79. Rujitanaroj, P.; Wang, Y.C.; Wang, J.; Chew, S.Y. Nanofiber-mediated controlled release of siRNA complexes for long term gene-silencing applications. *Biomaterials* **2011**, *32*, 5915–5923. [CrossRef] [PubMed]
80. Saraf, A.; Baggett, L.S.; Raphael, R.M.; Kasper, F.K.; Mikos, A.G. Regulated non-viral gene delivery from coaxial electrospun fiber mesh scaffolds. *J. Control. Release* **2010**, *143*, 95–103. [CrossRef]
81. Mickova, A.; Buzgo, M.; Benada, O.; Rampichova, M.; Fisar, Z. Core/Shell Nanofibers with Embedded Liposomes as a Drug Delivery System. *Biomacromolecules* **2012**, *13*, 952–962. [CrossRef]
82. Pillay, V.; Dott, C.; Choonara, Y.E.; Tyagi, C.; Tomar, L.; Kumar, P.; Du Toit, L.C.; Ndesendo, V.M.K. A review of the effect of processing variables on the fabrication of electrospun nanofibers for drug delivery applications. *J. Nanomater.* **2013**, *2013*, 789289. [CrossRef]
83. Leung, V.; Ko, F. Biomedical applications of nanofibers. *Polym. Adv. Technol.* **2011**, *22*, 350–365. [CrossRef]
84. Srikar, R.; Yarin, A.L.; Megaridis, C.M.; Bazilevsky, A.V.; Kelley, E. Desorption-Limited Mechanism of Release from Polymer Nanofibers. *Langmuir* **2008**, *24*, 965–974. [CrossRef] [PubMed]
85. Gandhi, M.; Srikar, R.; Yarin, A.L.; Megaridis, C.M.; Gemeinhart, R.A. Mechanistic Examination of Protein Release from Polymer Nanofibers. *Mol. Pharm.* **2009**, *6*, 641–647. [CrossRef] [PubMed]
86. Khansari, S.; Duzyer, S.; Sinha-Ray, S.; Hockenberger, A.; Yarin, A.L.; Pourdeyhimi, B. Two-Stage Desorption-Controlled Release of Fluorescent Dye and Vitamin from Solution-Blown and Electrospun Nanofiber Mats Containing Porogens. *Mol. Pharm.* **2013**, *10*, 4509–4526. [CrossRef] [PubMed]
87. Zhang, Q.; Li, Y.; Lin, Z.Y.W.; Wong, K.K.Y.; Lin, M.; Yildirimer, L.; Zhao, X. Electrospun polymeric micro/nanofibrous scaffolds for long-term drug release and their biomedical applications. *Drug Discov. Today* **2017**, *22*, 1351–1366. [CrossRef] [PubMed]
88. Vacanti, J.P.; Langer, R. Tissue engineering: The design and fabrication of living replacement devices for surgical reconstruction and transplantation. *Lancet* **1999**, *354*, S32–S34. [CrossRef]
89. Diegelmann, R.F.; Evans, M.C. Wound healing: An overview of acute, fibrotic and delayed healing. *Front. Biosci.* **2004**, *9*, 283–289. [CrossRef] [PubMed]
90. Halim, A.S.; Khoo, T.L.; Yussof, S.J.M. Biologic and synthetic skin substitutes: An overview. *Indian J. Plast. Surg.* **2010**, *43*, S23–S28. [CrossRef] [PubMed]
91. Khil, M.-S.; Cha, D.-I.; Kim, H.-Y.; Kim, I.-S.; Bhattarai, N. Electrospun nanofibrous polyurethane membrane as wound dressing. *J. Biomed. Mater. Res. B. Appl. Biomater.* **2003**, *67*, 675–679. [CrossRef]
92. Saeed, S.M.; Mirzadeh, H.; Zandi, M.; Barzin, J. Designing and fabrication of curcumin loaded PCL/PVA multi-layer nanofibrous electrospun structures as active wound dressing. *Prog. Biomater.* **2017**, *6*, 39–48. [CrossRef]
93. Mohseni, M.; Shamloo, A.; Aghababaei, Z.; Vossoughi, M.; Moravvej, H. Antimicrobial Wound Dressing Containing Silver Sulfadiazine With High Biocompatibility: In Vitro Study. *Artif. Organs* **2016**, *40*, 765–773. [CrossRef]
94. Khodkar, F.; Ebrahimi, N.G. Preparation and properties of antibacterial, biocompatible core-shell fibers produced by coaxial electrospinning. *J. Appl. Polym. Sci.* **2017**, *134*, 44979. [CrossRef]
95. Online, V.A.; Du, L.; Xu, H.Z.; Li, T.; Zhang, Y.; Zou, F.Y. Fabrication of ascorbyl palmitate loaded poly(caprolactone)/silver nanoparticle embedded poly(vinyl alcohol) hybrid nanofibre mats as active wound dressings via dual-spinneret electrospinning. *RSC Adv.* **2017**, 31310–31318. [CrossRef]
96. Liu, X.; Lin, T.; Fang, J.; Yao, G.; Zhao, H.; Dodson, M.; Wang, X. In vivo wound healing and antibacterial performances of electrospun nanofibre membranes. *J. Biomed. Mater. Res. Part A* **2010**, 499–508. [CrossRef]
97. Wang, S.-D.; Ma, Q.; Wang, K.; Chen, H.-W. Improving Antibacterial Activity and Biocompatibility of Bioinspired Electrospinning Silk Fibroin Nanofibers Modified by Graphene Oxide. *ACS Omega* **2018**, *3*, 406–413. [CrossRef]
98. Fröhlich, M.; Grayson, W.L.; Wan, L.Q.; Marolt, D.; Drobnic, M.; Vunjak-Novakovic, G. Tissue engineered bone grafts: Biological requirements, tissue culture and clinical relevance. *Curr. Stem Cell Res. Ther.* **2008**, *3*, 254–264. [CrossRef] [PubMed]
99. Nooeaid, P.; Salih, V.; Beier, J.P.; Boccaccini, A.R. Osteochondral tissue engineering: Scaffolds, stem cells and applications. *J. Cell. Mol. Med.* **2012**, *16*, 2247–2270. [CrossRef] [PubMed]
100. Uma Maheshwari, S.; Samuel, V.K.; Nagiah, N. Fabrication and evaluation of (PVA/HAp/PCL) bilayer composites as potential scaffolds for bone tissue regeneration application. *Ceram. Int.* **2014**, *40*, 8469–8477. [CrossRef]

101. Su, Y.; Su, Q.; Liu, W.; Lim, M.; Venugopal, J.R.; Mo, X.; Ramakrishna, S.; Al-Deyab, S.S.; El-Newehy, M. Controlled release of bone morphogenetic protein 2 and dexamethasone loaded in core-shell PLLACL-collagen fibers for use in bone tissue engineering. *Acta Biomater.* **2012**, *8*, 763–771. [CrossRef]
102. Shafiee, A.; Soleimani, M.; Chamheidari, G.A.; Seyedjafari, E.; Dodel, M.; Atashi, A.; Gheisari, Y. Electrospun nanofiber-based regeneration of cartilage enhanced by mesenchymal stem cells. *J. Biomed. Mater. Res. Part A* **2011**, 467–478. [CrossRef]
103. Wu, S.; Wang, J.; Jin, L.; Li, Y.; Wang, Z. Effects of Polyacrylonitrile/MoS2 Composite Nanofibers on the Growth Behavior of Bone Marrow Mesenchymal Stem Cells. *ACS Appl. Nano Mater.* **2018**, *1*, 337–343. [CrossRef]
104. Qu, Y.; Wang, B.; Chu, B.; Liu, C.; Rong, X.; Chen, H.; Peng, J.; Qian, Z. Injectable and Thermosensitive Hydrogel and PDLLA Electrospun Nanofiber Membrane Composites for Guided Spinal Fusion. *ACS Appl. Mater. Interfaces* **2018**, *10*, 4462–4470. [CrossRef] [PubMed]
105. Guyton, A.; Hall, J. *Textbook of Medical Physiology*, 11th ed.; Elsevier Saunders: Philadelphia, PA, USA, 2006; pp. 802–804, ISBN 0721602401.
106. Bach, A.D.; Beier, J.P.; Stern-Staeter, J.; Horch, R.E. Skeletal muscle tissue engineering. *J. Cell. Mol. Med.* **2004**, *8*, 413–422. [CrossRef] [PubMed]
107. Bian, W.; Bursac, N. Engineered skeletal muscle tissue networks with controllable architecture. *Biomaterials* **2009**, *30*, 1401–1412. [CrossRef] [PubMed]
108. Liao, H.; Zhou, G.-Q. Development and Progress of Engineering of Skeletal Muscle Tissue. *Tissue Eng. Part B Rev.* **2009**, *15*, 319–331. [CrossRef]
109. Mckeon-fischer, K.D.; Flagg, D.H.; Freeman, J.W. Poly(acrylic acid)/poly(vinyl alcohol) compositions coaxially electrospun with poly (3-caprolactone) and multi-walled carbon nanotubes to create nanoactuating scaffolds. *Polymer* **2011**, *52*, 4736–4743. [CrossRef]
110. Gurunathan, K.; Murugan, A.V.; Marimuthu, R.; Mulik, U.; Amalnerkar, D. Electrochemically synthesised conducting polymeric materials for applications towards technology in electronics, optoelectronics and energy storage devices. *Mater. Chem. Phys.* **1999**, *61*, 173–191. [CrossRef]
111. Guimard, N.K.; Gomez, N.; Schmidt, C.E. Conducting polymers in biomedical engineering. *Prog. Polym. Sci.* **2007**, *32*, 876–921. [CrossRef]
112. Ateh, D.D.; Navsaria, H.A.; Vadgama, P. Polypyrrole-based conducting polymers and interactions with biological tissues. *J. R. Soc. Interface* **2006**, *3*, 741–752. [CrossRef]
113. Schmidt, C.E.; Shastri, V.R.; Vacanti, J.P.; Langer, R. Stimulation of neurite outgrowth using an electrically conducting polymer. *Proc. Natl. Acad. Sci. USA* **1997**, *94*, 8948–8953. [CrossRef]
114. Joo, S.; Kang, K.; Nam, Y. In vitro neurite guidance effects induced by polylysine pinstripe micropatterns with polylysine background. *J. Biomed. Mater. Res. A* **2015**, *103*, 2731–2739. [CrossRef]
115. Lee, J.Y.; Bashur, C.A.; Milroy, C.A.; Forciniti, L.; Goldstein, A.S.; Schmidt, C.E. Nerve growth factor-immobilized electrically conducting fibrous scaffolds for potential use in neural engineering applications. *IEEE Trans. Nanobioscience* **2012**, *11*, 15–21. [CrossRef] [PubMed]
116. Gomez, N.; Lee, J.Y.; Nickels, J.D.; Schmidt, C.E. Micropatterned polypyrrole: A combination of electrical and topographical characteristics for the stimulation of cells. *Adv. Funct. Mater.* **2007**, *17*, 1645–1653. [CrossRef]
117. Lee, J.Y.; Bashur, C.A.; Goldstein, A.S.; Schmidt, C.E. Biomaterials Polypyrrole-coated electrospun PLGA nanofibers for neural tissue applications. *Biomaterials* **2009**, *30*, 4325–4335. [CrossRef] [PubMed]
118. Depla, D.; Mahieu, S.; Greene, J.E. Chapter 5—Sputter Deposition Processes. In *Handbook of Deposition Technologies for Films and Coatings*, 3rd ed.; Martin, P.M., Ed.; William Andrew Publishing: Boston, MA, USA, 2010; pp. 253–296, ISBN 978-0-8155-2031-3.
119. Maurya, D.; Sardarinejad, A.; Alameh, K. Recent Developments in R.F. Magnetron Sputtered Thin Films for pH Sensing Applications—An Overview. *Coatings* **2014**, *4*, 756–771. [CrossRef]
120. George, J. *Preparation of Thin Films*; CRC Press: New York, NY, USA, 1992.
121. Gomathi, N.; Sureshkumar, A.; Neogi, S. RF plasma-treated polymers for biomedical applications. *Curr. Sci.* **2008**, *94*, 1478–1486.
122. Liston, E.M. Plasma treatment for improved bonding: A review. *J. Adhes.* **1989**, *30*, 199–218. [CrossRef]
123. Gao, A.; Hang, R.; Huang, X.; Zhao, L.; Zhang, X.; Wang, L.; Tang, B.; Ma, S.; Chu, P.K. The effects of titania nanotubes with embedded silver oxide nanoparticles on bacteria and osteoblasts. *Biomaterials* **2014**, *35*, 4223–4235. [CrossRef] [PubMed]

124. Huang, H.-L.; Chang, Y.-Y.; Chen, H.-J.; Chou, Y.-K.; Lai, C.-H.; Chen, M.Y.C. Antibacterial properties and cytocompatibility of tantalum oxide coatings with different silver content. *J. Vac. Sci. Technol. A* **2014**, *32*, 02B117. [CrossRef]
125. Wang, Q.; Wang, X.; Li, X.; Cai, Y.; Wei, Q. Surface modification of PMMA/O-MMT composite microfibers by TiO2coating. *Appl. Surf. Sci.* **2011**, *258*, 98–102. [CrossRef]
126. Lee, T.-J.; Kang, S.; Jeong, G.-J.; Yoon, J.-K.; Bhang, S.H.; Oh, J.; Kim, B.-S. Incorporation of Gold-Coated Microspheres into Embryoid Body of Human Embryonic Stem Cells for Cardiomyogenic Differentiation. *Tissue Eng. Part A* **2015**, *21*, 374–381. [CrossRef] [PubMed]
127. Bolbasov, E.N.; Rybachuk, M.; Golovkin, A.S.; Antonova, L.V.; Shesterikov, E.V.; Malchikhina, A.I.; Novikov, V.A.; Anissimov, Y.G.; Tverdokhlebov, S.I. Surface modification of poly(l-lactide) and polycaprolactone bioresorbable polymers using RF plasma discharge with sputter deposition of a hydroxyapatite target. *Mater. Lett.* **2014**, *132*, 281–284. [CrossRef]
128. Majumdar, A.; Butola, B.S.; Thakur, S. Development and performance optimization of knitted antibacterial materials using polyester-silver nanocomposite fibres. *Mater. Sci. Eng. C* **2015**, *54*, 26–31. [CrossRef] [PubMed]
129. Kalakonda, P.; Aldhahri, M.A.; Abdel-wahab, M.S.; Tamayol, A.; Moghaddam, K.M.; Ben Rached, F.; Pain, A.; Khademhosseini, A.; Memic, A.; Chaieb, S. Microfibrous silver-coated polymeric scaffolds with tunable mechanical properties. *RSC Adv.* **2017**, *7*, 34331–34338. [CrossRef]
130. Badaraev, A.D.; Nemoykina, A.L.; Bolbasov, E.N.; Tverdokhlebov, S.I. PLLA scaffold modification using magnetron sputtering of the copper target to provide antibacterial properties. *Resour. Technol.* **2017**, *3*, 204–211. [CrossRef]
131. Barbarash, L.S.; Bolbasov, E.N.; Antonova, L.V.; Matveeva, V.G.; Velikanova, E.A.; Shesterikov, E.V.; Anissimov, Y.G.; Tverdokhlebov, S.I. Surface modification of poly-ε-caprolactone electrospun fibrous scaffolds using plasma discharge with sputter deposition of a titanium target. *Mater. Lett.* **2016**, *171*, 87–90. [CrossRef]
132. Blantocas, G.Q.; Alaboodi, A.S.; Abdel-baset, H.M. Synthesis of Chitosan—TiO$_2$ Antimicrobial Composites via a 2-Step Process of Electrospinning and Plasma Sputtering. *Arab. J. Sci. Eng.* **2018**, *43*, 389–398. [CrossRef]
133. Beregoi, M.; Busuioc, C.; Evanghelidis, A.; Matei, E.; Iordache, F.; Radu, M.; Dinischiotu, A.; Enculescu, I. Electrochromic properties of polyaniline-coated fiber webs for tissue engineering applications. *Int. J. Pharm.* **2016**, *510*, 465–473. [CrossRef] [PubMed]
134. Goreninskii, S.I.; Bogomolova, N.N.; Malchikhina, A.I.; Golovkin, A.S.; Bolbasov, E.N.; Safronova, T.V.; Putlyaev, V.I.; Tverdokhlebov, S.I. Biological Effect of the Surface Modification of the Fibrous Poly(L-lactic acid) Scaffolds by Radio Frequency Magnetron Sputtering of Different Calcium-Phosphate Targets. *Bionanoscience* **2017**, *7*, 50–57. [CrossRef]
135. Dastjerdi, R.; Montazer, M. A review on the application of inorganic nano-structured materials in the modification of textiles: Focus on anti-microbial properties. *Colloids Surf. B* **2010**, *79*, 5–18. [CrossRef]
136. Murugesh Babu, K.; Ravindra, K.B. Bioactive antimicrobial agents for finishing of textiles for health care products. *J. Text. Inst.* **2015**, *106*, 706–717. [CrossRef]
137. Zhong, W. Efficacy and toxicity of antibacterial agents used in wound dressings. *Cutan. Ocul. Toxicol.* **2015**, *34*, 61–67. [CrossRef] [PubMed]
138. Ren, S.; Dong, L.; Zhang, X.; Lei, T.; Ehrenhauser, F.; Song, K.; Li, M.; Sun, X.; Wu, Q. Electrospun nanofibers made of silver nanoparticles, cellulose nanocrystals, and polyacrylonitrile as substrates for surface-enhanced raman scattering. *Materials* **2017**, *10*, 68. [CrossRef] [PubMed]
139. Ge, L.; Li, Q.; Wang, M.; Ouyang, J.; Li, X.; Xing, M.M.Q. Nanosilver particles in medical applications: Synthesis, performance, and toxicity. *Int. J. Nanomed.* **2014**, *9*, 2399–2407.
140. Yip, J.; Jiang, S.; Wong, C. Characterization of metallic textiles deposited by magnetron sputtering and traditional metallic treatments. *Surf. Coat. Technol.* **2009**, *204*, 380–385. [CrossRef]
141. Slawson, R.M.; Van Dyke, M.I.; Lee, H.; Trevors, J.T. Germanium and silver resistance, accumulation, and toxicity in microorganisms. *Plasmid* **1992**, *27*, 72–79. [CrossRef]
142. Zhao, G.; Stevens, S.E. Multiple parameters for the comprehensive evaluation of the susceptibility of Escherichia coli to the silver ion. *BioMetals* **1998**, *11*, 27–32. [CrossRef] [PubMed]
143. Schierholz, J.M.; Beuth, J.; Pulverer, G. Silver coating of medical devices for catheter-associated infections? *Am. J. Med.* **1999**, *107*, 101–102. [PubMed]
144. Mishra, V.K.; Kumar, A. Impact of Metal Nanoparticles on the Plant Growth Promoting Rhizobacteria. *Dig. J. Nanomater. Biostruct.* **2009**, *4*, 587–592.

145. Hamouda, T.; Baker, J.R. Antimicrobial mechanism of action of surfactant lipid preparations in enteric gram-negative bacilli. *J. Appl. Microbiol.* **2000**, *89*, 397–403. [CrossRef] [PubMed]
146. Pal, S.; Tak, Y.K.; Song, J.M. Does the antibacterial activity of silver nanoparticles depend on the shape of the nanoparticle? A study of the gram-negative bacterium Escherichia coli. *J. Biol. Chem.* **2015**, *290*, 1712–1720. [CrossRef] [PubMed]
147. Tian, J.; Wong, K.K.Y.; Ho, C.M.; Lok, C.N.; Yu, W.Y.; Che, C.M.; Chiu, J.F.; Tam, P.K.H. Topical delivery of silver nanoparticles promotes wound healing. *ChemMedChem* **2007**, *2*, 129–136. [CrossRef] [PubMed]
148. Demling, R.H.; Leslie DeSanti, M.D. The rate of re-epithelialization across meshed skin grafts is increased with exposure to silver. *Burns* **2002**, *28*, 264–266. [CrossRef]
149. Poon, V.K.M.; Burd, A. In vitro cytotoxity of silver: Implication for clinical wound care. *Burns* **2004**, *30*, 140–147. [CrossRef] [PubMed]
150. Irfan, M.; Perero, S.; Miola, M.; Maina, G.; Ferri, A.; Ferraris, M.; Balagna, C. Antimicrobial functionalization of cotton fabric with silver nanoclusters/silica composite coating via RF co-sputtering technique. *Cellulose* **2017**, *24*, 2331–2345. [CrossRef]
151. Chen, Y.H.; Hsu, C.C.; He, J.L. Antibacterial silver coating on poly(ethylene terephthalate) fabric by using high power impulse magnetron sputtering. *Surf. Coat. Technol.* **2013**, *232*, 868–875. [CrossRef]
152. Liu, X.; Gan, K.; Liu, H.; Song, X.; Chen, T.; Liu, C. Antibacterial properties of nano-silver coated PEEK prepared through magnetron sputtering. *Dent. Mater.* **2017**, *33*, e348–e360. [CrossRef] [PubMed]
153. Muzio, G.; Miola, M.; Perero, S.; Oraldi, M.; Maggiora, M.; Ferraris, S.; Vernè, E.; Festa, V.; Festa, F.; Canuto, R.A.; et al. Polypropylene prostheses coated with silver nanoclusters/silica coating obtained by sputtering: Biocompatibility and antibacterial properties. *Surf. Coat. Technol.* **2017**, *319*, 326–334. [CrossRef]
154. Grass, G.; Rensing, C.; Solioz, M. Metallic copper as an antimicrobial surface. *Appl. Environ. Microbiol.* **2011**, *77*, 1541–1547. [CrossRef]
155. Castro, C.; Sanjines, R.; Pulgarin, C.; Osorio, P.; Giraldo, S.A.; Kiwi, J. Structure-reactivity relations for DC-magnetron sputtered Cu-layers during E. coli inactivation in the dark and under light. *J. Photochem. Photobiol. A Chem.* **2010**, *216*, 295–302. [CrossRef]
156. Prakash, N.H.; Sarma, A.; Sarma, B. Antibacterial studies of copper deposited water hyacinth fiber using RF Antibacterial studies of copper deposited water hyacinth fiber using RF plasma sputtering process. *Mater. Technol.* **2018**, *33*, 621–633. [CrossRef]
157. Rino, M.; Nathalie, D.G.; Tim, D.; Peter, D.; Christophe, L. Plasma Surface Modification of Biodegradable Polymers: A Review. *Plasma Process. Polym.* **2011**, *8*, 171–190. [CrossRef]
158. Staszek, M.; Siegel, J.; Rimpelová, S.; Lyutakov, O.; Švorčík, V. Cytotoxicity of noble metal nanoparticles sputtered into glycerol. *Mater. Lett.* **2015**, *158*, 351–354. [CrossRef]
159. Tverdokhlebov, S.I.; Bolbasov, E.N.; Shesterikov, E.V.; Antonova, L.V.; Golovkin, A.S.; Matveeva, V.G.; Petlin, D.G.; Anissimov, Y.G. Modification of polylactic acid surface using RF plasma discharge with sputter deposition of a hydroxyapatite target for increased biocompatibility. *Appl. Surf. Sci.* **2015**, *329*, 32–39. [CrossRef]
160. Bolbasov, E.N.; Maryin, P.V.; Stankevich, K.S.; Kozelskaya, A.I.; Shesterikov, E.V.; Khodyrevskaya, Y.I.; Nasonova, M.V.; Shishkova, D.K.; Kudryavtseva, Y.A.; Anissimov, Y.G.; et al. Surface modification of electrospun poly-(L-lactic) acid scaffolds by reactive magnetron sputtering. *Colloids Surf. B* **2018**, *162*, 43–51. [CrossRef]
161. Pantojas, V.M.; Velez, E. Initial Study on Fibers and Coatings for the Fabrication of Bioscaffolds. *P. R. Health Sci. J.* **2009**, *28*, 258–265.
162. Bolbasov, E.N.; Antonova, L.V.; Stankevich, K.S.; Ashrafov, A.; Matveeva, V.G.; Velikanova, E.A.; Khodyrevskaya, Y.I.; Kudryavtseva, Y.A.; Anissimov, Y.G.; Tverdokhlebov, S.I.; et al. The use of magnetron sputtering for the deposition of thin titanium coatings on the surface of bioresorbable electrospun fibrous scaffolds for vascular tissue engineering: A pilot study. *Appl. Surf. Sci.* **2017**, *398*, 63–72. [CrossRef]

© 2019 by the authors. Licensee MDPI, Basel, Switzerland. This article is an open access article distributed under the terms and conditions of the Creative Commons Attribution (CC BY) license (http://creativecommons.org/licenses/by/4.0/).

Review

Biopolymers for Biomedical and Pharmaceutical Applications: Recent Advances and Overview of Alginate Electrospinning

Jolanta Wróblewska-Krepsztul [1], Tomasz Rydzkowski [1,*], Iwona Michalska-Pożoga [1] and Vijay Kumar Thakur [2,3,*]

1. Department of Mechanical Engineering, Koszalin University of Technology, Raclawicka 15-17, 75-620 Koszalin, Poland; jolanta.wroblewska-krepsztul@tu.koszalin.pl (J.W.-K.); iwona.michalska-pozoga@tu.koszalin.pl (I.M.-P.)
2. Enhanced Composites and Structures Center, School of Aerospace, Transport and Manufacturing, Cranfield University, Bedfordshire MK43 0AL, UK
3. Department of Mechanical Engineering, School of Engineering, Shiv Nadar University, Uttar Pradesh 201314, India
* Correspondence: tomasz.rydzkowski@tu.koszalin.pl (T.R.); Vijay.Kumar@cranfield.ac.uk (V.K.T.)

Received: 15 January 2019; Accepted: 6 March 2019; Published: 10 March 2019

Abstract: Innovative solutions using biopolymer-based materials made of several constituents seems to be particularly attractive for packaging in biomedical and pharmaceutical applications. In this direction, some progress has been made in extending use of the electrospinning process towards fiber formation based on biopolymers and organic compounds for the preparation of novel packaging materials. Electrospinning can be used to create nanofiber mats characterized by high purity of the material, which can be used to create active and modern biomedical and pharmaceutical packaging. Intelligent medical and biomedical packaging with the use of polymers is a broadly and rapidly growing field of interest for industries and academia. Among various polymers, alginate has found many applications in the food sector, biomedicine, and packaging. For example, in drug delivery systems, a mesh made of nanofibres produced by the electrospinning method is highly desired. Electrospinning for biomedicine is based on the use of biopolymers and natural substances, along with the combination of drugs (such as naproxen, sulfikoxazol) and essential oils with antibacterial properties (such as tocopherol, eugenol). This is a striking method due to the ability of producing nanoscale materials and structures of exceptional quality, allowing the substances to be encapsulated and the drugs/biologically active substances placed on polymer nanofibers. So, in this article we briefly summarize the recent advances on electrospinning of biopolymers with particular emphasis on usage of Alginate for biomedical and pharmaceutical applications.

Keywords: biopolymers; packaging; pharmaceutical; biomedical; electrospinning; alginate

1. Introduction

Modern packaging should protect products from external factors, extend the period of maintaining the quality of the product and above all, minimize impact on the environment [1–6]. Figure 1 summarizes some of the properties desired in common packaging materials. Detailed properties have been discussed in some excellent review articles, so we won't be going into more detail [1,2,7–9]. In this direction, public awareness is rapidly growing and attracting greater attention to ecology, as well as the use of natural substances for a range of applications from automotive to biomedical [10–16]. A tendency in the packaging industry to design innovative materials and introduce new solutions is gaining greater attention [17,18]. Active and modern packaging frequently relies on natural substances such as natural polysaccharides [9]. It is imperative to preserve all aspects of environmental safety during production, as well as the effectiveness and safety of use for patients [19–23]. Interestingly, the cyclic olefin copolymer (COC) is most commonly used for tamper-resistant packaging and is a remarkable alternative to a damped polyvinyl chloride (PVC).

Figure 1. General properties of commonly used packaging materials.

Packaging in medical and biomedical engineering is defined as a technique that enables the closure of a pharmaceutical product from its production to its end use [24]. The role of pharmaceutical packaging is to provide life-saving drugs, surgical devices, nutraceuticals, pills, powders and liquids, to name a few [7,25]. Pharmaceutical packaging influences the isolation and ensures the safety, identity and convenience of using the drug. The packaging must communicate well with the patient so that there are no adverse effects on the health of the patient. The key issue in packaging is also the issue of ecological safety [26,27]. Drug companies that pack drugs are among the industry leaders due to their technological advances. Current trends in industrial research on such new materials are the result of a continuous series of challenges being faced by the industry. The packaging industry is constantly evolving and is an important factor in the development of the pharmaceutical industry and biomedical sciences. Electrospinning of polymers/nanomaterials [28,29] is one of the potential methods in the packaging process that allows for the use of biopolymers [30]/natural substances for the production of medical packaging, dressings, biosensors, medical implants, and is a growing trend in biomedical sciences [10,31–39]. Figure 2 illustrates the schematic of a coaxial electrospinning setup. The inset shows an illustration of a coaxial jet under applied voltage [39]. Mercante et al in their excellent review articles have also shown the number of publications that's involved the use of electrospinning in sensor applications and the number of these publications is increasing day by day [10].

Different types of electrospun fibers are being produced for biomedical applications [32]. However, the electrospinning of biopolymers is a very challenging process. For example, in the case of chitosan, its neutralization lead to the loss of chitosan traits by using unsuitable substances. In the production of continuous filaments, suitable solvents should be used so that there is no interruption. In addition, during chitosan nanofiber synthesis, the value of the electric field generated is also crucial. Too high of an electric field causes repulsion between ionic groups of the polymer backbone, which disturbs the formation of continuous filaments. The use of chitin and chitosan nanofibers in biomedical and other applications has been recently reviewed [32–34]. Jayakumar et al, in their article, had primarily focused on the properties, preparation and biomedical applications [32]. Electrospinning of silk has also been reported [35]. It is used because of beneficial properties such as non-toxicity and biocompatibility.

Due to its positive mechanical properties, it can be used in various temperature and humidity ranges. However, in the case of silk, one of the protein components, sericin, should be removed before biological application because it can cause allergic reactions, for example, in the case of medical dressings [36–38].

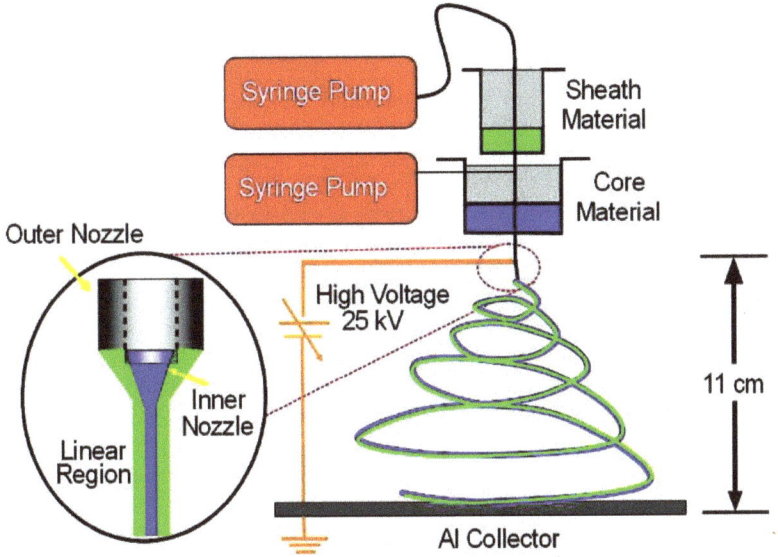

Figure 2. Schematic of coaxial electrospinning setup. The inset shows an illustration of a coaxial jet under applied voltage [39]. Reprinted with permission from Ref. [39]. Copyright American Chemical Society, 2010.

One of the organisms that is produced as a structural component of silk protein is Bombyx mori [40]. The protein it produces, silk fibroin, is a large crystalline macromolecule consisting of repeating units of amino acids, mainly alanine (A), glycine (G) and serine (S). The fibers that are formed with its participation exhibit good mechanical properties [41]. Researchers have also demonstrated that electrospinning of biopolymer blends of chitin and silk fibroin is possible [42]. In this process, nanofibrous membranes of blended chitosan and silk fibroin were successfully prepared using electrospinning in a HFIP/TFA spinning solvent. With the increment in the content of silk fibroin, the typical diameter of the as-prepared nanofibers was found to increase. The incorporation of silk fibroin was also found to contribute to the enhancement in the mechanical properties of nanofibrous membranes. Furthermore, with the increment in the chitosan content, the antibacterial activity became significantly suitable for wound dressings [42]. Electrospun collagen–chitosan nanofiber having a typical fiber diameter of 434–691 nm was also prepared as a biomimetic extracellular matrix (ECM) for endothelial cells and smooth muscle cells [43]. Different characterization techniques such as FTIR spectra analysis, XRD analysis, DSC and tensile testing were carried out to analyse the developed materials [43]. Double-network (DN) agarose/polyacrylamide nanofibers were prepared by electrospinning [44]. The DN of agarose/polyacrylamide (PAAm) nano fibers developed using simultaneous photo-polymerization and electrospinning. Different characterization techniques were used to confirm the realization of a crosslinked double-network. In comparison to the pristine agarose, the electrospun fibrous agarose/PAAm demonstrated 66.66% enhancement in the strength [44]. Similarly, electrospinning of alginate is being carried out for a number of applications [45]. Alginate is an important biopolymer with vast potential [46].

1.1. Synthetic Polymers

Different types of polymers ranging from natural to synthetic are rapidly becoming the most interesting subject of research in the sector of the biomedical industry [47]. They are often used in the packaging of medicines [48,49], as well as in the development of flexible ampoule/syringes that are more easy to use. However, adsorption and migration of the bioactive substance to the polymer changes in pH, permeability of oxygen, optical properties and the release of leached components affect their use and should be taken into account [50,51]. Interaction of the different outer components not only affects the drug but also the function of the polymeric container. Polyolefins, high-density polyethylene (HDPE) or polypropylene (PP) are some of the most common polymers used for the production of vials. Often, multilayer containers are developed to achieve such requirements as inertia, oxygen or UV protection. Polycyclic and olefinic polymers and copolymers (Daikyo Crystal Zeniths) have been used for filling polymer syringes [51,52]. Devices such as PVC tubes containing di-2- ethylhexylphthalate (DEHP) plasticizer are used in dialysis for blood supply or extracorporeal oxygenation. The bags containing the polymer are used to donate blood and store blood products. Due to lipophilicity, the plasticizer is transferred from the polymer surface to lipids and red blood cell membranes [53]. It has been found that the plasticizer in blood bags reduces haemolysis of red blood cells by about 50% compared to blood stored in non-plastic containers, which improves the quality of the blood product [50]. Tubes for extracorporeal circulation are often heparinized to reduce the coagulation, which causes intense contact with PVC and increases thrombogenicity [54].

For the storage of red blood cells, an alternative plasticizer such as butyryl-trihexyl-citrate (BTHC) or di-iso-nonyl-1, 2-cyclohexanedicarboxylate (DINCH) is used. Polyolefins as alternative polymers are used to store platelets [55,56]. Polyethylene and polyurethanes are used to create tubes. Tubes of positronic pumps are usually made of silicone [51]. Hemodialysis membranes are produced as bundles of hollow fibers with a surface in contact with blood. The technical requirements concern mainly the permeability for substances smaller than albumin, preventing the passage of impurities from the dialysate to the blood, and the compatibility of the membrane with blood. Previously, dialysis membranes were made of cellulose [51,57]. The hydroxyl groups were replaced with acetylene derivatives or other modified additives, preventing the activation of the complement system and the associated leukocyte activation and leukocyte sequestration in the lungs [51,57]. Synthetic membranes consist of a hydrophobic base material and hydrophilic components. The polyaryl sulfone co-precipitation membranes, polysulfone (PSf) and polyvinylprolidone (PVP) membranes are the most popular for a number of applications [58]. In addition, other membrane materials such as polyamide (PA), polycarbonate (PC) and polyacrylonitrile (PAN), PMMA, polyester polymer alloy (PEPA), ethylene vinyl alcohol copolymer (EVAL), and molecular thin nanoporous silicon diaphragms are also used. Poly (ethylene glycol) (PEG) is used in membranes to improve compatibility with blood [59]. Polymer stents used in the upper sections of the ureter are designed to overcome the problems of sperm infection. Silicone is the best biocompatible material with the lowest incrustation tendency. Its use is limited by low mechanical stiffness and high resistance. Therefore, polyurethane products with better mechanical properties than silicone were optimized [51,60]. The stents were coated with glycosaminoglycans (GAGs, heparin or pentosan polysulfone), phosphorylcholine, which increases the comfort of patients, reduces bacterial colonization and encrustation [51,60].

1.2. Biopolymers: Structure of Alginate

Recently, there has been a great thrust on the usage of biopolymers for a number of applications, especially in the biomedical and pharmaceutical [61–65]. The functional efficiency of the biopolymer molecules depends on the composition, physicochemical properties and structural features [66–68]. It is possible to rationally design the composition and structure of the biopolymer to obtain the appropriate functional attributes [23]. The internal structure of the polymer molecule determines many functional characteristics such as permeability, chargeability and integrity [69]. The stability of the biopolymer particles and their ability to aggregate is influenced by the electrical characteristics. Biopolymer particles with a high electric charge will repel and prevent aggregation. Molecules of biopolymers and their electrical properties influence the interaction with other molecules present in the surrounding environment [69]. Among natural biopolymers, alginate is one of the most popular and intensely studied [70,71]. It is an anionic biopolymer consisting of units of mannuronic acid and guluronic acid in irregular blocks [72]. Mannuronic acid and guluronic acid are linked by glycosidic linkages [73–75]. Mannuronic acid forms β (1 → 4) bonds and α bonds (1 → 4) with guluronic acid [76]. The stiffness of molecular chains is ensured by the rigid and bent conformations of guluronic acid [77,78]. Hadas and Simcha have recently reported their interesting work on the characterization of sodium alginate and calcium alginate with particular emphasis on their structure [79]. Different properties and applications of alginate have also been reviewed [80]. The properties of alginates used in biomedicine can be shaped by modifying the availability of their hydroxyl and carboxyl groups [81]. It affects the properties of alginates, such as solubility, hydrophobicity and their biological activity. Alginate hydrogels were created by crosslinking polymer chains [82]. The chemical properties of alginate hydrogels were found to depend on the cross-linking density of the chain [83]. One of the methods used in the design of alginate hydrogels is intermolecular cross-linking, in which only the alginate guluron groups react with the divalent cation most often the calcium used to gel the alginate [84]. Marguerite has summarized the applications of alginate especially for packaging in an excellent review article, so we won't go into detail [85].

2. Electrospun Antioxidant/Antibacterial Materials for Medical Packaging

Electrospinning is one of the most prominent techniques currently being used in the development of a number of products for biomedical applications, as discussed in the preceding section [86–88]. The electrospun creates ultrathin fibers collected in a random pattern. The mats produced in this way are used as filters, catalytic carriers, dressings and drug delivery systems [89]. Currently, scientific research focuses on the use of nanofiber properties and the focus on determining the parameters of electrospinning of biopolymers for medical and pharmaceutical applications [90–92]. In this technique, varying electrified fields are applied to produce polymer filaments, which can then be embedded in various device platforms (briefed in introduction section). The basic device for electrospinning consists of four parts: a syringe containing a polymer solution, a metallic needle, a power source and a metal collector (with variable construction) [93]. The polymer solution is extruded through the tip of the needle by forming a polymer stream that is leached out of the needle and expands as a result of deflection and dissipation of the electric charge voltage of the solution surface. The basic set for electrospun fibers is shown in Figure 3 [94]. Electrospinning of anisotropic fiber yarns (Biohybrid) was carried out via the extraction of microfibrils from bacterial cellulose networks [94]. In the equipment, a plastic syringe of 5 mL was used as the polymer reservoir in the continuous feed system. An aluminum reservoir that was filled with water was used and the fibers were subsequently collected on the surface of the water 120 mm from the needle tippet [94].

The polymeric solution stream flows towards the metal manifold with simultaneous evaporation of the solvent and then the nanofibers are deposited on the collector surface. When nanoparticles are deposited instead of nanofibers, the process takes place using electrospray [95,96]. Electrospinning is a fiber production process with a diameter of 0.01 to 10 μm by using electrostatic forces. In the electrospinning technique, a syringe filled with a polymer solution has a high potential source from 10 to 30 kV. The electrical voltage produces free electrons, ions or pairs of ions attracted to the electric field [97]. One of the elements is Taylor's cone, in which the polarity of the solution depends on the generator's voltage. The drive between similar charges in the electric space acts against the surface tension and fluid elasticity in the polymer solution, distorting the drop to the shape of the cone structure. In addition to the critical charge density, the Taylor taper becomes unstable and the liquid stream is released from the tip of the cone. Then, in the presence of the electric field, the polymer stream is formed, creating a light continuous form of the fiber. The fluid stream through continuous stretching becomes unstable because the number of spiral paths accumulated on the collector electrode increases [98]. This area is called whipping region. Polarized electrospun nanofibers move down to collision with less collector plate potential. The collector's morphology affects the straightening of the fiber. Different collectors are used, such as drum rotary collector, movable belt collector and straight mesh collector. To produce nanofibers with controlled morphology, parameters such as molecular weight of the polymer, surface tension, solution viscosity, solution conductivity, flow rate, temperature, collector morphology and distance of the collector apex are optimized [99]. It has been found that the diameter of the fibers increases with increasing concentration and viscosity. Increasing the molecular weight reduces the risk of interference on fibers such as balls and drops. It was determined that nanofibers can be formed with a variety of secondary structures. Thus, nanofibers with a core-shell structure, with an empty interior or with a porous structure may be formed [99,100]. As a drug delivery system, a mesh made of nanofibres produced by the electrospinning method offers a number of advantages. This is an attractive method due to its ability to produce nanoscale materials and structures of exceptional quality [101]. This allows substances to be encapsulated and drugs and biologically active substances placed on polymer nanofibers [102]. Biologically active substances can be attached to nanofibers like a mesh in the electrospinning process. In this way, antibacterial and antioxidant materials are created for medical applications [103,104]. Coating materials containing chitosan formulations with antioxidant and antifungal properties were also formed [88,105].

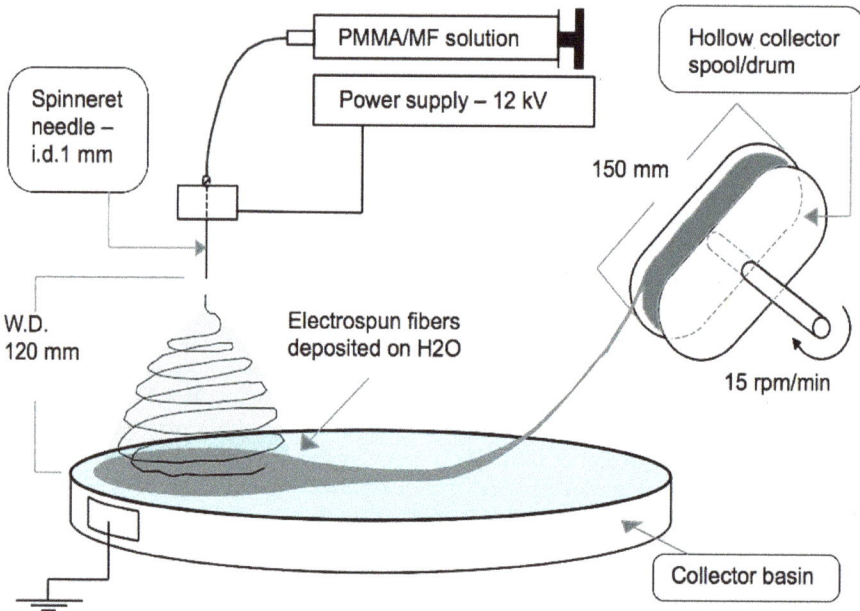

Figure 3. Electrospinning setup with a spinneret–water surface working distance of 120 mm and a spool running at 15 rpm/min. The spool (hollow) was designed to collect fibers with minimal spool–fiber yarn contact [94]. Reprinted with permission from Ref [94]. Copyright American Chemical Society, 2010.

By choosing a base mesh polymer, it can be easily designed to have improved mechanical properties, biocompatibility and cellular response, which makes mesh a good medical product in the nanocomposite materials sector. The electrospinning creates a macroporous scaffolding containing randomly oriented or aligned nanofibres on which the drug is placed. Electrospun nanofiber scaffolds provide the optimal environment for vaccinated cells [88]. Therapeutic compounds such as lipophilic and hydrophilic drugs, proteins, antimicrobials, etc. can be incorporated into the polymer nanofiber mesh by using monoaxial or coaxial electrospinning [106].

Nanofibers can help treat skin damage and can be considered a substitute for skin tissue [107]. The drug enclosed in the nanofiber mesh is released by various mechanisms when the nanofibrils mesh is swollen, biodegradable or absorbed by the human body. As an effective dressing, it inhibits bacterial growth during wound healing. The lower drug release rate allows for non-wound healing of wounds from a few days to several weeks [100,105]. Polymer after electrospinning acts as a barrier controlling the release of loaded molecules. The advantage of this method is the ability to close almost all drugs (especially hydrophobic) in the core, regardless of the drug-polymer interaction. Drugs, proteins, uptake factors and genes can be included in nanofibers [105].

Silver nanoparticles exhibit the capability to interact with bacteria. This allows the size of silver nanoparticles 1–10 nm in diameter. The smaller size provides better antibacterial activity. Scientific research has proven the antibacterial activity of electrospun nanofibers containing polylactic acid and silver nanoparticles (AgNPs) against Staphylococcus aureus and *Escherichia coli*. Fibers of silver nanoparticles and polyethylene oxide were mixed with polyurethane fibers exhibiting antibacterial activity on *Escherichia coli* [103,105]. Antimicrobial efficacy increases with an increase in the concentration of silver. The activity can also be increased by reducing silver nanoparticles. Smaller silver nanoparticles have the ability to disperse the bacterial membrane and inhibit bacterial growth. Antimicrobial activity of electrospun nanofibers results from the distribution of silver nanoparticles to electrospun nanofibers. Increased access of silver nanoparticles to electrospun nanofibers increases microbial capacity [108–110]. Polyacrylonitrile fibers combined with silver nanoparticles showed antibacterial activity on gram positive Bacillus cereus and gram negative *Escherichia coli*. The fibers formed from the combination of silver nanoparticles and Nylon 6 have an antibacterial effect on the gram of negative *Escherichia coli* and gram positive on Staphylococcus aureus [111].

Antioxidative effects of electrospun nanofibers have also been proven by obtaining multifunctional biomaterials, especially through the use of vitamin E and natural biopolymers [105]. Fibers resulting from the combination of polylactic acid, silver nanoparticles and vit. E inhibited the growth of *Escherichia coli* up to 100%. The release time of silver ions from nanofibres immersed in water lasted up to 10 days. It was proved that the combination of polylactic acid nanofibres, silver nanoparticles and vitamin E showed antioxidant activity in studies on fresh apple juice [112].

It has been proven that nanofibers act as a membrane that actively reduces polyphenol oxidase activity. Such materials can be used for preservation in the food industry in fruit and juice packaging [105,112]. Nanofibers can be coated with biocompatible polymers (hydrolysed collagen, elastin, hyaluronic acid, chondroitin sulfate). In this way, the electrospun fibers are coated with pure polyurethane to use the substance to improve the antibacterial effect of urinary catheters. The membranes thus obtained, have good antimicrobial activity on *Escherichia coli*, Salmonella typhimurium, Listeria monocytogenes [105].

Much research has been focused on the learning and operation of plant extracts using electrospun fibers [113,114]. Researchers have managed to encapsulate with the use of electrospun fibers several raw plant extracts such as Centella asiatica, baicalein, green tea, Garcinia mangostena, Tecomella undulata, aloe vera, Grewia mollis, chamomille, grape seed, Calendula officinalis, Indigofera aspalathoides, Azadirachta indica, Memecylon edule and Myristica andamanica [112]. Also, essential oils such as linalool, pinene, eugenol and cymene are used [115–117]. Antibacterial action with essential oils is determined by their hydrophobic nature. Bioactivity of essential oils combined using electrospun fibers to create scaffolding of fibers with antibacterial properties and thermo-mechanically controlled mobility. Electrospinning of chitosan nanoparticles and cinnamon essential oils in the ratio 1:1 has been demonstrated [112]. Fibers with a diameter of 38-55 nm were formed. The fibers were prepared from an aqueous solution containing 5% w/v acetic acid and various concentrations of essential oil (0.5 and 5.0% by volume). After fabrication, the fibers were cross-linked with glutaraldehyde to increase the stability of their chemical properties. The activity of chitosan nanofibres and ether oil particles has been proven in action against *P. aeruginosa, E. coli*. Efficacy of the essential oil was demonstrated in a study in which cellulose acetate was used as a polymer matrix [118]. Fibers containing essential oils of peppermint and lemongrass were also analyzed. The generated scaffolds inhibited *E. coli* proliferation and were non-toxic to fibroblasts and keratinocytes [116]. The Shikonin component of Lithospermum erytrorhizon root dried was also used in the electrospinning technique [116]. It has anticancer, antioxidant, anti-inflammatory and antibacterial effects. Shikonina was encapsulated and placed on PCL/poly (trimethylene carbonate) fibers. Note that the drug was released in the first hour at an increased rate and then for 48 h at a fixed dose. The shikonin-laden fibers showed a profound effect against *E. coli and S. aureus*. The healing properties were also tested in combination with alkannins, which are naturally occurring hydroxynaphthoquinone. The connection was used for the preparation of topical and transdermal patches. Cellulose acetate, poly (lactic acid) (PLA) and two different poly (lactic-glycol) mixtures were used as a matrix. Inclusion complexes of cyclodextrins (CD-IC) with plant extracts and complexes with eugenol (EG) were also formed [119]. Eugenol has a bactericidal effect and, in combination with inclusion cyclodextrins, has been used in electrospun fibers. The use of cyclodextrins complexes, increases the solubility of natural extracts in water affects antioxidant and antibacterial activity. Chitosan-based dressings were also developed for biomedical applications. A mixture of chitosan, TiO_2, poly (N-vinyl pyrrolidone) a synthetic polymer with good biocompatibility was prepared [120]. The membrane thus formed showed high activity against the microorganisms *S. aureus, B. subtilis, E. coli, P. aeruginosa*. The use of a membrane of chitosan, poly (N-vinylpyrrolidone) and silver oxide showed a similar effect, but with the added advantage of film transparency, allowing for observation of the wound [120].

Sodium alginate based nanofibers were also synthesized using polyethylene oxide (PEO) as "carrier polymer" [121]. It was concluded from the study that the sodium alginate on its own could not be electrospun. However, with the addition of a suitable polymer, it can be easily electrospun [121] and the PEO–PEO interactions with high molecular-weight entangled PEO were the key to "carrying" the alginate from the prepared solution to synthesise the fibers using electrospinning (Figure 4).

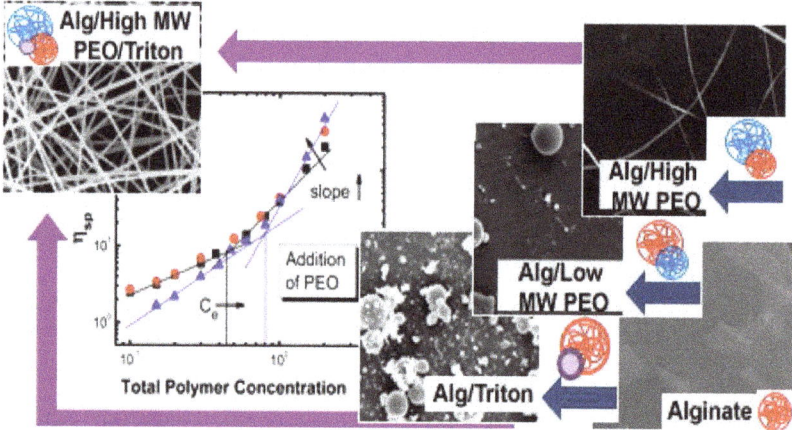

Figure 4. Schematic for the synthesis of alginate–polyethylene oxide blend nanofibers and the role of the carrier polymer in electrospinning. Reprinted with permission from Ref [121]. Copyright American Chemical Society, 2013.

In other studies, it was proved that nanofiber mats containing silver ions were more active compared to nanofibre mats without silver nanoparticles [122,123]. To monitor the human breath, smart fabrics were synthesized via electrospinning of the in situ assembly of well-dispersed Ag nanoparticles [124]. In this work, lightweight and flexible Ag/alginate nanofiber sensor were successfully developed that have the capability to sensitively monitor human breath (Figures 5 and 6).

Figure 5 shows the schematic for the fabrication of Ag/alginate nanofibers and Figure 6 shows the SEM images of different alginate nanofibers.

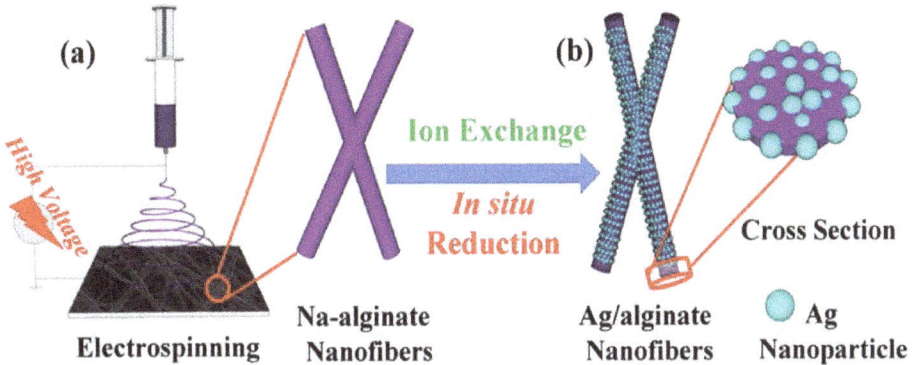

Figure 5. Schematic diagram illustrating the fabrication process of Ag/alginate nanofibers. (**a**) Na-alginate nanofibers prepared by electrospinning. (**b**) Ion-exchange and in situ reduction processes. Reprinted with permission from Ref. [124]. Copyright American Chemical Society, 2018.

Figure 6. SEM images of (**a**) as-electrospun Na-alginate nanofibers and (**b**–**d**) Ag/alginate nanofibers obtained at different reduction times: (**b**) 10, (**c**) 20, and (**d**) 30 min. Reprinted with permission from Ref. [124]. Copyright American Chemical Society, 2018.

Other synthetic polymers such as polyurethane, polyacrylonitrile, poly (acrylamide), poly (nitrilesulfonic acid sodium salt, poly (sulphobetaine methacrylate) were also used to develop wound dressings. Fluoroquinolones and norfloxacin were attached to polyphosphates by chemical modification using amino acid esters (alanine, glycine, and phenylalanine) as chain extenders, and these components were then used to create nanofibers by electrospinning [125].

Antimicrobial biodegradable multilayer systems were developed using electrospinning, especially the active multilayer structures based on natural polymers [126]. The system was developed in such a way that it consisted of different layers; for example, alginate-based film as outer layer; using a PHBV8 film as outer layer or no outer layer. Different characterizations such as oxygen and water vapour permeabilities, intermolecular arrangement, transparency and thermal properties were evaluated. In addition the antimicrobial activity was also evaluated. Alginate based coatings for Food Packaging Applications have been recently reviewed by Tugce et al. [127]. In this article, authors have summarized the recent advances on the usage of alginate for recent edible coatings.

3. Biopolymers for Biomedical and Pharmaceutical Packaging

Active and modern packaging biomaterials contain natural substances that are abundantly found in nature [1,11,128]. Biomaterials are often based on natural polysaccharides [129–133]. Among polysaccharides, Alginates have found applications in the food sector, water purification, biomedicine and packaging [73,134–139]. Algae contain nutrients such as vitamins, salts, iodine and sterols. Organisms containing large amounts of alginate in the cell walls are the brown algae Phaeophyceae such as Fucus, Laminaria, and Aseophyllum. The amount of alginates obtained generally depends on the species of algae and the extraction methods used [140]. They are linear polymers composed of (1→4)-α-L-guluronic acid blocks (GG) β-D-mannuronic acid blocks (MM) additionally, of heteropolymeric sequences of M and G (MG blocks) [74,80]. In biomedicine, alginates are used for controlled drug release, encapsulation, scaffolds in ligaments, tissue engineering and in dentistry for the preparation of forms in the presence of slow-release calcium salt [141]. The pharmaceutical industry uses purified alginates for dispersion or stabilization of substances. In biomedicine, alginates are used for controlled drug release, encapsulation, scaffolds in ligaments, tissue engineering and in dentistry for the preparation of forms in the presence of slow-release calcium salt. Figure 7 shows the application areas of alginate hydrogels. The alginate produces edible coatings with good barrier and mechanical properties allowing the protection of active ingredients by encapsulation [3]. Garlic oil is often added as a natural antibacterial agent in such coatings. Alginate is partially dusted with calcium and mixed with starch to obtain high water retention in the paper coating. This is important in order to obtain a uniform mass and coating by pressing to improve its rheology [85]. Alginates have found a number of applications in biomedical sciences as wound dressing materials [142]. Especially sodium alginate used in the form of a hydrogel has stimulated more and more scientific interest due to its physicochemical properties. Materials made of alginate are considered to be friendly to humans due to tissue biocompatibility, which allows for their use in biomedical engineering [135,140]. Highly absorbent dressing materials are formed by the production of wet spinning fibers. With the addition of calcium and sodium, high-absorbency sodium and calcium fibers were produced. Antimicrobial fibers were also formed by adding alginic acid or silver. By adding zinc, the fibers that generate the immune system were created. Fibers for immobilizing or supporting bioactive molecules were readily prepared [143]. Antimicrobial properties were imparted to cotton fabrics employing alginate–quaternary ammonium complex nanoparticles [144]. Using the ionic gelation method, a new type of nanoparticle (average size of 99 nm) that was composed of sodium alginate (SA) and 3-(trimethoxysilyl) propyl-octadecyldimethylammonium chloride (TSA) was synthesized. Fibers exhibited an efficient antimicrobial activity that was even maintained after 30 laundry cycles (non-leaching antimicrobial agent) [144]. Ionic gelation was used to develop a new class of nanoparticles that consists of sodium alginate (SA) and 3-(trimethoxysilyl) propyl-octadecyldimethylammonium chloride (TSA). The ratio of SA/TSA was found to exhibit a significant effect on the average size of the SA–TSA nanoparticles. Nanoparticles having an average size of 99 nm were selected for the study and, after using a pad-dry-cure method, were loaded onto cotton fabrics. Different characterization techniques were used to analyse the treated fabrics. It was concluded from the study that the SA–TSA nanoparticles exhibit high potential to be used as a non-leaching agent imparting robust antimicrobial characteristics to the studied cotton fabrics [144].

The new generation of medical textiles marks the field of expansion for scientists and researchers [145]. The current dressings are non-toxic, bacteriostatic, antiviral, non-allergic, hemostatic, highly absorbent and, above all, biocompatible. It is possible to modify them so that they contain medicines with some mechanical properties. Current textile materials in modern packaging are also highly diverse. These include tapes, fabrics, non-woven fabrics, knitted fabrics, composite materials [16,146]. Lignin and cellulose [147,148] are the most abundant natural polymers available as by-products of various industries. Recent scientific reports show that lignin is increasingly used as an essential component of hydrogels [149]. This allows for the creation of various types of materials, especially in the medical or pharmaceutical sector [150]. It is also possible to add additives to dressings such as odor absorbing, soothing pain and irritation.

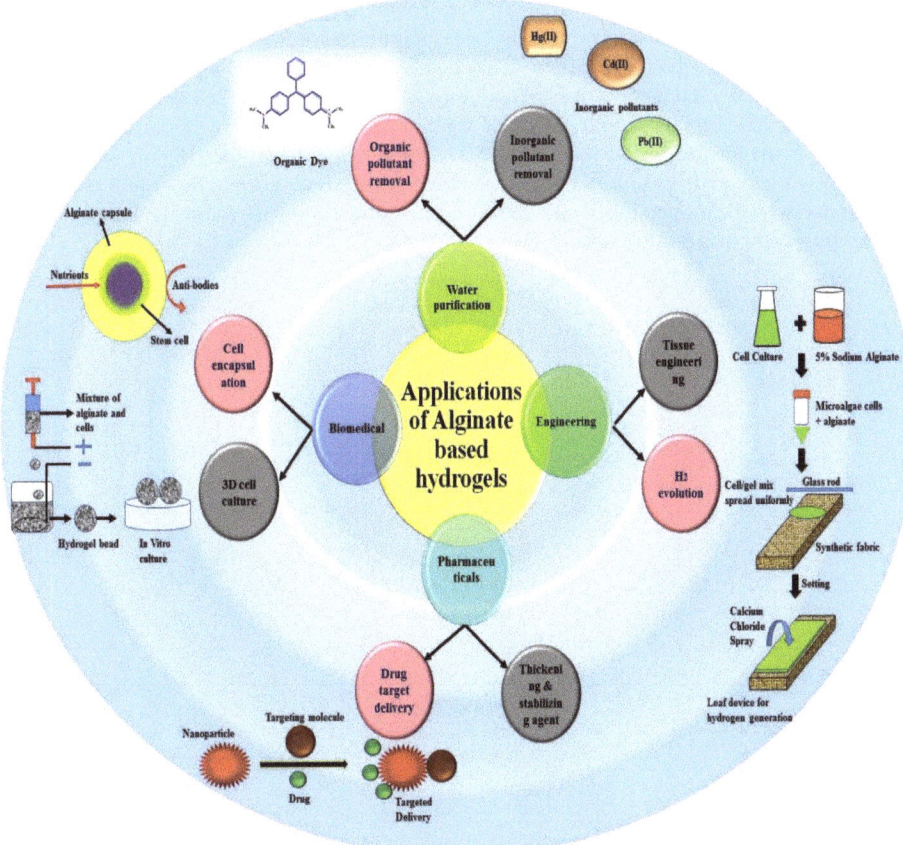

Figure 7. Various applied applications of alginate based hydrogels [135]. Reprinted with permission from Ref. [135]. Copyright Elsevier, 2018.

Nanofibres are being used in the treatment of wounds due to interesting properties such as fiber diameter at the nanoscale, porosity, low weight. Polymers from various natural resources are attracting the attention of more and more scientists for exploring their use in wide range of application [151–153].

Nanofibers of alginates were created by electrospinning in the presence of various synthetic polymers and surfactants [154]. The electrospinning process has been used inexpensively by an efficient technique for the production of nanofibers with the use of biopolymers and other organic substances used in medicine and pharmaceuticals [93,155]. This technique is used to create dressings, drug delivery systems and scaffolds in tissue engineering. The electrospinning creates ultrathin fibers, collected in a random set [156]. The mats that are created during this process are used as wound dressings, catalytic carriers and nanocomposites with many applications [99]. The versatility of the electrospinning technique is so developed that it allows for its use in drug delivery systems and the creation of poly (lactide-co-glycolide) scaffolds, which allows the permanent release of the drug from nanofibres while maintaining their structure and biological activity [86,157]. Nanofibres have also been used in the chemical field as catalysts, sensors and chemical and physical adsorbents. Still, most popular is the use of unique properties of nanofibers prepared using different biopolymers via electrospinning [158]. Several materials have been commercialized under different trademarks such as Coalgan from Brothier Laboratories (France), which is sold as an haemostatic fibers pad for nosebleeds; Algosteril® or Sorban® are also non-woven dressing absorbing rapidly and retains wound fluid resulting in haemostasis and accelerates wound healing. Alginate based hydrogels were also synthesized and used as novel platform for in situ preparation of metal–organic framework [MOF] [159]. In order to establish the feasibility of the synthesis technique, the Hong Kong University of Science and Technology-1 [HKUST-1] –alginate composite was selected as a model system. Figure 8 shows the schematic for the preparation of the MOF–alginate composite and Figure 9 displays the cross-section backscattered electron images. It was concluded from the study that MOF particles can be incorporated into the alginate substrates and for in situ MOF growth, the metal ion cross-linked alginate hydrogels provided outstanding templates.

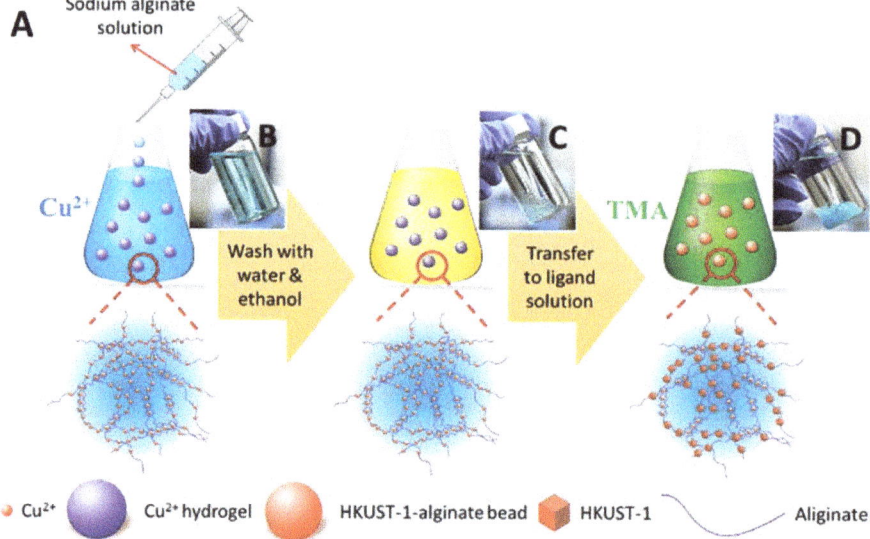

Figure 8. (**A**) Schematic of the preparation of the MOF–alginate composite. Photographs of (**B**) alginate hydrogels cross-linked by Cu^{2+} right after the addition of a sodium alginate aqueous solution to a Cu^{2+} aqueous solution, (**C**) alginate hydrogels cross-linked by Cu^{2+} after being washed with water and ethanol, and (**D**) HKUST-1–alginate hydrogels [159]. Reprinted with permission from Ref. [159]. Copyright American Chemical Society, 2016.

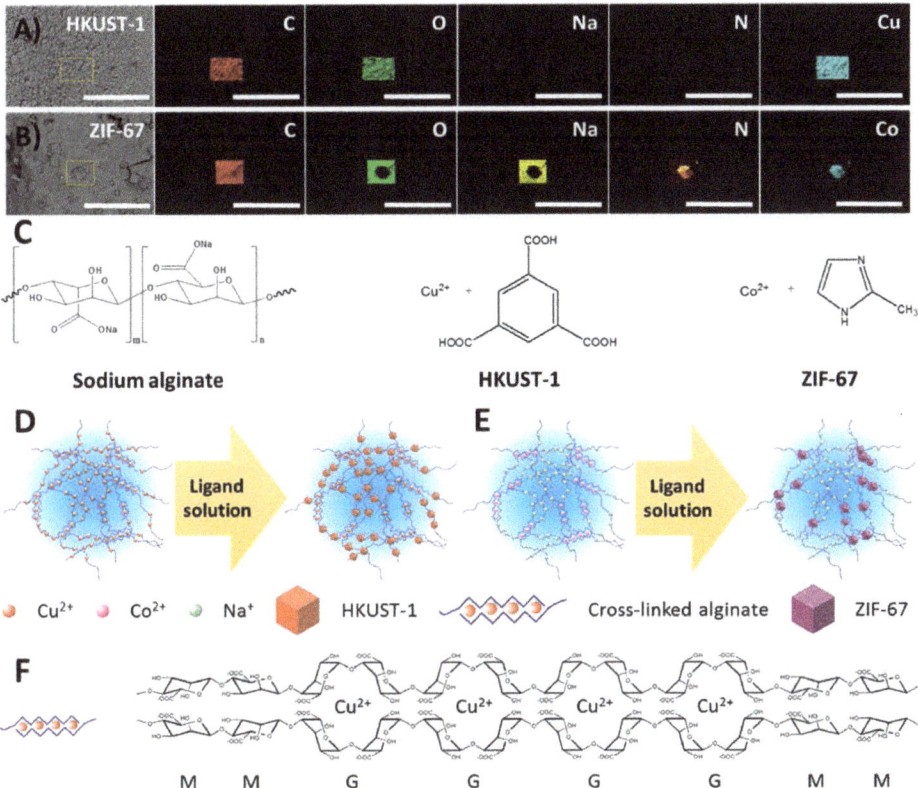

Figure 9. Cross-section backscattered electron images of (**A**) HKUST-1– and (**B**) ZIF-67–alginate composites and their corresponding EDX elemental maps (scale bar of 60 μm). (**C**) Chemical composition of sodium alginate, HKUST-1, and ZIF-67. Schematic of the formation of (**D**) the HKUST-1–alginate composite, (**E**) the ZIF-67–alginate composite, and (**F**) the "egg-box" model of metal ion cross-linked alginate [159]. Reprinted with permission from Ref. [159]. Copyright American Chemical Society, 2016.

Smart Biopolymers

Intelligent biopolymers are becoming more and more attractive in biotechnology and medicine, as well as in packaging [160,161]. They are used in biomedicine and tissue engineering. Intelligent biopolymers have been reported to play an important role in drug delivery [22,162,163]. They are easy to produce, are a good carrier of nutrients and maintain the stability of the drug [150,164]. It is possible to inject them in vitro as a liquid and they can form a gel at body temperature. Thermosensitive polymers are used to solubilize hydrophobic drugs and are used in the production of preparations with low solubility drugs as a drug carrier. The use of intelligent polymers for drug delivery is promising in view of the mechanism of drug release observed [165]. External and internal stimuli that affect the mechanism include temperature, light irradiation, electric current and intelligent hydrogels that can immobilize enzymes have the ability to gel by phase transition [165]. The chemical signal translates into a mechanical signal causing shrinking or swelling of the gel. This phenomenon is used for the controlled release of the drug. The diffusion of the drug from the beads depends on the condition of the gel [166]. The intelligent polymer is generally integrated with a wall microcapsule or a liposome lipid bilayer. The conformational transition of the polymer affects the integrity of the microcapsule or liposome and allows controlled release of the drug incorporated into the microcapsule or liposome [167].

Thanks to the use of drugs released in hydrogels, pharmaceutical aspects can increase productivity, profitability and wide range of applications [70]. One example of an insulin delivery system is a hydrogel comprising an insulin-containing reservoir within a poly (methacrylic acid-g-ethylene glycol) copolymer in which glucose oxidase was immobilized [168,169]. Currently, there is a limited research on fluorinated polymers as a drug carrier in drug delivery applications. Polymer membranes have been used as passive materials for drug release due to their ability to be hydrophilic to hydrophobic in surface wettability. The characteristics of the process are reversible and can go to the initial state [170,171]. The use of lipid based biopolymers for cancer therapy has been recently reviewed [172]. In this article, authors have focused on the advantageous biologic and physicochemical characteristics including controlled drug release, long circulatory half-lives and facile targeted therapy of the natural and synthetic lipid.

4. Summary and Future Prospective

Packaging plays an imperative role in supply chain and has received great attention in number of industries. However, the existing packaging systems are primarily based on synthetic polymers/plastics from fossil resources. Seeing the issues associated with the synthetic polymers/plastics, biomedical and pharmaceutical industries in particular all around the globe are looking for new bio-based sustainable packaging materials that can extend the shelf life of these materials due to the environmental and health issues associated with traditional packaging waste. Increasing the use of sustainable polymeric materials in packaging could address such concerns. The main goals of polymer researchers in the development of new medical and pharmaceutical materials are associated with a reduction in the risk of disease transmission and the spread of infections. The search for new effective antimicrobial dressing materials is constantly growing. Electrospinning is being used in medicine for the production of non-woven structures at the nanoscale. In the electrospinning process, the chemical and physical parameters of the obtained nanofibre mats can be adjusted by dressing the appropriate materials and parameters of the process. The electrospinning process is also influenced by nanoparticles of various natural materials and biopolymers. The addition of natural nanomaterials affects the morphology and size of the electrospun fibre. Along with alginate, the other biopolymers being used in the electrospinning technique include hyaluronic acid, cellulose, silk, gelatin and collagen, to name a few. The mixture of biopolymers and synthetic polymers is also used to create the novel biomaterials with specific properties such as mechanical resistance, thermal stability and barrier resistance. Mixtures with natural polymers may also affect the structural, morphological and subsequent degradation properties of the electrospun fibers

The treatment of disease states with the help of biologically active materials created with the help of electrospun nanofibers is part of modern regenerative medicine. The use of natural materials such as plant oils and extracts means that the nonwovens created in the electrospinning process find many applications with safety of use. It is of great importance that it is possible to optimize process conditions and create nanofibers that meet specific requirements. Among the various materials studied, alginate has very high potential for packaging applications and is being seen as the future of packaging materials. However, a very limited study has been carried out on the use of Alginate based electrospun mats for packing application, and to realize its real potential an extensive study in this direction is needed.

Author Contributions: Conceptualization, J.W-K., T.R., I.M-P., and V.K.T.; investigation, J.W-K., T.R., I.M-P., and V.K.T.; writing-original draft preparation, J.W-K., T.R., I.M-P., and V.K.T.; visualization, J.W-K., T.R., I.M-P., and V.K.T.

Funding: V.K.T. wish to acknowledge and thank the British Council for the support of this work through a UKIERI grant (No. DST /INTUK/P-164/2017).

Conflicts of Interest: The authors declare no conflict of interest.

References

1. Cazón, P.; Velazquez, G.; Ramírez, J.A.; Vázquez, M. Polysaccharide-based films and coatings for food packaging: A review. *Food Hydrocoll.* **2017**, *68*, 136–148. [CrossRef]
2. Gomez-Estaca, J.; Gavara, R.; Catala, R.; Hernandez-Munoz, P. The Potential of Proteins for Producing Food Packaging Materials: A Review. *Packag. Technol. Sci.* **2016**, *29*, 203–224. [CrossRef]
3. Hambleton, A.; Debeaufort, F.; Bonnotte, A.; Voilley, A. Influence of alginate emulsion-based films structure on its barrier properties and on the protection of microencapsulated aroma compound. *Food Hydrocoll.* **2009**, *23*, 2116–2124. [CrossRef]
4. Khalil, H.P.S.A.; Davoudpour, Y.; Saurabh, C.K.; Hossain, M.S.; Adnan, A.S.; Dungani, R.; Paridah, M.T.; Sarker, M.Z.I.; Fazita, M.R.N.; Syakir, M.I.; et al. A review on nanocellulosic fibres as new material for sustainable packaging: Process and applications. *Renew. Sustain. Energy Rev.* **2016**, *64*, 823–836. [CrossRef]
5. da Silva, F.T.; da Cunha, K.F.; Fonseca, L.M.; Antunes, M.D.; Mello El Halal, S.L.; Fiorentini, A.M.; da Rosa Zavareze, E.; Guerra Dias, A.R. Action of ginger essential oil (*Zingiber officinale*) encapsulated in proteins ultrafine fibers on the antimicrobial control in situ. *Int. J. Biol. Macromol.* **2018**, *118*, 107–115. [CrossRef] [PubMed]
6. Kuntzler, S.G.; Vieira Costa, J.A.; de Morais, M.G. Development of electrospun nanofibers containing chitosan/PEO blend and phenolic compounds with antibacterial activity. *Int. J. Biol. Macromol.* **2018**, *117*, 800–806. [CrossRef] [PubMed]
7. Kwon, H.; Kim, D.; Lee, K.; Seo, J.; Lee, H.-J. The Effect of Coating Process and Additives on EVA Coated Tyvek (R) for Gas Sterilizable Medical Packaging Applications. *Packag. Technol. Sci.* **2017**, *30*, 195–208. [CrossRef]
8. Lauriano Souza, V.G.; Fernando, A.L. Nanoparticles in food packaging: Biodegradability and potential migration to food—A review. *Food Packag. Shelf Life* **2016**, *8*, 63–70. [CrossRef]
9. Wróblewska-Krepsztul, J.; Rydzkowski, T.; Borowski, G.; Szczypiński, M.; Klepka, T.; Thakur, V.K. Recent progress in biodegradable polymers and nanocomposite-based packaging materials for sustainable environment. *Int. J. Polym. Anal. Charact.* **2018**, *23*, 383–395. [CrossRef]
10. Mercante, L.A.; Scagion, V.P.; Migliorini, F.L.; Mattoso, L.H.C.; Correa, D.S. Electrospinning-based (bio)sensors for food and agricultural applications: A review. *TrAC Trends Anal. Chem.* **2017**, *91*, 91–103. [CrossRef]
11. Thakur, S.; Chaudhary, J.; Sharma, B.; Verma, A.; Tamulevicius, S.; Thakur, V.K. Sustainability of bioplastics: Opportunities and challenges. *Curr. Opin. Green Sustain. Chem.* **2018**, *13*, 68–75. [CrossRef]
12. Shao, P.; Niu, B.; Chen, H.; Sun, P. Fabrication and characterization of tea polyphenols loaded pullulan-CMC electrospun nanofiber for fruit preservation. *Int. J. Biol. Macromol.* **2018**, *107*, 1908–1914. [CrossRef] [PubMed]
13. Ochoa, M.; Collazos, N.; Le, T.; Subramaniam, R.; Sanders, M.; Singh, R.P.; Depan, D. Nanocellulose-PE-b-PEG copolymer nanohybrid shish-kebab structure via interfacial crystallization. *Carbohydr. Polym.* **2017**, *159*, 116–124. [CrossRef] [PubMed]
14. Depan, D.; Misra, R.D.K. Hybrid Nanoscale Architecture of Wound Dressing with Super Hydrophilic, Antimicrobial, and Ultralow Fouling Attributes. *J. Biomed. Nanotechnol.* **2015**, *11*, 306–318. [CrossRef] [PubMed]
15. Singha, A.S.; Thakur, V.K. Synthesis and Characterization of Pine Needles Reinforced RF Matrix Based Biocomposites. *J. Chem.* **2008**, *5*, 1055–1062. [CrossRef]
16. Thakur, V.K.; Singha, A.S. Mechanical and Water Absorption Properties of Natural Fibers/Polymer Biocomposites. *Polym.-Plast. Technol. Eng.* **2010**, *49*, 694–700. [CrossRef]
17. Antunes, M.D.; da Silva Dannenberg, G.; Fiorentini, A.M.; Pinto, V.Z.; Lim, L.-T.; da Rosa Zavareze, E.; Guerra Dias, A.R. Antimicrobial electrospun ultrafine fibers from zein containing eucalyptus essential oil/cyclodextrin inclusion complex. *Int. J. Biol. Macromol.* **2017**, *104*, 874–882. [CrossRef] [PubMed]
18. Kurd, F.; Fathi, M.; Shekarchizadeh, H. Basil seed mucilage as a new source for electrospinning: Production and physicochemical characterization. *Int. J. Biol. Macromol.* **2017**, *95*, 689–695. [CrossRef] [PubMed]
19. Kohsari, I.; Shariatinia, Z.; Pourmortazavi, S.M. Antibacterial electrospun chitosan-polyethylene oxide nanocomposite mats containing ZIF-8 nanoparticles. *Int. J. Biol. Macromol.* **2016**, *91*, 778–788. [CrossRef] [PubMed]

20. Miculescu, F.; Maidaniuc, A.; Voicu, S.I.; Thakur, V.K.; Stan, G.E.; Ciocan, L.T. Progress in Hydroxyapatite–Starch Based Sustainable Biomaterials for Biomedical Bone Substitution Applications. *ACS Sustain. Chem. Eng.* **2017**, *5*, 8491–8512. [CrossRef]
21. Pandele, A.M.; Comanici, F.E.; Carp, C.A.; Miculescu, F.; Voicu, S.I.; Thakur, V.K.; Serban, B.C. Synthesis and characterization of cellulose acetate-hydroxyapatite micro and nano composites membranes for water purification and biomedical applications. *Vacuum* **2017**, *144*, 599–605. [CrossRef]
22. Corobea, M.C.; Muhulet, O.; Miculescu, F.; Antoniac, I.V.; Vuluga, Z.; Florea, D.; Vuluga, D.M.; Butnaru, M.; Ivanov, D.; Voicu, S.I.; et al. Novel nanocomposite membranes from cellulose acetate and clay-silica nanowires. *Polym. Adv. Technol.* **2016**, *27*, 1586–1595. [CrossRef]
23. Pandele, A.M.; Neacsu, P.; Cimpean, A.; Staras, A.I.; Miculescu, F.; Iordache, A.; Voicu, S.I.; Thakur, V.K.; Toader, O.D. Cellulose acetate membranes functionalized with resveratrol by covalent immobilization for improved osseointegration. *Appl. Surf. Sci.* **2018**, *438*, 2–13. [CrossRef]
24. Van de Walle, E.; Van Nieuwenhove, I.; Vanderleyden, E.; Declercq, H.; Gellynck, K.; Schaubroeck, D.; Ottevaere, H.; Thienpont, H.; De Vos, W.H.; Cornelissen, M.; et al. Polydopamine-Gelatin as Universal Cell-Interactive Coating for Methacrylate-Based Medical Device Packaging Materials: When Surface Chemistry Overrules Substrate Bulk Properties. *Biomacromolecules* **2016**, *17*, 56–68. [CrossRef] [PubMed]
25. Rogina-Car, B.; Grgac, S.F.; Katovic, D. Physicochemical Characterization of the Multiuse Medical Textiles in Surgery and as Packaging Material in Medical Sterilization. *Autex Res. J.* **2017**, *17*, 206–212. [CrossRef]
26. Rowson, J.; Sangrar, A.; Rodriguez-Falcon, E.; Bell, A.F.; Walton, K.A.; Yoxall, A.; Kamat, S.R. Rating Accessibility of Packaging: A Medical Packaging Example. *Packag. Technol. Sci.* **2016**, *29*, 607. [CrossRef]
27. Kale, R.D.; Gorade, V.G.; Madye, N.; Chaudhary, B.; Bangde, P.S.; Dandekar, P.P. Preparation and characterization of biocomposite packaging film from poly(lactic acid) and acylated microcrystalline cellulose using rice bran oil. *Int. J. Biol. Macromol.* **2018**, *118*, 1090–1102. [CrossRef] [PubMed]
28. Nasajpour, A.; Mandla, S.; Shree, S.; Mostafavi, E.; Sharifi, R.; Khalilpour, A.; Saghazadeh, S.; Hassan, S.; Mitchell, M.J.; Leijten, J.; et al. Nanostructured Fibrous Membranes with Rose Spike-Like Architecture. *Nano Lett.* **2017**, *17*, 6235–6240. [CrossRef] [PubMed]
29. Muhulet, A.; Miculescu, F.; Voicu, S.I.; Schütt, F.; Thakur, V.K.; Mishra, Y.K. Fundamentals and scopes of doped carbon nanotubes towards energy and biosensing applications. *Mater. Today Energy* **2018**, *9*, 154–186. [CrossRef]
30. Thakur, S.; Sharma, B.; Verma, A.; Chaudhary, J.; Tamulevicius, S.; Thakur, V.K. Recent approaches in guar gum hydrogel synthesis for water purification. *Int. J. Polym. Anal. Charact.* **2018**, *23*, 621–632. [CrossRef]
31. Nasajpour, A.; Ansari, S.; Rinoldi, C.; Rad, A.S.; Aghaloo, T.; Shin, S.R.; Mishra, Y.K.; Adelung, R.; Swieszkowski, W.; Annabi, N.; et al. Tissue Regeneration: A Multifunctional Polymeric Periodontal Membrane with Osteogenic and Antibacterial Characteristics (Adv. Funct. Mater. 3/2018). *Adv. Funct. Mater.* **2018**, *28*, 1870021. [CrossRef]
32. Jayakumar, R.; Prabaharan, M.; Nair, S.V.; Tamura, H. Novel chitin and chitosan nanofibers in biomedical applications. *Biotechnol. Adv.* **2010**, *28*, 142–150. [CrossRef] [PubMed]
33. Thakur, V.K.; Thakur, M.K. Recent Advances in Graft Copolymerization and Applications of Chitosan: A Review. *ACS Sustain. Chem. Eng.* **2014**, *2*, 2637–2652. [CrossRef]
34. Thakur, V.K.; Voicu, S.I. Recent advances in cellulose and chitosan based membranes for water purification: A concise review. *Carbohydr. Polym.* **2016**, *146*, 148–165. [CrossRef] [PubMed]
35. Xi, H.; Zhao, H. Silk fibroin coaxial bead-on-string fiber materials and their drug release behaviors in different pH. *J. Mater. Sci.* **2019**, *54*, 4246–4258. [CrossRef]
36. Voicu, S.I.; Condruz, R.M.; Mitran, V.; Cimpean, A.; Miculescu, F.; Andronescu, C.; Miculescu, M.; Thakur, V.K. Sericin Covalent Immobilization onto Cellulose Acetate Membrane for Biomedical Applications. *ACS Sustain. Chem. Eng.* **2016**, *4*, 1765–1774. [CrossRef]
37. Wenk, E.; Wandrey, A.J.; Merkle, H.P.; Meinel, L. Silk fibroin spheres as a platform for controlled drug delivery. *J. Control. Release* **2008**, *132*, 26–34. [CrossRef] [PubMed]
38. Valentini, L.; Bon, S.B.; Pugno, N.M. Combining Living Microorganisms with Regenerated Silk Provides Nanofibril-Based Thin Films with Heat-Responsive Wrinkled States for Smart Food Packaging. *Nanomaterials* **2018**, *8*, 518. [CrossRef] [PubMed]
39. Bedford, N.M.; Steckl, A.J. Photocatalytic Self Cleaning Textile Fibers by Coaxial Electrospinning. *ACS Appl. Mater. Interfaces* **2010**, *2*, 2448–2455. [CrossRef]

40. Montalban, M.G.; Coburn, J.M.; Abel Lozano-Perez, A.; Cenis, J.L.; Villora, G.; Kaplan, D.L. Production of Curcumin-Loaded Silk Fibroin Nanoparticles for Cancer Therapy. *Nanomaterials* **2018**, *8*, 126. [CrossRef] [PubMed]
41. Zhou, C.-Z.; Confalonieri, F.; Jacquet, M.; Perasso, R.; Li, Z.-G.; Janin, J. Silk fibroin: Structural implications of a remarkable amino acid sequence. *Proteins Struct. Funct. Bioinform.* **2001**, *44*, 119–122. [CrossRef] [PubMed]
42. Cai, Z.; Mo, X.; Zhang, K.; Fan, L.; Yin, A.; He, C.; Wang, H. Fabrication of Chitosan/Silk Fibroin Composite Nanofibers for Wound-dressing Applications. *Int. J. Mol. Sci.* **2010**, *11*, 3529–3539. [CrossRef] [PubMed]
43. Chen, Z.G.; Wang, P.W.; Wei, B.; Mo, X.M.; Cui, F.Z. Electrospun collagen–chitosan nanofiber: A biomimetic extracellular matrix for endothelial cell and smooth muscle cell. *Acta Biomater.* **2010**, *6*, 372–382. [CrossRef] [PubMed]
44. Cho, M.K.; Singu, B.S.; Na, Y.H.; Yoon, K.R. Fabrication and characterization of double-network agarose/polyacrylamide nanofibers by electrospinning. *J. Appl. Polym. Sci.* **2016**, *133*. [CrossRef]
45. Hu, W.-W.; Lin, C.-H.; Hong, Z.-J. The enrichment of cancer stem cells using composite alginate/polycaprolactone nanofibers. *Carbohydr. Polym.* **2019**, *206*, 70–79. [CrossRef] [PubMed]
46. Wang, J.; Salem, D.R.; Sani, R.K. Extremophilic exopolysaccharides: A review and new perspectives on engineering strategies and applications. *Carbohydr. Polym.* **2019**, *205*, 8–26. [CrossRef] [PubMed]
47. Iqbal, N.; Khan, A.S.; Asif, A.; Yar, M.; Haycock, J.W.; Rehman, I.U. Recent concepts in biodegradable polymers for tissue engineering paradigms: A critical review. *Int. Mater. Rev.* **2019**, *64*, 91–126. [CrossRef]
48. Clark, C.M.; Bottomley, L.; Jackson, P. The development of a training package for community pharmacists focused on medicines use review in common skin diseases. *Pharm. World Sci.* **2008**, *30*, 377–378.
49. Lorenzini, G.C.; Hellstrom, D. Medication Packaging and Older Patients: A Systematic Review. *Packag. Technol. Sci.* **2017**, *30*, 525–558. [CrossRef]
50. Jenke, D. Evaluation of the chemical compatibility of plastic contact materials and pharmaceutical products; safety considerations related to extractables and leachables. *J. Pharm. Sci.* **2007**, *96*, 2566–2581. [CrossRef] [PubMed]
51. Maitz, M.F. Applications of synthetic polymers in clinical medicine. *Biosurf. Biotribol.* **2015**, *1*, 161–176. [CrossRef]
52. Jenke, D.R. Extractables and leachables considerations for prefilled syringes. *Expert Opin. Drug Deliv.* **2014**, *11*, 1591–1600. [CrossRef] [PubMed]
53. Sampson, J.; de Korte, D. DEHP-plasticised PVC: Relevance to blood services. *Transfus. Med. Oxf. Engl.* **2011**, *21*, 73–83. [CrossRef] [PubMed]
54. Frank, R.D.; Müller, U.; Lanzmich, R.; Groeger, C.; Floege, J. Anticoagulant-free Genius haemodialysis using low molecular weight heparin-coated circuits. *Nephrol. Dial. Transplant.* **2006**, *21*, 1013–1018. [CrossRef] [PubMed]
55. Hornsey, V.S.; McColl, K.; Drummond, O.; Macgregor, I.R.; Prowse, C.V. Platelet storage in Fresenius/NPBI polyolefin and BTHC-PVC bags: A direct comparison. *Transfus. Med. Oxf. Engl.* **2008**, *18*, 223–227. [CrossRef] [PubMed]
56. Kretschmer, V.; Marschall, R.; Schulzki, T.; Härtel, B.; Neumann, H.J. New polyolefin foil for 5-day storage of platelet concentrates (PC) collected by apheresis. *Transfus. Med. Hemother.* **1992**, *19*, 141–144. [CrossRef]
57. Sunohara, T.; Masuda, T. Cellulose triacetate as a high-performance membrane. *Contrib. Nephrol.* **2011**, *173*, 156–163. [PubMed]
58. Miculescu, M.; Thakur, V.K.; Miculescu, F.; Voicu, S.I. Graphene-based polymer nanocomposite membranes: A review. *Polym. Adv. Technol.* **2016**, *27*, 844–859. [CrossRef]
59. Johnson, D.G.; Khire, T.S.; Lyubarskaya, Y.L.; Smith, K.J.P.; Desormeaux, J.-P.S.; Taylor, J.G.; Gaborski, T.R.; Shestopalov, A.A.; Striemer, C.C.; McGrath, J.L. Ultrathin silicon membranes for wearable dialysis. *Adv. Chronic Kidney Dis.* **2013**, *20*, 508–515. [CrossRef] [PubMed]
60. Venkatesan, N.; Shroff, S.; Jayachandran, K.; Doble, M. Polymers as ureteral stents. *J. Endourol.* **2010**, *24*, 191–198. [CrossRef] [PubMed]
61. Fernando, I.P.S.; Kim, D.; Nah, J.-W.; Jeon, Y.-J. Advances in functionalizing fucoidans and alginates (bio)polymers by structural modifications: A review. *Chem. Eng. J.* **2019**, *355*, 33–48. [CrossRef]
62. Bayer, I.S. Thermomechanical Properties of Polylactic Acid-Graphene Composites: A State-of-the-Art Review for Biomedical Applications. *Materials* **2017**, *10*, 748. [CrossRef] [PubMed]

63. Gunathilake, T.M.S.U.; Ching, Y.C.; Ching, K.Y.; Chuah, C.H.; Abdullah, L.C. Biomedical and Microbiological Applications of Bio-Based Porous Materials: A Review. *Polymers* **2017**, *9*, 160. [CrossRef]
64. Gigli, M.; Fabbri, M.; Lotti, N.; Gamberini, R.; Rimini, B.; Munari, A. Poly(butylene succinate)-based polyesters for biomedical applications: A review. *Eur. Polym. J.* **2016**, *75*, 431–460. [CrossRef]
65. Dubey, S.P.; Thakur, V.K.; Krishnaswamy, S.; Abhyankar, H.A.; Marchante, V.; Brighton, J.L. Progress in environmental-friendly polymer nanocomposite material from PLA: Synthesis, processing and applications. *Vacuum* **2017**, *146*, 655–663. [CrossRef]
66. Madhumitha, G.; Fowsiya, J.; Roopan, S.M.; Thakur, V.K. Recent advances in starch–clay nanocomposites. *Int. J. Polym. Anal. Charact.* **2018**, *23*, 331–345. [CrossRef]
67. Coscia, M.G.; Bhardwaj, J.; Singh, N.; Santonicola, M.G.; Richardson, R.; Thakur, V.K.; Rahatekar, S. Manufacturing & characterization of regenerated cellulose/curcumin based sustainable composites fibers spun from environmentally benign solvents. *Ind. Crop. Prod.* **2018**, *111*, 536–543.
68. Thakur, V.K.; Singha, A.S. Rapid Synthesis, Characterization, and Physicochemical Analysis of Biopolymer-Based Graft Copolymers. *Int. J. Polym. Anal. Charact.* **2011**, *16*, 153–164. [CrossRef]
69. Jones, O.G.; McClements, D.J. Functional Biopolymer Particles: Design, Fabrication, and Applications. *Compr. Rev. Food Sci. Food Saf.* **2010**, *9*, 374–397. [CrossRef]
70. Reakasame, S.; Boccaccini, A.R. Oxidized Alginate-Based Hydrogels for Tissue Engineering Applications: A Review. *Biomacromolecules* **2018**, *19*, 3–21. [CrossRef] [PubMed]
71. Aguero, L.; Zaldivar-Silva, D.; Pena, L.; Dias, M.L. Alginate microparticles as oral colon drug delivery device: A review. *Carbohydr. Polym.* **2017**, *168*, 32–43. [CrossRef] [PubMed]
72. Draget, K.I.; Taylor, C. Chemical, physical and biological properties of alginates and their biomedical implications. *Food Hydrocoll.* **2011**, *25*, 251–256. [CrossRef]
73. Zia, K.M.; Zia, F.; Zuber, M.; Rehman, S.; Ahmad, M.N. Alginate based polyurethanes: A review of recent advances and perspective. *Int. J. Biol. Macromol.* **2015**, *79*, 377–387. [CrossRef] [PubMed]
74. Venkatesan, J.; Bhatnagar, I.; Manivasagan, P.; Kang, K.-H.; Kim, S.-K. Alginate composites for bone tissue engineering: A review. *Int. J. Biol. Macromol.* **2015**, *72*, 269–281. [CrossRef] [PubMed]
75. Saltz, A.; Kandalam, U. Mesenchymal stem cells and alginate microcarriers for craniofacial bone tissue engineering: A review. *J. Biomed. Mater. Res. A* **2016**, *104*, 1276–1284. [CrossRef] [PubMed]
76. Pawar, S.N.; Edgar, K.J. Alginate derivatization: A review of chemistry, properties and applications. *Biomaterials* **2012**, *33*, 3279–3305. [CrossRef] [PubMed]
77. You, J.-O.; Park, S.-B.; Park, H.-Y.; Haam, S.; Chung, C.-H.; Kim, W.-S. Preparation of regular sized Ca-alginate microspheres using membrane emulsification method. *J. Microencapsul.* **2001**, *18*, 521–532. [PubMed]
78. de Vos, P.; Faas, M.M.; Strand, B.; Calafiore, R. Alginate-based microcapsules for immunoisolation of pancreatic islets. *Biomaterials* **2006**, *27*, 5603–5617. [CrossRef] [PubMed]
79. Hecht, H.; Srebnik, S. Structural Characterization of Sodium Alginate and Calcium Alginate. *Biomacromolecules* **2016**, *17*, 2160–2167. [CrossRef] [PubMed]
80. Lee, K.Y.; Mooney, D.J. Alginate: Properties and biomedical applications. *Prog. Polym. Sci.* **2012**, *37*, 106–126. [CrossRef] [PubMed]
81. Yang, J.-S.; Xie, Y.-J.; He, W. Research progress on chemical modification of alginate: A review. *Carbohydr. Polym.* **2011**, *84*, 33–39. [CrossRef]
82. Bhat, S.D.; Aminabhavi, T.M. Pervaporation separation using sodium alginate and its modified membranes—A review. *Sep. Purif. Rev.* **2007**, *36*, 203–229. [CrossRef]
83. Franklin, M.J.; Ohman, D.E. Mutant analysis and cellular localization of the AlgI, AlgJ, and AlgF proteins required for O acetylation of alginate in *Pseudomonas aeruginosa*. *J. Bacteriol.* **2002**, *184*, 3000–3007. [CrossRef] [PubMed]
84. El-Sherbiny, I.M. Enhanced pH-responsive carrier system based on alginate and chemically modified carboxymethyl chitosan for oral delivery of protein drugs: Preparation and in-vitro assessment. *Carbohydr. Polym.* **2010**, *80*, 1125–1136. [CrossRef]
85. Rinaudo, M. Biomaterials based on a natural polysaccharide: Alginate. *TIP* **2014**, *17*, 92–96. [CrossRef]
86. Soares, R.M.D.; Siqueira, N.M.; Prabhakaram, M.P.; Ramakrishna, S. Electrospinning and electrospray of bio-based and natural polymers for biomaterials development. *Mater. Sci. Eng. C-Mater. Biol. Appl.* **2018**, *92*, 969–982. [CrossRef] [PubMed]

87. Lutz, M.; Engelbrecht, L.; Laurie, A.; Dyayiya, N. Using CLEM to investigate the distribution of nano-sized antimicrobial agents within an EVOH matrix. *Int. J. Polym. Anal. Charact.* **2018**, *23*, 300–312. [CrossRef]
88. Miguel, S.P.; Figueira, D.R.; Simoes, D.; Ribeiro, M.P.; Coutinho, P.; Ferreira, P.; Correia, I.J. Electrospun polymeric nanofibres as wound dressings: A review. *Colloids Surf. B-Biointerfaces* **2018**, *169*, 60–71. [CrossRef] [PubMed]
89. DeFrates, K.G.; Moore, R.; Borgesi, J.; Lin, G.; Mulderig, T.; Beachley, V.; Hu, X. Protein-Based Fiber Materials in Medicine: A Review. *Nanomaterials* **2018**, *8*, 457. [CrossRef] [PubMed]
90. Jakobsson, A.; Ottosson, M.; Zalis, M.C.; O'Carroll, D.; Johansson, U.E.; Johansson, F. Three-dimensional functional human neuronal networks in uncompressed low-density electrospun fiber scaffolds. *Nanomed. Nanotechnol. Biol. Med.* **2017**, *13*, 1563–1573. [CrossRef] [PubMed]
91. Venugopal, J.; Zhang, Y.Z.; Ramakrishna, S. Electrospun nanofibres: Biomedical applications. *Proc. Inst. Mech. Eng. Part N J. Nanoeng. Nanosyst.* **2004**, *218*, 35–45. [CrossRef]
92. Hassan, M.K.; Abukmail, A.; Hassiba, A.J.; Mauritz, K.A.; Elzatahry, A.A. PVA/Chitosan/Silver Nanoparticles Electrospun Nanocomposites: Molecular Relaxations Investigated by Modern Broadband Dielectric Spectroscopy. *Nanomaterials* **2018**, *8*, 888. [CrossRef] [PubMed]
93. Cheng, J.; Jun, Y.; Qin, J.; Lee, S.-H. Electrospinning versus microfluidic spinning of functional fibers for biomedical applications. *Biomaterials* **2017**, *114*, 121–143. [CrossRef] [PubMed]
94. Olsson, R.T.; Kraemer, R.; López-Rubio, A.; Torres-Giner, S.; Ocio, M.J.; Lagarón, J.M. Extraction of Microfibrils from Bacterial Cellulose Networks for Electrospinning of Anisotropic Biohybrid Fiber Yarns. *Macromolecules* **2010**, *43*, 4201–4209. [CrossRef]
95. Reneker, D.H.; Yarin, A.L. Electrospinning jets and polymer nanofibers. *Polymer* **2008**, *49*, 2387–2425. [CrossRef]
96. Alehosseini, A.; Ghorani, B.; Sarabi-Jamab, M.; Tucker, N. Principles of electrospraying: A new approach in protection of bioactive compounds in foods. *Crit. Rev. Food Sci. Nutr.* **2018**, *58*, 2346–2363. [CrossRef] [PubMed]
97. Akhgari, A.; Shakib, Z.; Sanati, S. A review on electrospun nanofibers for oral drug delivery. *Nanomed. J.* **2017**, *4*, 197–207.
98. Kitsara, M.; Agbulut, O.; Kontziampasis, D.; Chen, Y.; Menasche, P. Fibers for hearts: A critical review on electrospinning for cardiac tissue engineering. *Acta Biomater.* **2017**, *48*, 20–40. [CrossRef] [PubMed]
99. Jesus Villarreal-Gomez, L.; Manuel Cornejo-Bravo, J.; Vera-Graziano, R.; Grande, D. Electrospinning as a powerful technique for biomedical applications: A critically selected survey. *J. Biomater. Sci.-Polym. Ed.* **2016**, *27*, 157–176. [CrossRef] [PubMed]
100. Pham, Q.P.; Sharma, U.; Mikos, A.G. Electrospinning of polymeric nanofibers for tissue engineering applications: A review. *Tissue Eng.* **2006**, *12*, 1197–1211. [CrossRef] [PubMed]
101. Asghari, F.; Samiei, M.; Adibkia, K.; Akbarzadeh, A.; Davaran, S. Biodegradable and biocompatible polymers for tissue engineering application: A review. *Artif. Cells Nanomed. Biotechnol.* **2017**, *45*, 185–192. [CrossRef] [PubMed]
102. Khorshidi, S.; Solouk, A.; Mirzadeh, H.; Mazinani, S.; Lagaron, J.M.; Sharifi, S.; Ramakrishna, S. A review of key challenges of electrospun scaffolds for tissue-engineering applications. *J. Tissue Eng. Regen. Med.* **2016**, *10*, 715–738. [CrossRef] [PubMed]
103. Sharma, J.; Zhang, X.; Sarker, T.; Yan, X.; Washburn, L.; Qu, H.; Guo, Z.; Kucknoor, A.; Wei, S. Biocompatible electrospun tactic poly(methyl methacrylate) blend fibers. *Polymer* **2014**, *55*, 3261–3269. [CrossRef]
104. Cheng, H.; Yang, X.; Che, X.; Yang, M.; Zhai, G. Biomedical application and controlled drug release of electrospun fibrous materials. *Mater. Sci. Eng. C-Mater. Biol. Appl.* **2018**, *90*, 750–763. [CrossRef] [PubMed]
105. Munteanu, B.S.; Vasile, C. Antioxidant, antibacterial/antifungal nanostructures for medical and food packaging applications. *J. Nanosci. Nanomed.* **2017**, *1*, 15–20.
106. Al-Enizi, A.M.; Zagho, M.M.; Elzatahry, A.A. Polymer-Based Electrospun Nanofibers for Biomedical Applications. *Nanomaterials* **2018**, *8*, 259. [CrossRef] [PubMed]
107. Hassiba, A.J.; El Zowalaty, M.E.; Nasrallah, G.K.; Webster, T.J.; Luyt, A.S.; Abdullah, A.M.; Elzatahry, A.A. Review of recent research on biomedical applications of electrospun polymer nanofibers for improved wound healing. *Nanomedicine* **2016**, *11*, 715–737. [CrossRef] [PubMed]
108. Son, W.K.; Youk, J.H.; Park, W.H. Antimicrobial cellulose acetate nanofibers containing silver nanoparticles. *Carbohydr. Polym.* **2006**, *65*, 430–434. [CrossRef]

109. Xu, X.; Zhou, M. Antimicrobial gelatin nanofibers containing silver nanoparticles. *Fibers Polym.* **2008**, *9*, 685–690. [CrossRef]
110. Sheikh, F.A.; Ju, H.W.; Lee, J.M.; Moon, B.M.; Park, H.J.; Lee, O.J.; Kim, J.-H.; Kim, D.-K.; Park, C.H. 3D electrospun silk fibroin nanofibers for fabrication of artificial skin. *Nanomed. Nanotechnol. Biol. Med.* **2015**, *11*, 681–691. [CrossRef] [PubMed]
111. Tijing, L.D.; Ruelo, M.T.G.; Amarjargal, A.; Pant, H.R.; Park, C.-H.; Kim, C.S. One-step fabrication of antibacterial (silver nanoparticles/poly(ethylene oxide))—Polyurethane bicomponent hybrid nanofibrous mat by dual-spinneret electrospinning. *Mater. Chem. Phys.* **2012**, *134*, 557–561. [CrossRef]
112. Munteanu, B.S.; Aytac, Z.; Pricope, G.M.; Uyar, T.; Vasile, C. Polylactic acid (PLA)/Silver-NP/VitaminE bionanocomposite electrospun nanofibers with antibacterial and antioxidant activity. *J. Nanopart. Res.* **2014**, *16*, 2643. [CrossRef]
113. Motealleh, B.; Zahedi, P.; Rezaeian, I.; Moghimi, M.; Abdolghaffari, A.H.; Zarandi, M.A. Morphology, drug release, antibacterial, cell proliferation, and histology studies of chamomile-loaded wound dressing mats based on electrospun nanofibrous poly(ε-caprolactone)/polystyrene blends. *J. Biomed. Mater. Res. B Appl. Biomater.* **2014**, *102*, 977–987. [CrossRef] [PubMed]
114. Suganya, S.; Ram, T.S.; Lakshmi, B.S.; Giridev, V.R. Herbal drug incorporated antibacterial nanofibrous mat fabricated by electrospinning: An excellent matrix for wound dressings. *J. Appl. Polym. Sci.* **2011**, *121*, 2893–2899. [CrossRef]
115. Cowan, M.M. Plant products as antimicrobial agents. *Clin. Microbiol. Rev.* **1999**, *12*, 564–582. [CrossRef] [PubMed]
116. Zhang, W.; Ronca, S.; Mele, E. Electrospun Nanofibres Containing Antimicrobial Plant Extracts. *Nanomaterials* **2017**, *7*, 42. [CrossRef] [PubMed]
117. Liakos, I.; Rizzello, L.; Hajiali, H.; Brunetti, V.; Carzino, R.; Pompa, P.P.; Athanassiou, A.; Mele, E. Fibrous wound dressings encapsulating essential oils as natural antimicrobial agents. *J. Mater. Chem. B* **2015**, *3*, 1583–1589. [CrossRef]
118. Kontogiannopoulos, K.N.; Assimopoulou, A.N.; Tsivintzelis, I.; Panayiotou, C.; Papageorgiou, V.P. Electrospun fiber mats containing shikonin and derivatives with potential biomedical applications. *Int. J. Pharm.* **2011**, *409*, 216–228. [CrossRef] [PubMed]
119. Kayaci, F.; Ertas, Y.; Uyar, T. Enhanced Thermal Stability of Eugenol by Cyclodextrin Inclusion Complex Encapsulated in Electrospun Polymeric Nanofibers. *J. Agric. Food Chem.* **2013**, *61*, 8156–8165. [CrossRef] [PubMed]
120. Archana, D.; Singh, B.K.; Dutta, J.; Dutta, P.K. In vivo evaluation of chitosan-PVP-titanium dioxide nanocomposite as wound dressing material. *Carbohydr. Polym.* **2013**, *95*, 530–539. [CrossRef] [PubMed]
121. Saquing, C.D.; Tang, C.; Monian, B.; Bonino, C.A.; Manasco, J.L.; Alsberg, E.; Khan, S.A. Alginate–Polyethylene Oxide Blend Nanofibers and the Role of the Carrier Polymer in Electrospinning. *Ind. Eng. Chem. Res.* **2013**, *52*, 8692–8704. [CrossRef]
122. Kayaci, F.; Sen, H.S.; Durgun, E.; Uyar, T. Electrospun nylon 6,6 nanofibers functionalized with cyclodextrins for removal of toluene vapor. *J. Appl. Polym. Sci.* **2015**, *132*. [CrossRef]
123. Álvarez-Paino, M.; Muñoz-Bonilla, A.; Fernández-García, M. Antimicrobial Polymers in the Nano-World. *Nanomaterials* **2017**, *7*, 48. [CrossRef] [PubMed]
124. Zhang, J.; Wang, X.-X.; Zhang, B.; Ramakrishna, S.; Yu, M.; Ma, J.-W.; Long, Y.-Z. In Situ Assembly of Well-Dispersed Ag Nanoparticles throughout Electrospun Alginate Nanofibers for Monitoring Human Breath—Smart Fabrics. *ACS Appl. Mater. Interfaces* **2018**, *10*, 19863–19870. [CrossRef] [PubMed]
125. Unnithan, A.R.; Gnanasekaran, G.; Sathishkumar, Y.; Lee, Y.S.; Kim, C.S. Electrospun antibacterial polyurethane-cellulose acetate-zein composite mats for wound dressing. *Carbohydr. Polym.* **2014**, *102*, 884–892. [CrossRef] [PubMed]
126. Cerqueira, M.A.; Jose Fabra, M.; Lorena Castro-Mayorga, J.; Bourbon, A.I.; Pastrana, L.M.; Vicente, A.A.; Lagaron, J.M. Use of Electrospinning to Develop Antimicrobial Biodegradable Multilayer Systems: Encapsulation of Cinnamaldehyde and Their Physicochemical Characterization. *Food Bioprocess Technol.* **2016**, *9*, 1874–1884. [CrossRef]
127. Parreidt, T.S.; Müller, K.; Schmid, M. Alginate-Based Edible Films and Coatings for Food Packaging Applications. *Foods* **2018**, *7*, 170. [CrossRef] [PubMed]

128. Goudarzi, V.; Shahabi-Ghahfarrokhi, I. Development of photo-modified starchikefiran/TiO2 bio-nanocomposite as an environmentally-friendly food packaging material. *Int. J. Biol. Macromol.* **2018**, *116*, 1082–1088. [CrossRef] [PubMed]
129. Oner, M.; Kizil, G.; Keskin, G.; Pochat-Bohatier, C.; Bechelany, M. The Effect of Boron Nitride on the Thermal and Mechanical Properties of Poly(3-hydroxybutyrate-co-3-hydroxyvalerate). *Nanomaterials* **2018**, *8*, 940. [CrossRef] [PubMed]
130. Zakuwan, S.Z.; Ahmad, I. Synergistic Effect of Hybridized Cellulose Nanocrystals and Organically Modified Montmorillonite on kappa-Carrageenan Bionanocomposites. *Nanomaterials* **2018**, *8*, 874. [CrossRef] [PubMed]
131. Huang, Y.; Mei, L.; Chen, X.; Wang, Q. Recent Developments in Food Packaging Based on Nanomaterials. *Nanomaterials* **2018**, *8*, 830. [CrossRef] [PubMed]
132. Wardhono, E.Y.; Wahyudi, H.; Agustina, S.; Oudet, F.; Pinem, M.P.; Clausse, D.; Saleh, K.; Guenin, E. Ultrasonic Irradiation Coupled with Microwave Treatment for Eco-friendly Process of Isolating Bacterial Cellulose Nanocrystals. *Nanomaterials* **2018**, *8*, 859. [CrossRef] [PubMed]
133. Yang, C.-L.; Chen, J.-P.; Wei, K.-C.; Chen, J.-Y.; Huang, C.-W.; Liao, Z.-X. Release of Doxorubicin by a Folate-Grafted, Chitosan-Coated Magnetic Nanoparticle. *Nanomaterials* **2017**, *7*, 85. [CrossRef] [PubMed]
134. Nagar, P.; Goyal, P.; Gupta, A.; Sharma, A.K.; Kumar, P. Synthesis, characterization and evaluation of retinoic acid-polyethylene glycol nanoassembly as efficient drug delivery system. *Nano-Struct. Nano-Objects* **2018**, *14*, 110–117. [CrossRef]
135. Thakur, S.; Sharma, B.; Verma, A.; Chaudhary, J.; Tamulevicius, S.; Thakur, V.K. Recent progress in sodium alginate based sustainable hydrogels for environmental applications. *J. Clean. Prod.* **2018**, *198*, 143–159. [CrossRef]
136. Li, J.; He, J.; Huang, Y. Role of alginate in antibacterial finishing of textiles. *Int. J. Biol. Macromol.* **2017**, *94*, 466–473. [CrossRef] [PubMed]
137. Pan, L.; Wang, Z.; Yang, Q.; Huang, R. Efficient Removal of Lead, Copper and Cadmium Ions from Water by a Porous Calcium Alginate/Graphene Oxide Composite Aerogel. *Nanomaterials* **2018**, *8*, 957. [CrossRef] [PubMed]
138. Song, W.; Su, X.; Gregory, D.A.; Li, W.; Cai, Z.; Zhao, X. Magnetic Alginate/Chitosan Nanoparticles for Targeted Delivery of Curcumin into Human Breast Cancer Cells. *Nanomaterials* **2018**, *8*, 907. [CrossRef] [PubMed]
139. Tang, S.; Wang, Z.; Li, P.; Li, W.; Li, C.; Wang, Y.; Chu, P.K. Degradable and Photocatalytic Antibacterial Au-TiO2/Sodium Alginate Nanocomposite Films for Active Food Packaging. *Nanomaterials* **2018**, *8*, 930. [CrossRef] [PubMed]
140. Liu, Z.; Li, J.; Zhao, X.; Li, Z.; Li, Q. Surface Coating for Flame Retardancy and Pyrolysis Behavior of Polyester Fabric Based on Calcium Alginate Nanocomposites. *Nanomaterials* **2018**, *8*, 875. [CrossRef] [PubMed]
141. Zhao, L.; Wang, Y.; Zhao, X.; Deng, Y.; Li, Q.; Xia, Y. Green Preparation of Ag-Au Bimetallic Nanoparticles Supported on Graphene with Alginate for Non-Enzymatic Hydrogen Peroxide Detection. *Nanomaterials* **2018**, *8*, 507. [CrossRef] [PubMed]
142. Qin, Y. The characterization of alginate wound dressings with different fiber and textile structures. *J. Appl. Polym. Sci.* **2006**, *100*, 2516–2520. [CrossRef]
143. Qin, Y. The gel swelling properties of alginate fibers and their applications in wound management. *Polym. Adv. Technol.* **2008**, *19*, 6–14. [CrossRef]
144. Kim, H.W.; Kim, B.R.; Rhee, Y.H. Imparting durable antimicrobial properties to cotton fabrics using alginate–quaternary ammonium complex nanoparticles. *Carbohydr. Polym.* **2010**, *79*, 1057–1062. [CrossRef]
145. Thombare, N.; Jha, U.; Mishra, S.; Siddiqui, M.Z. Guar gum as a promising starting material for diverse applications: A review. *Int. J. Biol. Macromol.* **2016**, *88*, 361–372. [CrossRef] [PubMed]
146. Singha, A.S.; Kumar Thakur, V. Saccaharum Cilliare Fiber Reinforced Polymer Composites. *E-J. Chem.* **2008**, *5*, 782–791. [CrossRef]
147. Singha, A.S.; Thakur, V.K. Synthesis, Characterisation and Analysis of Hibiscus Sabdariffa Fibre Reinforced Polymer Matrix Based Composites. *Polym. Polym. Compos.* **2009**, *17*, 189–194. [CrossRef]
148. Singha, A.S.; Thakur, V.K. Mechanical, Thermal and Morphological Properties of Grewia Optiva Fiber/Polymer Matrix Composites. *Polym.-Plast. Technol. Eng.* **2009**, *48*, 201–208. [CrossRef]

149. Thakur, S.; Govender, P.P.; Mamo, M.A.; Tamulevicius, S.; Mishra, Y.K.; Thakur, V.K. Progress in lignin hydrogels and nanocomposites for water purification: Future perspectives. *Vacuum* **2017**, *146*, 342–355. [CrossRef]
150. Thakur, V.K.; Thakur, M.K. Recent advances in green hydrogels from lignin: A review. *Int. J. Biol. Macromol.* **2015**, *72*, 834–847. [CrossRef] [PubMed]
151. Singha, A.S.; Shama, A.; Thakur, V.K. X-ray Diffraction, Morphological, and Thermal Studies on Methylmethacrylate Graft Copolymerized *Saccharum ciliare* Fiber. *Int. J. Polym. Anal. Charact.* **2008**, *13*, 447–462. [CrossRef]
152. Singha, A.S.; Thakur, V.K. Fabrication and Characterization of H. sabdariffa Fiber-Reinforced Green Polymer Composites. *Polym.-Plast. Technol. Eng.* **2009**, *48*, 482–487. [CrossRef]
153. Zhang, Y.; Li, Y.; Thakur, V.K.; Gao, Z.; Gu, J.; Kessler, M.R. High-performance thermosets with tailored properties derived from methacrylated eugenol and epoxy-based vinyl ester. *Polym. Int.* **2018**, *67*, 544–549. [CrossRef]
154. Kim, G.; Park, K. Alginate-nanofibers fabricated by an electrohydrodynamic process. *Polym. Eng. Sci.* **2009**, *49*, 2242–2248. [CrossRef]
155. Mokhena, T.C.; Jacobs, V.; Luyt, A.S. A review on electrospun bio-based polymers for water treatment. *eXPRESS Polym. Lett.* **2015**, *9*, 839–880. [CrossRef]
156. Mendes, A.C.; Stephansen, K.; Chronakis, I.S. Electrospinning of food proteins and polysaccharides. *Food Hydrocoll.* **2017**, *68*, 53–68. [CrossRef]
157. Qasim, S.B.; Zafar, M.S.; Najeeb, S.; Khurshid, Z.; Shah, A.H.; Husain, S.; Rehman, I.U. Electrospinning of Chitosan-Based Solutions for Tissue Engineering and Regenerative Medicine. *Int. J. Mol. Sci.* **2018**, *19*, 407. [CrossRef] [PubMed]
158. Lotfian, S.; Giraudmaillet, C.; Yoosefinejad, A.; Thakur, V.K.; Nezhad, H.Y. Electrospun Piezoelectric Polymer Nanofiber Layers for Enabling in Situ Measurement in High-Performance Composite Laminates. *ACS Omega* **2018**, *3*, 8891–8902. [CrossRef]
159. Zhu, H.; Zhang, Q.; Zhu, S. Alginate Hydrogel: A Shapeable and Versatile Platform for in Situ Preparation of Metal–Organic Framework–Polymer Composites. *ACS Appl. Mater. Interfaces* **2016**, *8*, 17395–17401. [CrossRef] [PubMed]
160. Hashimoto, T.; Suzuki, Y.; Suzuki, K.; Nakashima, T.; Tanihara, M.; Ide, C. Review—Peripheral nerve regeneration using non-tubular alginate gel crosslinked with covalent bonds. *J. Mater. Sci.-Mater. Med.* **2005**, *16*, 503–509. [CrossRef] [PubMed]
161. Miculescu, F.; Maidaniuc, A.; Miculescu, M.; Dan Batalu, N.; Cătălin Ciocoiu, R.; Voicu, Ş.I.; Stan, G.E.; Thakur, V.K. Synthesis and Characterization of Jellified Composites from Bovine Bone-Derived Hydroxyapatite and Starch as Precursors for Robocasting. *ACS Omega* **2018**, *3*, 1338–1349. [CrossRef] [PubMed]
162. Thakur, M.K.; Thakur, V.K.; Gupta, R.K.; Pappu, A. Synthesis and Applications of Biodegradable Soy Based Graft Copolymers: A Review. *ACS Sustain. Chem. Eng.* **2016**, *4*, 1–17. [CrossRef]
163. Trache, D.; Hussin, M.H.; Haafiz, M.K.M.; Thakur, V.K. Recent progress in cellulose nanocrystals: Sources and production. *Nanoscale* **2017**, *9*, 1763–1786. [CrossRef] [PubMed]
164. Thakur, V.K.; Kessler, M.R. Self-healing polymer nanocomposite materials: A review. *Polymer* **2015**, *69*, 369–383. [CrossRef]
165. Thakur, V.K.; Thakur, M.K. Recent trends in hydrogels based on psyllium polysaccharide: A review. *J. Clean. Prod.* **2014**, *82*, 1–15. [CrossRef]
166. Li, H.; Tan, C.; Li, L. Review of 3D printable hydrogels and constructs. *Mater. Des.* **2018**, *159*, 20–38. [CrossRef]
167. Wong, C.Y.; Al-Salami, H.; Dass, C.R. Microparticles, microcapsules and microspheres: A review of recent developments and prospects for oral delivery of insulin. *Int. J. Pharm.* **2018**, *537*, 223–244. [CrossRef] [PubMed]
168. Peppas, N.A.; Keys, K.B.; Torres-Lugo, M.; Lowman, A.M. Poly(ethylene glycol)-containing hydrogels in drug delivery. *J. Control. Release* **1999**, *62*, 81–87. [CrossRef]
169. Pattanashetti, N.A.; Heggannavar, G.B.; Kariduraganavar, M.Y. Smart Biopolymers and their Biomedical Applications. *Procedia Manuf.* **2017**, *12*, 263–279. [CrossRef]

170. Barbosa, J.; Correia, D.M.; Gonçalves, R.; Ribeiro, C.; Botelho, G.; Martins, P.; Lanceros-Mendez, S. Magnetically Controlled Drug Release System through Magnetomechanical Actuation. *Adv. Healthc. Mater.* **2016**, *5*, 3027–3034. [CrossRef] [PubMed]
171. Cardoso, V.F.; Correia, D.M.; Ribeiro, C.; Fernandes, M.M.; Lanceros-Méndez, S. Fluorinated Polymers as Smart Materials for Advanced Biomedical Applications. *Polymers* **2018**, *10*, 161. [CrossRef]
172. Kim, M.W.; Kwon, S.-H.; Choi, J.H.; Lee, A. A Promising Biocompatible Platform: Lipid-Based and Bio-Inspired Smart Drug Delivery Systems for Cancer Therapy. *Int. J. Mol. Sci.* **2018**, *19*, 3859. [CrossRef] [PubMed]

© 2019 by the authors. Licensee MDPI, Basel, Switzerland. This article is an open access article distributed under the terms and conditions of the Creative Commons Attribution (CC BY) license (http://creativecommons.org/licenses/by/4.0/).

MDPI
St. Alban-Anlage 66
4052 Basel
Switzerland
Tel. +41 61 683 77 34
Fax +41 61 302 89 18
www.mdpi.com

Nanomaterials Editorial Office
E-mail: nanomaterials@mdpi.com
www.mdpi.com/journal/nanomaterials

www.ingramcontent.com/pod-product-compliance
Lightning Source LLC
LaVergne TN
LVHW070049120526
838202LV00101B/1968